The Enigma of Evolution
and
the Challenge of Chance

Denise Carrington-Smith

First published 2018 by
Storixus Media and Publishing
Australia

ISBN - 9780648364016 (Paperback Edition)
ISBN - 9780648364023 (eBook Edition)

S

www.storixus.com

Acknowledgements

Sincere thanks are extended to the following people for their help and support in the preparation of this book.

Firstly, I must acknowledge the opportunity afforded me by James Cook University, Cairns, to undertake my programme of study and my two supervisors, Associate Professor John Campbell and Dr. Elizabeth Tynan, for their patient and supportive supervision of my work.

I am grateful for the help of Professor Derek Partridge in my bringing together the chapters relating to the work of Alfred Wallace..

I truly appreciate the assistance I received from the Weaver family, Roy and Neil, Ann and Sally, who helped and guided me as this book went through the first stages of preparation for publication.

Finally, family and friends - what would one do without them? There was an abundance of emotional support, supplemented at times by practical assistance.. Whatever its nature, the help and support received from them all is deeply appreciated. I would particularly like to thank Caroline Lee for the illustration on page 377, the artist who drew the fascinating picture which graces the front cover, who wishes to remain anonymous, and the two of my children who gave me hours of practical help, but who also do not wish to be named. They know who they are although they will never truly know how much their help was appreciated.

Thank you, each and every one.

Author's note

Much of the literature upon which this book is based was written long before gender neutral language was introduced and the generic use of male gender terminology is an intrinsic part of the history of the English language. It was felt that to change what was said to what could have been said, should have been said, or even would have been said, would be to change the essence of the discourse being analysed. For this reason, the language used by an author, be it neutral or gender biased, will be the language used when discussing that particular author's work.

For convenience, the masculine pronoun has been retained when referring to 'God' and the feminine when referring to 'Nature'.

Foreword

When I took up the study of Archæology, I was delighted to find that the subject encompassed more than just artifacts; it included the study of the people who made them. Artifacts (artefacts) are objects not found in nature, which have been artificially produced by humans. Some confirm written history, others throw new light upon the written record, yet others reach far beyond the scope of the written word, back in time for, not thousands, but millions of years. I became fascinated by what we could learn about our earliest ancestors.

I came across a report in a book by Roberts and Manchester (1997), entitled *The Archæology of Disease*, of the remains of a Neanderthal which gave evidence of him having suffered from a life-style related disease, diffuse idiopathic skeletal hyperostosis, appropriately abbreviated to DISH since it is the result of overeating! I realized that such pathology provided another window of opportunity to glean information, not just about the lifestyle of these early people, but about their nature. I scoured the literature for other lifestyle related pathologies diagnosed from skeletal remains and the conclusions which I drew about the nature of these people were presented in my Masters thesis.

My Ph.D. thesis was quite different. It was a voyage of exploration - I had no idea where my journey would take me. I was inspired to embark upon it by the reading of Paul Jordan's book, published in 1999, *Neanderthals: Neanderthal Man and the Story of Human Origins*. The conclusions he drew about the Neanderthals, their nature and their lifestyle, were so different from those of Ralph Solecki, who, in 1971, had published his book about the Neanderthals, entitled *Shanidar: The First Flower People*. Both men reached different conclusions based upon much the same evidence. Why?

Yet do we not see this happening every day in our Courts? Same evidence, yet two sides drawing such different conclusions from that evidence, feeling so strongly about their position, that they are prepared to disrupt their lives, and spend tens, if not hundreds, of thousands of dollars, arguing for the case in which they so passionately believe. Sometimes the people presenting the case in Court feel truly passionate about the position for which they are arguing; other times they argue the case they have been paid to argue - although, if they did not have at least some sympathy for the position they represent, it is unlikely that they would have agreed to accept the case.

Here it is 'evolution' which is on trial. (This actually happened in the Courts of Law in the United States of America on two occasions - with two different results!) However, this book covers more than the Courts of Law. It covers the Court of Academia and the Court of Public Opinion, of which you form a part. It starts with a brief look at the opinion of two Greek philosophers, since their views had a profound effect on European thinking at the start of the Age of Enlightenment, then skips forward in time to the 18th century, when the Biblical story of Creation, of Adam and Eve, first came seriously to be questioned. I made every effort, as I worked my way forward, to study each era chronologically, not to allow, for example, later re-evaluations of the age of fossils, to influence my understanding of the circumstances at the earlier time, which caused one person to think one way and another person to think another.

Why does one person believe in God and another not? I have no idea, but of one thing I am certain. People endeavour to interpret the evidence in accordance with their pre-existing beliefs, rather than adopt beliefs based on the evidence. Occasionally, a person does change their core belief system, but that is a rare occurrence. Yet opinions do change over time. Why? Always there is a dominant belief system (dominant discourse) which is not necessarily the majority opinion, which I found interesting.

When I started out on my journey, I had no idea what books I would be led to read. I had no idea that I would find myself reading that great classic of the economist, Adam Smith, published in 1776, *An Inquiry into the Nature and Causes of the Wealth of Nations*, nor did I expect to read Jonathan Swifts' *Gulliver's Travels*, which I did to track down the origin of the word 'yahoo'! I had read some of Darwin's work, but not Lamarck's, whose work I was to find incredibly interesting. Each book I read led me to others. Some were truly fascinating, others truly disturbing. Not all of that which I found either fascinating or disturbing was able to be included in my thesis and not all that I found either fascinating or disturbing has been able to be

IV

included here. It is a large subject, far beyond the scope of one book.

I hope you will find this book interesting and enjoyable. Will it change your outlook? Possibly. Will it change your core beliefs? Probably not. But then, it might. After all, miracles do happen sometimes. Don't they?

Denise Carrington-Smith

2018

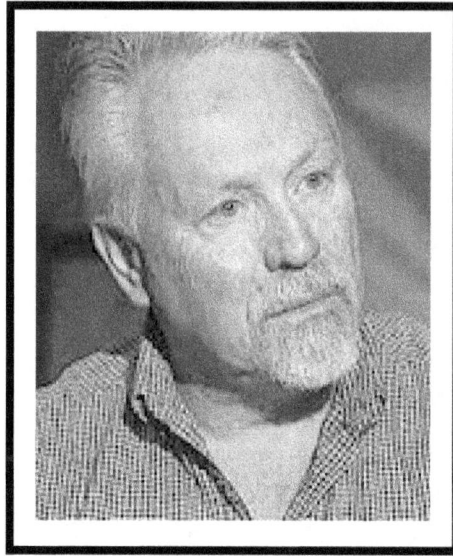

Prof. Alan Thorne

(1939 - 2012)

Canberra

This book is dedicated to

Alan Thorne

who never gave up on what he believed to be right

Periods of Time

The time periods given below are approximations. They may be subject to revision from time to time.

Mya (millions of years ago)	Epoch	

Tertiary Period

65 - 53	Paleocene	
53 - 35	Eocene	
35 - 25	Oligocene	
25 - 5	Miocene	Hominoids
5 - 2	Pliocene	Hominids

Quaternary Period

1 - 0.5	Lower Pleistocene	Homo
0.5 - 0.05	Middle Pleistocene	
0.05 - 0.01	Upper Pleistocene	Homo sapiens
>0.1	Holocene	

Ages

Stone Ages

1,000,000 – 150,000	Lower Pleistocene	Acheulian
		Clacton
		Levalloisian
150,000-40,000	Middle Pleistocene	Mousterian
50,000-10,000	Upper Pleistocene	Chatelperronian
		Aurignacian
		Gravettian
		Solutrean
		Magdalenian
10,000 –	Neolithic	Azilian

Contents

Illustrations

The greatest obstacle to the progress of science

is not ignorance;

it is the illusion of knowledge.

David J. Boorstein

Part I

ANTICIPATION

Before 1850

Thoughts on Evolution before Darwin

Chapter 1

Turning the Universe

Upside Down

A journey through time takes no time at all – one only has to think! A journey through thought takes far longer, for the same reason. One has to think!

I have never questioned the fact of evolution. However, when I took up the study of archæology (which I did in my retirement) I started to question Darwin's theory of evolution by natural selection. As I was to find, I was not alone in accepting that, while a process such as natural selection could account for *variation* within species, it could not be considered the sole and sufficient cause of all evolution of life on this Earth.

This book is based on the research I undertook in the course of completing both my Master's and my Doctor's degrees. For the first, I researched pathologies found in fossil remains to learn what I could about ancient people, such as the Neanderthals, and even *Homo erectus*. I came to the conclusion that, in many ways, they were not so very different from us. This led me to ponder why, in so much of the literature, there appears to be an underlying assumption that, while human beings *(homo sapiens)* are 'wise', all our forebears were decidedly stupid! Some accounts portrayed ancient 'humans' as exhibiting far less common sense than the animals.

Some of the material included in this book was omitted from my thesis, either because my original exceeded the word limit (which it definitely did!) or because the material was deemed too controversial. My supervisors supported me where they could, but did not agree with all the conclusions to which I came.

It is often said that the best place to start is at the beginning and, where the West is concerned, that is very definitely with the Greeks. This is true in relation to art, philosophy and politics. The influence of the Middle East came later, during and after the time of the Romans, following the spread of Christianity and, later, Islam, as the Moors pushed north into Europe via Spain. It is with the resurgence of Greek thinking that we are concerned here, with the rediscovery of the philosophical works of Plato and Aristotle during the Enlightenment. Plato's concept of God and Aristotle's concept of 'The Great Chain of Being', with humans at the top, infiltrated and underpinned Christian thinking for centuries and, in many cases, continues to do so to this day. For this reason, it will seem very familiar. The concepts are simple and easy to understand, so please do not be put off by the appearance of these two names in this opening chapter. Without understanding their contribution, it is impossible to understand the later debate which, at times, became quite controversial and even bitter.

As I researched the literature on evolutionary thought in Europe, from the time of the Ancient Greeks until the dawning of the Age of Science in the eighteenth century, one book, *The Great Chain of Being*, by Arthur Lovejoy (1936), was referenced so frequently that it became clear that the knowledge acquired by Professor Lovejoy in this field was so extensive that it was generally accepted by later writers. I was happy to do the same. As I moved forward in time, the literature became more extensive until it had expanded to such a degree that it would be impossible for one person to read and evaluate it all, at least within the time-frame allowed for the completion of a thesis. I make no pretense that this book covers all that has been written on the subject of evolution, especially post Darwin, but, again, certain books were repeatedly referenced and it was clear that these were the seminal volumes which provided the basis for the opinions which permeated this field of thought. It was upon these I concentrated, but, of course, I was more than happy to include lesser known works when they made material contributions, often by putting forward an opposing idea.

While Eastern philosophers had embraced the concept of æons of time, of ages succeeding ages, of universes being created and dissolved, of universal change and evolution, Western thought had for hundreds of years been encased in the more limiting Judaic tradition. This held that the world had been created but a few thousand years ago and that it had taken but six days for its Creator to complete His Creation. Apart from the destruction caused by the Flood, the world today was very much as it had been at the time of Adam and Eve. No new living forms had been created and, some people maintained, none had been lost.

The imperfection of Man, his potential for improvement in this world, with the possibility of perfection in the next, was accepted Church doctrine during the first millennium A.D. The Earth was assumed to be flat, the Sun,

4

Moon and stars occupied the heavenly sphere, which was assumed to be a 'sphere' because of the rotatory motion of the Sun and Moon around the Earth.

It was the rediscovery of the works of the ancient Greek philosophers, particularly Plato, Socrates and Aristotle, but also of Ptolemy, which revolutionized Western thought during the second millennium, resulting in the questioning of previously accepted religious and secular thought and ushering in the Ages of Enlightenment, Reason and Science.

That the Earth was but a small speck in the cosmos had been accepted in earlier times. The Greek astronomer, Ptolemy, had spoken of the Earth as a mere dot in comparison with the heavens (Lovejoy 1936/1964: 100):

> It has been shown that the distance between the center of the earth and the summit of the sphere of Saturn is a journey of about eight thousand seven hundred years of 365 days, assuming that one walks forty leagues a day (i.e. the distance in round numbers is 125 million miles) … And if the earth is thus no bigger than a point relatively to the sphere of the fixed stars, what must be the ratio of the human species to the created universe as a whole?

As the philosophies of the ancient Greeks, especially Plato, filtered their way through the thoughts of philosophically minded Europeans, it was inevitable that comparisons would be made between the teachings of the ancient Greeks and those of the ancient Jews. Some attempted to merge the two positions, others saw the differences as irreconcilable. Initially, the Church was not much disturbed by the ponderings of the few, but as these ponderings became more widespread, attempts were made to suppress them, at least in part, as Galileo and Copernicus discovered to their cost. Nevertheless, alternative Greek thought gradually percolated through the upper levels of European society.

The 'Goodness' of Plato

The Jews believed in a very personal concept of God. Not only was he very intimately concerned with the affairs of the Jewish people, he exhibited some very human characteristics. For example, he was a 'jealous' God, who wreaked vengeance on his enemies and those of 'His People' – the Jews. Plato's concept of God was quite different. God was complete, perfect, self-sufficient, in need of nothing, not even praise or worship, from humans who could add nothing to his ever-present completeness – 'Goodness'. No Earthly creature was self-sufficient. All were in need of sustenance, of companionship. Being incomplete, they were imperfect, not 'Good'. This did not mean that humans, or any other parts of Creation, were 'evil', merely not 'perfect', not complete. This concept of the completeness of God was to play a large part in shaping the philosophical and religious thought of the 17th, 18th and 19th centuries.

If God was 'complete' and 'perfect', what need had he of Creation? Plato hypothesized that God, being 'Good', had no envy. Therefore, He could not deny existence to anything capable of being created and had, therefore, 'allowed' creation to take place, Self-Sufficing Perfection becoming Self-Sufficing Fecundity. While nothing incomplete was beautiful, and no part of creation was complete in itself, taken together, creation was 'Good', i.e. perfect, complete and beautiful. From this concept came the thought that if any creature were to disappear from the Earth, i.e. become extinct, the world would be less than complete, less than perfect. Since the completeness of the Earth mirrored the completeness of God, such a notion was incompatible with the perfection of God (Mayr 1982).

Plato had envisaged a God complete in itself, yet which had caused (or allowed) to be created the entire Universe, not seeing these two concepts as being in any way irreconcilable. From Plato's concept of plenitude (everything that could be, was) came Aristotle's concept of continuity. Aristotle argued that there was no true boundary between organic and inorganic, between plants and animals, or between different types of animals. Some marine forms attached themselves to rocks, as plants were attached to the Earth, and died if separated therefrom; bats were mammals, yet flew like birds. Otters and seals were mammals as much at home in the water as fish, while some fish flew through the air, if only for short distances. Apes were neither true quadrupeds nor true bipeds.

Aristotle acknowledged the possibility of classification while upholding the possibility of continuity. It was these two distinct concepts which were to prove so troublesome to naturalists in the 18th and 19th centuries. Aristotle did not himself arrange creatures in a single Chain of Being, but he did suggest an arrangement based on the 'powers of soul' possessed by the various organisms, the rational soul of Man being the most superior. Each higher order possessed all the powers of those below it, with an additional differentiating one of its own.

While agreeing with Ptolemy regarding the comparative size of the Earth, Bacon 'Westernized' Plato's philosophy by maintaining that, as Man had been made to serve God, so the world had been made to serve Man, without whom the world would be without aim or purpose.

According to Lovejoy, this attitude was portrayed by many 19th and 20th century writers as if it were the attitude of Europeans throughout the Middle Ages and into modern times, instead of being a minority view. The

Mediæval view of the Universe was not that the Earth took the centre, i.e. the most important, place. In fact, it was quite the reverse. The Earth was placed in the most lowly position, as far removed as possible from the heavenly spheres, the abode of God and His angels. The actual centre of the Universe was Hell, with Earth, in its sub-lunary position, scarcely any more elevated. The sub-lunar sphere was corruptible; the heavenly spheres beyond the Moon were perfect, complete, incorruptible. Earth was described as "the worst, lowest, most lifeless part of the universe, the bottom of the house" (Lovejoy p. 102).

Aristotle based his system on pairs of opposites: hot/cold, moist/dry. These 'essences' manifested, in various proportions, in earth, air, fire and water, all of which had their natural 'place'. The place of Earth was the centre, which, as stated above, was held to be the worst, most lowly place. The Sun, being hot, had the greatest tendency to rise and was thus placed in the heavens, not merely physically, but in essence. Placing the Earth in the heavens, in an equal place with the Sun, elevated it from its humble (sub-lunar) position to a celestial one, and it was this elevation which was the cause of the opposition by the Church to the Copernican system.

Aristotle's Chain of Being extended from God, down through the Heavens and its angels, to Man, then to the animals, plants and the Earth itself. This was compatible with the Christian belief that humans held a unique position; they were the only creatures which partook of two natures, being half spiritual and half material. The animals had no souls and the angels had no bodies. The human position, while an important link in this chain, was nevertheless a lowly one. Humans were the lowest beings who could aspire to enter the heavenly regions, but only by great effort, having led a pure life, and even then not without the Saviour's Grace.

The people of the Middle Ages believed that the Heavens were filled with spiritual, angelic beings, far outnumbering souls on Earth. Although there were some differences between the spiritual cosmologies of the Eastern and Western branches of the Christian Church, and between them and the spiritual cosmology of Islam, all were based on Judaic philosophy and found their way into European thought *via* tradition, not the direct teaching of the Bible. One tradition postulated seven heavens inhabited by the Seraphim, Cherubim, Thrones, Virtues, Principalities, Powers and Dominions, the first two of these being the realms of the archangels and angels. (Archangels are lower than angels, giving them the ability, on rare occasions, to communicate with humans, something which the angels cannot do.) Another tradition was slightly different: archangels, angels, Seraphim, Cherubim, Principalities, Dominions and Powers. Another combined the two, raising the number of Heavens from seven to nine, which some held to be the more 'complete' number. The Devil was formerly an angel. Like Adam, Lucifer committed the sin of Pride, which came before his Fall, as it did before Adam's. Pride, therefore, was the initial, pre-eminent, most deadly sin. Today, pride is usually seen as a cold, aloof emotion. Then it was seen to possess passion, a passionate belief that one's own views were right to the extent of wishing to impose these views on others.

Aliens?

The Copernican Revolution, far from reducing the stature of the Earth from the centre of the Universe to an insignificant speck in the Heavens, had, in fact, elevated it from its lowly sub-lunar position into the Heavens, where it circled with the other celestial spheres, as one of them. The atheist, or materialist, who removed not only God from His heaven, but all the angelic beings as well, did not demote human beings from a spiritual state to a lower one, but rather elevated them to the position of the most intelligent beings in the Universe.

Of course, there could be other forms of life on other planets. This was a possibility earnestly debated then and the debate continues to this day. The incredible amount of life on this planet, from the most simple to the most complex, made it statistically unlikely that there was no life anywhere else in the Universe. Indeed, some saw it as inconceivable that the power of God had been exhausted in the production of so "insignificant and wretched a being as Mankind" (Lovejoy p. 115).

Fascinating as this problem is to modern astrobiologists and astronomers, it was of even greater interest to Mediæval philosophers. If there were beings on other planets, were they perfect, or had they also suffered a 'fall'? If 'fallen', had they also been saved by Christ? Had Jesus incarnated on other planets? Had he suffered and died many times over? Which was worse: to believe that we humans were the only 'fallen' people in the Universe, or to believe that Christ must suffer again and again, because he could not leave a single 'fallen' person unsaved?

An hypothetical connection between humans and apes was readily accepted. Natives of Africa and South East Asia considered the chimpanzee and the orang-utan to be degenerate humans, this view being adopted by early European explorers. Later writers, such as Huxley and Darwin, were to consider humans as advanced apes. This reversal had serious philosophical implications. The Chain of Being was seen as descending from God. Humans had fallen from a state of Grace. To suggest that humans were improved apes was to suggest that the original creation had been less than perfect, an heretical view unacceptable to the Church.

There was far more at stake than a scientific debate as to whether the Universe was formed suddenly or over a period of time. 'Creationists' held their views, not merely because of the book of *Genesis*, but because of a belief in the 'Goodness' of God. Fossil finds showing that other forms had existed in the past which were not alive now, and forms alive now that had not existed in the past, were difficult to explain within the concept of 'Goodness'. It started to be suggested that all forms had been conceptualized by God from the very beginning, but that they took shape over a period of time, changing gradually from one shape to another, rather as does a baby *en route* to becoming an adult.

The concept of a continuous Chain of Being, descending from God down to the lowest of his creations was seen by some to imply that every creature that could be created had been created A creature existed in every niche in which it was possible for a creature to exist. Nevertheless, it did seem that there was space available for more creations and during the Middle Ages it was credible to suggest that these lived in parts of the Earth as yet unexplored. As European exploration expanded, new creatures were indeed discovered. Some, like the kangaroo and the duck-billed platypus, were quite unusual and did seem to fill another niche. By the nineteenth century, it became clear that the number of creatures yet unsighted must be diminishing. Only a favoured few could travel the world with the intention of finding new species, but anybody could scour their own countryside for undescribed species. The 19th century was the heyday of the amateur naturalist. Birds, beetles, butterflies, fish, plants, to say nothing of rocks and fossils, were sought, watched, caught, collected and mounted by the amateur naturalist. Museums, zoological gardens and conservatories all sprang into being and were heavily patronized.

Charles Darwin's early interest in entomology was part of this culture. Professional naturalists were swamped with specimens from at home and abroad, as Darwin was to find out on his return from his voyage on the *Beagle*, when the museums were reluctant to accept his boxes because they were so inundated with specimens.

It was not only in relation to animals that reports of extraordinary new finds were received in Europe. Reports arrived of giants, after the discovery of an outsized footprint in Patagonia (the name of the country itself means 'big foot'), and the legend of the Yeti/Abominable Snowman lingered on well into the 20th century, and even into the 21st. Apes were generally unknown to Europeans before the era of exploration and the discovery of the orang-utan provided a much needed link between monkeys and humans, particularly in view of the natives' claims that they were, in fact, degenerate humans. Reports came in from Africa of more 'orangs', later known as chimpanzees and gorillas, also believed by their local human neighbours to be degenerate humans. While the gorillas were 'giants', the chimpanzees were 'pygmies'. Lo and behold! 'Pygmy' humans were also found in Africa. Pieces of the Divine puzzle were gradually falling into place!

Early contact with the black races of Africa had left nothing but favourable impressions (Fernádez-Armesta 2004). African kings were depicted in European literature as being as rich and powerful as any European monarch. Seeing the black Africans as connecting links with the apes, which had degenerated from them, was in no way derogatory. The size of the smallest African people, being closer to that of the chimpanzees, merely illustrated the completeness of God's creation in which no gap that could possibly be filled had been left vacant. It did, indeed, take all sorts to make a world.

Natura non facit saltum

The dictum *Natura non facit saltum* (Nature makes no leaps) was the fundamental premise of the Great Chain of Being. From the point of view of classification, everything graded into or overlapped with something else, so that all were interlinked. There was a steady upward gradient from the simplest form of life to the most complex, from the lowest to the highest – Man! Upon this one point there seemed to be no dispute. There was dispute, however, as to whether or not humans had souls – the majority opinion was that they did. Did animals have souls? The majority (Christian) opinion was they did not, but there were some who considered that all living things had souls. And what about plants? Were they not living too? During the 19th century, there was a growing movement, with its roots in Germany, to consider all living things as having some form of consciousness, however small. This was a resurgence of the old Celtic belief and it experienced another revival in the 1960s and 1970s, with the publication of books such as *Supernature* by Lyall Watson (1973). Talking to the trees, to the plants in your garden, became quite fashionable. The dominant position had always been that humans were in some way different from all other forms of life. Whether that difference was that humans alone had souls and would continue to live in another form and in another place after death, was, and still is, a matter of debate.

Also debated was whether there had been but one creation of life, which remained virtually unchanged, or whether there had been one act of creation from which the forms living today had evolved. Had there been continual creations? Why would God expend all His creative force in one almighty Act and then do nothing?

The debate continues to this day.

Chapter 2

The Birth of Evolutionary Theory

It must not be thought that just because the new cosmology gradually became generally accepted that it became universally accepted. We learn very early in our lives that, contrary to what appears to be so obvious, the Sun does not circle the Earth but the Earth circles the Sun. We learn to accept that, contrary to what also appears to be so obvious, the Earth is not stationary but is hurtling through space at an incredible speed, even when there does not seem to us to be so much as a breath of air moving on our skin.

These were not concepts readily accepted. It took time for them to become the majority belief. Even into the second half of the twentieth century, the Flat Earth Society remained in existence. The advent of satellites and the images they beamed back to Earth saw its final demise.

The time we are about to consider, when the concept of evolution began to be discussed in European scientific circles, falls midway between the time of Copernicus and the present day. Having re-organized their thinking about the Heavens (space), scientists were about to re-organize their thinking about time, the time the Earth had been in existence, the time when life first came into being on this planet - and *how*? The Old Testament account, accepted not only by the Christian majority in Europe at that time, but also by Jews and Muslims, taught that organic life first came into being in the form of vegetation on the third day of Creation - the same day that dry land was separated from the sea. The Sun and the Moon (two great lights) were created on the fourth day, moving creatures, both in the waters and in the air, on the fifth day, earth-bound creatures, including humans on the sixth. The biblical account, therefore, allowed some difference in time between the origin of plants and animals, but not evolution of one life form from another.

The first murmurings within the scientific community on the subject of evolution did not appear until the middle of the eighteenth century. They were to precipitate an argument every bit as bitter and profound as had the teachings of Copernicus, for science would require a change in philosophical thinking which ran counter to prevalent religious thinking. The battle was originally little more than a skirmish but erupted into a full scale war. While the battle of the 'Heavens' (space and its occupants, the stars) would seem to have been won, the battle in regard to life on Earth is still far from over. The increasing number of Muslims migrating into Europe and other Western countries, such as America and Australia, bringing with them their Islamic belief in the account of Creation as related in the Old Testament, is resulting in an *increase* in the percentage of people *rejecting* the concept of evolution in Western countries (Dawkins 2009).

The gentleman scientist of the eighteenth century was in a fortunate position. The invention of the printing press had made information more readily available than ever before and gentlemen's libraries were lined with books on diverse subjects to stimulate the minds and interests of their readers, yet the number was not yet so large that the task of absorbing all relevant knowledge on any one subject was nigh impossible. The educated amateur gentleman scientist, or natural philosopher as he was then known, was versed in mathematics, the emerging sciences of physics and chemistry, as well as astronomy, and was likely well versed in botany and zoology as well. All fields were open to his enquiring mind and, being financially independent, he could follow his interests along whichever path he chose.

Another invention, that of the microscope, had revealed new worlds to the human eye. Cells had been recognized for the first time, but their structure was as yet unknown. Quite how the coming together of cells from the male and female parent of any species, be it plant or animal, produced an offspring "after its own kind" was becoming more of a mystery, not less, as scientists sought to understand the mechanism of reproduction.

Pierre Louis Moreau de Maupertuis (1698-1759)

The first natural philosopher to propose a scientific (albeit brief) theory of evolution was the Frenchman, Pierre Louis Moreau de Maupertuis. Not a strong child, being spoilt by his mother has been suggested as the reason why, in adult life, he found it difficult to accept criticism, but many a person, spoilt or not, finds criticism difficult to accept when it is perceived to be unjustified. If Maupertuis was convinced that he was right, and his critics wrong, perhaps that was because he was!

As a young man, he spent time travelling, as was to be expected, and also enrolled for two years' service in the army, also to be expected. Military service in peacetime was not onerous. With a military uniform and the title of Captain, he was able to mix freely in society, while at the same time continuing his study of mathematics, which was his first and most enduring discipline.

In 1728, the year that Newton died, Maupertuis had visited London where he had learned of, and accepted, Newton's laws of gravitation, which he introduced into France on his return. In his *Introduction* to his translation of Maupertuis' *Venus Physique*, Simone Boas cited a letter written by Voltaire giving a light-hearted summary of the differences in thought at that time between France and England. These included that the French attributed the tides to pressure from the Moon, while the English (following Newton) attributed them to the attraction of the Moon; in Paris the Earth was thought to be the shape of a melon while in London it was thought to be squashed at the poles. This helps us to understand, at least to some extent, how new ideas, which we now take for granted, had to battle for acceptance. Maupertuis did accept Newtonian physics, and tried to promote it on his return to Paris, but was considered by some to be a 'traitor'.

In 1736 Maupertuis became part of a team that set out to prove that the Earth was not a perfect sphere. While another team took measurements from near the equator in Peru, Maupertuis led an expedition to Lapland, which was as close to the North Pole as was then feasible, to take comparative measurements. His team spent a year in the Arctic, during which time his health appears to have been quite good. On his return, these measurements, combined with those of the team from Peru, were able to show that the Earth was indeed slightly flattened at its pole.

Today we may admire the mathematical and physical effort which was necessary to produce this result, but consider it of nothing more than passing interest, but, at that time, the finding had important philosophical implications.

It had been a shock to the collective Western system to take the Earth from its lowly sub-lunar position and place it in the Heavens. Some were still finding it impossible to accept that the Earth was not flat, but was a sphere, like other heavenly bodies. It had been difficult to accept that the Earth moved through the heavens, that it circled the Sun. Two things remained constant: the Moon circled the Earth and the Sun, Moon and other heavenly bodies were perfect spheres. Now it was being shown that the Earth was not a perfect sphere. It was squashed!

Unlike the Copernican revolution which had shattered the collective peace of the Western mind, the finding of Maupertuis and his colleagues caused barely a ripple among the general population, because so few people had access to their work and it was little known, except among the close scientific community and, of course, the mariners who relied on accurate latitudinal and longitudinal measurements.

His findings were controversial, not least because they were inspired by Newtonian physics, and his fellow Frenchmen were determined not to accept a shape for the Earth which "had been conceived by an Englishman" (Maupertuis 1752/1966: xvii). That the Earth was not a perfect sphere was one more nail in the coffin of Plato's 'Goodness', of the concept of the perfection of God and of all his creation (except humans). Philosophical idealism was giving way to humanistic practicality. For the first time, people were openly questioning the validity of parts of the Bible, mostly the Old Testament, although as the reproductive process became more fully understood, the fundamental principle of Christianity, that of the Virgin Birth, was also increasingly questioned.

Although at one time close friends, Maupertuis and Voltaire had a falling out of opinion and their relationship became increasing acrimonious. This may have influenced Maupertuis' decision in 1740 to move to Berlin, at the invitation of Frederick the Great, to re-organize the Berlin Academy of Sciences, of which he was to be President for fourteen years (1742-1756).

Glass (1955) referred to Maupertuis as 'a forgotten genius', although it was not Maupertuis' astronomical work to which Glass was referring, but his early proposal of a theory of evolution, which encapsulated in its simplicity the basics, not only of the theories which later became known as Darwinian Natural Selection, but also of Mendelian genetics. In his work, *Essai de cosmologie*, published in 1750, Maupertuis said (Glass 1955: 103):

> May we not say that, in the fortuitous combination of the productions of Nature, since only those creatures *could* survive in whose organization a certain degree of adaptation was present, there is nothing extraordinary in the fact that such adaptation is actually found in all those species which now exist? *Chance*, one might say, turned out a vast number of individuals; a small proportion of these were organized in such a manner that the animals' organs could satisfy their needs. A much greater number showed neither adaptation nor order; these last have all perished … Thus the species which we see today are but a small part of all those that a *blind destiny* has produced. (*Emphasis added by Glass 1955*).

And again, a year later in *Systems of Nature*, Maupertuis wrote (Glass et al. 1959: 77):

> Could one not explain by that means [mutation] how from two individuals alone the multiplication of the most dissimilar species could have followed? They could have owed their first origination only to certain fortuitous productions, in which the elementary particles failed to retain the order they possessed in the father and mother animals; each degree of error would have produced a new species; and by reason of *repeated deviations* would have arrived at the *infinite diversity* of animals that we see today. *(Emphasis added by Glass et al. 1959)*.

Maupertuis made no specific mention of the factor of time, which was to play such an important part in the development of the theories of both Lamarck and Darwin. However, he clearly suggested that repeated chance or random change (deviation/mutation) had led to the development of the millions of species we know today by a process of the elimination of the least 'fit' or, as Herbert Spencer was famously to say "the survival of the fittest". His claim that destiny was blind was in stark contrast with the accepted belief, not only in France but throughout the then known world, that creation was the work of God.

The term 'evolution' was not then used to refer to the birth of species or genera, but to that of individuals. How did the individual, human or animal, unfold (evolve) in the womb? Contemporary thinking favoured 'preformation', the idea that either the mother or the father carried within their reproductive organs a minute form which was to unfold and expand through its embryonic stage until it was capable of sustaining its individual life at birth. The minute form was not believed to be a miniature of the adult form that simply grew. Rather they thought of the miniature form as a bud which gradually unfolded (evolved) into a form sufficiently mature for the new life to be able to sustain itself outside the womb (or egg, in the case of oviparous creatures). The changing form and shape of embryos was well known. Many observations were being made, at the expense of unfortunate female animals, of the state of the uterus, its condition and contents at varying times after mating, in an attempt to ascertain the very earliest time at which it could be said that a new life form had begun to 'evolve'.

At that time it was much debated whether it was the female who provided the miniature form which would grow into the infant, the male ejaculate merely providing nourishment (ovism) or whether it was the male spermatozoon which provided the miniature, the female form providing the nourishment (animalculism). Maupertuis not only held that both parents contributed, but that they contributed equally. Offspring clearly bore resemblances to both parents. While the resemblance to the one furnishing the 'miniature' was understandable, resemblance to the other parent was harder to explain. It was suggested that the 'nutrition' supplied by the other parent was the source of such resemblance. Maupertuis rejected this concept on the grounds that, if nutritive material could so influence the growing embryo that it would bear a resemblance to the nutrient source, why did the human infant not resemble the fruit, vegetables and other nutritive substances eaten by the mother during her pregnancy?

The concept that semen may have a purely nutritive purpose, or may serve merely to stimulate the female 'form' into growth, was not as outlandish as it may at first appear. Geneticists today are aware of *thelytokous*. Species of animals exist in which males are either totally absent or are very rare and genetically non-functional. Many are simply all-female species, although whether the fact that they are not an interbreeding population precludes them from being described as a 'species' at all is a debatable point (White 1973a). The small freshwater fish, *Poecilioa formosa*, from south Texas and northern Mexico, is genetically all-female, but its eggs need to be activated by the sperm of a male from a closely related species, either *P. latpinna* or *P. shenops*, before they will begin to develop (White p. 154). With the spider beetle, *Ptinus mobilis*, the situation is slightly different; development of the egg in this all-female species does not proceed beyond first metaphase until the egg is penetrated by a sperm from the related bisexual species, *P. clavipes* (White p. 155). In neither case is there any genetic contribution from the male.

By the middle of the eighteenth century, when Maupertuis was undertaking his work, the microscope had revealed the presence of what were claimed to be 'animalcules' in the male semen. Did the female 'semen' produce anything similar? Graaff believed that he had discovered the female egg, but what Graaff had seen was the Graaffian follicle where eggs develop. At times, Maupertuis suggested that the female discharged a multitude of eggs, possibly as many as the male 'animalcules', but they were so small that they had yet to be seen, even under magnification. He even suggested that these might emanate from the lining of the womb. However, at other times he spoke of the 'egg' in the singular. For example, in *The Earthly Venus* (1752: 9) he wrote:

> Fallopius found two tubes whose floating extremities, in the abdomen, have a kind of fringe which may reach over the ovary, retrieve the egg, and then propel it to the uterus into which these tubes open.

It seemed logical that, no matter how many eggs the female produced, usually only one would grow in the womb, although twin, or even triple, births were, of course, known. What was not known was where or how the egg interacted with the spermatozoa and whether one or numerous spermatozoa were involved in the production of the fœtus. Maupertuis believed that the egg and the 'animalcules' each contained particles responsible for the formation of parts of the body but did not know whether one, several, or many spermatozoa were involved.

Using the principle of 'attraction' (gravitation) as propounded by Newton, Maupertuis believed that each of these particles would gravitate towards, and join with, their most closely related particle, so that all the body part particles would form a whole, which would then take on embryonic form. Once two related particles had combined, any other particle which might have been suitable for that role would have no place and would be discarded.

Particles from either parent had equal chance of forming any particular part of the embryo. The child would thus resemble both parents in some degree, although the proportion of representation would vary, some children resembling one parent much more than the other. If the process worked perfectly, a 'perfect child' would result. If some particle was missing, or became damaged, a child deformed by deficiency would be born. If some particle in excess of that required somehow managed to attach itself, then extra body parts would result, even to the extent of producing conjoined twins, infants with two heads or with one head and two bodies.

In 1745, an albino Negro had been brought to Paris. Maupertuis learned that albinoism occurred sporadically among Negroes. Indeed, in Senegal there were whole families of Negroes who were 'white' and there were known to be albino 'Indians' in Panama (Glass 1955: 106). It seemed to Maupertuis that the potential for this to occur must be contained within the hereditary particles, albeit present in less quantity than normal particles.

He reasoned that something similar happened with polydactyly (the presence of a sixth digit on one or more hand or foot). Maupertuis studied the history of a family from Berlin who exhibited this trait. It was apparent that this condition could be passed on by either the male or the female, that it was an inherited characteristic rather than a congenital deformity.

Study of a second family, named Kelleia, gave similar results as far as male/female transmission was concerned, but the appearance of the trait seemed to have been more frequent.

While certain facts, such as the transmissibility of features to the offspring by both parents, seemed to Maupertuis to be beyond dispute, he remained uncertain of the precise process of reproduction. Maupertuis made it clear to his readers that he was not writing for the purpose of giving answers to problems, rather to point them out, to share his thoughts in the hope of inspiring others to ponder the same problems and, hopefully, make further contributions to scientific thought on this fascinating topic.

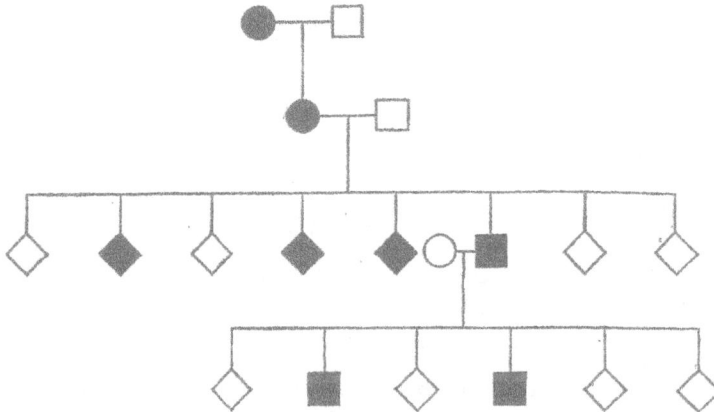

Fig 2.1: Polydactily in the Ruhe family shown in black:
black squares = males; diamonds = females; circles = sex unknown
(Glass 1955: 106)

More than a century before Darwin published *On the Origin of Species* in 1859, all the basic principles of heredity, monstrosities/sports, those which were "one-off" deformities against those which were variations subject to inheritance, adaptability and survivability, were the subject of scientific investigation. Until Glass rediscovered Maupertuis in the 1950s, his work remained virtually unknown. Even now, it receives scant attention and little has been translated into English. It will be argued throughout this book that there was a concerted attempt by many prominent writers on the subject of (human) evolution to downplay the role of anyone whose work might be seen to detract from the uniqueness and greatness of Darwin's contribution.

Georges-Louis Leclerc, Comte de Buffon (1707-1788)

Count Buffon was the pre-eminent naturalist of the 18th century, being 'Keeper' of the prestigious *Jardin du Rois* in Paris. He wrote extensively, his works appearing intermittently, but regularly, as a series of more than forty 'volumes' between 1749 and 1804, some of which were quite short, but together they comprised a sizeable amount of work. In 1781, Buffon's work was translated into English by William Smellie. In 1834 this appeared as a two volume work under the title *Natural History, General and Particular, including the History and Theory of the Earth, a General History of Man*, published by the Chambers family of Edinburgh.

Buffon believed the Earth to be very old and to be but a small planet in an immense Universe. These are concepts with which we have grown up but which were quite revolutionary in the second half of the eighteenth century when he was writing, at least as far as the concept of the immensity of time past was concerned. No doubt remembering Copernicus and Galileo, Buffon took care not to antagonize the Church and, amazingly, lived a long and successful life - unpersecuted.

Buffon believed that the Earth, and the other planets, had been formed from a portion of the Sun, which had become detached. This hot vaporous mass had gradually cooled, becoming first molten and then solid rock. The Moon, Buffon believed, had in turn been formed from a portion of the Earth, probably torn from the region of the Pacific Ocean by some unknown force, possibly that of a passing comet or other heavenly body.

By now, hopefully, you will be mentally taking yourself back in time and appreciating quite how revolutionary were his ideas, formed in a society in which people were raised with a literal belief in the account of Creation as given in *Genesis*, as well as of the Flood and Noah's Ark. Again, as with Maupertuis, it must be remembered that little of Buffon's work would have become known among the general populace but it made a lasting impression among the intelligentsia. In England, as we shall shortly see, Erasmus Darwin was thinking in a similar manner (and mentions Buffon's work in his own writing) but it was not until 1834 that Buffon's work became available to the people of Britain in their own language.

The science of geology was in its infancy but at that time there was an acceptance, at least among natural philosophers, that there had been changes upon the surface of the Earth as it was formed, although the amount of time involved was not generally thought to be extensive, at least by today's standards. In this, again, Buffon and Erasmus Darwin were in the forefront of thinking. Some philosophers thought that water, particularly after the Flood, had been the main instrument in the shaping of the Earth's surface. The people adhering to this school of thought were referred to as 'Neptunists' (and they included Erasmus Darwin). The 'Neptunists' believed that the Earth had once been completely covered with water, much of which had subsequently disappeared, partly through evaporation but mostly by being engulfed in great chasms within the Earth. Buffon belonged to the opposing school of thought, being a 'Vulcanist'. They were called after Vulcan, the God of Fire. That the Earth's centre was hot, molten, was evident from the lava thrown up by erupting volcanoes, as well as by the bubbling water found at hot springs. Buffon allowed that many of the superficial features of the Earth were due to the action of volcanoes, earthquakes and erosion by wind, weather and water, he attributed much of the irregularity of the Earth's crust to the cooling, and therefore 'shrinking', of the Earth itself. Volcanic eruptions and earthquakes may make abrupt changes, but erosion by wind and weather would take time. As with the work of Maupertuis, expanses of time, far greater than that 'allowed' by biblical history, were being invoked.

Buffon estimated that the solid matter of which the Earth was comprised weighed some four times that of the matter of the Sun, great changes having taken place as the Earth cooled and solidified. However, what most impressed Buffon was the formation of sedimentary rock, which had been laid down under water over considerable periods of time, but which was now elevated above sea level, in some cases to a great height. He realized that the Biblical account of the deluge was totally inadequate to account for these formations and instead proposed that the account in *Genesis* referred not to 'days' of twenty-four hours duration but rather 'Days' as in 'Ages'. The Biblical deluge occurred in the last, seventh Age, after the creation of Man.

Buffon claimed that the land beneath the sea was the same as the land above it - composed of mountains, hills, valleys, even 'rivers', as underwater currents flowed more rapidly in the vicinity of hills and valleys. He claimed that the same factors that operated to wear down rocks in one place and to cause sediment to accumulate in another, operated under the sea as well as on dry land. While waves were a superficial phenomenon caused by the operation of the wind, the tides were caused by the Moon, whose effects would be felt throughout the ocean, no matter its depth. Great volumes of water were constantly being dragged across the bottom of the ocean, wearing down hills and mountains, depositing sediment, etc.

Buffon was aware that the climate of some parts of the world appeared to have changed. He attributed this to the movement of land across the surface of the globe. Over immense amounts of time, as rock was worn away in one place and sediment deposited in another, land masses would gradually have 'moved' into warmer or cooler climates.

Buffon's theory of the history of the Earth was one of change over immense periods of time, but of change that was always occurring in conformity with the basic principles of physics, including gravity, which was the new and dominant concept of the time, thanks (in France) to Buffon's older contemporary, Maupertuis.

Linnæus' scheme for the classification of all living things, published between 1735 and 1774, while Buffon was a young man, was to have profound consequences for European thought. It became necessary for the serious naturalist to label all plants and animals, however large or small, within the Linnæan binomial system of classification. General acceptance of nomenclature within the community of naturalists was essential if naturalists were to understand each other's writings, yet the task of classification was fraught with difficulties and caused much controversy, none more so than which plants and animals were varieties and which were distinct species.

Initially, Buffon thought that all living things graded themselves imperceptibly, making classification difficult, but he changed his mind. He came to recognize species as definite entities, the ability to interbreed being the defining attribute, although he also realized that this definition could never be anything but a 'working definition', due to the difficulty in establishing the breeding potential of all creatures.

Buffon used a system of argument and counter argument, which left the reader to decide which point of view to accept and the author free from any accusation of promoting heresy. For example, Buffon suggested that the horse and the donkey might be regarded as a single family having a common origin. The same logic would suggest that humans and apes had a common origin and that all animals may have descended from a single animal, but Buffon went on to say: "But no! Certain from revelation that all animals have participated equally in the grace of direct creation" (Lovejoy 1959: 98). His first suggestion was a clear pronunciation of evolutionary theory, more thoroughly enunciated and more widely known than the brief ponderings of Maupertuis. It was the point of view taken by his younger colleague, Lamarck, whose work will be discussed shortly. However, there was one condition which Buffon considered must be accommodated before evolution could be accepted as a fact, that condition being the production of a new species by direct descent.

Like many others, Buffon was intrigued by mules, not only those produced by the mating of the horse and the ass but those of other animals, such as the dog and the wolf, the term 'mule' being applied to species crosses, whether they were fertile or not. Some animals from different species could mate and reproduce, resulting in a healthy animal that could live many years but which was not itself able to reproduce. Buffon carried out his own experiments and studied reports of cross-breedings in an attempt to ascertain whether it was possible to produce fertile offspring from different species. He had no success (Buffon 1781/1834, vol.1: 28-36).

Dogs were able to vary so much, yet still constituted one species (vol.1: p. 357). This observation led Buffon to conclude that variation within species, no matter how great, was not sufficient for the formation of new species. Lovejoy (p. 99) questioned why Buffon should consider the sterility of hybrids proof of separate descent and quoted Buffon's claim that it would need to be assumed that:

> ... two animals, male and female, had not only so far departed from their original type as to belong no longer to the same species ... that is to say, to be no longer able to reproduce by mating with those animals which they formerly resembled — but had also both diverged to exactly the same degree, and to just that degree necessary to make it possible for them to produce only by mating with one another.

Lovejoy went on to say: "The logic of this is to me, I confess, a trifle obscure; but it is evident that Buffon conceived that the evolution from a given species of a new species infertile with the first could come about only through a highly improbable conjunction of circumstances". Unlike Lovejoy, I find Buffon's logic quite compelling. For a new species to come into existence, the evolved specimen must be *unable* (not merely unwilling) to breed with its own parent, or any other member of its parents' species. Furthermore, to continue as a new species, it must find a mate which has evolved in a manner similar to itself, also unable to breed with other members of its parents' species.

It was as if Buffon wanted to believe in evolution, but his study of hybrid animals and their sterility stopped him fitting the facts to the theory.

Buffon followed Maupertuis in rejecting 'preformation', believing that reproduction was the result of the coming together of particles from the male and the female. He suggested that in infancy and childhood these particles were used for nutrition and growth. When growth had been completed there were surplus particles, which could then be used for reproduction (Buffon 1781/1834, vol.1: 186):

> There exists, therefore, a living matter, universally distributed through all animal and vegetable substances, which serves alike for their nutrition, their growth and their reproduction ... reproduction takes place only through the same matter's becoming superabundant in the body of the animal or plant. Each part

of the body then sends off the organic molecules which it cannot admit. Each of these particles is absolutely analogous to the part by which it is thrown off, since it was destined for the nourishment of that part. Then, when all the molecules sent off by all the parts of the body unite, they necessarily form a small body similar to the first, since each molecule is similar to the part from which it comes ... these two seminal fluids are extracts from all parts of the body; and a mixture of them is all that is necessary for the formation of a certain number of males and females.

There is a clear similarity between Buffon's ideas and those of Charles Darwin when the latter enunciated his theory of *pangenesis* (1868), although it is not suggested that Darwin was aware of this similarity of thought.

James Hutton (1726-1797)

James Hutton was little known in his life time and his work may well now be forgotten were it not for one of his colleagues, John Playfair, who took it upon himself, not merely to republish Hutton's work after his death, but to append an 'explanation' which was longer, and a lot more readable, than the original work. It was this work which inspired Charles Lyell (see below) who, in turn, inspired Charles Darwin.

A recluse, whose writings were little known outside his immediate circle, James Hutton was a member of the Royal Society of Edinburgh. In 1795, he published a two volume work entitled *Theory of the Earth*, a small edition of no more than 500 copies. It made difficult reading. The book has never been reprinted and only a few of the original copies survive. In 1802, five years after Hutton's death, Playfair published an abridged version (140 pages) of Hutton's work, together with 'Explanatory Notes and Additions', which ran to 385 pages, a facsimile reproduction of which was published in 1956. The facsimile reproduction of Hutton's book also contained facsimile reproductions of three papers presented by Hutton to the Royal Society of Edinburgh in 1785, 1788 and 1794.

A subsequent edition also included a facsimile presentation of the *Abstract* presented by Hutton to the Society in March and April 1785, summarizing his first paper and a biography of Hutton written by Playfair.

Hutton was much interested in coal and other similar natural products such as bitumen and petroleum. He maintained that all were of vegetable origin, although some of his colleagues argued that some grades of coal and oil were mineral based. It was one thing to accept that the bones and shells of sea creatures had become deposited on the sea floor, been compressed into chalk, then raised from beneath the sea to form chalk hills, it was quite another to explain how vegetable matter had become compressed beneath rock and transformed into coal or oil. Clearly the vegetable matter must have grown above ground at one time. Hutton envisaged cycles of land being raised above the surface of the sea, bearing vegetation, then becoming compressed again beneath it. At the minimum he postulated three such cycles, but believed there were probably far more. By extension, he believed that these cycles would continue into the future. Hutton was formulating his theories at approximately the same time as Buffon but it is unlikely that he was aware of Buffon's work, which did not become available in Scotland until 1834, when it was translated into English and published in Edinburgh by the Chambers Publishing House.

It was not only coal which took Hutton's interest. He was fascinated with the origin of veins of metals and minerals which were 'injected' into the rocks of other formations. The 'Neptunists' claimed that these veins were the result of matter which had been held in solution during the Universal Flood which had covered the Earth before any land rose above sea level. As the Earth dried out, these minerals, which had been held in solution, gradually dried out to provide the minerals and metals mined today. Hutton argued that many of these substances were not water soluble. Furthermore, there was no known solvent which could carry all of the known metals and minerals. They could never have been held in solution.

Like Buffon, Hutton was a 'Vulcanist'. He did not deny that water had played an important part in the erosion of land, but he denied that it had played any part in its formation. He argued that organic matter on the sea floor was subject to immense pressure, as also was any non-organic matter beneath the sea. This pressure would cause the temperature to be raised. While it was true that coal, heated to combustion temperature on the surface of the Earth, would burn and turn to ash, in the absence of air such combustion would not be possible. There was no way of knowing at that time exactly what would be the result of heat being applied, under pressure, to matter, but Hutton postulated that it was this which formed metals, minerals and substances such as coal and oil.

In the same way that air and water circulated under the influence of temperature differentiation, so, Hutton believed, did the substance of the Earth. The movement would be far slower, but it would happen. In some places the heat would cause the surface of the Earth above it to 'bulge'. In this way, great land masses would be raised above sea level. This raising of great masses to the surface of the Earth would cause fractures, which would allow molten matter to run as fluids into other structures, forming the metal and mineral veins in which Hutton was so interested. This expansion and fracturing would also explain the tilting of layers of sedimentary rock, which must have been laid down horizontally. Hutton did not rely on volcanoes and earthquakes alone to account for the formation of land, although he did acknowledge that they had a role. Hutton was not prepared to speculate on time, simply stating that, in comparison with human life, it was eternal.

In 1774, Hutton visited the salt mines in Cheshire, accompanied by 'his friend Mr. Watt, of Birmingham' (Playfair 1802/1970: 151). This 'Mr. Watt' was presumably the engineer, James Watt, who was also a friend of Erasmus Darwin, Charles Darwin's famous grandfather. Erasmus Darwin recorded conducting an experiment with Dr. Hutton of Edinburgh in the early 1770s (King-Hele 1963). Darwin had studied medicine for two years in Edinburgh and he may have made Hutton's acquaintance at that time or his participation in this experiment may be explained by their mutual acquaintance with James Watt, who was also from Scotland.

The experiment was related to Erasmus Darwin's theory that as air rose, it expanded, because there was less pressure at a higher altitude. The same amount of heat was now distributed over a greater area, causing the air to cool, without transfer of heat to any other substance. This became known as adiabatic expansion. The now cooler air would be less able to retain moisture and rain would fall. Clearly, if this were the case, then gases (or any other substance) would increase in temperature when compressed, without addition of heat from any other source. Upon this latter application of the principle much of Hutton's theory depended.

Hutton did not speculate as to First Causes, the origin of the Solar System or the formation of the Earth itself, as did Buffon. Whatever laws or forces operated to bring the Earth into existence were not those operating in its maintenance. Whatever laws or circumstances would operate at some time in the future to bring the Earth to an end were not operating now in the day to day maintenance of the Earth as we know it. Hutton was concerned only with tracing, as accurately as possible, what had occurred since the Earth was formed by applying the laws of physics and chemistry, which operated now as they had done ever since the Earth began. It was through the interpretation of these laws that scientists could draw valid conclusions about the past.

Chapter 3

Darwin's 'Great' Grandfather

There is one more great mind, that of Erasmus Darwin (1731-1802), whose ideas need to be studied in this review of early evolutionary theorists of the 18th century. To King-Hele (1963, 1968, 1999) the modern reader is indebted for a fascinating look at the life and work of Dr. Erasmus Darwin, now best known for being the grandfather of Charles Darwin, but known to his contemporaries as the finest physician and most knowledgeable man in all England. It was the 'Darwin name' which drew attention to the young graduate, Charles, when a naturalist was being sought as a companion for Captain Fitzroy on the voyage of the *Beagle*. Erasmus Darwin married twice, had fourteen children, most of whom survived into adulthood, and, like his grandson, Charles, stuttered, an affliction which he did not allow to be a deterrent in his social or professional life.

Erasmus Darwin was a compulsive inventor and enjoyed the company of others of similar mind, such as Josiah Wedgwood (potter), James Keir (chemist), James Watt (scientist), Matthew Boulton (manufacturer) and William Small, Professor of Natural History from America. Meeting regularly at around the time of the full moon (which gave them light by which to drive home), they became known as the Lunar Society of Birmingham, or, more simply, the Lunatics. This small, illustrious group was later to be joined by others, such as Joseph Priestley (chemist) and Samuel Galton (manufacturer). Darwin's son, Robert, married Susannah, daughter of Josiah Wedgwood, Charles Darwin being the grandson of both Erasmus Darwin and Josiah Wedgwood.

King-Hele (1963) listed some seventy five subjects in which Erasmus

Darwin might justifiably be considered a pioneer, ranging from artesian wells and artificial insemination to speaking machines and submarines. Darwin assisted James Watt with his work on steam engines and proposed a steam-carriage to replace the horse-drawn variety. He was sure that Priestley's discovery of oxygen would lead to the development of vehicles able to travel underwater.

> Led by the Sage, Lo! Britain's sons shall guide
> Huge SEA BALLOONS beneath the tossing tide.
> The diving castles, roof'd with spheric glass,
> Ribb'd with strong oak, and barr'd with bolts of brass,
> Boy'd with pure air shall endless tracks pursue,
> And PRIESTLEY'S hand the vital flood renew ...
> Deep, in war, waves beneath the Line that roll,
> Beneath the shadowy ice-isles of the Pole ...
>
> *(The Botanic Garden, Pt.I:IV)*

Darwin wrote a number of quite lengthy books in rhyming couplets and *The Botanic Garden* was also the medium in which he predicted the development of aeroplanes for both civil and military use:

> Soon shall they arm, UNCONQUERED STEAM! Afar
> Drag the slow barge, or drive the rapid car;
> Or on wide-waving wings expanded bear
> The flying-chariot through the fields of air;
> Fair crews triumphant, leaning from above,
> Shall wave their fluttering kerchiefs as they move;
> Or warrior-bands alarm the gaping crowd,
> And armies shrink beneath the shadowy cloud.
>
> *(The Botanic Garden, Pt.I:IV*

Mention has already been made of Erasmus Darwin's interest in adiabatic expansion. In a paper published in the *Philosophical Transactions of the Royal Society* in 1788, he gave his account of his work with Hutton and Edgeworth during the mid-1770s. He made the claim that air cooled as it expanded without any exchange of heat with an external source, causing 'devaporation' – the formation of clouds and rain. Although it would seem that they only met on the one occasion, Erasmus Darwin and Hutton corresponded over many years.

Darwin invented for the sheer joy of inventing. Outgoing by nature, he shared his ideas and seemed to have had no interest in making money from them. In this he seems to be the complete antithesis of his grandson, Charles, who was by nature introverted and who kept his thoughts and ideas secret, at times even from his few close friends.

Erasmus Darwin's podgy face stares out from the page of many a book on evolution, belying the physical and mental stature of this great man. His somewhat overweight condition may perhaps be explained by the fact that, having realized that the body and brain operated on sugars, he was a great believer in this substance. He campaigned successfully for sugar beet to be grown commercially in England.

Renowned as the best physician in England, as much for his compassion towards the poor and the mentally ill as for his medical results, Darwin's life-time of medical experience was contained in his *Zoonomia* or *The Laws of Organic Life*, the first volume of which was published in 1794. This was followed by a second volume in 1796, both volumes being considerably expanded in later editions.

Zoonomia was a medical text, intended by Darwin to contain all the medical knowledge at his disposal, including his own observations and recommendations. His suggested treatments, given under M.M. (Materia Medica) are extensive because they contain all remedies advocated by the medical fraternity, as well as by himself. He established a system of classification of disease, using the same format as that used for plants and animals. He proposed four main classes: *Irritation, Sensation, Volition* and *Association. Irritation* referred to disorders resulting from external irritation, *Sensation* to disorders resulting from internal irritation (pleasure/pain), *Volition* to disorders excited by desire or aversion and *Association* those diseases associated with external causes, such as alcohol, to which Darwin was implacably opposed.

Darwin was even more opposed to Christianity than he was to alcohol! He was appalled by the 'hell fire' preaching adopted by many Christian clergy, which he believed caused great distress to many people. Some people became so consumed with the idea that they would burn in hell for even trivial misdemeanors that they were driven to suicide, even though this would precipitate them into the very hell they so dreaded. Darwin advocated education, learning to think logically and rationally, as the best means for preventing people from falling under the spell of religious and/or hysterical mania, such education being as important for girls as for boys.

Darwin was well aware that much of the suffering which he was called upon to alleviate was self-inflicted, but he also realized that even those who led exemplary lives might still endure years of suffering in old age. Even this, to some extent, was 'self-inflicted'. Before humans became civilized, he believed, old age was unknown, at least in its incapacitating aspect. Darwin did not see Nature as cruel since Nature did not tolerate sickness or old age.

Evolution

Erasmus Darwin's thoughts on evolution were contained in *Zoonomia*, published in 1794, under the section dealing with *Generation*. It is remarkable how closely the elder Darwin's thoughts anticipated those of his now more famous grandson.

Darwin pointed out changes which had occurred in animals due to domestication. He made specific reference to horses, dogs, cattle, sheep, camels, rabbits, hares, pigeons and partridges and finished by referring to humans. He spoke of the mode of life of those who laboured at the anvil, oar or loom, which increased their strength, as well as circus performers, who increased their agility and the shapeliness of their limbs. He referred also to the adverse effects of alcohol. All these factors he considered capable of affecting future generations, for better or worse. Darwin drew attention to the great changes which took place during the lives of some animals. A crawling caterpillar transformed into an aerial butterfly, an aqueous tadpole into an air-breathing frog. Less dramatic, but equally important, were the changes which transformed the boy into the man, the girl into the woman. These changes (*Zoonomia* II: 233-40):

> ...which are in part produced by their own exertions in consequence of their desires and aversions, of their pleasures and their pains, or of irritations, or of associations; and many of these acquired forms or propensities are transmitted to their posterity.

From consideration of the similarity of structure between quadrupeds, birds and amphibious animals, Erasmus Darwin concluded, as had Buffon, that "all alike have been produced from a similar living filament".

Darwin addressed three basic 'desires': lust, hunger and security. The first was to become Charles Darwin's 'Sexual Selection', the other two were the basis of 'Natural Selection', being developed according to the species' needs as predator or prey, for defense, security or attack. He concluded (*Zoonomia* II: 240):

> ... would it be too bold to imagine that in the great length of time since the earth began to exist, perhaps millions of ages before the commencement of the history of mankind, would it be too bold to imagine that all warm-blooded animals have arisen from one living filament, which THE FIRST GREAT CAUSE endued with animality, with the power of acquiring parts, attended the new propensities, directed by irritation, sensations, volitions and associations; and thus possessing the faculty of continuing to improve by its own inherent activity, and of delivering down those improvements by generation to its posterity, world without end?

In the third edition of *Zoonomia*, published in 1801, Darwin outlined a theory of reproduction which was virtually identical with that of Maupertuis, although he may well not have been aware of Maupertuis' work in this area. In suggesting that small particles are produced by bodily organs which are carried by the blood to the sex organs, where particles from both parents unite to form the new offspring, Erasmus Darwin was also anticipating his grandson's theory of *pangenesis*.

The Temple of Nature, published posthumously in 1803, contained the following verses:

> Organic Life beneath the shoreless waves
> Was born and nurs'd in Ocean's pearly caves;
> First forms minute, unseen by spheric glass,
> Move on the mud, or pierce the watery mass;
> These, as successive generations bloom
> New powers acquire, and larger limbs assume;
>
> *(Canto i: 296-301)*

Darwin followed evolution through various forms of flora and fauna until he finally reached Man:

> Imperious man, who rules the bestial crowd,
> Of language, reason, and reflection proud,
> With brow erect, who scorns this earthy sod,
> And styles himself the image of his God;
> Arose from rudiments of form and sense,
> An embryon point, or microscopic ens!
>
> *(Canto 1: 310-315)*

In the same way that Erasmus Darwin was concerned about the lack of a systematic theory in relation to the human body, so, too, was he concerned about the lack of a systematic theory in relation to plants, their physiology, and their maintenance in a healthy condition. He set out to remedy this deficiency in a comprehensive work entitled *Phytologia* or *The Philosophy of Agriculture and Gardening*, published in 1800.

Darwin felt that the dividing line which had been drawn between plants (immobile) and animals (mobile) was an artificial one. Within the animal kingdom, there was a great deal of variation with regard to mobility. Some animals were capable of moving very swiftly, others were capable of almost constant movement, while others, such as barnacles, barely moved at all. Darwin observed that a 'bud' from a polyp could grow into a complete animal in the same way that might a bud from a tree or the branching cells of the coral 'insect'.

Darwin proved that plants respire, although by means different from that of animals. He attributed to leaves the functions of both respiration and circulation. Darwin endeavoured to show that plants also have 'muscles', 'nerves' and 'brains' by pointing out that many plants open and close their petals at dawn and dusk, some also in response to changes in weather. Some animals, such as shell fish, move less in a day than many plants, supporting his arguments with special reference to orchids, *Drosera* (Sundew) and *Mimosa* (Sensitive Weed). Charles Darwin also wrote papers on movement in plants such as *Mimosa*, on the fertilization of orchids by insects, and on the ability of certain insectivorous plants, such as *Drosera*, to consume insects by trapping them in their sticky digestive juices.

The Botanic Garden was written as 1,944 lines of rhyming couplets. Darwin pointed out that, while it may be the destiny of plants to be devoured, many lived far longer lives than many animals, during which time they were able to enjoy the benefits of the warmth of the sun, the fresh air, refreshing rain, the absorption of nutrients from the soil and, of course, reproduction. Darwin was convinced that plants experienced *joy* in living. Within this work Darwin took the opportunity to include an incredible variety of topics, from the Solar System, the outer atmosphere, shooting stars, lightning, rainbows, refraction, the aurora, phosphoric lights, glow-worms, volcanoes, gunpowder, cannon, steam engines, flying machines, submarines, electricity, electrically driven machines, Wedgwood pottery, coal, America, the Spanish Conquest, the French Revolution – and these are but some!

From his sometime colleague, James Hutton, Darwin took the knowledge that much of the surface of the Earth owed its existence to previously living matter. While Hutton viewed this fact with the cold eye of the mineralogist, Darwin saw everywhere beneath his feet the evidence of past life – the testimony of past *joy*. The rocks and the soil were part of the cycle of life, the cycle from soil to vegetable to animals and, of course, to human. Darwin had a profound feeling for the joy of life, that life was Good.

Erasmus Darwin died seven years before Charles was born and, therefore, had no direct influence on his grandson. Charles Darwin made no mention of the work of his grandfather in the first edition of *On the Origin of Species* (1859/1998). The third edition of *The Origin*, published in April, 1861, included an addendum of some eight pages acknowledging the work of some thirty previous writers on the subject of evolution. In a footnote to his comments on Geoffroy St Hilaire, Darwin added (p. 371):

> It is curious how largely my grandfather, Dr. Erasmus Darwin, anticipated the erroneous growth of opinion, and the views of Lamarck, in his *Zoonomia*, (I, 500-10), published in 1794... It is a rather singular instance of the manner in which similar views arise at about the same period, that Goethe in Germany,

> Dr. Darwin in England, and Geoffroy Saint Hilaire... in France, came to the
> same conclusion on the origin of species, in the years 1794-5.

While most modern historical sketches of evolutionary thought give place
to both Buffon and Erasmus Darwin, in the latter case it is more Darwin's
relationship to his grandson which is of interest, rather than his own
accomplishments.

Hutton had no interest in the evolution of biota. However, his work gave
credence to the concept of the Earth being of great age, far greater than
that indicated by the Bible, thus supporting the work of his contemporary,
Buffon.

Erasmus Darwin also argued that the Earth was of great age. He
embraced the concepts of both 'evolution' and 'continuity'. It is of interest to
note that he referred to 'The First Great Cause', rather than 'God'. He made
no secret of his anti-Christian views and avoided the use of a word which
carried with it connotations regarding Divine Attributes associated by
Christians with the (presumed) Nature of God. However, Erasmus Darwin
was a God-fearing man and his view of evolution was clearly religious.

As the result of the wide interest of the works of Maupertuis, Buffon and
Erasmus Darwin, by the time the 19th century dawned there was already
an interest in the concept of living forms becoming changed over time. This
concept was replacing that of the sudden appearance and subsequent
stability of created forms as portrayed in the Biblical account of Creation
given in *Genesis*. As will be seen in the next chapter, in the 1790s Lamarck
was already giving public lectures embracing the theory of evolution.
Evolutionary theory had started its journey towards becoming the
established (dominant) theory.

Chapter 4

Lost in Translation

It was a chance reading of an English translation of Lamarck's *Philosophie Zoologique,* first published in France in 1809, that opened up to me a completely new understanding of the history of evolutionary thought in the Western world. Lamarck's great work was based on lectures he had given during the 1790's, thus his thought and writings straddled the turn of the eighteenth/nineteenth centuries.

I would not have pulled the book down from the library shelf had I not already been familiar with Lamarck's name. Every text book on the subject of evolution (which I had been delighted to find included in the course on archæology in which I had enrolled) had commenced with an overview of evolutionary thought preceding Darwin and Lamarck's name had been prominent. The brief overview was usually accompanied by a picture of a giraffe stretching upwards its long neck in an endeavor to reach leaves higher on a tree, supposedly illustrating Lamarck's belief in the inheritance of acquired characteristics: parent giraffe stretches its neck, baby giraffe is born with a (slightly) longer neck. This concept was the subject of polite academic ridicule and I expected some light and amusing reading.

How wrong I was!

I found myself confronted with the most scholarly work on evolution that I had yet read and several years later, after reading hundreds of books on the subject while completing my thesis, I still consider Lamarck's work to be the most logical, the most scholarly.

Before studying Lamarck's ideas, it is time to study the man himself.

Jean-Baptiste-Pierre-Antoine de Monet Chevalier de Lamarck (1744-1829)

Jean-Baptiste Lamarck was born in France, at a place called Bazantin in Picardy, on 1st August, 1744, the youngest of eleven children. He came from an old established family, his father, Philippe Jacques de Monet, being lord of a manor, but they were of limited means, not members of the wealthy French aristocracy. He married late. In 1777, at the age of 32 or 33, he formed a lasting relationship with Marie-Françoise de Fontaines de Chuignolles, who bore him six children. However, they were not married until 1792, the ceremony being performed when Marie-Françoise was on her deathbed. While it was not uncommon for peasant people to live in *de facto* relationships, for a person of Lamarck's standing, the following of this path must have been a deliberate decision. Perhaps his antipathy towards the Church (see below) discouraged him from entering its portals and his principles prevented him from participating in a ceremony endorsed by a Church in which he did not believe. As will soon become apparent, Lamarck early showed an independent streak and all his life followed his own direction, popular or not.

Having taken the plunge once, Lamarck must have found the experience not as traumatic as he had anticipated, for he married a further three times, fathering two more children, although the legitimacy of his final relationship is not certain.

His three older brothers having made careers in the Army, it was his father's decision that Jean-Baptiste should enter the Church and he was placed in the Jesuit College at Amiens in 1755, while still only eleven years old. He was to spend four years there. His father died in 1759 and Jean-Baptiste took the opportunity to make a career change. He ran away to join the Army! At that time, the French were fighting in Germany, it being close to the end of the Seven Years' War. He purchased a horse, acquired a letter of introduction from a friend and, thus equipped, joined the French Grenadiers on the eve of the Battle of Fissinghausen. He was then 16 years of age. The French were soundly defeated, the officers of Lamarck's company being killed. It is reported that he took charge and showed courage under fire, which resulted in his immediately being appointed an officer. In a footnote to his *Introduction* to his 1914 translation of Lamarck's work, Elliot questioned the accuracy of this account, which was based upon a letter written to Cuvier in 1830, shortly after Lamarck's death, by one of his sons. Elliot felt that the letter magnified Lamarck's achievements. However, I find nothing untoward in the account of this part of his life. Teenage boys are notoriously hot-headed and feel themselves to be invincible. Lamarck's family had a long history of military involvement and his three older brothers had been allowed to follow this path. That a boy of his age should prefer a military, rather than a Church, career is quite understandable. The officers being killed, it would have been imperative that a substitute be appointed with all speed. In those days, officers did not work their way up through the ranks; they were appointed from among the ruling classes, not merely because of an assumed ability to lead, but because they were educated. An ability to read and write far greater than that acquired by most peasant children at the local village school, was essential. The arrival of a young man thus qualified must have been seen as very fortuitous by the Field Marshall who commissioned him. Today, we think of 16 year-olds as school children. Such was not the case then, when children entered the work force, took apprenticeships or entered their vocational training at a College or University, away from home, by approximately twelve years of age. Lamarck had spent four years at College before his 'escape' and would have been considered a man.

Lamarck was discharged from military service on medical grounds at the still young age of twenty-two. He suffered an enlargement of the cervical glands which condition, it was suggested, had resulted from 'horseplay' in the barracks, during which Lamarck had been severely pulled by his hair, stretching his neck. By way of footnote, Elliot gave a caveat, explaining that this was the account given by Cuvier, which differed from that given by his son in the above cited letter. While such 'horseplay' may have aggravated a pre-existing condition, it is unlikely that it was the primary cause of an enlargement sufficiently severe to require an operation and subsequent discharge from the army. Whatever the condition from which Lamarck suffered, it either resolved itself, or the operation performed was successful, because no mention is made of any further related problem during the remaining sixty-three years of his life.

Lamarck had now to decide upon a further career. The Church, the Army, Medicine or Law were acceptable paths open to younger siblings of the aristocracy. Lamarck had already made an 'escape' from the Church and his chosen career in the Military was now closed to him. With his capacity for detailed and logical thought, it may be supposed that Lamarck would have been well suited to the Law, but he chose medicine instead. He moved to Paris, where he lived in a top floor apartment (garret) for a short while, before moving in with his brother. For a year, he supported himself by working as a bank clerk; then he took up his medical studies, which lasted four years (1767-1771). At this time, medical treatment was largely based on the prescription of herbal remedies, although the use of chemicals, such as arsenic and mercury, was becoming increasingly popular. Botany formed a major part of Lamarck's study and it was in this discipline he was to specialize, never practicing as a doctor. The *Jardin du Roi* (later known as *Le Jardin des Plantes*) was not merely a botanical garden where one could view plants from around the world, as were the Royal Botanical Gardens at Kew in England. Because of the close connection between herbs and medicine, the botanical gardens in Paris were also the centre for medical education and biological research.

It was during this time that Lamarck made the acquaintance of Buffon, who became his patron. After ten years' study and work, Lamarck, with the assistance of Buffon, published *Flora Française* (1781), a comprehensive account of the flowering plants of France. This resulted in Lamarck being admitted to the French Academy of Science.

Buffon assisted Lamarck in other ways. On his recommendation, Lamarck was appointed 'Botanist to the King'. The education of both Lamarck and of Buffon's son was extended when Lamarck was chosen by Buffon to accompany his son on a two year tour of Germany, Hungary and Holland (1781-1782), where they studied rare plants and had the opportunity of meeting other eminent botanists. On his return, Lamarck was appointed keeper of the Herbarium at the *Jardin*, writing his *Dictionaire de Botanique* and *Illustrations de Genres*. After Buffon's death in 1788, Lamarck continued his work at the *Jardin*. He recommended its re-organization, submitting a proposal to the *Assemblée Nationale* which was accepted, in large part, when, in 1793, the *Jardin* became the *Museum d'histoire Naturelle*. The new complex was extended to cover twelve different scientific fields, each overseen by its own professor. No longer was it merely a 'Garden'. It now encompassed the study of animals. There were two chairs of zoology and Lamarck was appointed Professor of the department dealing with 'inferior' animals, the 'insects' and 'worms' of Linnæus. The Chair for 'superior' animals (mammals, birds, reptiles and fish) was awarded to Geoffroy St. Hilaire.

It might have been supposed that the eminent Lamarck would have been appointed to the chair of botany. Quite why he transferred from botany to biology is not known. It seems possible that, during his years of botanical study, Lamarck became interested in the beetles and bugs, butterflies and bees, as well as all the other many, varied and *incredible* species with which our plants have such a close relationship. It seems possible that Lamarck, after having devoted more than twenty years to the study of plants, requested a change, a chance to break new ground. Little was known of the 'inferior' animals compared with the 'superior' animals, yet their anatomy, physiology, method of reproduction and overall way of life was so much more varied and interesting than that of the 'superior' animals, now known as vertebrates. Which was the greater challenge? That Lamarck may have chosen the invertebrates must remain speculation, but so, too, must the idea that he was allocated their study as some form of 'consolation prize' after the Chair of botany had been awarded to someone else as suggested by Elliot: " ... there now seemed nothing suitable remaining for him except this chair of zoological remnants".

Honeywill (2008) supported Elliot's claim that the Chair for the study of inferior animals had been awarded to Lamarck almost as a 'consolation prize' because it was not wanted by anybody else. However, Honeywill added that Lamarck later purchased a modest summer house with the proceeds of the sale to the Government of '"one of his spectacular shell collections". Quite when Lamarck started his shell collections is not recorded, but it would seem possible that it was Lamarck's interest in the 'inferior beings' which had led to his suggestions for the *Jardin's* re-organization, not the other way around. It was Lamarck who had recommended that the *Jardin* should be, not merely reorganized, but extended to encompass animals as well as plants. It was he who wrote the recommendations which were submitted to the *Assemblée* in 1793. He was at that time held in high regard on account of his botanical work, his controversial ideas on evolution not yet having been enunciated, let alone published. Why should he not have been awarded a Chair, acceptable to him, on the grounds of his work?

One of the first things which Lamarck did, after taking up his new position, was introduce the terms 'vertebrate' and 'invertebrate' to replace the terms 'superior' and 'inferior'. There is nothing in Lamarck's work to suggest that he considered any life form to be 'superior' or 'inferior' to any other, including humans, merely different, more or less complex.

The Empress Josephine took a great interest in the creatures being newly discovered in previously little known lands and when Baudin arrived back in Paris in 1804, after sailing as far as Australia, the Empress Josephine laid claim to the emus and the kangaroos, which she kept in her extensive gardens. To me, the thought of travelling for months in what was by today's standards quite a small ship, with emus and kangaroos on board, is mind-boggling, but they did! Lamarck would have had access to extraordinary specimens of creatures previously unknown to the northern hemisphere, other than possibly by drawings, some alive, others preserved, but at least 'in the flesh'. This occurred five years before Lamarck published his *Philosophy Zoologique,* although, of course, he was constantly publishing other material.

Lamarck lived in turbulent times. When he was growing up France was ruled by a King. When he fought in the Army, he did so in the service of the King. He relocated to Paris in time to experience the trauma of the French Revolution, an event without precedence in European history.

Lamarck was from a minor aristocratic family, albeit the youngest of eleven children. Nevertheless, to live, work and survive in Paris during this time was an achievement in itself! Lamarck does seem to have been a genuine supporter of Napoleon, possibly because Napoleon overthrew the Catholic Church. Napoleon permitted a belief in a 'Supreme Being' but would not tolerate Christianity, showing his contempt by using *Nôtre Dâme* cathedral to stable his horses! Remembering that the young Lamarck had 'escaped' from the Jesuits at the first opportunity, it is reasonable to assume that he genuinely embraced Napoleon's vision.

Part of Lamarck's duties at the *Jardin* was the giving of lectures. It was during this time, the 1790s, that Lamarck first formulated his ideas on evolution by the writing of some early papers. In 1809, he published the ideas he had been formulating as the result of his work. His *Philosophie Zoologique* outlined his theory of evolution and was the first book ever to be published devoted solely to this topic. Lamarck's later work, *Histoire naturelle des Animeaux sans vertebres,* comprised a number of volumes which appeared between 1815 and 1822.

The acme of Lamarck's career had coincided with the acme of Napoleon's power. By the time Lamarck's later work was published, Napoleon had been defeated at the Battle of Waterloo. As Napoleon's fortunes faded, so did those of Lamarck. The Catholic Church became re-established and his younger colleague, Georges Cuvier (1769-1832), a devout Catholic, completely rejected any theory of evolution which could not be reconciled with the Biblical account of Creation and the establishment of Man on Earth as told in the book of *Genesis.*

Cuvier became known as a brilliant anatomist. He was the first person to subject fossil bones to scientific scrutiny and to attempt to classify them into families and species. Cuvier acknowledged that there were fossil remains of creatures which no longer existed and that these may have become extinct as a result of a natural catastrophe, such as a flood. He even acknowledged that the Flood recorded in *Genesis* may have been the last of a series, but he completely rejected the concept of change through the process of evolution. The high regard in which Cuvier was held, the falling out of favour of ideas associated with the Napoleonic regime, as well as the re-establishment of Christian views, all combined to counter Lamarck's revolutionary concept, which was rejected and derided. Cuvier campaigned against Lamarck's evolutionary ideas with a religious fervor – and won! Lamarck was discredited during his lifetime and Cuvier's attacks continued after his death. How different the history of evolutionary theory may have been had Geoffroy St. Hilaire not surrendered his Chair to Cuvier when he chose to accompany Napoleon on his expedition to Egypt. According to Honeywill, St. Hilaire and Lamarck were like-minded; St. Hilaire supported Lamarck's ideas and the two men were life-long friends.

I was only able to read Lamarck's work in English because in 1914 Hugh Elliot had taken the time and trouble to translate it. Perhaps he had been inspired to do so in 1909, the centenary of the book's original publication? Who knows – but the translation must have taken much time and effort to complete. One would have assumed that Elliot knew more about Lamarck's theory than anybody else at that time, but his *Introduction*, in which he attempted to explain Lamarck's theory, contains gross errors and misrepresentations, which are difficult to comprehend.

Elliot correctly stated Lamarck's two conditions for change: the effect of environment and an innate tendency of the living force to develop (expand). He also correctly stated Lamarck's thesis that, tracing the ancestry of living creatures backwards from the mammals to infusoria, decreasing levels of complexity were found. From this Elliot (1914/1963: xxiv) concluded that Lamarck believed:

> All existing animals are on the road of development from Monas to man, and man's ancestors include every existing species of animal. Not only had he birds, reptile and fish ancestors, but also arachnid, insect, worm, starfish, etc. He passed through the stage of being a scorpion and a spider. He traversed in turn every known species of insect. He was a tapeworm, a sea-anemone, a polyp and an amœba.

No wonder Elliot (page xxxv) stated that Lamarck's doctrine was "wholly absurd". Lamarck maintained that there was a progressive degree of organizational complexity throughout the series of classes of animals, but he never suggested, for example, that a reptile was evolved from a crab, or that an insect gave rise to a scorpion. Lamarck did believe that each order connected with another above and below it in order of complexity, and that if no obvious connecting link was to be seen, that was because it had yet to be identified. He was encouraged by the discovery of new fauna in Australia, especially the monotremes, which he saw as being a connecting link between birds and mammals.

Elliot also seemed to misrepresent Lamarck's theory of spontaneous generation. He condemned this as being based on a false premise, that infusorians die in cold weather, yet return with the warm weather, hence having been spontaneously generated (page xxxvii). Lamarck gave this illustration *in support* of his premise, but it was not the premise itself. Lamarck believed that the laws operating in this world, indeed in the Universe, were constant. He argued that if it were possible at some time in the past for inert matter to develop into living matter, then it must be possible for the same thing to happen again (and again), provided the same conditions existed. These conditions Lamarck believed to be the presence of a gelatinous or mucilaginous substance, water, warmth and some form of energy, which Lamarck likened to electricity. In this respect, Lamarck was more consistent in his beliefs than Darwin who believed that life started but once, millions of years ago. To Lamarck, this was 'Special Creation' in which he did not believe. It seemed logical to him that if the process could occur once, it could occur again.

Elliot (page xxxviii) also seemed to misunderstand Lamarck's theory of increasing complexity of organs:

> The series is full of anomalies. Birds, for instance, are lower in the scale than animals. Yet their pectoral muscles and sternum are developed out of all proportion to their proper place in the series.

Lamarck based his classification on internal systems, such as the digestive, circulatory, respiratory and nervous system, etc. At no time did he use anything as trivial as the development of a set of muscles as the basis for classification.

Elliot recapitulated Lamarck's contribution to the scientific study of zoology, giving credit to Lamarck for a number of innovations. However, he criticized Lamarck for having no conception of, for example, the carnivora (page lxii). Lamarck was interested in the developing complexity of systems, starting with digestion, but did not see the type of nutrient ingested to be of significance for classification. Lamarck grouped animals according to their anatomy. To him feet, hooves, claws and fins were more important for classification purposes than food consumed.

Elliot was committed to the materialistic philosophy. By contrast, Lamarck accepted a Creative Force as being constantly active in nature, working gradually in conformity with well-defined laws. Lamarck's belief that this living, creative force, was constantly pushing the boundaries of creation, rather like a flood of water making its way down a dry river bed, bursting its own bounds as it sought out new avenues, was in some ways similar to that of Teilhard de Chardin in the 20th century, whose work will be considered later.

Lamarck was the first person to publish a book which clearly postulated, not only evolution as a natural phenomenon, but a theoretical basis as to the order in which evolution had taken place. His book was intended to discredit the doctrine of 'Special Creation' as portrayed in the Bible.

While acknowledging Lamarck's vision, Elliot concluded: "Then came the *Origin of Species*, a work which naturally and immediately superseded every earlier publication". According to Elliot, Lamarck's work was "slowly entering upon the final stage of oblivion". It was this belief that stimulated Elliot to translate Lamarck's work, a labour which he said was undertaken with the view of vindicating Lamarck's memory. While we owe Elliot a debt of gratitude for having made available to the English-speaking world so important a book, nevertheless it is to be regretted that Elliot should have sought to undermine the very work which he so laboriously translated with the intention of preserving.

What were Lamarck's theories and teachings? I have read, and re-read, Lamarck's book, as translated by Elliot, and my conclusions are completely different from those of Elliot.

I believe Lamarck was right!

Chapter 5

Lamarck's Aim

Although the topic of evolution had been broached by Buffon, Lamarck was the first person to write a complete book on the subject. It was his aim to show how all life on this Earth had evolved from single cells. Lamarck did not believe that all life on Earth had evolved from *one* single cell which mysteriously became alive and from which all other life forms had evolved; Lamarck did not believe in 'one off' events. His aim was to show how a 'life force' was everywhere present, how, under its influence, life on this Earth had evolved from single cells into the myriad of forms that we see today. Change had followed a steady path, sometimes branching, but always following natural laws. Progressive change could be traced, not merely by fossil evidence, but by the evidence of the anatomy and physiology of current life forms. Lamarck's book ran to some 800 pages. It was the fashion at that time to allow wide margins around the text for notes. When printed after today's fashion by Elliot in his translation into English, it ran to just over 400 pages, which will give you some idea of the book's length and thoroughness.

Lamarck's book fell into three parts. The first part was concerned with what he called *artificial devices*, such as the division of living things (both plant and animal) into species. Humans decided what constituted a 'species' and what was merely a 'variety'. There was often dispute as to the validity of a suggested nomenclature. Then, starting with the most complex animals, mammals, he showed the *degradation* (simplification) which ran through the animal kingdom, as one studied the 'lesser' animals. This, he believed, showed the influence of the environment and habit on the organs of animals, the resultant 'use' or 'disuse' influencing development or arrest of development. He concluded the first part of his work by suggesting a far more extensive and thorough arrangement and classification of the animal kingdom than any previously put forward. Lamarck retained Linnæus' four categories of vertebrates: mammals, birds, reptiles, fish, but divided Linnæus' last two classes, insects and worms, into ten: molluscs, cirrhipedes, annelids, crustaceans, arachnids, insects, worms, radiarians, polyps and infusorians. He explained how each class resembled the one previous to it in certain ways but differed in others – by not having some attribute of the more complex class just described. (Remember, in this part he was working *backwards*).

In the second part of his book, Lamarck discussed what constituted the essence of animal life and the conditions necessary for "the existence of this wonderful natural phenomenon". He then endeavored "to ascertain the exciting cause of organic movements ... the properties of cellular tissues ... the sole conditions under which spontaneous generation can occur", i.e. the conditions necessary for life to take hold. In this section of the book, he worked forwards. First he established the difference between inorganic and organic matter, then considered how each of these differences (for example, the ability to digest food, the ability to reproduce) had influenced growth and the evolution of different systems and forms.

The third and final part of his work discussed "the physical causes of feeling, of the power to act, and the acts of the intelligence found in certain animals". This included:

- origin and formation of the nervous system;
- the nervous fluid;
- physical sensibility ... sensation;
- the reproductive power of animals;
- the origin of will and the faculty of willing;
- ideas and their different kinds;
- understanding, attention, thought, imagination, memory, etc.

The current chapter is intended to give an overview of Lamarck's thinking. The following chapter will outline how Lamarck presented his material, how his line of logic extended, page by page, throughout his book.

As already mentioned, Lamarck's mentor, Buffon, had written extensively on his theories of how the Earth had been formed – and reformed – over the millennia. Geology was a very new science, studied in depth by very few people. It is perhaps difficult to put oneself back in time and to appreciate quite how ground-breaking was Buffon's work. Until the eighteenth century, it had generally been accepted that the Earth was much the same then as it always had been. Minor changes, such as the diversion of a river, possibly following a lava flow or a flood, the rising and falling of sea levels, these were well accepted, as were the rising above the sea of small islands as the result of volcanic activity, but major land masses, mountains and their associated valleys, were all supposed to have been there 'since time immemorial'. The idea that large masses of land, such as the chalk Downs of the south coast of England, had been formed over unimaginable periods of time *beneath* the sea from the skeletal remains of small sea creatures, and then, by some unimaginable force, lifted up to their present height *or more*, before being weathered to their present condition, was a major shift in thinking and one which we can but admire in those brave and far-sighted individuals whose minds first imagined and grasped the concept.

Although Buffon was careful not to antagonize the Church by openly supporting the evolution of *living* things, be they plant, animal or human, he did discredit the biblical account of Creation as far as the Heavens, the stars, the Sun, the Moon and the Earth were concerned. For Buffon, the creation of the Universe was something which had taken place over æons of time. The Earth had been shaped and reshaped by wind and tide. Following in the footsteps of his master, Lamarck wrote that (p. 45):

> ... nothing on the surface of the earth remains permanently in the same state ... elevated ground is constantly denuded by the combined action of the sun, rain-waters and yet other causes; everything detached from it is carried to lower ground; the beds of streams, of rivers, even of seas, change in shape and depth, and shift imperceptibly; in short, everything on the surface of the earth changes its situation, shape, nature and appearance, and even climates are not more stable.

Although he had turned his back on Christianity, Lamarck retained a belief in a Supreme Creator. He saw this Being as having little direct influence on the development of life, other than having established matter and the laws by which it acted and interacted. These laws, having been established by the Creator, were eternal and unchanging. All earthly 'change' was but a re-arrangement of matter brought about by the action of these immutable laws.

While Lamarck had set himself the task of trying to understand these changes, as they affected both living and non-living material, he recognized the danger of 'jumping to conclusions' by drawing inferences from perceived facts, which inferences may later be found to be false as more facts came to light. It had formerly seemed so obvious that the Earth was stable, unmoving, that it was the Sun, Moon and stars which moved around the Earth, but that inference from a perceived 'fact' had been shown to be false. Lamarck was determined to avoid making a similar mistake (p. 7):

45

> There are then few positive truths on which mankind can firmly rely. They include the facts which he can observe, and not the inferences that he draws from them; they include the existence of nature ... also the laws which regulate the movements and changes of its parts. Beyond that all is uncertain.

Lamarck realized that in order to understand in the most basic way what constituted life, how life itself had originated, how living matter differed from inert, inorganic matter, it was necessary to study the most basic form of life, the cell. Lamarck was the first person clearly to state that cells were the fundamental structure of all living things, both plant and animal (p. 230):

> It was no new discovery that all organs of animals are invested by cellular tissue, even down to their smallest parts ... Yet ... nothing more seems to have been said than the mere facts themselves; no one that I know of has yet perceived that cellular tissue is the universal matrix of all organisations, and that without this tissue no living body could continue to exist.

All cells were constituted of two parts: one solid (the cell membrane), the other fluid, contained within the membrane. Plant cells were comprised of mucilaginous matter, that of animals, gelatinous matter. These two forms later became known as protoplasm.

Lamarck held that both warmth and water were necessary for life. Ancient philosophers had noted that life was more abundant in tropical areas where there was greater warmth. However, in desert areas, there was much warmth, but little or no water and little or no life. At the poles, there was plenty of water, but not much warmth and not much life. Lamarck pointed out that temperatures at the poles were above 'absolute zero', so some degree of heat was available for potential life, even there. However, warmth, water and protoplasm alone were not enough. For a cell to be living, there had to be movement, which Lamarck believed was supplied by a vital energy (aura vitalis), somewhat analogous to electricity (then but recently discovered) which would permeate mucilaginous or gelatinous matter, if conditions were right. From his microscopic studies, Lamarck concluded that over time, as cells increased in size and complexity and experienced the need to increase their ability to absorb nutrients, they tended to make pathways which (a good deal) later developed into the organs of digestion, respiration, circulation, etc. For Lamarck, the life force was an active presence which helped to mould its surrounding cells according to its needs.

No body that was perfectly dry could be alive. If all moisture were withdrawn from a cell (i.e. if it were desiccated), the potential for life could still be retained and revived when the cell once again received water. In some cases, this could occur after considerable lapses of time, even months.

It was from facts such as these that Lamarck drew his conclusion that the *aura vitalis* was present everywhere in nature and could invade/invest simple matter at any time, if conditions were right. Here it may be noted that Lamarck did, in fact, make an inference, although that inference carried a very important rider: *if conditions were right.*

Lamarck made quite clear that his work had been written with the intention of disproving the account of Creation given in *Genesis* (p. 129). He argued in favour of spontaneous generation on the grounds that there could be only two methods by which life had come into being: by Special Creation or by Natural Law. Lamarck rejected Special Creation. Having adopted 'Natural Law' as his guide, he reasoned that if life had become established in protoplasm at one point in time, then it must be able to become established at another, *if the same conditions existed,* since the same laws would always be in operation. Lamarck did not advocate spontaneous generation in a sterile environment any more than he supported the spontaneous generation of 'imperfect' animals, as will be discussed below. Lamarck recognized that Nature had virtually unlimited time in which to change and adapt: "for her [Nature] time has no limits" (p. 114).

Lamarck introduced the terms 'vertebrate' and 'invertebrate' and initiated the use of the two divisions, organic and inorganic, to replace the three divisions of animal, vegetable and mineral which had previously been used. If inorganic matter could be incorporated into the organic sphere by ongoing spontaneous generation, why had not all matter become organic over the millions of years of Earth's history? Lamarck believed that on death, all organic matter returned to an inorganic form: "... from the remains left by each of these bodies after death, have sprung the various minerals known to us" (p. 239). Chemical reactions took place in a living body, forming new compounds which were not known to form outside organic matter. Some of these compounds survived the death of their 'host' and contributed to the matter which made up the surface of the Earth. By 'minerals' Lamarck did not mean 'metals'. He did not specify any minerals in particular but we know that some occur in an organic as well as an inorganic form (iron, calcium, phosphorous, sodium, etc.) so his proposition was not as unreasonable as it may at first sound. Lamarck's understanding of chemistry may have been a little uncertain, but remember, he formed his theories in the 1790s, before there was the certainty about chemistry that there is today.

The degree to which earth, soil and rock were composed of matter which had previously been organic (living) was being increasingly recognized by writers such as Buffon, Erasmus Darwin and Hutton. It was part of the cycle of life and death. Today we are used to the idea that the massive 'White Cliffs of Dover', as well as miles upon miles of Downs behind them which form the southern coast of England, are formed from the skeletal remains of earlier sea-living creatures. Reading Lamarck's work, I felt that he had taken the time to attempt to visualize the Earth as it once must have been, devoid of any organic matter on its surface. By 'visualizing', I do not mean holding the concept briefly before the mind, but meditating upon it. Try for yourself – it is not easy! It is not only those chalk cliffs of Dover which have gone – and the miles of Downs behind them. It is any soil, clay or rock which contains any organic material at all! What would be left of our Earth today if all organic material suddenly disappeared? For centuries, Europeans had used marble to beautify their dwellings, as pillars, as statues. Marble is chalk/limestone, which has been compressed under the weight of the sea, which process subjected it to heat. It isn't really a 'rock' at all! Had I been living in the eighteenth century, I am not at all sure that I would have embraced the new ideas!

Lamarck did not try to imagine some 'natural law' which would turn inorganic matter (rock) into organic matter (living cells). Life itself was the direct gift of the Supreme Creator, as was the Universe of matter.

Lamarck was reluctant to accept that any species might have become extinct, other than recently by the 'hand of man', believing rather that fossil creatures no longer existing had evolved into other forms. Believing as he did that *all* change, including climate and physical environment, took place very slowly, he held that life forms would have time gradually to adjust and evolve in keeping with the circumstances they were experiencing.

Lamarck established the single-celled infusoria as the most basic of all living forms and claimed that these extremely small and transparent animals and plants, of gelatinous and mucilaginous substance, had changed over long periods of time so as to increase their complexity, giving rise to all later plants and animals. Lamarck declined to speculate how or when these changes in complexity had taken place, only stating that they were the result of a combination of life's inner urge to grow and the changing external environment in which the living form, simple or complex, found itself.

The concept of an inner urge, of 'volition' or 'desire', was in line with thought of the 18th and 19th centuries. Desires and aversions were seen as an expression of an entity's inner nature which controlled and shaped its life and, to some extent, its destiny. As a natural therapist, I specialized in homœopathy and read many early texts written in the late eighteenth and nineteen centuries. I became accustomed to the use of the words 'wants', 'needs', 'desires' and 'aversion' with their somewhat archaic meanings, that of *driving forces*, virtually impossible to resist. I found myself somewhat taken aback when I realized that others, not so trained, had understood Lamarck's references to 'wants', 'needs', desires (*besoin*), as meaning something more in the nature of a 'wish' or a 'dream'. It is far more than that. A person may 'desire' warmth. They will never be completely happy if forced to live in a cold environment, where they feel the need to 'rug up', to huddle close to the fire. I left my home, my family, my friends, everything I had ever known, to travel half way around the world to live in Australia, because I had a deep-seated 'desire' for warmth. These desires are driving forces which shape peoples' lives. The inner urge, the 'need', to which Lamarck was referring was irresistible and in response to these inner drives and needs, life forms changed and adapted.

Lamarck's views on the process of reproduction followed those of Maupertuis and Buffon, but they were infused with his own evolutionary thinking. He accepted excess of nutrition after growth had been completed as being the means by which material for a new being with a similar organization was engendered. However, if evolution was to be a reality, then it must be possible for improvements to be passed on. This was quite within the scope of the particulate theory of reproduction and inheritance outlined by Maupertuis and accepted by Buffon and, at first sight, appears similar to Darwin's theory of *pangenesis*. However, Lamarck made it clear that he was referring to modifications and developments acquired and attained by *populations*, not by individuals.

Speaking of adaptations which seemed to him to have come about due to use/disuse of particular body parts (such as poor vision in moles which lived underground), Lamarck stated (p. 124):

> I can, in short, cite a multitude of instances ... which bear witness to the differences that accrue ... from the use or disuse of any ... organs, although *these differences are not preserved in the new individuals which arise by reproduction*: for if they were their effects would be far greater. *(Italics added)*

Later, on the same page, he re-iterated his understanding that change was only passed on according to the laws of inheritance when it had become common to the species:

> Now every change that is wrought in an organ through habit or frequently using it, is subsequently preserved by reproduction, *if it is common to the individuals who unite together in fertilization for the propagation of their species* ... In reproductive unions, the crossing of individuals who have different qualities or structures is necessarily *opposed to the permanent propagation* of these qualities and structures. Hence it is in man who is exposed to so great a

diversity of environment, the accidental qualities or defects which he acquires
are not preserved. (Italics added)

Lamarck made it quite clear that he did not believe that a change which came about in an individual animal, no matter how fortuitous, would be passed on to its progeny because, were that to happen, there would be so much change there would not be enough stability for species to become established, let alone continue in a recognizable form century after century. Fossils were providing evidence of change over what Lamarck believed to be vast amounts of time, but they were also showing stability over the centuries. The Egyptians had not only mummified humans; they had also mummified animals, such as cats and dogs. It was accepted that these species had changed little over the last three thousand years. Similarly, artistic depictions of horses and cattle showed no evidence of noticeable change. How change became sufficiently common in groups of individuals to warrant becoming an inherited characteristic, Lamarck did not know, and upon that which he did not know, he did not speculate.

The belief that nature had endowed her creations with the ability, not only to reproduce other individuals similar to those of the parents, but also to preserve progress towards more complex entities, formed the basis of the doctrine of the inheritance of acquired characteristics, so associated with Lamarck's name. Nevertheless, these two passages show quite clearly that Lamarck did not hold that changes or developments occurring during the life of an *individual* were passed on to the next generation. The often repeated references to the 'Lamarckian principle of the Inheritance of Acquired Characteristics' is a misrepresentation of the truth, whether accidental or intentional. Some authors, such as de Beer (1963: 7), Dawkins (1986: 287-289) and Gould (2002: 170-197) acknowledged that Lamarck had been treated unfairly, but little, if any, acknowledgement was made of the fact that Lamarck did not believe in the inheritance of characteristics acquired by the *individual*, but Darwin did! Even Gould, whose lengthy discussion of Lamarck's work is far superior to that of most other authors, does not mention these passages.

Lamarck recognized that animals long in an environment adapted to that environment. He cited many pages of examples, including those of domesticated animals, such as ducks and geese which had become too heavy to fly long distances. Towards the end of his examples, he cited that of the giraffe (p. 22):

... the giraffe ... is obliged to browse on the leaves of trees and to make
constant efforts to reach them. From this habit long maintained in all its race, it
has resulted that the animals fore-legs have become longer than its hind legs,

and that its neck is lengthened to such a degree that the giraffe, without standing up on its hind legs, attains a height of six metres.

It will be seen from the above quotation that Lamarck makes reference to 'effort' and a 'habit long maintained in all its race' but not to 'longing' or 'willing'. If there was any 'volition' involved, it was on the part of the *aura vitalis* of the species, not the individual animal. Lamarck explained (p. 107):

> *The environment affects the shape and organization of animals*, that is to say that when the environment becomes very different, it produces in course of time corresponding modifications in the shape and organization of animals ... whatever the environment may do, it does not work any direct modification whatever in the shape and organization of animals. But great alterations in the environment of animals lead to great alterations in their needs and these alterations in their needs necessarily leads to others in their activities. Now if the new needs become permanent, the animals then adopt new habits which last as long as the needs that evoked them. *(Italics in original)*

Lamarck believed that function preceded form – that changed form followed changed requirements (needs). Darwin believed the opposite, that form changed by chance variation and if the new form was useful, it would be selected.

It was Lamarck's view that only the simplest of organisms could be utilized by nature to give rise to life. He was aware that his opinions were opposed by others, who held that life could be generated in many basic worms, fungi, moulds, etc. 'Worm' was a general term applied to grubs and insects which the ancients thought were generated spontaneously through the action of heat, fermentation and decomposition of organic matter. Lamarck rejected this view, arguing that all insects were either oviparous or viviparous and that 'worms' only appear on putrid meat after flies had laid eggs there, claiming that all animals, however imperfect, had the power of reproducing their own species.

Lamarck established the single-celled infusoria as the most basic of all living forms and claimed that "these extremely small and transparent animals and plants, of gelatinous and mucilaginous substance" had changed over long periods of time so as to increase their complexity, giving rise to all later plants and animals. He did not speculate how or when these changes in complexity had taken place, only stating that they were the result of a combination of life's inner urge to grow and the changing external environment in which the living form, simple or complex, found itself.

Lamarck (p. 239) held that "Nature only establishes life in bodies that are at the time in a gelatinous or mucilaginous state". It was Lamarck's opinion that an error had been committed when rejection of the idea of the spontaneous generation of "imperfect animals" had led to rejection of the concept of spontaneous generation at all. It was Lamarck's belief that life only came into existence in the simplest cell that gave him his belief in evolution.

On page 179 Lamarck illustrated his theory of the origin of various animals by a diagram. In the accompanying text, Lamarck gave his opinion that the animal scale had been started by at least two separate branches and that as it proceeded it appeared to terminate in several twigs in certain places. The line which gave rise to 'worms' was distinct from that which gave rise to the infusorians, polyps and radiarians. The 'worm' line itself divided very early, giving rise to both the insects, including arachnids and crustaceans, and to the marine annelids, cirrhipedes and mollusks, from whom all later fish, reptiles, birds and mammals evolved.

The diagram referred only to Worms. This was made clear by Lamarck having listed the groups 'Infusoria', 'Poplyps' and 'Radiarians' separately in the upper right-hand corner of the diagram. Unfortunately, Elliot omitted this listing in his reproduction of Lamarck's illustration, then claimed that is was Lamarck's teaching that all advanced organisms passed through *all* other forms on their way to their present form, as discussed in the previous chapter.

It is my understanding of Lamarck's teaching that he perceived Nature *as a whole* to be a living, evolving 'Being', both organic and inorganic matter having a specific role in this evolution. Living matter continually died and returned to its inorganic form, to be incorporated once again into living, organic matter. Although he accepted a 'Sublime Author' of all things, he thought (p. 180):

> Nature ... should be regarded as a whole made up of parts, with a purpose that is known to its Author alone, but at any rate not for the sole benefit of any single part.

Here Lamarck was making clear that he did not believe that the world had been created for the convenience or pleasure of humans alone, but that each and every part of creation was equally important to its Author. Add to this his hypothesis that humans had evolved from apes (see next chapter), and it will be seen by how much Lamarck anticipated Darwin.

TABLE

SHOWING THE ORIGIN OF THE VARIOUS ANIMALS.

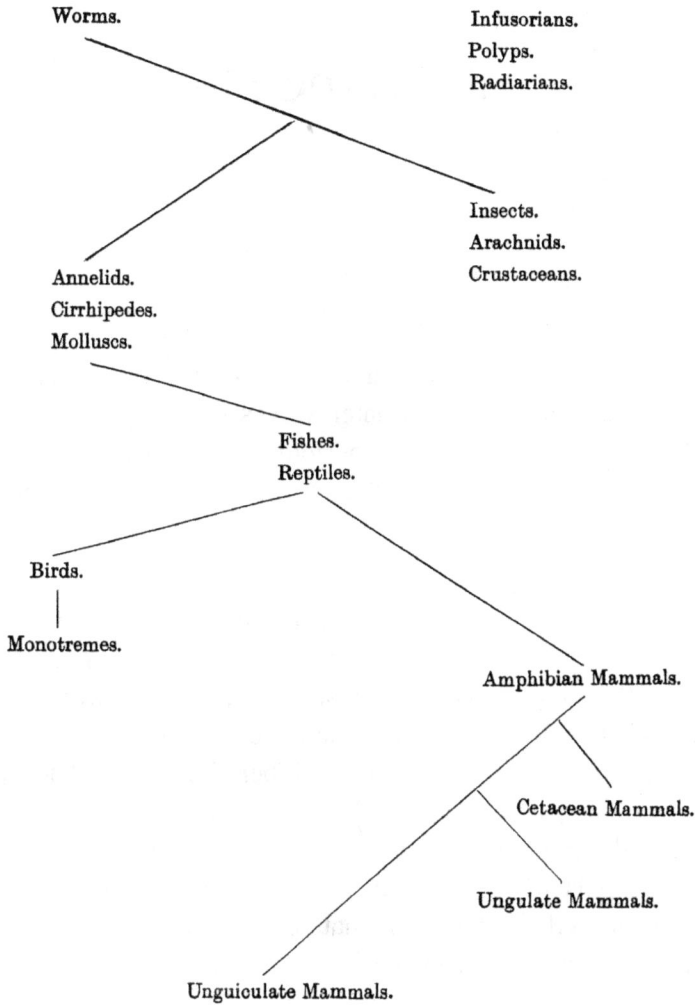

Worms.   Infusorians.
                                                Polyps.
                                                Radiarians.

                                                Insects.
                                                Arachnids.
Annelids.   Crustaceans.
Cirrhipedes.
Molluscs.

                        Fishes.
                        Reptiles.

Birds.

Monotremes.

                                    Amphibian Mammals.

                                        Cetacean Mammals.

                                        Ungulate Mammals.

            Unguiculate Mammals.

Fig 5.1: Diagram of the 'Chain of Being' branching out from Worms
(Lamarck 1914/1963: 179)

Chapter 6

Hitting the Mark

There were many points of similarity, as well as difference, between the theories of Lamarck and the later works of Charles Darwin. Both saw evolution happening over immense periods of time, by imperceptible steps. Both saw the environment in which all animals lived as being subject to slow, but constant, change.

Lamarck traced physical evolution backwards from the most complex creatures to the most simple. Mammals had the most complex organization. Besides internal organs such as kidneys, liver and spleen not found in many lower orders, they had a complete diaphragm, a heart with four chambers and they suckled their young. Birds and monotremes were the only other animals which had hearts with four chambers, they cared for their young but not by suckling, and their diaphragm did not completely separate the chest cavity and abdominal areas. Reptiles had only three chambers to their heart, fish had gills instead of lungs, and so on.   (Apparently, crocodiles have four-chambered hearts;   if that is the case, it would seem that Lamarck never dissected a crocodile.)

Worms were the lowest creatures in this line of evolution. Radiarians had taken a completely separate line of evolution, radiating outwards from a centre or axis. Polyps were another evolutionary line, being gemmiparous, with homogenous bodies, no nervous system and a single opening which served both to ingest food and expel waste. The final category was the infusoria, which fed by absorption, and reproduced by simple cell division. It was this group, and this group only, which Lamarck believed able to be reproduced by direct spontaneous generation.

Lamarck also noted the appearance of organs of special sense. For example, all creatures that had ears also had eyes, but many that had eyes did not have ears, showing that eyes had evolved first. The sense of touch was the most primitive sensory perception, Lamarck believed, and digestion the most primitive system, preceding circulation, respiration and so on.

Lamarck started by mapping the path of evolution of physical forms in reverse, from that of the most complex to the least, noting the disappearance of first one, then another, of their systems of organization. He then, in the third part of his book, turned his attention to the 'mental', or as he called them the 'moral', characteristics of living beings. Here Lamarck worked forward in time, following the progression of forms of animal life from the 'least' perfect to the 'most'.

Lamarck conceded that even the most inferior forms of animal life may, and in many cases did, possess some sort of 'inner feeling', a sensation, an awareness of being alive, perhaps that which we would today call 'consciousness'. However, Lamarck believed logic demanded that we accept the fact that inferior forms, which do not possess a nervous system, could not experience feelings or pain in any way analogous to that experienced by higher forms. The more complex the nervous system, the more feelings, physical, emotional and intellectual, were experienced.

No muscular movement would be possible without the involvement of a system of nerves. Simple radiarians were developing nerve/muscle fibres and some, such as the starfish, were also developing nerve ganglia allowing for independent movement. As they had as yet no centre of control (brain), they reacted to their environment, rather than made 'conscious' decisions. Insects, however, were capable of complicated muscular movement, of their legs, wings and so forth, and must, therefore, possess a nervous system. Furthermore, they possessed one of the organs of special sense, sight, which meant that they must have the beginnings of an area of central control (brain) and, in fact, a very small 'bulb' was present at the anterior end of the nervous system. Lamarck considered the development of the nervous system in this class of animal to be too simple to permit the production of 'ideas', but it was sufficient to enable the creatures to make simple decisions – when and where to fly, for example.

Lamarck recognized that the nervous system throughout the body consisted of two distinct parts, connected by a central area of control, the brain. The first division activated the movement of muscle tissue, the second the sensory system, responsible for the perception of feeling or pain. Lamarck knew that it was possible to have movement without feeling and feeling without movement, so he realised that these two activities of the nervous system must be separate from one another, although connected through the brain. Therefore, he believed that animals with an imperfectly evolved nervous system could be capable of movement without necessarily being capable of feeling pain, although they might be aware of touch.

Despite the presence of a rudimentary brain, Lamarck did not believe that any invertebrate animal had the power of independent thought (idea) or action, outside that for which they were programmed. They could 'decide' the exact moment at which to take flight, upon which flower to alight, but their choices were limited to these pre-programmed activities. The lives of all invertebrate creatures were controlled solely by instinct, which he believed to have come about over immense periods of time, during which pathways had been forced by the nervous system throughout the body and brain, directing movement of the nervous fluid into pre-determined avenues.

The evolution of vertebrate animals was accompanied by the appearance of a new organ, which was present in no invertebrate creature: the hypocephalon. This was the name given by Lamarck (p. 320) to the two soft, wrinkled hemispheres, adjacent to (superimposed upon), but separate from, the true brain. Because these two wrinkled hemispheres were soft, he believed them to be inactive, capable merely of receiving impressions. The hypocephalon was therefore capable of receiving impressions, which were stored in minute chambers, impossible to see, but which he deduced to be present at the termination of each tiny canal present in the soft tissue. Over and through these flowed the nervous fluid, which he believed to be a form of electrical impulse.

Even fish, Lamarck believed, had the power of some independent thought. He made this claim because in fish there first appeared the rudiment of the hypocephalon. It was very small and the cranium itself was not completely filled, so the amount of 'thought' or 'learning' possible to a fish must be very limited. The situation was similar with reptiles. They, too, had a small hypocephalon and an incompletely filled cranium. It was in the birds that Lamarck saw the first real progress in intellectual activity, i.e. the ability to choose to vary their actions.

The body was constantly being exposed to many stimuli, sights and sounds, to which no attention was paid. These made no impression on the hypocephalon. Only when attention was being paid was an impression made. These impressions could later be reactivated by the nervous fluid being redirected to them, either deliberately or randomly. Deliberate recall was the basis of memory. Simple 'ideas', according to Lamarck, were merely impressions consciously received. When these were recalled and compared one with another, they gave rise to complex ideas, which were the basis of both reason and judgment, contributing to what we know as intellect.

Lamarck maintained that 'moral' (mental) characteristics were subject to the same laws of use and disuse as physical characteristics. Therefore it followed that the more the 'moral' faculties were used, the greater they would become. He stressed the importance of early education, since early impressions were those which carved out the first neural pathways. He deplored those parents or caregivers who were too lenient with young children, forgiving poor behaviour in the fond hope that it would improve as they became older. Making new pathways in older brain tissue was difficult (p. 370). The more people were educated to pay attention to many impressions and ideas, to recall these impressions and ideas, to compare them, to bring reason to bear and to form sound judgments, the more would the intellect of the human race be elevated (pp. 383-384).

Lamarck deduced that the bigger the hypocephalon, the greater would be the animal's ability to receive impressions, to think, reason and, to some extent, form judgments. He saw dreaming in animals, such as cats and dogs, as evidence for imagination, which resulted from the coming together of previously impressed ideas. However, animals lacked the ability that humans had of directing their attention for a sustained period and were, therefore, limited in their ability to formulate 'new' ideas, which were really new formulations of old impressions.

Lamarck claimed that a separate family for humans (*bimana*) was justified on the grounds that only humans had opposable thumbs on their hands, but not on their feet. Apes with the latter physiology had been named *quadrumana* (p. 169). Lamarck (pp. 170-171) believed that if some race of *quadrumanus* were to abandon tree-dwelling for ground dwelling, needed to stand upright for the purpose of seeing long distances, gave up using its jaws for cutting, tearing, grasping, etc., then this race would evolve in a way strikingly different from its fellow apes.

Lamarck further pointed out that apes were quite capable of assuming an upright position for short periods of time, as were humans. Humans alone, of all creatures, were unable to maintain their position of locomotion for extended periods of time while stationary. After a surprisingly short time, humans must either walk, sit or lean for support in order to remain upright. This showed Lamarck (pp. 172-173) that the upright position was not the original position of humans, but that it had been developed as a result of evolution. Lamarck held back from making a definitive statement in his text that humans were evolved from the same stock as apes, but he left his reader in no doubt as to his opinion, since that was how *bimanus* was shown in the relevant table. Lamarck made it clear that he considered humans part of the animal kingdom, both physically and mentally.

In his life-time, Lamarck's ideas were rejected by his peers, largely due to the opposition of his colleague, Baron Cuvier. The rejection of Lamarck's views on evolution by so many later writers is hard to understand since Lamarck's views accorded with the materialistic thinking which became increasingly popular during the 19th century. His belief that thought, even reasoned thought, was but a chance operation of the brain, not the product of a reasoning mind under the influence of an immortal soul, was definitely materialistic.

Recent interpretations of Lamarck's theory have differed greatly. As already pointed out, Elliot (p. xxiv) held that Lamarck believed:

> All existing animals are on the road of development from Monas to man, and man's ancestors include every existing species of animal. Not only had he bird, reptile and fish ancestors, but also arachnid, insect, worm, starfish, etc. He passed through the stage of being a scorpion and a spider. He traversed in turn every species of insect. He was a tapeworm, a sea-anemone, a polyp and an amœba.

Bowler (1984: 80-81) had a different interpretation. He believed Lamarck's teaching to be that each line came into existence as the result of a separate act of spontaneous generation, that the most simple forms of life existing today were the most recently generated, more advanced forms, such as human, having evolved over great expanses of time from the earliest spontaneously generated forms. It is unclear what simple forms Bowler believed to be in existence in earlier times that later gave rise to humans. Were they the same as the simple forms of life existing today which were the most recently generated? If so, was nature continually recreating the same simple forms?

My understanding is that Lamarck saw evolution as a great tree. All came from the same 'root' – or should that be 'trunk'? At some point, a new 'branch' would be formed, this event coming about due to changed environment, combined with the urge, the need (*besoin*) of the *aura vitalis*. For example, at some point in evolutionary history, insects and spiders undoubtedly had a common ancestor, but once the arachnid branch had split from the insect branch, they were forever after separate in their evolution. Insects had six legs and went through a larval stage before developing into their final form whereas spiders had eight legs and developed their adult form directly within an egg, to mention but two differences. No insect species had ever subsequently given rise to a species of spider and, conversely, no spider species had ever given rise to a new species of insect.

In the same way, once a branch had evolved which was warm blooded, it henceforth never gave rise to a line of cold blooded creatures. Lamarck never mentioned dinosaurs. For him, the first warm blooded creatures to evolve were birds. All birds help regulate their body warmth by means of feathers. All are bipedal – none grew any further limbs, being restricted to the basic skeletal pattern of the vertebrates which preceded them. All laid eggs. The second line of warm blooded creatures, which also branched away from the reptiles, used hair/fur/wool for insulation. Those who had lost most of their hair developed sub-cutaneous fat, but never did any mammal grow feathers. Most were quadrupedal, a few bipedal, but none had grown an extra set of limbs which could evolve into wings. All gave live (placental) birth, with the offspring developing *in utero*. Lamarck made no suggestion how this major evolutionary branch managed to reproduce itself while in the process of converting from oviparous to placental birth; neither did Darwin, or any other evolutionary theorist whose work I have read.

One last example – some reptiles, such as turtles and crocodiles, have returned to the water, if not completely, then for a major part of their lives. They still return to land to lay their eggs and they are still air breathing. Similarly, some birds, such as penguins, also spend much of their lives in water, but still return to land to breed. The list of mammals which now spend much, if not all, of their lives in water is much longer: whales, dolphins, sea lions, seals, walruses, hippopotami – the list goes on. Some now give birth in water – but all are air breathing. Some can stay under water for extraordinarily long periods of time (by our standards) but none have shown any tendency to revert to gills, or develop some new way of extracting oxygen from the water.

Some regressions do appear to have occurred. Snakes have lost their limbs, as have whales; some fish and moles, which live underground, have lost their sight, but their anatomy still bears witness to their evolution. Once a branch of evolution has broken away, for whatever reason and by whatever means, it is committed to that path. Countless species and varieties may evolve, but all will carry the tell-tale pattern of their evolutionary history.

Evolution is a never ending process. Who is to say, when we discover a new species or variety, how long it has been in existence, especially if it is an insect with no fossil history? It may have evolved quite recently – and who is to say what is 'recent' when it comes to evolutionary time? There is no fossil evidence of any new class of animal coming into existence for millions of years - but that does not mean that it will never happen again, although the chances are that it will not be in our life time!

Lamarck's old age was one of poverty, exacerbated by blindness. Lamarck died alone and received a pauper's burial - a far cry from the honour accorded to Darwin half a century later, namely burial in Westminster Abbey. It is said that at his burial, one of his daughters cried out:

"History will remember you, my Father."

I hope she was right!

Chapter 7

French Resistance

Before returning to England to follow the story of the emergence and acceptance of evolutionary theory in that country, there is one more Frenchman whose work needs to be considered, not because he contributed anything towards the theory, but because he mounted an influential and effective campaign against it.

Lamarck was already 25 years old when George Cuvier was born and was just embarking upon his illustrious career at *Le Jardin*, which was to last some forty years. As already mentioned, his patron, Count Buffon, and his older colleague, Geoffroy St. Hilaire, both supported his work and his ideas, but everything changed for Lamarck after Buffon died and St. Hilaire left *Le Jardin* to accompany Napoleon on his quest to conquer Egypt. St. Hilaire's place was taken by Cuvier.

Georges Cuvier (1769–1832)

Baptised Jean Léopold Nicolas Frédéric, Cuvier took the name Georges after the death of his older brother. Born in Montebéliard, in the Duchy of Württenberg, Cuvier was brought up a devout Lutheran, unlike Buffon and Lamarck, who were both French and both of whom had been educated by the Jesuits. Cuvier moved to Paris in 1793, becoming a French citizen. His first position, held between 1793 and 1795, was as a secretary/clerk, during which time he made an intensive study of natural history. He held a number of public appointments, but his appointment to the position of Professor of Comparative Anatomy at the *Muséum National d'Histoire Naturelle* in 1802 ensured his place in history. Together with Etienne Geoffroy St Hilaire, he had published a short work in

1795 on the classification of mammals. This was followed three years later by a zoology textbook, *Essays on the Theory of the Earth* (Cuvier 1798/1817/1978) based on a series of lectures he had given on comparative anatomy, which ran to several editions. In 1816, Cuvier published *The Animal Kingdom according to its Organization* (Cuvier 1863/1969).

Cuvier was undoubtedly the greatest comparative anatomist of his time. It was said that Cuvier could reconstruct a whole animal from a single bone, such was his understanding of the correlation of parts necessary to maintain a complete whole. As Cuvier continued his work, he became more and more convinced that other naturalists were placing too much emphasis on comparatively unimportant external features in determining whether two or more animals were of the same species. Cuvier pointed out that superficial characteristics, such as colour or thickness of fur, were the most variable. By contrast, the skulls of the wolf and the fox were virtually the same throughout their territories, although there were superficial differences (Cuvier 1798: 117).

Cuvier argued that if species were evolving over time as a result of gradual variation, there ought to be evidence of this fact. Cuvier studied mummified remains of animals from Egypt: cats, ibises, birds of prey, dogs, monkeys, crocodiles and the head of a bull. He was aware that these creatures were but two or three thousand years old, yet could discover no sign of major variation or change over that time. Remember, it was then generally accepted that the Earth was a mere six thousand years old. Suggestions by people such as Buffon, Erasmus Darwin, Hutton and Lamarck that the Earth was of inestimable age were not accepted by the Church or the general body of scientists and educated people, including Cuvier. If evolution had taken place, then, so it was generally thought, it had taken place in a mere six thousand years. Cuvier's claim that there was no discernible change over three thousand years was a very powerful counter-argument. Cuvier claimed the onus of proof was on those who proposed that change was continually taking place to substantiate their position with concrete evidence.

It is ironic that it should be Lamarck's name which has become so indelibly associated with the picture of the giraffe, with its upwardly stretching neck, when it was Cuvier who used this animal, in this pose, as the frontispiece for his book *The Animal Kingdom According to Its Organization*, which was first published in 1816.

Cuvier was the first comparative anatomist to extend his art to the examination of fossils. He endeavoured not only to reconstruct past life

Fig.7.1: Frontispiece: The Animal Kingdom According to Its Organization.
(Cuvier 1863/1969)

forms, but to place them into families, orders, genera and species, where possible. The more he studied, the more he was struck by the fact that none of the fossils he examined had exact living counterparts, and some were completely different from anything now living. This was true of mammals, birds and fish. Only among the invertebrates, such as snails, was it possible to find fossil shells which seemed to bear close resemblance to their living counterparts today, although Cuvier did not consider external characteristics, such as colour and size, particularly relevant. In some ways, this countered his own argument. Clearly there had been change over time. Creatures living in the past were not living now. Creatures for which no ancient fossil counterpart could be found were now living.

Cuvier accepted that many species had become extinct, *contra* Lamarck who believed that they had but changed. Cuvier attributed these extinctions to geological catastrophes. As an example, he suggested that if New Holland (Australia) were to sink beneath the sea, kangaroos, and other species unique to Australia, would be lost forever. If, at the same time, land were to rise above the ocean north of Australia, it might be possible for animals, such as elephants and lions, to migrate from Africa or Asia to what are now the islands of South-East Asia, and thence to Australia, giving the impression that they were new creations (Cuvier 1798: 125-126).

Cuvier regarded the history of the world as documented by fossils as of equal importance with the history of the world as documented in the Bible. Nevertheless, Cuvier's difficulty in reconciling these two histories is clear to the reader. While Cuvier followed Buffon in conceding that the biblical account referred only to events which had occurred since the last geological revolution, he was unable to follow Buffon regarding the immense length of time which Buffon believed had elapsed since creation.

To Cuvier, each fossil was an historical document. Without fossils, we would be unaware that different creatures had ever existed. As industrialization progressed, more and deeper excavations were being made and fossils were being found deep inside the Earth belonging to creatures which clearly must once have lived upon its surface. Cuvier was not satisfied with the concept of fossils merely being covered by sediment. To him, the great depth at which some of these fossils were being found indicated that great upheavals must have taken place, not once but two or three times. Cuvier followed Buffon in believing that there had been a series of epochs and, like Buffon, equated the last with the Biblical Flood. The Age of Exploration had shown just how much of the Earth's surface was currently under water. That catastrophes may have raised some areas of land up, while drowning others and sending untold numbers of species into oblivion, seemed plausible. New creations occurred after each catastrophe. Once created, species were immutable.

Cuvier noted that the Hebrew texts placed the creation of the world 5817 years before the year in which he was writing (1813), the Sumerian texts placed it at 6513 and the Septuagint at 7685 years before 1813. To this extent, Cuvier was prepared to allow for some inaccuracy in the biblical account, but that was all.

Cuvier recognized four great divisions within the animal kingdom: vertebrates, mollusks, articulates and radiates. In this he was far more conservative than Lamarck, who recognized fourteen divisions. Cuvier acknowledged a general similarity of skeletal structure among the vertebrates but did not consider these similarities to be proof of common descent. He argued that if evolution had happened as Lamarck suggested, there should be far greater evidence of divergence and change in the fossil record. Deviation, Cuvier believed, would have adverse consequences. Two, three, four chambered hearts were all perfect for their hosts. How could a species gradually develop another chamber to their heart? How could they survive during such a change? What benefit would there be during the process of change suggested by Lamarck? Cuvier concluded that species must be fixed for the duration of their existence.

One thing remained a mystery – the origin of the human species. As yet, no human fossil bones had been found. Examination of battle grounds had shown that human bones degenerated at a rate equivalent to that of horses, so the lack of human fossil bones could not be attributed to their more ready decay (Cuvier 1798). Cuvier could only suggest that humans had lived on some part of the globe now under water, that a few had survived some cataclysmic event. Cuvier did not accept Lamarck's inference that humans had evolved from apes and even Darwin was to hesitate before this suggestion.

Cuvier found himself deeply at odds with the views of Lamarck. Cuvier held strongly to his Christian faith, a faith which Lamarck had abandoned. Lamarck's views on extended periods of time, of gradual creation, of constant change, were diametrically opposed to the teachings of the Bible as recorded in *Genesis*. Cuvier opposed Lamarck's views with fervor. He was brilliant at his work, very highly regarded and was reaching the peak of his career as Lamarck was in the twilight of his.

If you would like to learn more about Cuvier and his work, *George Cuvier: Zoologist,* (Coleman 1964) would be worth reading.

Chapter 8

On Paley's Watch

It is time now to return to England, which was to take centre stage in the unfolding drama of evolutionary theory. That is not to say that others on the Continent, particularly in France and Germany, were not also entering the field and making their own contributions, but the spread of knowledge and opinion was not as fast (or as vast) then as it is now. There were journals, and these were widely read. Books were becoming more accessible to the general public, as printers made an effort to produce cheaper editions in an attempt to tap the larger market. All educated English people were able to read French, if not speak it fluently, and people on the Continent were conversant with English, although they may well have been more fluent in some other European language. Nevertheless, books and journals written in one's native tongue were the ones most read and English theorists were more influenced by other English writers than they were by writers in foreign tongues, if for no other reason than one of national pride! Edinburgh, in Scotland, was a renowned seat of learning but only the reclusive Hutton, whom we have already discussed, made any early contribution to the advancement of evolutionary theory from 'across the border' and his contribution was mostly geological, not biological.

The three people we are about to consider, influenced the spread of evolutionary theory 'by default'. None wrote about evolution. One wrote on economics, another on population and the third on natural philosophy – of the wonders of God's creation. The writings of all three were well known in Victorian society and Darwin is known to have been familiar with the writings of at least two of them. His family adhered to the political doctrine of the third – the first one we will discuss – so Smith's ideas had penetrated his thought pattern, even if never specifically referred to in Darwin's writing.

Adam Smith (1723-1790)

Born in Kirkcaldy, Scotland, Adam Smith attended first Glasgow University and then, upon receipt of a scholarship, Balliol College, Oxford. After returning to Scotland, Smith supported himself by giving lectures on Moral Philosophy and Logic, until he was awarded a professorship, first in Logic and later in Moral Philosophy. In addition to theology and ethics, this course included general principles of law and government and of economics. A friend of Hutton and Hume, teacher of Boswell, Smith acquired a reputation for efficiency (Raphael 1985).

Smith's first book, *The Theory of Moral Sentiments,* was published in 1759, and was well received both in Britain and on the Continent. However, it was his second book, *An Inquiry into the Nature and Causes of the Wealth of Nations*, for which Smith is remembered today (Smith 1776/1937). Despite its length (900 pages), its basic premise was clear and easily able to be understood by any educated person.

There was a natural system of supply and demand which would ensure that necessary items would be produced in approximately the correct quantity to enable them to be supplied at their 'natural' price.

Smith asserted that the original human society, that of the hunter/gatherer, had been purely a consumer society, which generated no permanent wealth. The adoption of a pastoral lifestyle allowed the accumulation of wealth in the form of livestock, but this also was of no use unless consumed. When livestock increased beyond a number which could easily be fed on tribal lands, there was no option other than to kill surplus animals. These necessary deaths would have been sufficient to supply the tribe with any wool/skins necessary for their nomadic existence.

The next stage in the evolution of human societies occurred when humans abandoned their traditional nomadic lifestyle for a settled one. Food was no longer merely gathered, it was grown. It was with the advent of agriculture that the acquisition of wealth truly commenced. It was necessary to make claim to certain amounts of land, which must be defended against intrusion by other people. This less nomadic lifestyle encouraged the building of more permanent living quarters. For the first time there was division of labour, which became more pronounced as societies became larger.

Nevertheless, these societies were still predominantly consumer oriented. Until quite recent times, the nominal owner of large parcels of land in Britain, and other parts of Europe, built large halls in which they sometimes fed hundreds of people each day. The recipients of this 'free' food were not overly concerned that the land did not 'belong' to them, since they were receiving their share of its bounty. However, if adverse conditions, such as crop failure or war, reduced the amount of provisions available, the lower classes suffered more privation than did the upper, although it was in the interest of the Lord of the land to ensure that his workers did not starve, since he was dependent upon their labour for his wealth.

Smith contended that the labour of the farmer upon the land was the only productive labour. Mining, while apparently increasing wealth, depleted the land from which it came. Trees could be replaced, coal could not. However, coal was a consumer item and did not result in the long term economic problems that did other mined items, especially precious metals and stones. As these increasingly entered the market, large land owners for the first time were enticed to spend their excess wealth, not on their people, but on themselves. Smith's description of the peaceable farmers tending the soil and the rapacious industrialist consuming the product of their toil, brings to mind the two great natural divisions of the animal kingdom – carnivore and herbivore, predator and prey. Smith saw both as being necessary to produce balance in society, if society was to proceed beyond the stage of the hunter/gatherer or simple pastoralist.

The fourth stage of economic development was that of industry. The division of labour which had resulted in farmers and artisans producing more goods than they could consume within their family or immediate neighbourhood had allowed labour to be released from the country areas to the towns, where it was employed in producing manufactured items in excess of that which could be consumed by the factory owners, the factory workers and their neighbouring town dwellers.

Smith argued that specialization was far more productive than generalization. If a person grew his own food, wove his own cloth, sewed his own clothes, made his own shoes, time would be lost moving from one task to another. It was more economical for one person to specialize in tailoring, another in shoe making, etc. Trade networks would necessarily evolve and these must include some non-consumable items, otherwise industry would not survive.

Manufacturers needed consumers to buy their goods, which could not be done unless workers earned sufficient wages, not only to buy food for themselves and their families, but to have some left over to purchase other products. If the manufacturers preyed too heavily upon the workers, enriching themselves too much at the workers' expense by paying wages that were too low, the workers would starve and the manufacturers would be deprived, not only of their workers, but also of their consumers. Thus, claimed Smith, there was a natural minimum beyond which wages would not fall.

Smith was writing at a time when world trade was increasing at an unprecedented rate, especially for Britain. Trade routes had been opened to the Near and Far East. The American colonies had been a profitable export market and, although exports were adversely affected by the War of Independence, this would not prevent trade in the long term.

It was a major part of Smith's thesis that the actions of Governments were always a hindrance, even when they were intended to help. Subsidies might temporarily reduce the cost of an item in short supply and encourage its greater production, but this would result in oversupply and a subsequent drop in price, which might be so severe that production was no longer sustainable. This would result in a further, possibly even steeper, rise in price. No item was imported unless the importer saw evidence of a market. Tariffs designed to deter the import of certain items would raise, not only the price of the imported items, but also those of similar products available internally, to the detriment of the poorer portions of society. This might encourage further production, which would lead to oversupply, etc., etc.

Spain and Portugal had been among the poorest nations of Europe before the discovery and taking over of South America. As much gold and silver as possible was brought back to the home countries and its export forbidden. This had the effect of increasing the amount of gold and silver in circulation, but not of increasing the amount of food. More gold and silver was exchanged for food, but the people were no more wealthy. The gold and silver could not be traded for more corn, because export of these metals was forbidden. Paper money was of no use if it could not be redeemed for the requisite amount of gold or silver. The prohibition had to be relaxed, yet another example of Government interference which was counterproductive.

Smith argued that the attitude of Governments towards economics should be liberal. Trade should be left free to find its own natural level between supply and demand, between labour and wages. This attitude became known as *laissez faire* and was promoted by the Liberal (Whig) parliamentary party, to which the Darwin family adhered. Although Darwin made no mention of Smith's work, nevertheless he was profoundly influenced by the concept that everything worked out for the best if it was subjected to the minimum of interference. There was a natural balance, a natural order, which produced the best result. Malthus expanded upon Smith's ideas, especially in relation to population, and Darwin did acknowledge a debt to Malthus.

Robert Malthus (1766–1834)

Robert Malthus was yet another graduate of Cambridge University whose influence is evident in the unfolding saga of human history. Although ordained as a minister, he never took up a 'living', instead spending his life lecturing on political economy, principally at Haylesbury College.

In 1820, Malthus published a two-volume work, *Principles of Political Economy*, based upon these lectures, a second edition of which was published posthumously in 1836. It is for an earlier work, *An Essay on the Principles of Population: A View of Its Past and Present Effects on Human Happiness*, published in 1798, that Malthus is most remembered today (Malthus 1798/1816/1890). This work went through six editions during Malthus' lifetime (as did the major works of Smith before him and Darwin after him) and is the only work by which Darwin claimed to have been influenced in the development of his theory of evolution by natural selection.

Although a Minister of the Church, Malthus wrote in direct opposition to the teaching of Christ as given in *The Sermon on the Mount*, that humans should take no thought for the morrow, that God, who took care of the lilies of the field and the birds of the air, would also take care of them (Matthew 6: 25–34). Before his marriage at the age of thirty-eight, Malthus travelled widely in Europe and Asia and, thanks to his interest in Smith's views on economics, took particular notice of the state of the poor, not only across Europe, including Russia, but in India and China as well. In his opinion, the European country in which the poor suffered most from their poverty, was England, despite the fact that England had the most generous Poor Laws of any European country. Malthus argued that the Poor Laws enabled the poor to become so destitute by allowing them to reproduce beyond their 'natural' economic capacity.

Malthus' basic proposition was that, unchecked, the human population would increase at a far faster rate than would (or could) food supplies. It had to be assumed that the best land for cultivation had been utilized first and that remaining uncultivated land would prove less fertile and/or the climatic conditions would be less favourable. Even if all possible land were to be cultivated to the greatest possible degree, there was a limit beyond which food production could not progress, but there was no mathematical limit beyond which the human population could not proceed.

Population was known to be controlled by war, famine and pestilence. The number of people killed by war (mostly males), although appearing at times to be high, had very little permanent effect on population levels, which were controlled by female fertility. 'Predation/war' on females generally took place at birth, more females being subject to infanticide than males. Pestilence generally struck most severely when population levels were high and the general health and nutrition of the population low and was, therefore, associated with famine, which struck when population levels exceeded available food supply. No amount of productive human effort could do more than postpone the inevitable – starvation for large numbers of the population. The degree of misery associated with poverty and starvation was immense and it was unconscionable simply to allow this to happen without protest. No responsible government should encourage population increase by means of monetary incentives, such as State Poor Relief given on the basis of the size of the family or, worse still, an undertaking by the State that it would be responsible for the care of any child (or adult) unable to provide for itself. (Many people think that the Welfare State was an invention of the 20th century. Reading Malthus' account of the Poor Laws which operated in Britain in the 18th century, based on practices which had been passed down from previous centuries, soon disillusions one of that misconception!) The two countries which Malthus believed best exemplified his philosophy were those of China and Norway, the one with a large population, the other with a small.

Populate – and perish

The Chinese custom was for land to be passed on from father to all sons, not just the eldest, as was common in Europe. As a result the parcel of land owned by each peasant family had become so small that, despite the Chinese people being the most peaceable and industrious of any country that Malthus knew, they lived in the most abject poverty and, in many cases, a state of actual starvation.

Norway, by contrast, was the only country, as far as Malthus could ascertain, whose people were aware of the perils of overpopulation and which took active steps to prevent the problem. Young men were required to give ten years military service, longer than the requirement of any other country. When their military service was complete, the men returned home and waited for their chance to obtain a job with a house. This only happened when somebody died. The fortunate person who obtained the position became a 'houseman' and could then marry and raise a family. Any male unable to obtain a house was unable to marry and reproduce. No new houses were built and this resulted in a stable population, all of whom could be maintained at a level of subsistence above that of poverty.

Paley (a contemporary of Malthus whose work we will be considering next) saw an increasing population as a sign of happiness. Malthus argued that it might be a sign of past happiness but it was also a sure sign of future distress. Malthus disagreed with the view that every human being had a 'right' to subsistence. He argued that no person had the 'right' to be provided for by others, that each man had the duty to provide, not only for himself, but for his family, should he choose to have one. He cited St. Paul's admonition that if a man will not work, then neither should he eat. The only exception was the child, who had a 'right' to be provided for by its own parents. Parents had no right to expect other people to bear the expense of raising their offspring. Malthus knew of instances where the father of a large family had left, believing that this was the best way to ensure that his family was provided for – by the Parish.

Malthus' solution was twofold. He proposed that young people should be encouraged to delay marriage until they were in their thirties. Along with this, Malthus proposed that no people born after a certain date (twelve months after the Proclamation of a new law) should ever be entitled to Poor Relief. Malthus believed that if people knew that no other means of support was available, they would be far more circumspect in procreation. Procreation may be an instinct, but so was eating, and Malthus believed the one should be controlled in the same way as the other.

Notwithstanding the six editions of his book which were published in his lifetime, it has to be said that Malthus had very little influence on nineteenth century England. The Victorian family was the largest in recorded European history.

Of the people mentioned by Bettany in his *Introduction* to the republished 6th edition as having been important in forming and/or disseminating Malthus' views, only one is female: Harriet Martineau (Malthus 1816/1890: ix). Harriet Martineau travelled widely in Britain and America promoting Malthus' teaching. She was a close friend of Erasmus Darwin, the older brother of Charles Darwin, being a frequent visitor to the Darwin family home, and it is inconceivable that Charles was not well aware of Malthus' teaching long before his reported reading of Malthus' book. Indeed, it is more likely that Darwin's familiarity with Malthus' teaching was the reason for his late reading of this well-known work, the majority of the pages of his personal copy remaining uncut (Desmond and Moore 1991).

By the time Darwin published *On The Origin of Species* in 1859, the nineteenth century passion for the study of nature had convinced most people that, while God might 'care' for the fowls of the air in general, he did not 'care' for fowls in *particular*, as individuals. Victorian England embraced the Tennyson's concept of 'Nature red in tooth and claw' and Herbert Spencer's concept of 'survival of the fittest'. Aided by such thinking, Darwin was far more successful in establishing the truth of the balance of populations in nature by referring to the animal kingdom than Malthus had been by referring to the human kingdom.

William Paley (1743–1805)

Although William Paley was more than twenty years older than Robert Malthus, his work is being considered last because Paley had the most direct effect on the life and thinking of Charles Darwin. The study of Paley's work was compulsory when Darwin attended Cambridge University, was accepted by him as a young man, although later questioned.

Paley was born in Peterborough, where his father held a minor position in the Church, later moving to the North Riding of Yorkshire when his father took the position of headmaster at the local Grammar School. Paley performed well academically, graduating from Cambridge University and later returning there, lecturing in metaphysics, morals and the Greek Testament. In 1782, Paley was appointed Archdeacon of Carlisle, a position he held until his death.

In 1785, Paley published the first of his major works, *Principles of Moral and Political Philosophy* (Paley 1785/1833). Then in 1794, he published *View of the Evidence of Christianity* (Paley 1794/1833), a book directed against the growing tendency towards theism, evinced in some of his contemporaries - such as Erasmus Darwin. A few years later, in 1799, he published a smaller work, *Reasons for Contentment* (Paley 1799/1833). It was not until after the publication of *Evidence of Christianity* in 1794 that Paley took his degree of Doctor of Divinity at Cambridge. His last work, *Natural Theology*, published in 1802, was the most influential. It became a compulsory text at Cambridge

University for anyone studying there with the intention of being ordained within the Church of England, which was Charles Darwin's anticipated career path while he was studying at Cambridge.

At the time (1802) that Paley wrote *Natural Theology*, he was suffering ill health. Unable to perform his preaching duties, he committed his thoughts to paper in this major work, the first part of which is still frequently cited today.

Paley's *Evidence of Christianity* had been directed against theism - the belief in a God or a Supreme Power but not in Christianity - the position held by both Lamarck and Erasmus Darwin. *Natural Theology* was directed against atheism, the denial of the existence of any God whatsoever. That he felt such a book to be necessary bespeaks the growing influence of atheism towards the turn of the 18th/19th centuries.

Paley asked his reader to imagine that while out for a walk he knocked his foot against a stone. If asked from where the stone had come, the reader would be entitled to respond that for all he knew the stone had been there forever. Such a response would not be acceptable if the object which had been struck had not been a stone, but a watch. A watch is an organized object, in a way that a stone is not. If there was a watch, argued Paley, there must have been a watch-maker. Where there was design, there was a designer. Suppose the watch was somehow able to reproduce itself, the watch the reader had kicked might not be the original one, but one which had resulted from generations of reproduction. That would not eliminate the watch maker. Paley argued that no one would suppose that the watch was the result of 'the law of metallic nature', so why should anyone suppose that animals were the result of 'the law of animal nature'?

Paley then compared another 'man-made' object, the telescope, with a natural object, the eye. After discussing many types of eyes, mammal, bird, fish, etc., Paley moved on to the other senses and from there to other systems of the body. Referring to evolutionary theory, Paley (1802/1833: 448) wrote:

> ... every other animal, every plant, every organized body which we see, are only so many out of the possible varieties and combinations of being, which the lapse of infinite ages has brought into existence; that the present world is the relic of that variety; millions of other bodily forms and other species have perished, being by the defect of their formation incapable of preservation, or of permanence by generation ... The hypothesis teaches that every possible variety of being, hath at one time or other, found its way into existence (by what cause or in what manner is not said) and that those which were badly formed, perished; but how or why those which survived should be cast, as we see that plants and animals are cast, into regular classes, the hypothesis does not

explain; or rather, the hypothesis is inconsistent with this phenomenon.

Paley did not name the author(s) of the hypothesis to which he referred. He may have been referring to Buffon, or Erasmus Darwin, or both. He may have been referring to some of Lamarck's early writings, but not to his later ones, not just because he died before Lamarck's great work was published, but because Lamarck went to great lengths to explain why plants and animals (especially animals) were cast into regular classes. Paley challenged the atheist to explain the cause or manner in which plants and animals had come into being without the aid of a designer. It was this challenge that Darwin set out to meet. Countering Paley's arguments, overthrowing his well-known and much loved theory that living entities, by their very existence, proved the existence of a Divine Designer, was Darwin's greatest challenge and 'change by chance' was to be the foundation of Darwin's theory.

Paley (pp. 490-491) addressed the issue of teleology (planning or fore-thought) in nature, considering "*preparation*, i.e. the providing of things beforehand, which were not to be used until a considerable time afterwards, as the most certain proof of design". He cited the fact that milk was the only excretion of the body which was nutritious: "neither cookery nor chemistry have been able to make milk out of grass". Milk was produced in anticipation of the needs of the offspring at a time (end of pregnancy) when it might be thought that the female body would have no nourishment to spare. Also the eye was formed in the womb at a time when it had no function to perform: "an optical instrument made in a dungeon".

Paley cited many other examples of what he claimed were anticipatory design, arguments which Darwin never directly countered. However, Richard Dawkins (1986: 5) took up the challenge on Darwin's behalf:

> Paley's argument is made with passionate sincerity … but it is wrong, gloriously and utterly wrong … A true watchmaker has foresight; he designs … with a future purpose in his mind's eye. Natural selection, the blind, unconscious, automatic process which Darwin discovered, and which we now know is the explanation for the existence and apparently purposeful form of all life, has no purpose in mind. It has no mind and no mind's eye. It does not plan for the future. It has no vision, no foresight, no sight at all.

This statement, made equally passionately by Dawkins, does not *disprove* Paley's argument; it merely derides it.

Paley covered the anatomy and physiology of the human body with a degree of knowledge that might be expected of a physician, not a priest. He stressed the perfect adaptation and correlation of all body parts, and extended this adaptation and correlation to all other living things. His writings showed an extraordinary knowledge and understanding of the anatomy and physiology of creatures of all shapes and sizes. Like Erasmus Darwin, Paley (pp. 490, 496) argued that the pleasures of living far outweighed the pains:

> It is a happy world after all ... The air, the earth, the water, teem with delighted existence ... At this moment, in every given moment of time, how many myriads of animals are eating their food, gratifying their appetites, ruminating in their holes, accomplishing their wishes, pursuing their pleasures, taking their pastimes?

Chapter 9

Moving Blythly Forward

In the same way that we owe to King-Hele a resurrection of interest in the work and ideas of Erasmus Darwin, so it is to Loren Eiseley, that we owe a rediscovery of the work and ideas of Edward Blyth. Both a literary writer and an anthropologist, Eiseley held the Chair of Professor of Anthropology and the History of Science at the University of Pennsylvania. Two works by Eiseley on Darwin were published during his life time, in 1959 and 1961. After his death in 1977, a compilation of pieces was published, all but one of which had been written by Eiseley for scientific publication or prepared to be read to a gathering of scientists (Eiseley 1979: x).

Eiseley's first essay was a summary of Darwin's life and work in which he tackled the thorny issue of just how prevalent were theories of evolution in the last part of the 18th and the beginning of the 19th centuries and why Darwin had made so little acknowledgement of the work of his predecessors. While drawing attention to the works of people such as Maupertuis, Buffon, Erasmus Darwin, Lamarck, Cuvier, Hutton, Lyell, de Cabdille, St Hilaire, Wells, Matthew and Chambers, as well as a number of others, Eiseley nevertheless remained very loyal to Darwin inasmuch as he was prepared to forgive the great man any 'oversight' in making proper acknowledgment as part of the privilege owed to genius.

By the time he wrote his paper on Blyth, Eiseley's attitude had changed.

Edward Blyth (1810–1873)

Edward Blyth was born but one year after Darwin although he died nine years earlier. Neither man enjoyed good health throughout their life. Like Darwin, Blyth had shown a great interest in natural history since his childhood and was still in his twenties when he started contributing articles to journals and corresponding with other leading naturalists. He was part of a team which produced an illustrated translation of Cuvier's work, which was published in England in 1840. It was his study of the comparative anatomy of birds which lead Blyth to ponder whether all species had perhaps been derived from one common ancestor (Eiseley 1979). Blyth was interested in changes which had come about under domestication. He noted in particular the Ancorn sheep, which had been bred from one deviant animal, born with shorter legs than the rest of the flock. Being less able to jump fences, this ram was used to produce the Ancorn breed.

Blyth's health not being good, he was advised to move to warmer climates and accepted a position as curator of a museum in Calcutta, arriving there in 1841. Part of his duties included presenting monthly reports to the Asiatic Society of Bengal, which he did for twenty years, thus becoming one of the best known writers of his day. Blyth also contributed articles to other journals and his work was frequently cited by Darwin in *On the Origin of Species* and later works. His health failing, Blyth returned to cooler climes, visiting Darwin at Down and corresponding with him. He died 27th December 1873, a few days after his 63rd birthday.

Eiseley's findings in regard to the connection between the theories of natural selection as put forward by both Blyth and Darwin were published in an article in the *Proceedings of the American Philosophical Society* in 1959. It is perhaps unfortunate that his paper, which quite clearly accuses Darwin of plagiarism, should have appeared in that particular year since it was the centenary of the publication of *The Origin*. While the rest of the scientific world was celebrating Darwin's great achievement, Eiseley was casting doubt upon its originality.

Eiseley examined Darwin's works with Blyth in mind and found that Blyth was referred to extensively, especially in *Variation in Animals and Plants under Domestication* (1868) and *The Descent of Man* (1871). Blyth was cited more often than any other authority in *Variation*, not in relation to the central thesis, merely in relation to varieties and species. Darwin made no reference to Blyth in his 'Historical Sketch' included in the 3rd edition of *The Origin* (1861), which sketch had been added by Darwin in a belated attempt to satisfy criticism of his lack of proper acknowledgement of the work of others in the first two editions of his book. True to form Eiseley chose 1961, the centenary of the publication of Darwin's 3rd edition, to draw attention to this further 'oversight' of Darwin, which he did in a work entitled *Darwin's Century: Evolution and the Man who Discovered It.*

In 1842 and 1844, Darwin had written two unpublished essays, which Eiseley found to be very similar to some of Blyth's articles. He referred in particular to three articles by Blyth published in the *Magazine of Natural History* in 1835, 1836 and 1837, which he reprinted in full. It is known that Darwin read this journal, receiving copies while on his voyage on the *Beagle*. The January 1837 copy of the journal carried notation by Darwin on Blyth's article, as well as notes pinned to the back cover. Eiseley considered it significant that Darwin started his notebooks on the 'species question' a short time later. As mentioned above, Eiseley held the prestigious position of Professor of Anthropology and the History of Science at a respected University. No doubt he had been instructing students on the history of evolutionary theory for many years and the discovery that he may have been mistaken in some of the material he had presented to his students over that time would not have sat well with him. His 'mistake' may have become clear to him while undertaking specific research with the intention of writing articles to commemorate the two centenaries.

Eiseley pointed out that Blyth discussed natural variation as well as the tendency for variation to be passed on to future generations, the ability for naturally occurring variations to be artificially maintained and increased through domestic breeding, producing new varieties in domestic animals and crops, the possibility that something similar also occurred in nature, the tendency in herd animals for the strongest male to produce the greatest number of offspring, while the sickly and ill-adapted disappeared, 'selection under the struggle for existence in wild nature'. More particularly, Eiseley drew attention to specific examples used by Blyth and Darwin: naturally occurring mutations such as the Ancorn sheep, donkey-footed swine, tailless cats, back-feathered, five toed and rumpless fowls, domesticated cattle being grazed on poorer mountain pastures being smaller (more degenerate) than cattle grazed in valleys, domesticated fowl regularly supplied with food becoming more bulky and lazy and losing some of their ability to fly. While Blyth used the metaphor 'grouse are brown heather' and referred to ptarmigan 'as snow in winter', Darwin wrote of 'red grouse the colour of heather' and referred to 'ptarmigan white in winter'. Blyth discussed at length the role of protective colouration, using as his example the roles of the falcon and its prey. Darwin referred to the hawk and its prey, but both agreed that any ground dwelling bird, or small mammal, whose colouration differed from the optimum would be more subject to destruction. Protective colouration would be kept constant through natural selection.

The most convincing piece of evidence that Darwin directly copied from Blyth's work was thought by Eiseley to be the use by Darwin of the obsolete word 'inosculate', which meant *to join, to have a connection with or to be interwoven.* It was used by Blyth in his papers of 1836 and 1837. Darwin used the less strong word 'osculate', to touch or to join, once in *The Origin* (Darwin 1859/1997: 324). However, Barlow (1967: 62) pointed out that during his voyage on *HMS Beagle*, Darwin wrote a letter in which he referred to a bird that appeared to be a mix between a lark pigeon and a snipe: "Mr. MacLeay himself never imagined such an inosculating creature". The reference here was to the Quinary System of MacLeay, which saw nature, not as a ladder or chain, but as a series of connecting (osculating) circles. Barlow was probably correct in suggesting that Blyth and Darwin both drew on the same source for their use of this word.

Not all colouration in nature is designed as camouflage. Some is intended to attract attention, especially that of the opposite sex. Eiseley saw Blyth as foreshadowing Darwin in his discussion of sexual selection and its possible diversifying effect.

It was in his paper on *Psychological Distinctions Between Man and Other Animals* (1837) that Blyth raised the possibility of common descent:

> ... as man, by removing species from their appropriate haunts, superinduces changes on their physical constitution and adaptations, to what extent may not the same take place in wild nature, so that, in a few generations, distinctive characters may be acquired, such as are recognized as indicative of specific diversity? ... May not, then, a large proportion of what are considered species have descended from a common parentage?

Blyth continued by saying that he need not spend much time considering this possibility since "able writers have so often taken the subject in hand".

It was well known that the establishment of a new variety under domestication, be it plant or animal, took constant vigilance. Care had to be taken that reproduction took place only between carefully selected specimens. Domestic breeders were only able to achieve the results they did by strict segregation of breeding stock, something which did not happen in nature. Strict segregation had resulted in very distinct varieties but without constant human interference the establishment of new varieties would not be possible. At that time, there was much debate about what constituted a variety and what constituted a species. Plants were particularly difficult to classify since many reproduce asexually, by means of runners, bulbs, etc. Indeed, with a plant such as couch grass, which spreads by runners, it was at times difficult to tell where one *plant* ended and another began, let alone a variety or species! If, for example, a bird or a mouse was slightly different in colouring or shape in the north of England from that in the south, were they merely different varieties or different species? It was not always possible to capture these creatures and confine them to see if they would produce young. Even if they did not, was that because they were a different species or because they were in captivity? As zoological gardens became established, it was found that many animals, such as lions, would not breed in captivity. Even interbreeding was not a sure sign of sameness of species, because the horse and the donkey could interbreed and produce a healthy mule but they, surely, were different species?

Whereas Darwin was to be interested in how creatures came to differ from one another, Blyth was struck by their sameness. In the wild, thousands of creatures, birds, mammals, fish, would be born year after year, century after century, all looking the same to the human eye, even if the birds, in particular, did seem to be able to distinguish the one from the other. Domesticated animals varied in colour and shape, but in the wild such differences were unknown. Blyth was aware that occasionally a plant or animal would appear that differed in some way from its peers. Blyth concluded that these differences were disadvantageous, making that individual more subject to predation, particularly if the difference was in its colouring, which usually had a camouflaging effect. If differences were allowed by Nature to survive, then, over time, species individuality would break down. This had not happened. On the contrary, even the smallest differences were maintained by the species boundary.

Blyth argued that if flora and fauna were 'self-adapting', then they would adapt as they reached the outer limits of their natural habitat. Blyth noted the lack of intermediate types at habitat boundaries. Infinitely variable species, Blyth argued, would breed infinitely and would thus become less and less distinct. They would reach their full – adapted – potential at their habitat boundary, if, indeed, such a boundary existed. There would be no centre, no radiation, but a steady change as adaptation occurred. A species infinitely adaptable would be able to inhabit an infinite area – at least in its natural environment be that land, sea/fresh water, or air.

Blyth also discussed the possibility of sea currents transporting seeds. This latter possibility Blyth rejected on the grounds that water would germinate the seed, which would then die before reaching suitable soil, if the seed had not already been destroyed by the salt, an issue which Darwin was to take up in *The Origin*. Blyth appeared to be unaware that this subject had been addressed by Erasmus Darwin.

Blyth's argument was essentially the same as that of Lamarck. Lamarck also recognized that *individuals* occasionally varied in some way from other individuals of the same species. He also argued that if such changes were passed on to subsequent generations, there would be far more variety in nature than there was. His argument was framed in such a way that he appeared to be suggesting that deviant individuals would be prevented in some way from reproducing whereas Blyth suggested that 'selection by Nature' would eliminate the deviant individual by predation. Presumably both men would also have factored in the possibility that the deviant individual would be rejected by its peers for mating.

Blyth eventually rejected the idea of common descent, seeing 'selection by nature' as a conservative force. Darwin was to accept it, seeing 'Natural Selection' as a diversifying force, but he always struggled with the problem of 'blending'.

Eiseley's study of Blyth's work had made him question Darwin's ethics. There are three sentences in the papers of Blyth which Eiseley reproduced, which were rather surprisingly not specifically mentioned by Eiseley. The first occurred in Blyth's (1835) *On Varieties of Animals* in which he discussed the definition of species and varieties:

> The above is confessedly a hasty and imperfect sketch, a mere approximation towards an apt classification of 'varieties', but if it chance to meet the eye, and be fortunate enough to engage the attention of any experienced naturalist, who shall think it worth his while to follow up the subject, and produce a better arrangement of these diversities, my object in indicting the present article will be amply recompensed.

The second is somewhat similar and occurred in the final paragraph of the same article:

> Properly followed up, this subject might lead to some highly interesting and important results.

The third sentence of relevance occurred at the beginning of Blyth's (1836) paper on *Seasonal and Other Changes in Birds*:

> The subject is both extensive and complicated, and involves a number of other recondite inquiries. I could have wished that some naturalist better qualified than myself had taken it in hand.

Did Darwin take these sentences as both a challenge and a disclaimer? Having brought certain facts and thoughts to the attention of his fellow naturalist, did Darwin assume that Blyth was handing over these thoughts and ideas to anyone who wished to follow them through, without entailment or citation? Bearing in mind Eiseley's earlier work in which he defended Darwin's apparent plagiarism, it is surprising that this possible explanation was not offered by Eiesley.

We know that Blyth not only corresponded with Darwin, but met him at Down in Kent and never does Blyth seem to have made any attempt to claim priority. Why should he, when he knew that so many 'able writers' before him had so often 'taken the subject in hand' ?

Gould's dismissal

Stephen J. Gould was the pre-eminent commentator on evolutionary theory during the second half of the 20th century. He wrote extensively. Gould (2002: 137) completely dismissed Eiseley's theory. He agreed that Blyth had discussed natural selection but claimed that Eiseley had made a common mistake by not realizing that "all good biologists did so in the generations before Darwin". Considering Eiseley's academic credentials, and the considerable number of other people's work he also cited as having anticipated that of Darwin, this dismissal is somewhat insulting to Eiseley's memory. If natural selection was such a common subject for discussion, why is Darwin's work held today to have been so ground-breaking? Gould was a staunch 'Darwinist' and his dismissal of Blyth and Eiseley's work, while at the same time upholding that of Darwin, seems to me to be a double standard.

Chapter 10

Vestiges of Unity

There is one more writer whose ideas on evolutionary theory need to be considered before moving on to the better known ones of Charles Darwin, namely Robert Chambers, and one more person, who had a greater influence, not merely on the thinking of Charles Darwin but on the successful launching and promotion of his ideas than anyone else, and that person was Sir Charles Lyell.

Robert Chambers (1802–1861)

Although Hutton's book had made little impact on the thinking of Georgian Britain, Playfair's (1802/1956) *Explanations of the Huttonian Theory of the Earth* had been considerably more popular. Lyell's three volume work on the geology of the Earth, published in the 1830s, was also widely read and the educated people of Britain became accustomed, not only to the idea of movement in the Heavens, but also to the idea of change taking place upon the Earth. They were well prepared for *Vestiges of the Natural History of Creation* which was published anonymously in 1844 (Chambers 1844/1994), anonymity being a practice not uncommon in Victorian times.

It will be remembered that it was the Chambers brothers whose firm published the English translation of Buffon's works in 1834. Perhaps their interest in Buffon's work had been stimulated by a knowledge of that of Hutton, who, like them, came from Edinburgh? Whatever the reason for their undertaking, it meant that they were the first people in Britain to have the opportunity to study Buffon's vast work in their own language. It clearly made a good impression, for a decade later one of the brothers, Robert, published his own work.

In times past, great libraries had been the proud possession of the wealthy (and educated) few, but times were changing. Industrialization and commerce had meant that an increasing number of people in society needed to be conversant with the "Three R's". The great British Middle Class had come into being. The Chambers brothers ran a publishing house which specialized in publishing economical books and pamphlets designed to inform the non-University class of reader about a wide variety of topics. As will shortly be seen, Lyell also promoted his work by the publishing of a cheaper edition, following the first leather-bound edition, intended for the general reader. The Chambers Publishing House also published the *Edinburgh Journal,* which was well respected and widely read. In the process of publishing the works of other people, Robert Chambers no doubt became very well read.

The work undertaken by the Chambers brothers in establishing this business was such that by 1842 Robert Chambers was obliged to take time off to recover from a breakdown. Between the years 1842 and 1844 Chambers wrote his book, co-incidentally the same years during which Darwin made his first two 'sketches'. Eleven editions of *Vestiges* were published during Chambers' lifetime, all anonymously, and the tenth edition included an *Autobiographical Preface* in which Chambers gave an (anonymous) account of how he came to write his book.

Being convinced that the world operated under law, suggestions of 'fiats', 'special miracles' and 'interferences' were unsatisfactory to him. Chambers was aware of the transcendental approach to anatomy and physiology taken by Cuvier, of which he approved, but he dismissed Lamarck's theory as inadequate since it seemed to deny God any active role in the process of evolution. Chambers wrote from a theistic perspective, rather than an overtly Christian one. He saw God as actively involved in the process of both creation and evolution. Although acknowledging Lamarck as a man of the highest character, Chambers considered his work 'among the follies of the wise'. He rejected not only Lamarck's concept of evolution as a result of the 'wants' of the animal, but also that these changing 'wants' were precipitated by changing environmental conditions. Chambers' own theory of evolution relied solely on Design set in motion by the First Cause.

Although Robert Chambers never admitted authorship, he was soon on the list of possible 'suspects'. It was left to his friend, Alexander Ireland, to disclose his authorship after his death, lengthy correspondence between these two co-conspirators confirming the veracity of the claim. To protect his anonymity, Robert's wife, Anne, had copied out his work so that his handwriting would not be recognized and the manuscripts were sent to Ireland for printing by the London firm, John Churchill. The book was an immediate success, its popularity being fuelled by the criticism it attracted and discussion it generated. First published in October 1844, by January 1846 it had run to five editions, which increased to eleven during Chambers' lifetime. The twelfth edition was published posthumously and contained the introduction by Ireland confirming Chambers' authorship.

In response to criticisms of his work, in 1845 Chambers published (again anonymously) his *Explanations*. Both these books were included in the 1994 facsimile reproduction of Chambers' work, along with three reviews of Darwin's *The Origin*, published by Chambers, together with other additions. Because Chambers' two books were reproduced facsimile, there was a duplication in page numbering. In the Index, the editor used the letters 'V' and 'E' before the page number to indicate whether the reference was from *Vestiges* or *Explanations*. That system has been adopted here.

Chambers' work followed closely the example of Buffon. He aimed to bring together the stories told by the fossils and by geology and to meld these with contemporary scientific experiments and opinion, making one grand exposition. He commenced his book with an account of that which was known: the position of the Earth as one of a number of 'satellites' orbiting the Sun, forming the solar system; the Sun was but one of a myriad of suns, which we term stars, forming what Chambers called the astral system; the size of the Solar System (3,600,000,000 miles across); the shape of the Solar System (a flat oval); the Milky Way containing a myriad of stars; the existence of multitudes of stars not visible to the naked eye, but only by telescope. Taking the reader through the nebular hypothesis, Chambers proceeded from the conception of the Earth, its geological formation, the commencement of organic life in the ocean, development from fish to reptiles, to birds, to mammals, and eventually to Man.

The organization and extent of the Heavens led Chambers to one conclusion – Creation was the work of a Great Being. This was no pandering to political expediency, but a firm acclamation of Chambers' personal belief. He was writing after the manner of Paley, drawing attention to the wonders of the Universe.

Chambers' solution to the problem of evolution was simple, and in accord with the slow process of development throughout the Universe in general and of individual growth in particular. The changes which took place at puberty, Chambers argued, were predetermined before the child was born, the process being put in motion at the time of conception. So it was with the Universe. Divine Wisdom had established a principle of gradual, progressive development, in accordance with which gradual advances had been, and were being, made (p. V203). Chambers postulated that all development (evolution, both geological and biological) was predetermined by the Creator, and was not in response to changing environmental conditions or the wants/needs of the animal, as had been suggested by Lamarck.

Chambers saw the point of change as occurring during the embryonic stage of growth and development. Only in recent times, said Chambers, had physiologists observed that each animal passed, in the course of its germinal history, through a series of changes resembling the permanent forms of the various orders of animals inferior to it (p. V198). By considering only the embryonic form, it was possible to trace the path of evolution. What, said Chambers, if embryonic development was extended slightly? Would not the embryo then be able to develop to the next stage, as predetermined by the Divine Author? Chambers called his theory 'the Universal Gestation of Nature' (p. E72).

Chambers supported spontaneous generation not only of the most simple cells, but of moulds, fungi and 'worms' as well. He held that there was present-day evidence of 'occasional workings of the life-creating energy' (p. V178). He cited the practice of farmers mixing together horse and cow dung to propagate mushrooms, rejecting the claim that mushroom seeds were carried in the atmosphere, unperceived, on the grounds that these postulated seeds had never been seen and were, therefore, merely an abstract theoretical formation. Mould and infusoria, Chambers pointed out, increased their numbers by cell division, not by ova, so where did the hypothetical wind-borne seed come from?

Chambers cited the well-known experiments undertaken separately by Crosse and Weekes, involving crystallization of silicate of potash and nitrate of copper in the presence of a 'powerful voltaic battery'. A gelatinous matter had formed from which Weekes had observed a tiny 'insect' emerge. This was believed to be a species of *acarus* 'minute and semi-transparent and furnished with long bristles, which could only be seen by the aid of the microscope' (pp. V185-187). Chambers also referred to the case of the alleged transmutation of oats into rye when oats were kept cropped. Chambers' authority, a Dr. Lindley, had asked: "How can we be sure that wheat, rye, oats, and barley, are not all accidental off-sets from some unsuspected species?" (pp. E111-112). This is close to the concept of polymorphism in plants, especially among cereals (see White 1937, 1973, Dobzhansky 1970).

Chambers had been impressed, not only by the size of the Universe, but also with its order. In the first chapter of his book, Chambers spoke of the mathematical spacing of the planets. It is not surprising, therefore, that Chambers was attracted to the Quinary System of Animated Nature, outlined by MacLeay, to which he devoted a whole chapter of his book. MacLeay was yet another person who wrote on the subject of evolution during the first half of the 19th century and it has already been noted that both Blyth and Darwin had apparently read his work, or were at least sufficiently aware of it to use MacLeay's word '(in)osculate'. MacLeay saw living things not as a chain, or ladder, but as a series of osculating circles, no one part of which was superior to another. The components of these circles were invariably five in number. For example, in the animal kingdom there were five sub-kingdoms: the vertebrates, annulosa, radiata, acrita and mollusca.

This artificial combining and splitting of groups in order to bring their number to five was to be the undoing of the theory, a process well underway when *Vestiges* was published. Adherence to this failing theory, as well as to the theory of spontaneous generation, drew the most criticism from the scientific community, although Chambers' chief opponents were the Establishment, especially the Church.

Although Chambers wrote at length about the vastness of space, he had surprisingly little to say about time. There were only two passages in which he referred to the vast/enormous space of time needed for the 'gestation' of the whole of creation (pp. V202, V210).

Unlike Buffon and Cuvier, Chambers made no attempt to equate any part of his theory with the biblical account of creation. Chambers went so far as to suggest that a whole phenomenon of evolution was taking place, not only on this sphere, but on other spheres in space as well (p. V203). Such ideas were not welcomed by the Church as they tended to diminish the uniqueness of humans. At the time of the publication of Chambers' book, MacLeay's theory was going out of favour. Subsequent editions placed less emphasis upon it and by the final edition, all reference to it had been deleted.

Chambers had postulated a process of gradual, progressive development, guided by Divine Wisdom. Critics argued that the fossil evidence showed a sudden appearance of new forms at different times, not a gradual evolution. Darwin was to face the same criticism of his theory, which he endeavoured to counter simply by claiming that the fossil record was not perfect.

That *Vestiges* went to twelve editions is testament to its enduring popularity with the general reader. Although constantly criticized, it undoubtedly paved the way for Darwin's work when this was eventually published. Chambers' book was published after Darwin had already put aside his own first essay. Its publication was not the reason why Darwin decided not to seek publication of his ideas at that time, but ongoing editions may have been a contributing factor to Darwin's constant delay. No doubt Darwin took careful note of all the criticisms levelled at Chambers' ideas in the presentation of his own. As it was, there can be no doubt that Chambers paved the way for Darwin's longer work which was published fifteen years later.

Charles Lyell (1797-1875)

The profound influence the work of Charles Lyell exercised over the mind of the young Charles Darwin as he poured over the pages of *Principles of Geology* (Lyell 1830-1833) during his five year voyage on the *Beagle* is an integral part of the unfolding story of Darwin's theory of evolution. Lyell and Darwin later became close friends. Darwin (1867) was to say that he saw more of Lyell than of any other man. Lyell became converted to the fact of evolution, even if he never completely accepted all of Darwin's theory. During his lifetime, Lyell was acknowledged in England as the founder of scientific geology, which included the study not only of rocks, but of fossils, and therefore of earlier forms of life. Today, earlier writers, such as Buffon and Hutton, are seen to have preceded him, but they were not as well known in England as Lyell, who brought the study of geology to popular attention. It was due to Lyell's influence that geology became a degree subject at University, rather than an extra-curricular subject or hobby for gentlemen. He was knighted in 1848.

While studying at Oxford, Lyell had been much impressed by the work of James Hutton, of which he had learned by reading Playfair's *Illustrations of the Huttonian Theory*. After completing his studies, Lyell travelled in Europe, meeting Cuvier while in Paris. He was deeply impressed by Cuvier and subsequently sided with Cuvier against Lamarck, even though Lamarck's theory was much closer to Lyell's own theory of Uniformitarianism. Lyell adopted a non-confrontational (diplomatic) approach, where possible making it appear as if he were in agreement with the point of view of others (Lyell 1830-1833/1997: xxxii-xxxiii).

Born a gentleman, Lyell was privately 'anti-establishment', particularly in reference to the Church, becoming a Unitarian rather than remaining a member of the Church of England. Lyell saw no purpose in 'preaching to the converted' and continued to move in Tory (Conservative) circles, where he hoped his less than conservative views might make an impression. Although he was by preference a Whig (Liberal), he continued to review for the Tory *Quarterly Review* rather than the Whig *Edinburgh Review*, believing he would have more opportunity that way to influence opinion. When it came time to publish his book, entitled *Principles of Geology* in imitation of Newton's *Principles of Physics*, Lyell published with Murray, the Tory publisher, rather than with Knight, the Whig publisher, aiming his work at the Upper Classes, churchmen and politicians, not just by the content of his work but by its presentation and price. These gave his ideas respectability. Four years later, a cheaper version of his work was published and this two-pronged strategy was successful in bringing his ideas to the attention of a wider public.

Having travelled widely in Europe and visited the sites of volcanic eruptions, earthquakes, etc., Lyell came to the conclusion that the geology of the Earth could be accounted for by assuming that present events were similar in type and frequency to past events and that no 'external' cause was necessary as partial explanation. By using the vast amounts of time which were now allowed to the geologists, thanks to Buffon and Hutton, Lyell was able to calculate that the wearing away of rocks, such as the chalk cliffs on the south coast of England and the Niagara Falls on the border between Canada and the United States, could account for the retreat of land in some areas, the laying down of soil in others, with a resultant shift in land masses. To this Lyell added volcanic lava flows, which not only increased the height of land and sometimes formed new islands, but which could also change the flow of rivers, etc. Earthquakes contributed to the raising and destruction of land, causing some to be lost beneath the sea. Alterations in land formation would change wind flow and sea currents. Events in one part of the world could have far reaching effects on another.

Among the increasing number of fossils being found, some were clearly of plants and animals no longer in existence. Two things became apparent: (1) the deeper the fossil was found, the more likely it was to be of a simple form and (2) there were breaks in the fossil record showing what appeared to be the sudden replacement of previous forms by newer forms.

Lyell attributed the fact that most of the earliest fossils were of fish or other water dwelling creatures to the fact that sedimentation made it far more likely that the remains of water (especially sea) dwelling animals would be preserved. Animals dying on land would be devoured very soon after their death and would not have time to fossilize. Even animals washed downstream in flooded rivers would be eaten very quickly, long before their remains could be fossilized. Just because their remains had not been found was no reason to assume that land animals had not existed in the very earliest times.

The appearance of different forms in the ascending strata of one place could be explained by alterations in environment and climate. Slight changes in conditions would be tolerated, since it was well known that both plants and animals were able to adjust to changed living conditions within certain limits. Lyell cited animals which varied their colour or the thickness of their fur with the seasons. He also cited animals which were known to have made similar changes after being taken by humans to a new environment. However, he claimed, once initial variation had taken place, often quite quickly, further change took place only very slowly, if at all. If conditions altered too much, the plant or animal would be unable to adapt and would perish.

Having justified extinctions, Lyell was left with the need to explain how new forms had replaced extinct ones. Lyell rejected Lamarck's concept of evolution in favour of continuous replacement by influx from other parts of the world, as had his friend Cuvier, along with special creation. Lyell saw no reason to suppose that Creation had happened only once, but held that 'Acts of Creation' had produced new forms to replace extinct ones.

Lyell pointed out that if a round figure of one million species be assumed to inhabit the Earth, then, if only one species a year became extinct somewhere, it would take a million years for a complete changeover to take place. Most of these extinctions would take place unobserved, as would the creation of their replacement. Lyell asked how could it be known that a species had always existed, if it had never been observed before and there was no fossil evidence of its earlier existence? It is ironic that both Cuvier and Lyell rejected Lamarck's theory of spontaneous generation yet themselves proposed the nearly constant creation (or appearance) of new species.

At the time Lyell was writing, the discovery of the Neanderthals was still nearly thirty years away. Although stone tools had been found, there was no fossil history for humans. Lyell concluded that 'Man' was a recent creation. If the 'First Cause' had created 'Man' but recently, was there any reason why it should not also still be creating other plants and animals?

Lyell (1863) later admitted that in his earlier work he had not fully understood or appreciated Lamarck's system. The main difference of opinion between Lamarck and Cuvier had related to the fixity of species. Lamarck had seen the species boundary as one created by humans to assist them in categorizing plants and animals and held that living forms had an unlimited capacity to change over time. By 'time', Lamarck meant millions of years, a point which Lyell was to admit he had not fully grasped. Lyell argued that, while humans had been instrumental in the appearance of countless new varieties, they had not been able to produce even one new species. Furthermore, although variation was easily accomplished early in the process of domestication, further change became increasingly more and more difficult to achieve. There was a limit beyond which a species could not be made to vary (pp. 232-233).

Eventually converted by Darwin to the concept of evolution, Lyell never fully accepted Darwin's theory of evolution by natural selection. Having reread Lamarck's work, Lyell came to understand the length of time Lamarck had been proposing and eventually retracted his opposition to Lamarck's theory (Lyell 1863: 3):

> In the concluding chapters I shall offer a few remarks on the recent modification of the Lamarckian theory of progressive development and transmutation, which are suggested by Mr. Darwin in the 'Origin of Species by Variation and Natural Selection' and the bearing of this hypothesis on the different races of mankind and their connection with other parts of the animal kingdom.

Bearing in mind Darwin's antagonistic attitude towards Lamarck, it is doubtful that Darwin was overly pleased to have his great work described as a modification of Lamarck's theory! By the 1860s, human fossils of great antiquity, as well as large numbers of stone tools, were being found. Lyell visited a number of sites and became convinced that humans were not a recent creation.

Lyell was one of a growing body of people who preferred to refer to a 'First Cause' rather than to 'God'. His position was not secular but he was part of a rising tide of people who were distancing themselves from Christian beliefs, especially those relating to the Old Testament account of Creation. Despite his interest in geological change, he initially resisted the concept of biological change. He gradually came to accept the idea of the gradual evolution of living forms but never fully accepted Darwin's theory that evolution took place through a process of 'chance' change 'naturally' selected, although he certainly accepted Darwin's right to hold his belief and did as much as he could to help and support Darwin in the dissemination of his theory. Both men were strongly anti-Christian and this was the tie which bound them, even though it would seem clear from the statement cited above that Lyell was in truth more comfortable with Lamarck's theistic position than with Darwin's increasingly atheistic one.

Part II

IMPACT

Before 1900

Darwin's theory and its reception

Chapter 11

Laying the Foundations

So dominant did Darwin's theory of evolution by natural selection become, that "Darwin's theory of evolution" is sometimes spoken of as though it were Darwin who first proposed evolution itself, something which clearly was not the case. This book is not about Darwinism, *per se*, but about the way in which peoples' preconceived ideas have prejudiced their interpretation of evidence in relation to evolution itself and the history of the ideas which surrounded it. Nevertheless, it is inevitable that Darwin and his theory will dominate the remainder of this discussion.

Charles Darwin (1809-1882)

Charles Darwin was born on 12th February, 1809, in the English country town of Shrewsbury, nestled in the beautiful county of Shropshire which borders Wales. He was the second of six children, his father, Robert, being a doctor, as had been his father, Erasmus Darwin, before him. Affluent, but not aristocratic, the family were well-respected members of the rising English Middle Class. Throughout his life, Darwin never needed to work. He and his wife inherited all the money they would ever need to raise their family in comfort, with the assistance of their domestic staff. Darwin records in his short autobiography that he retained very little memory of his early years, apart from a trip to the seaside, which seems to have made a great impression. Was it on that holiday, as a four-year-old, that he developed his passion for collecting things? Many young children enjoy collecting one thing or another but he records his early passion for collecting all types of different items as being very strong, far stronger than any similar interest exhibited by any of his siblings. His mother died when he was eight, but he remembers very little about her, possibly because children of those times were generally kept in the nursery quarters, under the care of their nanny, for most of their pre-school years.

According to his own account, he must have been a bit of a handful at times. It would seem he had a lively imagination, which would lead him to invent stories for the purpose of playing tricks on other people, a behaviour which today would be described as 'attention seeking'. The skill of which he was most proud was shooting – he loved shooting birds! Reading the journal which he kept throughout his famous journey on the *Beagle*, it is clear that Darwin never had any compunction about killing animals, a characteristic which was no doubt desirable, even necessary, for a naturalist intent on collecting specimens, especially ones which were new, unusual or rare.

Darwin was not without compassion. He records the joy and pride he experienced when, under the tutelage of his father, he visited patients, took notes, reported back to his father who made suggestions for further questions and for appropriate treatment and saw people's health improve under his care. The young Darwin's budding career in medicine was brought to an abrupt halt when he became a medical student at the prestigious Edinburgh University and found himself required to attend the operating theatre. Two visits were more than enough. He fled the scene on both occasions before the operations were completed. As he pointed out, this was well before the "blessed days of chloroform" and the memories of these two incidents haunted him for years. This dichotomy of feeling towards people and animals persisted throughout his life. He ardently supported the abolition of slavery but just as ardently opposed the abolition of vivisection,

Charles spent two years in Edinburgh, in the company of his older brother, Erasmus, with whom he had a close relationship. Erasmus, too, was studying medicine. It was after his brother completed his time at Edinburgh that the young Charles found his medical studies too much of an ordeal. Perhaps attending operations was not required in the first two years? Be that as it may, when Erasmus left, Charles left too! His father was not pleased when he learned what had happened. He accused Darwin of being interested in nothing but "shooting, dogs and rat catching, and you will be a disgrace to yourself and all your family". Having no 'career' might be acceptable among the aristocracy, but the Middle Class had a strong work ethic and Charles was expected to make a contribution to the welfare of society, whether he needed the money or not. At his father's suggestion,

Darwin embarked upon a course intended to guide him into the life of a country clergyman, which seemed more suited to his affable character. At first, Charles harboured doubts about the acceptance of all the Articles of Faith required by the Church of England, but having carefully read *Pearson on the Creed*, he set aside any doubts and after three months' private tutoring to brush up on his Greek, entered Cambridge University at the beginning of 1828. Here he spent three years, during which, Darwin tells us, he did the minimum amount of work necessary to complete his course. No doubt he was not alone in this, since he was placed 10th among "the crowd of men who did not go in for honours". Since these numbered some 178, it would seem that he could apply himself when he wanted. Darwin attributed his success to his thorough study of the works of Paley, which he seemed genuinely to have enjoyed and accepted.

At this time, the subjects studied for the degree of Bachelor of Arts included the classics, some mathematics and, of course, for those intending ordination, theology. Even had geology been available as a degree subject, it would not have been included in the B.A. curriculum. However there was a growing interest in this fledgling field, the first volume of Lyell's *Principles of Geology* being published in 1830. His interest in natural history had led Darwin to be a regular attendee at informal gatherings for the study of botany under the tutelage of Professor Henslow and he also participated in a geological field trip organized by Professor Sedgwick. It was these extra-curricular activities which were to inspire his future career and seal his place in history.

On his first expedition to survey and map the coast of South America, the young Captain Fitzroy had been struck by the fact that there may have been valuable metals in the Fuegian mountains but he had nobody on board his ship experienced either in mineralogy or geology to evaluate their potential. He resolved, if ever he was sent on another surveying expedition, to ensure that his ship was carrying such a person on board. Cambridge University having been asked to recommend a suitable candidate for such a position, Professor Henslow thought of the young graduate, Charles Darwin. From an established family, but with no family commitments yet of his own, with no settled career demanding his attention, interested in natural history, an attendee of Henslow's own informal botany gatherings as well as of Sedgwick's geological field trip, he was the ideal candidate. Furthermore, he was of an easy-going temperament, which characteristic was to stand Darwin in good stead throughout the long voyage. Darwin recorded in his autobiography that Captain Fitzroy nearly rejected him for the position of Naturalist because the shape of his nose indicated that Darwin would not

99

possess sufficient energy or determination to complete the voyage. Darwin proved him wrong, but Fitzroy's evaluation seems not to have been completely inaccurate in that the voyage appears to have depleted Darwin's energies and left him constantly sick and somewhat socially reclusive for the remainder of his life. (Recent DNA analysis has now shown that Darwin suffered from Crohns disease).

The *Beagle* was embarking on a trip around the world. Skirting the west coast of Africa, it sailed down the east coast of South America, round the American Cape, up the west coast to the Galápagos Islands on the Equator, then west across the Pacific to New Zealand and Australia, across the Indian Ocean, round the African Cape before heading northwards and home! *HMS Beagle* did not set out on its long journey alone. It was accompanied by *HMS Adventure*, under Captain King. As was required of any naval vessel, on the completion of their commission the two captains wrote up their journals for submission to the Navy and subsequent publication. Darwin, too, had kept a journal, which was dispatched home in sections to supplement his letters. A decision was made to append Darwin's journal as a third volume. This was very well received by the public. A facsimile copy of Darwin's journal was published in 1979, allowing the reader to enjoy Darwin's account of his voyage 'in his own hand', with erasures and corrections, but without any of the changes Darwin made in the year it took him to reshape his journal for publication. In his introductory remarks to this facsimile edition, G. P. Darwin commented that Captain Fitzroy had taken offence at what he believed to be Darwin's failure to make proper acknowledgment of the help he had received from his shipmates. Similar accusations of failure to acknowledge the work of others were to be made after the publication of *On the Origin of Species.*

From his journal, it is clear that Darwin's primary interest was in geology. To supplement the knowledge he had gained during his field trip with Sedgwick, Professor Henslow had given him a copy of the first volume of Lyell's *Principles of Geology*, this being all that had been published at the time Darwin set sail in 1832. The remaining two volumes were received by him during the course of the voyage. Darwin's task was to collect geological samples, any samples of flora or fauna which he collected were solely at his discretion and expense. Henslow had kindly offered to arrange storage of any such material which Darwin shipped back until his arrival home, when it was anticipated that the flora and fauna would be sent to various museums. Unfortunately, the museums were swamped with samples being sent back by travelers from all four corners of the globe and few of Darwin's samples found a ready home.

Darwin wrote to Henslow from Rio de Janiero on 18th May, 1832, (Barlow 1967: 53):

> The geology was pre-eminently interesting and I believe quite new ... One great perplexity to me is an utter ignorance whether I note the right facts & whether they are of sufficient importance to interest others ... Geology and the invertebrate animals will be my chief object of pursuit through the whole voyage ... It is exceedingly interesting observing the different genera & species from those which I know; it is however much less than I had expected.

In another letter to Henslow dated 19th August, 1832, (Barlow 1967: 58), Darwin apologized for the small size of the geological specimens but wrote that "no person has a right to accuse me till he has tried carrying rocks under a tropical sun". Explaining the paucity of plant samples, he said: "I cannot summon resolution to collect where I know nothing ... It is positively distressing to walk in the glorious forest amidst such treasures and feel they are all thrown away on me". One cannot but feel for the young Darwin, not knowing where to start!

There seems to be an impression that it was to his time upon the Galápagos Islands that Darwin owed his insight into the concept of evolution by natural selection. This was not the case. From his journal we know that Darwin hated the Galápagos Islands. Their volcanic nature was too obvious to present Darwin with any opportunity for 'geologising' and they were almost barren, with little flora or fauna of interest. Darwin particularly disliked the iguanas, which he considered ugly and 'evil'. In fact, these iguanas were of particular interest, being the only iguanas anywhere in the world which swam and spent extended periods of time feeding underwater.

The Galápagos Islands were named after the giant tortoises found there (galápagos *(Sp.)*: tortoise) but their arrival is still a mystery. Isolated from each other, the population of each island had developed its own unique shell shape and markings. Darwin recorded this curiosity in his journal but does not appear to have given it any thought. He did not ponder how this large land animal, unable to swim between islands, had reached the Galápagos, which had never been joined to the mainland, nor to each other, for that matter.

Steadman and Zousmer (1988) discussed the problem at some length. There is no evidence that these animals were introduced to the islands by humans, nor is there evidence of human occupation of the islands before their discovery by Europeans in 16th century. It is generally assumed that they, like the rest of the flora and fauna of the islands, came from the nearby continent of South America, where giant tortoises are known to have existed in past geological times. However, there is no evidence for their existence in South America since the time that the first of the Galápagos Islands erupted from the sea floor, possibly some 30 million years ago. The ages of the different islands vary because they erupted at different times. The area is geologically unstable. The Nazca Plate is moving under the South American Plate and at the same time there is constant activity on the sea floor as volcanic matter spews from the fault line. Only occasionally does this activity result in the emergence of a new island above the surface of the sea and some small islands undoubtedly erode over the millennia and disappear back beneath the waves.

Currently there are eighteen main islands and a number of islets, some of which are little more than rocks. Because of the constant geological activity in the sea bed, the islands 'drift' and it is impossible to say precisely how close, or far apart, each of the islands was from the other at any time in the remote past. One thing is certain. Because these islands erupted from the sea bed, they have never, at any time, been joined to, or been part of, South America. At the present time, the nearest of the islands is 1000 km (600 miles) from the mainland. It is difficult to explain how land animals, such as the tortoises, survived such a sea voyage, since they were not endowed with sharp claws with which to cling. Steadman and Zousmer concluded that the journey must have been made by baby tortoises, although whether they were of the 'giant' variety, or whether they became 'giants' during the course of their evolution on the islands, remains unclear. Did their ancestors all arrive at the same time, carried on various pieces of flotsam which fortunately landed them on their separate islands? Did they arrive at intervals and chance to land on different islands? Be that as it may, they were there when Darwin visited, each colony distinguished from that of other islands by its unique shell.

In *The Origins* (p. 303), Darwin commented that it was a surprising fact that many of the new species of various fauna formed on one island had not quickly spread to another. He attributed this to the fact that the currents in the area were rapid and swept across the archipelago so that the islands were effectively far more separated from each other than they appeared on the map. Nevertheless, many species were found on several islands, presumably making the crossing during times of calm, or perhaps the reverse – being swept across during a storm. Using one fact to support two sides of an argument was a tactic to which Darwin frequently resorted and was one which provoked much criticism. In this case he used the sea both to explain how animals were able to cross a large expanse and to explain how they were prevented from crossing,

Later much was to be made of the minor differences between the finches of each island, which were in some ways even more remarkable, since birds have no difficulty flying across even large expanses of water and the finches would have had no difficulty flying from island to island had they so desired. Not only did Darwin not theorize as to the differences between the finches on the various islands, he did not even recognize them as finches, labelling one a wren, another a blackbird. It was only when his collection was later sorted that his mistakes were identified. By the time *The Origins* was finally published more than a quarter of a century later, Darwin was able to include the differences between the Galápagos finches as part of his evidence.

Only when he reached Australia did Darwin begin to ponder the degree of similarity/difference between fauna in the two hemispheres. Australia's unique fauna and flora made a deep impression upon him, yet, while there were distinct differences, nevertheless, there were similarities. In New South Wales, Darwin watched a lion ant catch its prey in a conical trap and recorded in his journal "without doubt this Judacious Larva belongs to the same genus, but to a different species, from the European one". He marveled that this tiny creature used so simple, yet so artificial, a contrivance to catch its prey. He wondered whether but one Creator could be responsible for such variety - yet how could more than one Creator have "worked over the whole world"? It is perhaps to Australia, not the Galápagos, that we owe Darwin's later philosophizing?

Darwin returned home from his voyage in December, 1836. In 1837, he started keeping notes on his readings and thoughts relating to the 'species question', namely, whether species were invariable. Central to this question was that of inter-specific sterility, which had been used by Buffon as an important criterion in the definition of a species. If the same species was represented by a number of varieties, had they been so created or had variation appeared over time, that is, had they evolved?

It was not until six years later, in 1842, that Darwin wrote up his thoughts in "a very brief abstract in pencil in 35 [hand-written] pages; and this was enlarged during the summer of 1844 into one of 230 [hand-written] pages" (F. Darwin 1887/1969: 84). In fact, the 1844 manuscript had only 189 pages, the remaining number being made up of blank sheets, interleaved as though for notes and additions (Darwin 1909/1969: xvi). Darwin intended this book to be published at some future time. He wrote a letter to his wife, Emma, dated 5th July, 1844, asking that, in the event of his death, she set aside £400 for the publication of the book. She was asked to find some competent person who would be prepared to "take trouble in its

103

improvement and enlargement". In addition to the sum of money specified, this person was also to receive the profits from the book.

To assist this editor, Darwin bequeathed this person all his books on Natural History which were either scored or had references, and charged Mrs. Darwin with the task of also handing over "all those scraps roughly divided into eight or ten brown paper portfolios". The letter ended (p. xvii–xviii):

> If there should be any difficulty in getting an editor who would go thoroughly into the subject ... then let my sketch be published as it is, stating that it was done several years ago, and from memory without consulting any works, and with no intention of publication in its present form.

That the work may have been done 'several years ago ... with no intention of publication in its present form' would, no doubt, have been true, since Darwin was clearly not satisfied with his effort, but his request that his wife state that the sketch had been done 'from memory without consulting any works' clearly was not. This is yet another example of tendencies which Darwin exhibited throughout his life: of being liberal with the truth and of not giving due credit to others.

In 1909, these two essays were published by Darwin's son, Sir Francis Darwin, under the compound title *The Foundations of the Origin of Species*. In numerous editorial footnotes, Francis Darwin indicated the names of authorities and references which Darwin had put in pencil in the margin of his book. The 1844 essay contained all the arguments which were to form the basis of *On the Origin of Species*, but without the plethora of examples with which the later book was burdened. Consequently, the outline of Darwin's thinking was more easily seen.

Darwin commenced his *Essay* of 1844 (hereafter referred to as *Foundations*) with a discussion on variation under domestication. As early as his second sentence, Darwin declared his belief in the doctrine of the inheritance of acquired characteristics:

> Under certain conditions organic beings even during their individual lives become slightly altered from their usual form, size, or other characters: and many of the peculiarities thus acquired are transmitted to their offspring.

That acquired characteristics could be transmitted to offspring was the cornerstone of Darwin's theory, in *Foundations* (Darwin 1909/1969), in *On the Origin of Species* (Darwin 1859/1998) and in *The Descent of Man* (Darwin 1871/1908). Darwin postulated that the altered conditions of domestication acted on the constitution of animals and plants, making their reproductive systems more 'plastic'. Darwin, correctly, gave equal attention to plants and animals, unlike Lamarck who addressed only animal evolution. A general theory of evolution must cover *all* living things, including bacteria and other prokaryotes, although Darwin was not aware of these last categories.

For new varieties to be formed under domestication, both selection and separation were necessary. If crossing were allowed, the variation, even if very distinct, would disappear. Separation and selection must be continued for several generations before a 'true' breed could be considered to have been established. Careful selection would result in further new races/varieties. However, Darwin warned, if the two races were allowed to interbreed freely, the two original varieties would disappear, being replaced by one 'mongrel' race, which would become homogenous. In the wild there was rarely more than one race/variety of each species. If they could interbreed, they would (p. 71): "I conclude, then, that races of most animals and plants, when unconfined in the same country, would tend to blend together". This conclusion was similar to that arrived at by both Buffon and Blyth. The line of logic Darwin was following was one which would tend to limit variability in nature, not increase it. This was a point which gave Darwin much trouble.

Darwin held that humans could not create variation. They could only select from that which had occurred. Darwin tended to favour excess of nutrition as the most likely cause of the increased variation which occurred under domestication, but, contra his own argument, pointed out that sheep and cattle, domesticated for thousands of years, were still capable of variation, even though their food supply was not increasing. Whether there was a limit beyond which variation was not possible had yet to be determined.

After discussing variation under domestication, Darwin considered variation under natural conditions. Variation in nature was slight compared with that under domestication but could occur if organisms in nature were *occasionally* subject to changing geological, and therefore environmental, conditions. (These were the conditions for change put forward by Lamarck.) Lyell's theory had postulated very gradual geological change and Darwin stipulated that organisms would have to be isolated in some way to prevent them simply moving with the gradually changing conditions. This might happen if an isthmus was cut off by rising sea levels, or a new volcanic island appeared. Darwin presented his theory thus (p. 87):

> What if … there was a Being of great penetrative insight, who could evaluate the innermost as well as the outer characteristics of any organism?

> What if … that Being had forethought which could extend over future centuries?

> What if ... that Being, working on several islands, was deliberately to choose from the characteristics thrown up by the creature's plastic reproductive system?
>
> What ... could that Being not accomplish over vast amounts of time, seeing how much blind, capricious man has actually accomplished over a few centuries?
>
> What if ... there was a secondary means in nature which could accomplish all that this hypothetical Being could accomplish?
>
> I believe such a secondary means does exist.

Then followed the section which was to form the paper read jointly with Wallace's paper before the Linnean Society in July, 1858, the official launching of Darwin's theory. In this part, reference was made to Malthus' well-known work in relation to the tendency for populations to outstrip their supply of food and of the resultant struggle for survival. Darwin then reverted to speculation (p. 97):

> What if ... a small population was isolated on an island?
>
> What if ... the conditions on this island were gradually, but continually, changing, then would not the plastic reproduction system throw up variation?
>
> What if ... some offspring were favoured by this variation and some hindered? Would not the favoured offspring have a better chance of survival?

'Death' would discard unwanted stock. Over thousands of generations, change would occur. This change would be assisted, in some cases, by sexual selection. One, or a few, males being dominant would sire most of the offspring or the female of some species (mostly birds) would choose the mate which attracted their attention by their display.

Certain animals, such as elephants, bears and hawks, had failed to breed in captivity. These animals could be tamed, but never domesticated. Some 'constitutional peculiarity' made tame species incompatible with their new environmental circumstances and prevented them reproducing. Darwin drew upon this fact to explain why hybrid animals seemed to be sterile. Hybrid animals were not known in the wild but the mule, born of a horse/donkey mating, was strong, healthy and lived a full life. However, it was unable to reproduce. Darwin argued that the 'constitution' of a hybrid animal was different from that of either of its parents and it was this 'peculiarity' which prevented it from breeding. If sufficient time and effort were to be directed towards ascertaining exactly what were the most favourable breeding conditions, there was no reason why tame animals should not breed, nor hybrids, in the same manner as any 'mongrel'. Such breeding could produce new varieties or species. Darwin further supported his argument by drawing attention to the fact that certain plants, which never crossed in the wild, were known to be capable of crossing under domestication and of producing fertile hybrids. Darwin concluded that sterility could not be the factor by which species were differentiated from varieties/races. This conclusion was to be vital to the development of his theory even though it contradicted his earlier conclusion. He argued both that random, non-isolated interbreeding would result in varieties/species losing their separate identities and becoming one stock and that interbreeding would produce hybrids which differed from both parents and thus provided the origin of a new variety/species.

Darwin argued that, while it was necessary for domestic breeders to isolate breeding stock in order to establish a new variety/species, once established, varieties/species would breed true and the breeder need not fear the throwing of crosses. If 'establishment' meant that crosses would not be thrown, then it is not clear how the new varieties/hybrids would come into existence in the first place. In the same way that Darwin had argued that the sea could be both a bridge and a barrier, he argued that a hybrid mating could produce a new variety which could lead to the establishment of a new species and that inter-breeding would result in an homogenous population.

The reader can easily understand why Darwin put his manuscript aside at that time and why it was a further fifteen years before he finally published his theory. Darwin was clearly struggling to reconcile two different positions. It was the potential for diversity which distinguished his theory from that of Blyth, but it was the concept of diversification that Darwin found so difficult to uphold. Blyth had considered the same issues, had flirted with the idea of species diversification, but his deliberations had forced him to conclude that nature operated to maintain species stability and continuity.

Although it was generally assumed by most breeders that there was a limit to the amount of variation which could be produced, Darwin was "unable to discover a single fact on which this belief is grounded". It was Darwin's opinion that (p. 104):

> Until man selects two varieties from the same stock, adapted to two climates or to other different external conditions, and confines each rigidly for one or several thousand years to such conditions, always selecting the individuals best adapted to them, he cannot be said to have even commenced the experiment.

Were such careful selection to be made over this amount of time, Darwin argued, the constitution of the organic being, be it animal or vegetable, would be changed and the resulting varieties would be indistinguishable from species.

The third chapter of Darwin's script was devoted to mental characteristics, including instincts. He opened his chapter with a caution (p. 112):

> Let me here premise that, as will be seen in the Second Part, there is no evidence and consequently no attempt to show that *all* existing organisms have descended from any one common parent-stock, but that only those have so descended which, in the language of naturalists, are clearly related to each other. *(Italics in original)*

Darwin then postulated the inheritance of acquired mental characteristics on the same grounds as those he had put forward for physical characteristics. He concluded (p. 115):

> These facts must lead to the conviction … that almost infinitely numerous shades of disposition, of taste, of peculiar movements, and even of individual actions, can be modified or acquired by one individual and transmitted to its offspring.

As possible examples, Darwin offered birds able to build better nests than their companions having a better chance of survival and passing on their greater abilities; birds that fed grubs to their chicks when they themselves were granivorous might once have been omnivorous and the habits of the adult changed, or a shortage of the correct food might have caused the parent bird to offer alternative nourishment to its chicks. Those chicks able to digest the alternative nourishment, in this case grubs, would pass on this ability to their own chicks. Darwin stressed, by use of italics, that he was suggesting *possibilities*, not *probabilities*.

Darwin believed not only that the eye could have evolved very gradually from a more simple structure, but that some body parts may have evolved from parts originally formed for different functions. Some naturalists believed that part of the ear had metamorphosed into the swimming bladder in fish, one of the rare occasions when Darwin cites prior opinions of other naturalists supporting evolution. Creatures such as bats, seals and gliding squirrels exhibited characteristics and abilities more commonly associated with other families. The jaguar swam and caught fish. Perhaps in time the jaguar would become a water animal? Darwin asked (p. 132):

> Who will say what could thus be effected in the course of ten thousand generations? Who can answer the same questions with respect to instincts? If no one can, the *possibility* (for we are not in this chapter considering the *probability*) of simple organs or organic beings being modified by natural selection and the effects of external agencies into complicated ones ought not to be absolutely rejected. *(Italics in original)*

In the second part of his book, Darwin attempted to convert *possibility* into *probability*.

Darwin had based his theory of natural selection on three basic propositions: time, isolation and the inheritance of acquired characteristics. Time was the first issue he dealt with in the second section of his book. Darwin expounded the theories of Hutton and Lyell regarding the age of various geological formations, of sediments, of the raising and lowering of sea levels consequent upon the raising and subsidence of land masses, of inundation, of formation of islands, of glaciation, concluding that it was a surprise that as many fossils had survived as had been found. The imperfection of the fossil record accounted for the fact that there was so little evidence of intermediate species.

Darwin suggested that animals, such as tortoises and crocodiles, had remained basically unchanged because their conditions had changed little and used this as evidence to contradict Lamarck's theory that there was in animate nature an inherent tendency towards change and development. However, conditions would have changed little in the oceans, which support a multitude of different life forms, even in the 'deep', a point which seems to have received little or no attention in the literature on evolution.

Using mammals as his example, Darwin drew attention to the differences between Australia/New Guinea and the rest of the world. While all areas had their own flora and fauna, there were many resemblances between them, these resemblances/differences being difficult to explain under Divine creation. Why would the Creator bother with small changes? Surely one would expect the Creator either to place the same creatures all over the world, or completely different ones? Although a major barrier had caused Australia to be isolated from the rest of the world, and had allowed a completely different fauna to evolve, namely the marsupials, yet there was a connection with the rest of the world in that Australia was inhabited by vertebrates, mammals, birds, fish, etc. Furthermore, a few marsupials existed in South America.

From major barriers and major differences, Darwin moved to minor barriers and minor differences. Deserts, mountains, rivers, arms of the sea, all could have served to separate one variety or species from another. Fauna and flora on islands which differed from that on the nearest mainland were nevertheless nearly always of a closely allied type, although slight differences could be observed between islands of a group such as the Galápagos. Prior occupation of an area could be sufficient to prevent new arrivals from taking hold. However, not all flora and fauna living in a certain environment were perfectly adapted, as was shown by the rapidity with which introduced European species were out-competing some indigenous species. Darwin's argument that if these creatures had been made by God, they would be perfectly adapted to their environment and, therefore, able to hold their own against introduced species, failed to explain why species which had adapted naturally to their environment over millions of years should be unable to defend their territory against invaders.

Before proceeding further with his thesis, Darwin recapitulated his three main criteria:

(1) repeated changes in conditions over several generations;

(2) steady selection of these slight varieties with a fixed end in view;

(3) isolation.

Darwin's second point, "a fixed end in view", was to become anathema to later Darwinists. Darwin had based his theory on the assumption of the existence of a "Being of great penetrative insight" with "forethought that could extend over future centuries". This 'Being' became almost invisible in *The Origin* - but not quite!

Thus far, Darwin had attempted to establish the means by which variation could occur within a species, but had not yet shown how a new species might evolve. He argued that a large number of animals on a large area could be expected to produce a 'sport' more frequently than a small number of animals in a small area, such as an island. However, Darwin realized that more interbreeding would take place in a large population, making new characteristics which had appeared by chance more difficult to establish (p. 184):

> If, however, he [the breeder on the mainland] could separate a small number of cattle, including the offspring of the desirable 'sport', he might hope, like the man on the island, to effect his end. If there be organic beings of which two individuals never unite, then simple selection whether on a continent or island would be equally serviceable to make a new and desirable breed; and this new breed might be able in surprisingly few years from the great and geometrical powers of propagation to beat out the old breed.

Darwin moved from a 'sport' occurring within a population of interbreeding animals to two groups of animals which never united to postulate the start of a new breed. The characteristics of the 'sport' would produce the new race, without reversion to the original type due to interbreeding, simply because the 'sport' was part of a small, isolated population, which, for some unknown reason, would never unite with other inhabitants. Possibly with the volcanic Galápagos Islands in mind, Darwin drew the picture of an island, or group of islands, gradually emerging from the sea. To this island, plants and animals would gradually arrive, having been transported by a variety of means, such as hurricanes, floods, floating trees or roots, or as seed in the stomach of some animal. Conditions would be different, and 'it might also easily happen that some of the species *on an average* might obtain an increase of food, or food of a more nourishing quality' (pp. 185-186):

> We might therefore expect on our island that very many slight variations were of no use to the plastic individuals, yet occasionally in the course of a century an individual might be born of which the structure or constitution in some slight degree would allow it better to fill up some office in the insular economy and to struggle against other species ... and if (as is probable) it and its offspring crossed with the unvaried parent form, yet the number of individuals being not very great, there would be a chance of the new and more serviceable form being nevertheless in some slight degree preserved.

This last conclusion is contrary to the evidence Darwin presented in the first part of his book in which he reviewed change under domestication, where he showed that great care had to be taken to isolate the *individual* carrying the desired new trait and to ensure its careful mating.

Darwin then envisaged a geological situation such that several adjacent islands might be formed. These would be stocked by similar or different chance arrivals from the nearest mainland, or by some of the already changed inhabitants of the first island. The islands would grow, possibly merge, eventually forming a large body of land, even a continent. As this happened, conditions would constantly be changing and so would the inhabitants, although isolation would no longer be a factor. As land rose in one part of the world, so it would sink in another. As continents sank, their land would become inundated, mountains become islands, creatures become isolated, and so the pattern would continue (p. 190):

> Let the now broken continent, forming islets, begin to rise ... let the islands become reunited into a continent ... The oftener these oscillations of level have taken place ... the greater the number of species which would tend to be formed.

Since sedimentary materials were not deposited on oceanic islands, the absence of intermediate fossils showing transitional stages could easily be understood. Fossils were preserved more readily during times of subsidence. Darwin conceded that enormous lapses of time would be needed for an island to be converted into a continent.

Darwin asked his reader to imagine that a certain species had become divided into six different regions. Conditions being different, they would develop into six different, but related, species. Some might not survive, but suppose half did. Suppose these three related species also became divided into several separate regions, they would develop into further species. These 'cousins' would form a genus. If the process were to be repeated, the future generations would form a family. The older forms would probably become extinct, but *families, genera, species, varieties* would continue. In comparing this scenario with that of domestic breeding, Darwin glossed over the fact that all varieties of domesticated species continue to be inter-fertile, they do not become species, genera or families. He had drawn an analogy which was not sustainable. Darwin attempted to correct his proposition by suggesting that intermediates had become extinct (Darwin pp. 212-213):

> It follows from our theory, that two orders must have descended from one common stock at an immensely remote epoch ... The existence of genera, families, orders, &, and their mutual relations, naturally ensues from extinction going on at all periods amongst the diverging descendants of a common stock.

Darwin here postulated for the first time that 'orders' have common descent. In the first part of his book, Darwin had stated that "as will be seen in the Second Part, there is no evidence and consequently no attempt to show that all existing organisms have descended from a common stock", yet this was clearly the direction in which his argument was leading him. In the final chapter of *Foundations*, Darwin referred to this point again (p. 252):

> No doubt the more remote the two species are from each other, the weaker the arguments become in favour of their common descent ... But if we once admit the general principles of this work ... we are legitimately led to admit their community of descent. Naturalists dispute how widely this unity of type extends: most, however, admit that the vertebrata are built on one type, the articulata on another; the mollusca on a third; and the radiata on probably more than one. Plants appear to fall under three or four great types. On this theory, therefore, all organisms *yet discovered* are descendants of probably less than ten parent forms. *(Italics in original)*

Lamarck and Chambers had assumed that evolution always occurred in the direction of greater complexity. Darwin correctly pointed out that this was not always the case. The 'Epizoic' Crustaceans were free swimming in early life, with articulated limbs and eyes, which they lost in maturity as they became attached to a fish upon which they preyed and which they never left. (The *complete* organism, throughout its whole cycle of life had moved towards increased complexity, compared with the earlier forms of life from which it was descended, as postulated by Lamarck.)

Associated with this tendency towards increased simplicity as well as increased complexity, Darwin saw vestigial and rudimentary organs, a circumstance which he claimed was completely incompatible with the doctrine of Special Creation. Why would God create a beetle with fused wings or a fish with useless eyes? Over time a 'station' occupied in other districts by less complicated animals might be left unfilled, and be occupied by a degraded form of a higher or more complicated class. The arguments Darwin made in this section are among the most logical of his entire essay.

Eiseley (1979) had shown, by similarity of thought and phrase, that Darwin had drawn from the work of Blyth when he wrote *The Origin*. Darlington (1959) made a similar examination, and reached a similar conclusion, in relation to Darwin's *Foundations* documents of 1842 and 1844.

Having completed his *Foundations* document to the best of his ability, Darwin turned his attention to the reworking of his *Journal*, the first edition of which had been published along with the report of the journey of *HMS Beagle* and *HMS Adventure*, prepared by Captain Fitzroy and Captain King. Publishing a second edition, retitled *Journal of Researches into the Natural History and Geology of the Various Countries Visited During the Voyage of HMS Beagle round the World* (Darwin 1845), allowed Darwin to fill out his account with thoughts and opinions which had occurred to him during the intervening years. Darlington studied Darwin's early writings and had been unable to find any hint of evolutionary thought in the notes and letters Darwin had written while on the *Beagle*, or, indeed, shortly thereafter. This led Davies (2008) to the opinion that Darwin had deliberately introduced his new thinking into the pages of the *Journal of Researches* in an attempt to deceive people into believing that these thoughts had occurred to him while on his voyage, thereby enabling him to be perceived as having anticipated the work of Edward Blyth which had been published 1835-1837. A careful comparison of the first and second editions of the *Journal* revealed that (p. 34-36):

> Throughout the second account, Darwin had inserted paragraphs dealing with evolutionary ideas that could only have been written by Darwin, the evolutionist, and not by Darwin the geologist, as he was on the Beagle voyage
> ...
>
> No one who bought either the 1839 or 1845 editions of the Beagle journal had the opportunity to check them against the original journals from the voyage which were still in Darwin's possession ...
>
> Gruber also showed how much further developed was Darwin's version of his Galápagos discoveries in the 1845 edition ... almost completely rewriting the natural history of the Galápagos for the new version of his journal.

Darwin undertook the task of revising his *Journal* immediately after he had completed his *Foundations* document, even changing its name to *Journal of Researches.* His ruse was eminently successful. To this day Darwin's name and that of the Galápagos Islands are inextricably intertwined and the deception would never have been discovered had not the original journal been made available to historians for study. It would seem that, realizing it may be some time before he was in a position to publish his ideas, Darwin took what steps he could to establish some priority. Darwin thereafter constantly referred in his letters to the length of time upon which he had been working on his big book. If it was Darwin's aim to deceive, there may have been some poetic justice in the fact that it was this second edition of his *Journal* which, in 1845, inspired the young Alfred Wallace to set out upon his journeys to distant lands to discover 'the origin of species'.

Chapter 12

On The Origin of Species

D arwin's major work, *On the Origin of Species by Means of Natural Selection or, The Preservation of Favoured Races in the Struggle for Life,* which Darwin finally published in 1859, was as much an extension of his 1844 Essay as it was an abstract of the longer work he was writing and which he had planned to call *Natural Selection. The Origin* was hastily assembled in one year following the receipt of a letter from Alfred Russel Wallace which clearly showed that Wallace had reached a similar conclusion to that of Darwin in relation to evolution.

Darwin's basic thesis remained very much the same as that put forward in *Foundations.* His first four chapters, which were devoted to explaining his theory, followed the same plan as the two parts of *Foundations.* Darwin once again commenced with a discussion of variation under domestication and from there he extrapolated to variation in nature.

In his first chapter, Darwin discussed the difficulty in deciding whether animals domesticated by humans thousands of years ago, such as dogs and cattle, had originated from one, or several, species. Today there are many more breeds of dog than there were in Darwin's day, but, even then, there were many breeds which differed greatly the one from the other, such as the greyhound and the bulldog. Interbreeding would produce an animal with a mixture, or blend, of the characteristics of the parents, not one with new features. Random ongoing breeding of such animals, as would occur in nature, would result in indistinct mongrels, not a distinct new breed. Only careful selection by the breeder over many generations would produce a new breed and were care to be relaxed and the new 'pure bred' animal be allowed to mate with another breed, the breed's new characteristic features would soon be lost. Darwin concluded that the many breeds of domestic dog must have originated from a number of distinctly different species. He did not explain why 'indistinct mongrels' would have resulted rather than distinct breeds (varieties/species) which would have been necessary if his theory of natural selection resulting from chance variation in the wild were to explain the millions of distinct varieties/species which have evolved in nature. Presumably the mongrel varieties in some way became isolated, allowing them to become distinct varieties or insipient species.

In *Foundations*, Darwin had postulated a Being of great forethought who planned ahead. That Being was no longer present. There was now no forethought, only random mating and chance variation, and from these two ingredients, Darwin was to argue, all the varieties of life on this Earth had emerged by the process he called "Natural Selection". In *Foundations*, Darwin had offered the concept of the island as the place where change was most likely to become established, believing that some form of isolation was necessary to prevent the new, favourable, variation from being 'swamped' by interbreeding with non-variant members of the breeding population. Darwin finished his first chapter on change under domestication by repeating this point – that new breeds/varieties could only become established by careful isolation of the breeding population.

The second chapter of *The Origins* was devoted to *Variation under Nature*. Following the format of *Foundations*, Darwin here discussed the difficulty in determining precise definitions of *'species'* and *'variety'*. Many naturalists had considered this problem and suggested solutions, none of which had proven completely satisfactory. Darwin declined to make a contribution to this on-going discussion, stating rather (p. 42):

> "... that I look upon the term species as one arbitrarily given for the sake of convenience to a set of individuals closely resembling each other, and that it does not essentially differ from the term variety."

The intricacies of the process of reproduction were still unknown at the time Darwin was writing. Chromosomes and DNA had yet to be discovered. Darwin believed that there was no insurmountable impediment to the interbreeding of any two creatures (or plants). Individuals simply *preferred* to mate with another individual similar to themselves. Species were not *essentially* stable; they were stable by convenience or convention, each generation simply following the custom of the previous generation in choosing its mate from among a restricted population. Darwin called varieties 'incipient species' while at the same time arguing that there was no such thing as a fixed species, or, indeed, a species at all. What were commonly called species were merely well-established varieties, which retained an ability further to vary. The essential instability of species was central to Darwin's theory.

As an example, Darwin referred to the current debate of the time – whether the Primrose *(Primula veris)* and the Cowslip *(Primula elator)* were varieties or separate species. While acknowledging that the term 'species' was generally applied to populations which appeared to be the result of distinct acts of creation and 'variety' to those which had resulted from common descent, such definitions were difficult to apply in practice. Under domestication, the history of the plant or animal was generally known. In nature, this was not the case. The supposed boundary could rarely be proven. Where the divide ran depended upon the individual opinion of various naturalists.

Darwin argued that the definitions offered were circular. To define species by inter-sterility and then claim that no species were inter-fertile with one another, or that all varieties were inter-fertile simply because that was the definition of 'variety', proved nothing.

Darwin's examples were drawn from plants, some insects and brachiopods, which also had polymorphic forms and/or were herma-phrodite. At this stage of his argument, Darwin made no reference to mammals. Later, in Chapter 8 where he discussed 'hybridism', Darwin stated (pp. 188, 189):

> The fertility of varieties, that is the forms known or believed to have descended from common parents, when inter-crossed, and likewise the fertility of their mongrel offspring is, on my theory, of equal importance with sterility of species; for it seems to make a broad and clear distinction between varieties and species …

> It is certain, on the one hand, that the sterility of various species when crossed is so different in degree and graduates away so insensibly, and, on the other hand, that the fertility of pure species is so easily affected by various circumstances, that for all practical purposes it is most difficult to say where fertility ends and sterility begins … It can thus be shown that neither sterility nor fertility affords any clear distinction between species and varieties.

In consecutive pages, Darwin claimed that fertility/sterility made a broad and clear distinction between varieties and species and that neither sterility nor fertility afforded any clear distinction at all.

The chief aim of Darwin's second chapter was to claim varieties, strongly marked and permanent varieties, sub-species and species were names of human contrivance, having no reality in nature. He was to return to these claims as he developed his thesis.

What humans could achieve under conditions of domestication, nature could achieve in the wild. The only difference was that change in the wild took longer, much longer.

Darwin's third chapter considered the struggle for existence. He considered the super-abundance of seed in both plant and animal kingdoms and the small likelihood of any one individual even reaching adulthood, let alone leaving numerous offspring. This was a common subject of discussion in Victorian England and this chapter was not intended to introduce new thought, rather to complete the groundwork for his theory, which he was to outline in the following chapter. Of most interest to us is probably the account Darwin gave of one of his own experiments, which gave readers their first glimpse of Darwin's obsessive care in his work (p. 54):

> ... on a piece of ground three feet long and two wide, dug and cleared, and where there could be no choking from other plants, I marked all the seedlings of our native weeds that came up, and out of the 357 no less than 295 were destroyed, chiefly by slugs and insects.

Throughout his life, Darwin conducted many experiments involving the counting of seeds, although at times it is not clear how they advanced his theory. There was no suggestion in this case that the slugs and insects selectively fed on the weaker or less favoured weeds so that the strongest, fittest survived. On the contrary, surely they would have fed first upon the most robust shoots, leaving the less robust with the best chance of growing to maturity and reproducing?

The fourth chapter of Darwin's book, entitled *Natural Selection*, in effect completed the reworking of the two parts of his *Foundations* essay. Here he finally outlines the basic principles of his theory (p. 64):

> ... can we doubt ... that individuals having any advantage, however slight, over others, would have the best chance of surviving and of procreating their kind? On the other hand, we may feel sure that any variation in the least degree injurious would be rigidly destroyed. The preservation of favourable variations and the rejection of injurious variations, I call Natural Selection.

Darwin then envisaged changing environmental and climatic conditions, pointing out that even the failure of one species to adapt, becoming extinct, would impact all other populations of plants and animals in that area, some because their food source no longer existed, forcing them to adopt new habits, others because they were no longer preyed upon. These changes would have a domino effect. Some changes might enable other species to change their habitat, moving into the changing area for the first time, which would add another 'domino'. He then returned to his favourite scenario – that of an island, or isolated area (p. 64):

> But in the case of an island, or of a country partly surrounded by barriers ... we should then have places in the economy of nature which would assuredly be better filled up, if some of the original inhabitants were in some manner modified ... every slight modification, which in the course of ages chanced to arise, and which in any way favoured the individuals of any of the species, by better adapting them to their altered conditions, would tend to be preserved and natural selection would thus have free scope for the work of improvement.

Darwin suggested that changed conditions would affect the physiology of all biota, including their reproductive systems, increasing the chance of modifications occurring in offspring. He conceded that "unless profitable variations do occur, natural selection can do nothing" (pp. 64-65). In the previous chapter on survival, Darwin had written at length on the fact that most destruction of living forms takes place at the stage of the seed/seedling, egg/newborn. The majority of potentially favourable chance improvements, it must be assumed, would be destroyed before they reached maturity and were able to reproduce. In view of the claim that Darwin was now making, that favourable chance variations were the progenitors of new species, I feel that Darwin was remiss in not, once again, drawing his reader's attention to this important point. He may have assumed that any serious reader would have absorbed this obvious fact from his previous discussion.

Without mentioning Paley's name, Darwin then demolished one of Paley's main arguments regarding the perfection of nature and, thus, of the Goodness of God. Darwin pointed out that there was not a country in the world where the introduction of foreign plants or animals had not seen at least some of the introduced species survive at the expense of the native inhabitants. If the natives were perfectly adapted to their environment, as they surely would have been if specially created by God, they would out-compete the introduced species which had evolved/been designed for different conditions. Darwin argued that nothing was completely, perfectly adapted to its environment, such that no future change was possible or desirable. Once again, as when making this same argument in *Foundations*, Darwin failed to address the fact that species evolved under natural selection were also vulnerable to competitioon. Should not they, too, have evolved in such a way that they would have the advantage over any introduced species, which had evolved under, and was, therefore, more suited to, different conditions?   If this situation is problematic for one argument, it is problematic for the other.

We then come to the most poetic paragraph of the entire chapter and, I believe, the most important one (p. 66):

> It may be said that natural selection is daily and hourly scrutinizing, throughout the world, every variation, even the slightest, rejecting that which is bad, preserving, and adding up all that is good, silently and insensibly working, whenever and wherever opportunity offers, at the improvement of each organic being in relation to its organic and inorganic conditions of life. We see nothing of these slow changes in progress, until the hand of time has marked the long lapse of ages, and then so imperfect is our view into long past geological ages, that we only see that the forms of life now are different from what they formerly were.

Gone is the Being of great penetrative insight and forethought, which the reader of *Foundations* would have associated with the Creator, or God. In His place, we now have Nature, or rather 'natural selection'. Later Darwinists, particularly Neo-Darwinists, were to claim that 'natural selection' was a process devoid of any thought or planning, since there was no Supreme Being, no Creator, no God, no entity capable of any planning or thought involved – all was chance. That was not the picture here painted by Darwin. He spoke of 'natural selection' as though she were a caring mother, watching over her children, ever mindful of their best interests, caring for and nurturing them every second of the day. Their welfare and improvement were her sole concerns. Darwin had rejected Lamarck's claim that there was an expansive force in nature actively seeking opportunities for change but the scenario he painted was not so very different, since his entity, natural selection, was constantly watching for opportunities to effect change, to make improvements. Natural Selection, however, was not in complete control. She could only effect change when opportunity presented itself – when changes occurred in the environment. Lamarck's Supreme Being orchestrated these environmental changes as well, although they all only occurred in accordance with the Laws of Nature which the Supreme Being had initiated.

Darwin's theory relied upon the fact that offspring were not identical reproductions of their parents, there were subtle differences, and it was upon these subtle differences that natural selection worked. What is not clear is whether Darwin understood these subtle differences as part of some plan or whether he perceived them as 'mistakes'. Neo-Darwinists now say that there is no plan, therefore there can be no mistakes, that all is chance. It is impossible to say now exactly what Darwin had in mind when he wrote that paragraph because it was clear from his later writings that he, himself, was uncertain. Be that as it may, the 'Being of great forethought' was gone, never to return. His departure opened the door for followers of Darwin, such as his great friend, Thomas Huxley, and other Humanists after him, to claim that 'natural selection' had obviated the need to postulate any form of Divine Creator, that there was no God. This was to have a profound effect on the thinking of the 20th century, as will be shown, and it is for this reason that I consider this paragraph to be one of the most important in Darwin's entire book.

Darwin gave examples of camouflage (grouse the colour of heather, etc.); he wrote of sexual selection, of male mammals competing with each other, of male birds enticing the female; he wrote of hypothetical examples, such as of the wolf which was more fleet, of a cub with a tendency to hunt a preferred prey passing on that tendency to its own offspring; he wrote of plants secreting juices, sweet or bitter according to whether they needed to attract insects or deter predators.

Having gently, but firmly, rejected the views of both Lamarck and Paley, Darwin now rejected those of Cuvier (pp. 74-75):

> ... as modern geology has almost banished such views as the excavation of a great valley by a single diluvial wave, so will natural selection, if it be a true principal, banish the belief of the continued creation of new organic beings, or of any great and sudden modification in their structure.

Darwin returned to the question of isolation, but here showed a change of thinking. Previously, he had seen islands as protected places, where new varieties could take hold. Now he realized that most islands had a more restricted biota than the larger land masses and had come to see islands more as restricted environments (p. 82):

> Although I do not doubt that isolation is of considerable importance in the production of new species, on the whole I am inclined to believe that largeness of area is of more importance ... Throughout a great and open area, not only will there be a better chance of favourable variations arising from the large number of individuals of the same species there supported, but the conditions of life are infinitely complex from the large number of already existing species; and if some of these many species become modified and improved, others will have to be improved in a corresponding degree or they will be exterminated ...

> I conclude that, although small isolated areas probably have been in some respects highly favourable for the production of new species, yet that the course of modifications will generally have been more rapid on large areas ...

Darwin offered in support of his theory a scenario which was completely unsupported by any evidence. He (p. 87) invited his reader to take the hypothetical case of a carnivorous quadruped. Suppose its population had grown to such an extent that its food source was now insufficient to support all its members. Darwin suggested that some individuals might feed on new kinds of prey, inhabit new places, climb trees, frequent water, perhaps become less carnivorous. Strictly speaking, 'less carnivorous' means 'omnivorous', including some vegetable matter in the diet, but it is possible that Darwin was envisioning this carnivorous quadruped eating insects?

Either way, there is no evidence to support this notion. We have all seen wild life documentaries showing lions, their ribs and flanks skeletally thin, panting beneath the shade of the trees as they wait for the coming of the rains, which will rejuvenate the grass and bring the return of their prey. They spend as little energy as they can while they wait. Some will survive, some will not. What they will not do is nibble at the bark of the trees, eat the fallen leaves, scratch the earth to uncover the roots, eat ants or insects, or anything which is not a normal part of their diet.

Omnivorous animals, such as rats, are more fortunate. They have a larger range of possible materials they can utilize as food, but even they have their limits. When there is insufficient food, the animal dies; it does not expand its form of diet. It will be remembered that Darwin had suggested that some granivorous birds might have offered their young grubs when there was a shortage of seed and postulated that this change of behavior had been passed on to their offspring. (There are some birds which are granivorous as adults which feed grubs to their young. Quite why this should be, no one knows.) Neither humans nor animals attempt to feed their starving young inedible material. People die wearing their clothes. They do not eat them in an attempt to stay alive. Humans, as well as animals, starve to death when their habitual food is unavailable. There was no evidence to support Darwin's conjecture, but plenty to contradict it. Over-population, with subsequent starvation, is not a valid scenario to explain dietary diversification.

In the remaining chapters of his book, Darwin expanded upon his theory. Two chapters each were devoted to the geological record and geographical distribution. His final chapter included discussion on embryology and rudimentary organs, which were probably his soundest arguments. Throughout he used numerous examples, some of which are more reasonable than others!   Below are some of the issues raised, with comments.

Hybridism

Darwin's fifth chapter was on the Laws of Variation. In it (pp. 125-128) Darwin suggested that all our domestic breeds of horse and donkey were descended from the zebra. He supported his argument by citing examples of unexpected stripes appearing, usually on the legs but sometimes on the flanks or back, on horses and donkeys. These stripes often disappeared as the animal matured.

The concept of common descent, particularly among the cat, dog and equine families, was appealing and now generally accepted. The problem was with showing how it had come about. Darwin was well aware that all our domestic breeds of cat, dog, horse, cattle, sheep, pigeons, geese, etc. etc., were considered 'varieties' because they could still interbreed and produce fertile offspring. Horses and donkeys could interbreed – but their offspring (mules) were sterile. Were they 'varieties' because they could interbreed or 'species' because their line was sterile? To Darwin, it did not matter. He considered the terms interchangeable.

In *Foundations*, Darwin had suggested changed conditions would have the effect of making the reproductive system more plastic, leading to the appearance of variation. Changed geological conditions which led to isolation, or enforced isolation under domestication, were the two situations most responsible. While it had been found possible to tame some wild animals held in captivity, such as lions, it had not been possible to domesticate them, i.e. they would not breed in captivity. Darwin (correctly) attributed this to wrong conditions and argued that when the right conditions were provided, these animals would breed in captivity. The implication was that hybrid animals similarly would subsequently be able to breed when provided with correct breeding conditions and that their progeny would be fertile, since this was one of the methods which he had previously suggested nature used to increase variation – but there was no evidence to support this assumption. All available evidence contradicted it.

Darwin returned to the problem of hybrids three chapters later, in Chapter VIII (pp. 187-211). Most of his examples were botanic, since it was far easier for plant breeders to cross varieties/species by artificial pollination, even those which would never have been able to cross in the wild for anatomical reasons, than it was to induce animals of different varieties/species to interbreed. The results of botanical crosses were varied but it had been found that new varieties often became less fertile, or completely sterile after a few generations. This, it was concluded, was due to 'inbreeding'. (Presumably, the further generations had also resulted from artificial pollination?) Domestic breeders, of both plants and animals, agreed that it was of benefit to their stock regularly to introduce 'new blood' from another breeder. Too much inbreeding caused sterility.

In discussing the infertility of hybrid crosses, Darwin cited not only crosses between horses and donkeys but also between canaries and finches, which were also sterile. Darwin believed the infertility of this latter cross could be attributed to the fact that the finches involved did not readily breed in captivity. Furthermore, attempts at further breeding from a first generation cross had generally been made between siblings, which were not fertile because they were too closely related.

Species and varieties

Darwin had previously argued that, in order to become established under domestication, new varieties/breeds had to be kept isolated, that breeding must be carried out between the most closely related of those exhibiting the desired characteristic, to prevent that new characteristic from being 'swamped'. Close interbreeding was essential to establish the new breed. Now he argued that close inter-breeding gave rise to sterility (p. 193): "... brothers and sisters have usually been crossed in each successive generation, in opposition to the constantly repeated admonition of every breeder."

European and Chinese geese, European and Indian cattle, each supposed by some to be separate species, interbred and produced fertile offspring. Some argued that the geese and cattle were each but varieties. Darwin (1859/1998: 194) did not take this view:

> ... most of our domestic animals have descended from two or more aboriginal species, since commingled by intercrossing. On this view, the aboriginal species must either at first have produced quite fertile hybrids, or the hybrids must have become in subsequent generations quite fertile under domestication. This latter alternative seems to me the most probable ... On this view of the origin of many of our domestic animals, we must either give up the belief of the almost universal sterility of distinct species of animals when crossed; or we must look at sterility, not as an indelible characteristic, but as one capable of being removed by domestication ...
>
> There is no reason to think that species have been specially endowed with various degrees of sterility to prevent crossing and blending in nature ... There is no fundamental distinction between species and varieties.

It seemed apparent to Darwin that certain groups of animals, such as the cats, dogs, equines and certain birds, were related and descended from a common ancestor. The problem lay in explaining how this had come about. His 'explanation' totally failed to account for the evolution of the several wild ancestors – they were just assumed to exist. The interbreeding which he suggested, did not, for some reason, result in a reduction in variation but to the establishment of another species. One can only assume that Darwin envisaged interbreeding producing a new (fertile) variety, some individuals of the new variety becoming isolated, giving rise to increased variation in nature, rather than homogeny.

Lamarck, also, had seen changed conditions as the precursor of variation. However, he never suggested that any established species crossed with another species, rather that individual populations changed. Those at the extremity of the original population's habitat would vary in different ways, as they would be experiencing different, changed, conditions. This is more in keeping with today's thought.

Darwin invited his reader to consider the many varieties of domestic pigeon as separate species (p. 128):

> Call the breeds of pigeons, some of which have bred true for centuries, species, and how exactly parallel is the case with that of the species of the horse-genus!

Having clearly stated that he saw no difference between a variety and a species, was Darwin entitled to call varieties of pigeons species? Despite the fact that in so doing he was disagreeing with every other naturalist, I believe the answer has to be 'Yes'. He was being internally consistent. Having thus shown that all closely-related domestic, 'species,' were inter-fertile, was he justified in assuming that all closely-related wild species were also inter-fertile? Clearly he thought so and the success of his book indicates that others agreed with him. I do not. To build his theory, even if only in part, on the assumption that wild animals would choose to interbreed with another species – and produce fertile offspring – was not justified and certainly not scientific!

The finches of the Galápagos Islands were something of an embarrassment to Darwin, who was well aware that his labelling of them had fallen well short of the standard expected. Lack (1974) published a study identifying thirteen different species, although his task was complicated by the large number of intermediate varieties. While most finches usually mated with their own 'species', about 10% were believed to mate with other 'species', making difficult a distinction between a 'species' and a 'well-marked variety'. Darwin made no reference in *The Origin* to the finches. They were a perfect example of evolution in progress – and Darwin missed it! Had Darwin spotted it, would he have considered the Galápagos Islands isolated, because of the distance between them and the South American mainland, or not isolated because of their proximity to each other?

In 1977, Steadman commenced research on the thousands of fossil bones to be found in the lava tubes of the Galápagos. It was his hope to trace part of the evolution of at least some of the islands' native inhabitants. He was seeking an answer to the question of whether the finches of the Galápagos Islands were in the process of diverging or converging. In another 100,000 years' time, would there be more species, or would such as now exist have merged into fewer 'mongrel' species (Steadman and Zousmer 1988)? He was unable to make a prediction.

The horns of the dilemma with which Darwin battled 150 years ago are still as piercing today!

The evolution of instinctive behaviour

Darwin's theory that individuals having an advantage over others, however slight, would survive and procreate and that variation in the least degree injurious would be rigidly destroyed applied to 'mental' characteristics as well as to physical. For example, under domestication, dogs had been trained to retrieve, herd, guard, etc. Those displaying most prominently the desired characteristic had been selected by breeders and passed on to their offspring the 'mental' pattern they had acquired by training during their life time. Gradually these patterns became so ingrained they became instinctive.

Instinctive behaviour had been developed in the wild in a similar manner. An advantageous behaviour practiced by a parent would be more easily acquired by its offspring until, after a number of generations, that advantageous behaviour became instinctive.

This scenario is still generally accepted by Darwinists and is a clear example of the inheritance of acquired characteristics, even over a small number of generations.

Social instincts

One cannot at times but admire the courage of Darwin. As is normal procedure in academic writing, he followed the chapters outlining his theory with one dealing with difficulties (pp. 132-158). In it, he made a very specific statement (p. 154):

> Natural selection will never produce in a being anything injurious to itself, for natural selection acts solely by and for the good of each. No organ will be found, as Paley has remarked, for the purpose of causing pain or for doing an injury to its possessor;

He then tackled the difficult problem of the sting of the wasp or the bee which could not be withdrawn after use, due to its backward facing serratures, causing the death of the insect. How could this situation have arisen under natural selection? Darwin's solution was to suggest that the sting had originally evolved as a boring instrument "originally adapted to cause galls" and subsequently utilized for other purposes. Thus the sting had not evolved with the intent that its possessor should die following its use; that was an unfortunate by-product of an adaptation. Perhaps Darwin foresaw the serratures disappearing over time? Nevertheless, injurious adaptations of this kind were not supposed to happen under natural selection. Anything unfavourable was supposed to be eliminated. Since the individual insect was dead, its 'experience' could not be passed on to the next generation, for better, for worse.

By including Paley's name, Darwin was pointing out that inexplicable things seemed to happen even in God's perfect creation, thereby pre-empting criticism of his theory by Creationists. Why these insects do not possess a retractable stinging apparatus is a problem for both Creationists and Darwinists.

Later, Darwin returned to the problem of social insects. He needed to explain the evolution of classes of sterile individuals (workers) in colonies of social insects. Again, he showed courage by telling his reader of (pp. 180, 181):

> ... one special difficulty, which at first appeared to me insuperable, and actually fatal to my whole theory. I allude to the neuters or sterile females in insect communities: for these neuters often differ widely in instinct and in structure both from the males and fertile females, and yet, from being sterile, they cannot propagate their kind.

> If ... it had been profitable to the community that a number should have been annually born capable of work, but incapable of procreation ... I can see no very great difficulty in this being effected by natural selection ... The great difficulty lies in the working ants differing widely from both the males and the fertile females in structure ... with the working ant we have an insect differing greatly from its parents, yet absolutely sterile; so that it could never have transmitted successively acquired modifications of structure or instinct to its progeny ... how is it possible to reconcile this case with the theory of natural selection?

Until this point, Darwin had maintained that variation occurred in individuals and it was *individuals* who survived in the struggle for life. Now he took an alternative position. It was the *community* which was all important – the individual was expendable (p. 182):

> This difficulty, although appearing insuperable, is lessened, or, as I believe, disappears, when it is remembered that selection may be applied to the family, as well as to the individual, and may thus gain the desired end.

> ... slight modification of structure, or instinct, correlated with the sterile condition of certain members of the community, has been advantageous to the community; consequently the fertile males and females of the same community flourished and transmitted to their fertile offspring a tendency to produce sterile members having the same modifications.

Darwin (pp. 182-183, 185), bravely, or foolishly depending upon the point of view taken, continued:

> ... the neuters of several ants differ, not only from the fertile females and males, but from each other, sometimes to an almost incredible degree ... It will indeed be thought that I have an overweening confidence in the principle of

natural selection, when I do not admit that such wonderful and well-established facts at once annihilate my theory ... I believe ... that by long continued selection of the fertile parents which produced most neuters with the profitable modification, all the neuters ultimately came to have the desired character.

I am bound to confess that, with all my faith in this principle, I should never have anticipated that natural selection could have been efficient in so high a degree, had not the case of these neuter insects convinced me of the fact.

Darwin's faith in his theory was unshakeable. All life had assumed its current form as the result of natural selection. We may not fully understand the process; what seemed incredible, almost impossible to us to be achieved by natural selection, *had* been achieved by natural selection, which only went to show just how powerful and wonderful natural selection was. *Nothing* could shake Darwin's belief!

Having reminded his readers of Paley's (outdated) views, Darwin could not resist taking a thrust at those of Lamarck (p. 185):

I am surprised that no one has advanced this demonstrative case of neuter insects against the well-known doctrine of Lamarck.

Lamarck, of course, held to a belief in a Supreme Being who orchestrated major changes. The major changes, which Lamarck suggested had taken place over long intervals of time, were inexplicable to Lamarck without this intervention.

The final sentence which closed this chapter has become very well known. Evolution had taken place, not as the result of special endowment or creation but following:

... one general law, leading to the advancement of all organic beings, namely, multiply, vary, let the strongest live and the weakest die.

Natura non facit saltum

Upon one point, Darwin remained steadfast, not only in writing *The Origin*, but for the remainder of his life: Nature made no leaps. Evolution was a gradual process; natural selection always acted with extreme slowness. This was a point upon which Darwin and his good friend, Thomas Huxley, did not agree. Huxley (1900) felt that Darwin had unnecessarily restricted his theory by denying any evolutionary role to 'sports' or 'monsters', terms given to flowers or animals which clearly, significantly and unexpectedly differed from their parent stock.

Darwin exhibited a very casual attitude toward the actual origin of any feature, the title of his book notwithstanding. Paley had taken the eye as his first example of Divine design. In Darwin's opinion how a nerve ending originally became sensitive to light "hardly concerns us" (!) but in this sensitivity once having been established, Darwin saw (pp. 144, 146):

> ... no very great difficulty ... in believing that natural selection has converted the simple apparatus of an optic nerve merely coated with pigment and invested by transparent membrane, into an optical instrument ...

> If it could be demonstrated that any complex organ existed which could not possibly have been formed by numerous, successive, slight modifications, my theory would absolutely break down. But I can find no such case.

One of the reasons why he could find no such case was that, when he did find one, such as the sterile worker ants and bees, he merely asserted that that case operated in accordance with natural selection by citing some mechanism which he imagined and then assuming that his position was proven.

It was not only individual organs in the process of transition which Darwin was called upon to explain, but species. How did a species of animal subsist in a transitional state? How did reptiles survive during the time when their two front legs were in the process of turning into wings? They would neither be able to run nor fly. How did a carnivorous land animal turn into an aquatic one, such as a whale? Darwin suggested (pp. 141-142):

> In North America the black bear was seen by Hearne swimming for hours with widely open mouth, thus catching, like a whale, insects in the water. Even in so extreme a case as this, if the supply of insects were constant, and if better adapted competitors did not already exist in the country, I can see no difficulty in a race of bears being rendered, by natural selection, more and more aquatic in their structure and habits, with larger and larger mouths, till a creature was produced as monstrous as a whale.

By the time he wrote *The Origin,* Darwin had been working on his theory for twenty years and had accumulated copious notes and examples. His choosing to include this example in *The Origin* cannot have been the result of haste in compiling his manuscript. It must have been included following much thought over many years and thus gives us a good indication of the process of his thinking during the development of his theory.

Darwin evinced a cavalier attitude towards 'difficulties', seeming to think that if he did not 'see' them, or consider them important, they either did not exist, or were of no consequence (p. 138):

> If a different case had been taken, and it had been asked how an insectivorous quadruped could possibly have been converted into a flying bat, the question would have been far more difficult, and I could have given no answer. Yet I think such difficulties have very little weight.

129

If there were two sides to an argument, Darwin at some point in his book adopted them both. Varieties gave rise to species, species gave rise to varieties; variation started with the individual and spread to the group; variation started with the group and spread to the individual; there were distinct differences between species and varieties, there were no differences between species and varieties; hybrids never gave rise to new species, hybrids were the start of new species; isolation was essential to prevent reversion to the original form, isolation caused inbreeding and led to infertility; nothing could evolve by natural selection which would harm the individual, the death of the individual was of no consequence if its fellows benefited; expanses of water acted as an isolating barrier, water acted as a means of dispersal.

It is little wonder that Darwin had been so hesitant to publish his ideas. He had a grand vision, but experienced great difficulty in working out the details.

At the beginning of his book, Darwin strove to convince his reader that there was no essential difference between a variety and a species. Now he denied the essential difference between hybrid/mongrel and fertile/infertile. This obfuscation allowed Darwin to use whichever meaning he needed to prove the point he was making. Without this freedom, Darwin would never have been able to formulate his theory.

The hoarder next door

Darwin was a collector, a hoarder even, of facts. The discriminating collector chooses carefully what items to add to his collection, arranges items carefully so that they are displayed to full effect and can easily be accessed. Some collectors lose their ability to discriminate; they start collecting anything and everything even remotely connected with their area of interest. Some end up by collecting - and hoarding - almost anything. Darwin started out as a collector and ended up as a hoarder. From his study at Down, Darwin wrote thousands of letters, asking other people to supply him with facts. These may have been neatly filed in his filing cabinets, but they became a muddled mess in his head. Instead of sorting through that which he already had, bringing his facts into order so that he might present them to the public, he did what so many other collector/hoarders do - he searched for more! Despite promises to himself, and hints to his friends, he was unable to bring his thoughts into sufficient focus to write his 'big book'. Eventually, he was so snowed under that his friends were forced to intervene, to help him sort through his material and produce something publishable. Without this 'intervention', it is unlikely that

Darwin would ever have published. More about this later.

Once he had been forced to clear his mind of some of his accumulated facts, Darwin was able to proceed with two further major publications, *The Variation of Plants and Animals under Domestication* (1868) and *The Descent of Man* (1871). These were better written than *On the Origin of Species* but still contained some muddled thinking as we shall see in the next chapter.

Chapter 13

Journey's End

The theoretical part of Darwin's work was contained in the first four chapters of *The Origin*, the remaining chapters being an extrapolation of his theory, illustrated by numerous examples. These four chapters were later expanded into a two volume work, *The Variation of Animals and Plants under Domestication* (Darwin 1868/1893), which was probably the most logically presented of all his theoretical treatises, and a second two volume work, *The Descent of Man and Sexual Selection* (1871/1908). While the remainder of Darwin's epic remained uncompleted, it could be said that Darwin did, in fact, complete the theoretical component of his great work.

In *The Variation*, Darwin covered an extraordinary amount of material in relation to both plants and animals under domestication. If there was one Darwin family trait which had been passed down from Erasmus to Charles, it was the ability to acquire vast amounts of information and to relay this information with great verbosity! Voluminous in detail as was this work, it was generally well constructed and most points were made clearly and logically. Writing of the ill-effects of inbreeding, Darwin told how there was an exchange of stags among the deer herds of England's Great Parks and how breeders of birds and dogs, etc., introduced a sire from another breeding stock to keep their line strong and healthy. Any breeder trying to develop a certain characteristic would endeavour to obtain an animal with characteristics as close as possible to those which were being developed. In *Foundations*, Darwin had favoured small, isolated populations for the establishment of new varieties; in *The Origin*, Darwin had been indecisive as to whether large or small populations allowed the greatest opportunity for variations to arise and become established. By the time he wrote *The Variation,* Darwin was quite clear that small, inbred populations were detrimental.

Darwin once again tackled the subject of hybrid sterility and what he had to say was interesting (pp. 170-171):

> But he who would take the trouble to reflect on the steps by which this first degree of sterility could be increased through natural selection to that higher degree which is common to so many species, and which is universal with species which have been differentiated to a generic or family rank, will find the subject extraordinarily complex. After mature reflection, it seems to me that this could not have been effected through natural selection. Take the case of any two species which, when crossed, produce few and sterile offspring; now, what is there which could favour the survival of those individuals which happened to be endowed in a slightly higher degree with mutual infertility and which thus approached by one small step towards absolute sterility?

> As species have not been rendered mutually infertile through the accumulative action of natural selection … we must infer that it has arisen incidentally during their slow formation in connection with other and unknown changes in their organization.

Darwin had encountered the same stumbling block as had Buffon. Later writers, such as Dawkins (1976 and elsewhere), Dobzhansky (1970 and elsewhere), Gould (2002 and elsewhere), and many more, were to consider natural selection as *all sufficient* in accounting for *all evolution,* citing Darwin as their authority. Darwin, himself, never made this claim. Even in *The Origin,* Darwin merely claimed natural selection to be the main factor. Here, after nine more years of thought, admidst a plethora of praise and criticism in relation to his thory, Darwin, twice, quite clearly stated that he had concluded that there were factors other than natural selection which had contributed to evolution. The crossing of wild species had been a prominent factor in Darwin's argument for the establishment of varieties and species which clearly appeared to be of common descent. Quite why or how these varieties subsequently became sufficiently differentiated to become mutually sterile, Darwin never really explained in *The Origin.* Now he had abandoned the idea. This did not destroy his theory since the idea of common descent from a single species could still be proposed.

Darwin did not re-address the problem of sterility among social insects.

Pangenesis

The latter part of *The Variation* saw Darwin return to the question of the inheritance of acquired characteristics. He had never doubted that new characteristics acquired during the lifetime of an individual could be inherited and now endeavoured to explain how this might come about. In both *Foundations* and *The Origin*, Darwin had put forward the idea that characteristics, such as smallness of stature due to adverse growing conditions and variations under domestication due to changed living conditions, could be passed on to future generations. Instincts, too, were patterns of behaviour repeated so frequently that they became inherited characteristics and Darwin had used this principle to explain how dogs had been bred to point, fetch, guard, herd, etc.

In *The Variation*, Darwin (pp. 369-370) introduced his theory of *pangenesis* thus:

> It is universally admitted that the cells or units of the body increase by self-division ... and ... become converted into the various tissues and substances of the body. But besides this means of increase, I assume that the units throw off minute granules which are dispersed throughout the whole system ... these ... multiply by self-division ... [and] may be called gemmules. They are collected from all parts of the system to constitute the sexual elements, and their development in the next generation forms a new being; but they are likewise capable of transmission in a dormant state to future generations and may then be developed.

The last stipulation was necessary to account for reversions or throwbacks. Darwin was, of course, well aware of the work of botanists who had crossed varieties of plants with well-defined characteristics, such as different coloured flowers, and was familiar with the concept that in the first hybrid generation one form dominated (was 'prepotent') over the other, but that when these hybrids were self-fertilized, the proportion of dominant (prepotent) forms decreased. Darwin concluded that 'prepotency', while an interesting phenomenon, did not contribute to long-term change, such as was needed to bring about evolution, because it was a force which expended itself over a few generations. (Darwin had himself experimented with *Antirrhinum* at about the same time that Mendel was carrying out his experiments with sweet peas.)

Other naturalists were studying inheritance in animals, endeavouring to find a 'law of inheritance', such that the characteristics contributed by the male were dominant over those contributed by the female (Vorzimer 1970). (This was Victorian England!) These experiments, and Darwin's theory of *pangenesis*, were concerned with how characteristics changed and how these changed characteristics were inherited, not with how these characteristics had appeared in the first instance.

Darwin gave no evidence for the existence of gemmules other than that he 'assumed' they existed. These gemmules, Darwin suggested, circulated throughout the body and were able to reproduce themselves. They might remain undeveloped during the early stages of life or during succeeding generations, their development depending on their union with other cells with which they had an affinity. Some creatures, such as salamanders, crabs and worms, were able to utilize 'gemmules' to reproduce lost parts; others, such as mammals, could use them only for repair.

According to Darwin's theory, at mating, gemmules from both parents would align themselves with similar gemmules from the other partner (and this could be a very refined process), but additional gemmules were somehow able to be transferred into the newly forming fœtus and continue to reproduce themselves, for generation upon generation, explaining the reappearance after many generations of some family trait.

As his example, Darwin suggested an animal subjected to a changed environment, either naturally or through domestication. If the climate were cooler, the animal might develop thicker fur and the gemmules from the increased hair follicles would circulate through the body, including the reproductive organs, and thus be available for incorporation into the offspring. The change could be individual. For example, the sons of blacksmiths would be expected to be born with a tendency to develop greater strength in their biceps than their peers. The change could be communal. For example, cave dwelling animals would gradually lose their power of vision. If all animals in a population were subject to the same change, it would not take long for the new gemmules to outnumber the old and the changed feature would become a common inherited characteristic. Changes in nerve pathways in the brain, brought about by repeated patterns of behaviour, would also be subject to the same law of inheritance, hence the establishment of instincts and of characteristics such as shown by the various breeds of domestic dog.

Darwin cited numerous examples of experiments carried out on animals, such as those where parts of the body had been excised or amputated to see if regrowth would occur, or where the spur of a cock had been inserted into the ear of an ox, growing to 24 cm. and weighing 396 grammes. Darwin was clearly fascinated by these experiments and this fascination helps to explain his antagonism towards the anti-vivisectionists, who were opposed to experimentation upon live animals (Romanes 1896: 61). None of these acquired 'characteristics' were inherited.

Darwin needed to explain how it was that dogs and sheep, which had had their tails docked for generations, were not born with docked tails, since his theory assumed that an altered body part produced altered gemmules. More than 3,000 years of circumcision had not resulted in the birth of a single Jewish man not in need of the operation. This, Darwin explained, was due to the original gemmules being inherited along with the altered gemmules, reproducing themselves more efficiently and therefore out-competing the altered gemmules. Changes made at a specific stage of life would manifest in future generations at the same stage of life. Darwin urged that young women should undergo intensive physical and mental training approaching the age of marriage, so that the benefits gained at that time might be transmitted to their offspring.

In the second edition of *The Variation*, Darwin inserted a footnote acknowledging that both Aristotle and Buffon had put forward theories similar to *pangenesis*. He also acknowledged the theory of parthogenesis published by Owen in 1840 and the *physiological units* proposed by Herbert Spencer (1863) in his *Principles of Biology*. None of these theories, Darwin claimed, were the same as his theory. Finally, he acknowledged the work of Mantegazza, which he admitted clearly foresaw the doctrine of *pangenesis*, but gave no further details.

Confusion continues to exist regarding Darwin's views on inheritance. Maynard Smith (1982: 91) explained that Darwin not only accepted Lamarck's views on the inheritability of acquired characteristics but that they were essential to his theory. Darwin's rejection of Lamarckism related to Lamarck's concept of progressive evolution driven by the 'needs' of the individual. Darwin was now placing less emphasis on the idea of evolution as a 'driven' process, or one resulting from 'forethought' by some natural force, than he had in *Foundations*, rather envisioning variation as happening randomly, with natural selection replacing God, or the Creative Force, in determining what would survive and what would perish. Unfortunately, more than thirty years later, Maynard Smith's elucidation of this point is ignored in the general literature, which still generally equates Darwin's rejection of 'Lamarckism' with a rejection of 'inheritance', not of 'direction'. For example, Fernandez-Armesto wrote (2004: 85):

> As he [Lamarck] formulated it in 1809, biota adapted to their environments and such adapted characteristics were passed on by heredity … Darwin — whose theory of evolution is now recognized to be incompatible with Lamarck's — actually endorsed his predecessor's views. In deference to Lamarck, Darwin advised young women to acquire 'many skills' before starting families. (Nonetheless, one of the advantages of his own account of evolution was that it did not rely on the dubious claim that acquired characteristics are heritable). *(Parentheses in original)*

Fernandez-Armesto was clearly well aware of Darwin's theory of *pangenesis*, since it was in that work Darwin made the suggestion relating to female education. His denial of Darwin's acceptance of the inheritance of acquired characteristics is difficult to explain. Wallace had no doubts: "Darwin always believed in the inheritance of acquired characteristics" (Wallace 1895, vol 2: 21)

As a result of his work as an immunologist, Steele (Steele 1979; Steele et al. 1998) came to the conclusion that separation of somatic and reproductive DNA might not be as absolute as thought. Steele specifically mentioned Darwin's theory of *pangenesis* as being an early example of this line of thought. Despite these explanations and tentative justifications, most Neo-Darwinists still misunderstand/misinterpret Darwin's theory, which they seem to regard as but the unfortunate aberration of an elderly man. Just how unimportant *pangenesis* is deemed to be by many modern writers is illustrated by Dennett (1995: 321) whose only reference to *pangenesis* comes in a sentence, itself in parentheses, suggesting that any reader interested in Darwin's "unconstrained imagination about mechanisms of inheritance" should consult "Desmond and Moore 1991, pp.531ff". Referral to Darwin's original work, rather than to the footnote of other authors, would have been more appropriate.

Descent of Man

In 1871, Darwin finally published his long awaited volume on human descent (Darwin 1871/1908). Darwin's apparent reluctance to address this issue has never been fully explained, since the idea was certainly not new. Indeed, he cited Lamarck, Wallace, Huxley, Lyell, Vogt, Lubbock, Buchner, Rolle and Häckel as being among those who had anticipated him, the one occasion upon which Darwin seemed happy for such precedence to have occurred! Although Darwin did stipulate that humans must have descended from a branch of the anthropoid stock, his book was predominantly about differences between current human races/species and how these had evolved. While Darwin thought Africa was the most likely continent upon which this anthropoid divergence took place, he pointed out that apes had lived in Europe during the Miocene era. As to whether humans were all of one species, that question could not be answered until a satisfactory definition of 'species' was forthcoming. However, Darwin tended to think that humans were all descended from the same anthropoid stock and were not different species.

Darwin went to great lengths to acquire information on differences between humans. For example, he had learned from the United States Commission that the legs of sailors employed in the recent war were longer by 0.327 of an inch than those of soldiers but their arms were shorter by

1.09 of an inch. "This shortness of the arms is apparently due to their greater use, and is an unexpected result: but sailors chiefly use their arms in pulling, and not in supporting weights" (p. 48). He seemed to be implying that sailor's arms actually became shorter during their lifetimes! He considered it probable that these modifications would become hereditary.

Humans could hardly have developed the manual skills necessary to make weapons or hurl them accurately had they not become bipedal (p. 77): "... from these causes alone it would have been an advantage to man to become bipedal". This comment implies forethought and planning on the part of evolution (teleology), an idea which he had clearly not completely abandoned, but which was not acceptable to his followers.

On the one hand, Darwin was certain that over time the 'civilized races' would exterminate, not only the apes, but the 'inferior races' as well, but on the other hand, he was forced to admit that, while 'savage races' eliminated their weakest, or recalcitrant members, 'civilized society' allowed these people to propagate. Furthermore, civilized societies enlisted their fittest men into their armed forces, leaving the less fit at home to procreate. These opposing forces led Darwin to two opposing conclusions. The first was that there was no reason why the tendency to do good and to do evil should not be inherited equally readily but, since the benefits of co-operative behaviour were so obvious (p. 192):

> Looking to future generations, there is no cause to fear that the social instincts will grow weaker, and we may expect that virtuous habits will grow stronger, becoming perhaps fixed by inheritance. In this case the struggle between our higher and lower impulses will be less severe, and virtue will be triumphant.

Darwin's second conclusion was less optimistic (p. 216):

> If the various checks specified ... do not prevent the reckless, the vicious and otherwise inferior members of society from increasing at a quicker rate than the better class of men, the nation will retrograde.

Post-Darwin social reformers tended to adopt Darwin's first conclusion, arguing that it was human nature to be co-operative and anti-social behavior, especially in the young, stemmed from the failure of society in general, and their parents in particular, to provide them with optimal conditions for their development.

Sexual selection

Darwin downplayed the influence of sexual selection in nature, which he concluded was limited. The role of the dominant male among group animals was already covered by the general principle of natural selection. Birds exhibited the most elaborate courtship rituals, yet many birds were monogamous, sometimes for life, so nearly all had equal chance to reproduce, the exception being surplus individuals unable to find a mate, and even these might mate if another individual were killed. However, in some cases the females did seem to exercise a decisive choice. Darwin concluded that secondary sexual characteristics attractive to the female would develop as a result of sexual selection. Since all females would eventually find a mate, leaving only a very few unpaired males (possibly none), it is difficult to see how the characteristics of the 'favoured' males who found mates first would be any more likely to survive the vicissitudes of life than those which mated a few hours, or days, later.

Darwin was interested in the relative numbers of male and female births. He was aware that some females gave birth exclusively to sons and others to daughters. He believed that "the tendency to produce either sex would be inherited like almost every other peculiarity" (p. 393), ignoring the fact that every human (and every other sexually reproducing species) has the same number of male and female progenitors.

Darwin was not alone in showing muddled thinking as far as the inheritability of male characteristics was concerned. Nearly a century later, Coon (1962: 93), referring specifically to human populations, wrote that natural selection tended to favour characteristics carried by males because a man can have several wives at one time but a woman can only be impregnated by one man at a time. This chauvinism apart, if one man is more sexually successful than another, his *individual characteristics* would have a greater opportunity of being passed on than those of a less compelling rival. However, the *ratio* of 'male' characteristics to 'female' characteristics passed on by inheritance is always the same: 1:1.

As to female inferiority, Darwin drew conclusions based on his theory of *pangenesis*. Characters acquired would be transmitted to the same sex at the same age. Therefore (pp. 860-861):

> ... the inherited effects of the early education of boys and girls would be transmitted equally to both sexes ... in order that woman should reach the same standard as man, she ought, when nearly adult, to be trained to energy and perseverance, and to have her reason and imagination exercised to the highest point, and then she would probably transmit these qualities chiefly to her adult daughters. All women, however, could not be thus raised, unless during many generations those who excelled in the above robust virtues were married, and produced offspring in larger numbers than other women ... men ... during manhood generally undergo a severe struggle in order to maintain themselves and their families; and this will tend to keep up or even increase their mental powers, and, as a consequence, the present inequality between the sexes.

This from a man who never had to work a day in his life! It is not clear why the struggle which men underwent should be presumed to increase their mental powers, rather than their physical, since many men undertook physical work, on the land, in the forge or at the carpentry bench, among others. Did not raising a family constitute 'a severe struggle' for women, particularly those raising the large numbers of children which were common in families of the Victorian era in England?

From the above, it will be seen that there were two aspects of his theory to which Darwin adhered unswervingly: gradualism and the inheritance of acquired characteristics. It may well be asked why it is that Lamarck's name is so indelibly associated with the doctrine of acquired characteristics, which he never supported as far as the individual was concerned, while Darwin, who expounded the idea again and again, from the beginning of *Foundations* through *The Origin, The Variation and The Descent of Man*, is continually absolved of heresy in this regard?

Chapter 14

Two Minds But a Single Thought

Alfred Russel Wallace (1823-1913)

Although all textbooks that deal with evolutionary theory acknowledge Wallace as co-discoverer of the theory of evolution by means of natural selection, his influence, compared with that of Darwin, is insignificant.

The two men travelled completely different paths to arrive at their common destination. Darwin was born to money and never needed to work to support himself, giving him the freedom to study and write in his own time. Wallace was born into an old family, but declining family fortunes forced him to earn his own keep. Wallace's first employment was as apprentice to a master builder. Subsequently he was apprenticed to his brother as a surveyor. After his brother died, he took a teaching position, which afforded him the opportunity to read widely on travel and natural history. He read *Vestiges*, and this book, along with the second edition of Darwin's *Journal*, published in 1845, inspired Wallace to travel overseas in search of 'the origin of species'.

Wallace's place in history is both certain – and uncertain. He will always be remembered as the man who, unbeknownst both to him and to Charles Darwin, was engaged in a race to complete a theory which explained the 'origin of species'. Currently, he is known as the person who narrowly lost the race, but whose 'charge' at the end spurred Darwin to victory. Some authors, such as Brackman (1980), Brooks (1984) and Davies (2008) are now questioning whether the race was, in fact, won by Wallace and whether Darwin, together with his friends Lyell and Hooker, conspired to falsify the record to ensure Darwin's claim to priority was upheld. Before visiting this complex debate in the next two chapters, it is appropriate to consider Wallace, the man and his work, apart from his unwitting entanglement with Darwin.

Wallace is usually portrayed as a person of lower class than Darwin, poor financially and poorly educated (see, for example Davies 2008; McCalman, 2009). This was not the case. It is clear from his autobiography, which Wallace published in 1905, that Wallace was proud of the fact that his family were descended from the Scottish hero, Sir William Wallace, their family crest supporting this contention. Among the family graves in the Churchyard of the small village of Hanworth, Middlesex, (population 750 in 1840) was that of Admiral Sir James Wallace, who died in 1803, although Wallace was not sure of his exact relationship.

One of the minor titles of the local Dukes of St. Albans was that of Baron Vere of Hanworth, the title being taken by a third son of one of the Dukes, although the Baron eventually inherited the Dukedom himself. Wallace's father's name was Thomas Vere Wallace and Wallace, rather modestly, suggested that his father was a tenant of the first Baron Vere. It was not customary in England for a tenant to take the name of the local aristocrat – and Wallace must have known this. It was, however, customary for daughters to give their family name to their sons as an intermediate name. Indeed, Wallace's own second name, Russel, was obtained in this way.

Wallace's father had qualified as a solicitor, the Law being an acceptable profession for minor branches of established families. He chose not to practice, being of independent means, but engaged in business enterprises, the first of which was not successful and the second of which resulted in him being defrauded by his partner. This led to a decrease in the family's fortunes. Nevertheless, the young Wallace boys all attended Grammar school. Grammar schools in England were fee-paying establishments, lower in prestige than the famous Public Schools, such as Eton, Harrow or Winchester, but nevertheless well regarded.

Wallace's mother was descended from the Greenells, believed to have escaped from France following the St. Bartholomew massacre of 1572. This family also had a family crest and Wallace's mother owned several oil-paintings of Greenell ancestors. The name 'Russell' came from this side of the family, although lacking the second 'l', due, Wallace thought, to an error at the registry. Wallace came from 'old blood', whereas the Darwin and Wedgwood families were *nouveau riche*.

Throughout his life, Wallace read extensively and was a prolific writer on many subjects. During his life-time, he published twenty-four books and over two hundred papers.

During his time as apprentice surveyor to his brother, Wallace became interested in geology. His interests were to grow from geology, with its fascinating fossils, through entomology to evolution (past), evolution (present) and possibilities for the future. Wallace became extremely involved in his later life with social issues, land ownership and inheritance, education, especially of females, child and female labour, social and medical issues, such as vaccination, pollution and 'junk food'.

As a young man, Wallace struck up a friendship with William Bates, an ardent entomologist, and the two set in place plans to travel to the Amazon. Wallace wrote to Bates that the prime objective of his proposed travel was to gather facts "towards solving the problem of the origin of species" (Wallace 1889a: iv). By 'origins', Wallace was referring to place, as well as time, the distribution of species being a subject of great interest to him. They needed to finance their own expedition but, fortunately, the railway companies were then in urgent need of surveyors. Wallace took up such a position and was able to save sufficient money from his wages by April 1848 to pay for his passage to the Amazon aboard a small trading vessel. Being self-funded, Wallace was able to determine his own agenda. He knew what he was doing, and why, unlike the unfortunate young Darwin who clearly embarked upon his voyage under-prepared, although through no real fault of his own.

In April, 1848, Wallace, together with his friend, William Bates, embarked for Brazil upon a journey which was not only to change Wallace's life forever, but possibly that of Darwin. This voyage nearly cost Wallace his life on more than one occasion. On 28th December, 1851, Wallace's friend, Richard Spruce, wrote to advise Mr. John Smith of the Royal Botanic Gardens at Kew, that Wallace was "almost at the point of death from a malignant fever, which has reduced him to such a state of weakness that he cannot rise from his hammock or even feed himself" (Brooks p. 28).

The 'Mr John Smith,' was my mother's great-uncle. It pleases me to know that my family had formed a friendship with the young Wallace before he travelled to Brazil, sufficient for Spruce to write this personal letter.

After three years' hard work, with many of his fascinating samples sent back to England eagerly purchased, and his interesting letters read informally at meetings, as had been Darwin's, Wallace established himself as a naturalist of note while yet in South America. Wallace's younger brother, who had joined him in the Amazon, was not so fortunate. He died of a fever.

This first voyage of the young Wallace was of three-and-a-half years' duration and ended in disaster. On the way home, the vessel on which he was travelling caught fire and sank, taking with it all his personal (duplicate) samples, so perilously collected. Wallace and his companions drifted in a lifeboat for ten days before being rescued, finally arriving back in England in October, 1852. Fortunately, Wallace's possessions were insured and he was able to set out in July, 1854, for the Malayan Archipelago, from which he would not return until 1862. During those two years in England, Wallace established himself further by giving talks on his experiences. He became sufficiently well-known and respected for Sir Roderick Murchison, President of the Royal Geographical Society, to make representations to the Government on his behalf which resulted in him being granted free passage to the Archipelago.

Having lived and worked for so many years among 'savage' people in two very different parts of the world, learning their languages and customs, he soon came to realize that a 'savage' child was no different from a 'civilized' child. Both needed to be taught such things as mathematics to acquire any sort of understanding of numbers beyond the most simple. He noted the native peoples' love of art, which he was later to conclude could not have resulted from natural selection. He considered native people to be "morally and intellectually our equals, if not superiors" and suggested that they needed to be protected from contamination by 'degraded' classes of civilized people.

Wallace's growing understanding of the distribution of species, both plant and animal, during his time on the Malayan Archipelago, which led to the formation of his theory of evolution, will be covered in the next chapter, which will also consider Wallace's influence upon Darwin and the role he played in the writing by Darwin of *On the Origin of Species*. This matter is quite controversial.

In 1870, Wallace put together a short anthology of what he then considered to be his most important contributions to the theory of natural selection. *Contributions* contained his first two papers of 1855 and 1858, together with other pieces. In 1889, he published his only full-length book on the topic, which he entitled *Darwinism* (Wallace 1889b). While many people were prepared to accept natural selection as an explanation for variation at specific, or even generic level, it was the wider groupings of family or order with which most people had problems. Wallace countered this criticism in the same way that he defended Darwin's position against those who asked how rudimentary organs could be of use to an animal while in the process of development (Wallace 1896: 128):

> ... [these objections] are really outside the question of the origin of all existing species from allied species not very far removed from them, which is all that Darwin undertook to *prove* by means of his theory ... To ask of a new theory that it shall reveal to us exactly what took place in remote geological epochs, and how it took place, is unreasonable. *(Italics in original)*

Wallace's use of italics for the word 'prove' does not change the fact that Darwin did extrapolate from species, to genera, to families, to orders, even if he did 'pretend' he was only 'hypothesizing'. Brackman believed Wallace's loyalty to Darwin could be explained in part by the fact that, until after Darwin's death when biographies, diaries and letters of his and his contemporaries started to be published, Wallace had believed that Darwin was well ahead of him, both in his theory and in his writing, and that it only gradually became apparent to Wallace the effect his papers had had on Darwin.

Wallace formed a close friendship with Lyell, in part because of his (Wallace's) interest in geological formations, particularly those formed by glaciers, to the understanding of which Wallace made original contributions. Wallace identified the geological and biological divide which runs northeast/southwest through the Malayan Archipelago, which is still known as the Wallace Line. Wallace records that he assisted Lyell by proof reading his later books, although he did also recall that he felt somewhat underpaid for his services! You may remember that Lyell had been somewhat guarded in his support for Darwin in his book on the *Antiquity of Man* which he published in 1863, referring to Darwin's work as a 'modification' of Lamarck's and one can but speculate as to whether Lyell's late 'conversion' to Darwin's ideas was brought about as much by his later friendship with Wallace as it was to his much longer friendship with Darwin. Wallace had only returned to England the previous year, by which time Lyell's book would have been well underway and their friendship not yet fully formed, so lack of recognition of Wallace's work in that book is not unexpected. Nevertheless, it does seem as if Lyell may have experienced a shift in his loyalties.

There is a point upon which the logic of both Wallace and Darwin appears to be open to question. Wallace (1870/1973: 298) stated:

> It is an essential part of Mr. Darwin's theory, that one existing animal has not been derived from any other existing animal, but that both are descendants of a common ancestor, which was at once different from either, but, in essential characters, intermediate between them both.

If evolution is an *ongoing*, gradual process, as suggested by Darwin, there is no reason to suppose that all divergence happened at some unspecified time in the (remote) past, nor is there any reason to suppose that the parent species did not co-exist with the daughter species for *at least some period of time*. Indeed, since it would be impossible for a 'daughter' species to evolve from an extinct 'parent' species, it is reasonable to assume that both the parent species and the new daughter species *must* have co-existed at some point in time, even if but briefly. The claim that no existing animal had been derived from any other existing animal lacked logic. If evolution is an ongoing process, taking place here and now, then there must be, somewhere on the face of this Earth, a daughter species in the process of evolving (separating) from a parent species. To suggest this not to be the case would be to deny the very essence of evolutionary theory as put forward by Lamarck, Darwin, Wallace and others.

In *The Descent of Man*, Darwin had suggested that birds and butterflies had acquired their bright colours as a result of sexual selection. Wallace disagreed. He (1889b) pointed out that several male butterflies will pursue the one female and the fittest and fleetest will mate – a simple case of the 'fittest' providing the greatest number of offspring. He concluded (p. 295):

> [The] … extremely rigid action of natural selection must render any attempt to select mere ornament utterly nugatory, unless the most ornamented always coincide with "the fittest" in every other respect; while, if they do so coincide, then any selection of ornament is altogether superfluous.

Wallace pointed out that the most highly coloured parts of the body were the peripheral teguments. A butterfly can survive with a torn wing if attacked by a bird in flight. Wallace believed that high colour was a form of protection, detracting potential attackers from vital areas of the body. They could also serve to intimidate, especially when displayed. As for birds, his observations had led him to decide that "female birds had unaccountable likes and dislikes in the matter of their partners".

In relation to the role of sex, Wallace also disagreed with another view which became increasingly popular as the 19th/20th centuries progressed – that the natural role of human males was one of polygamy/promiscuity. In all the time he had spent living among native people, both in South America and on the Malay Archipelago, he had never known any man to stray from his wife.

Another area where Wallace disagreed with Darwin was in relation to Darwin's theory of *pangenesis*, which Wallace (1905/1969: 422) originally accepted but later challenged. He preferred Weismann's theory of 'continuity of germ-plasm', which theory totally rejected the concept that any characteristic acquired during the life-time of the individual could be inherited by that individual's offspring. Weismann's theory was developed during the 1890s, after Darwin's death. Weismann's major work was published in 1904, the year before Wallace published his autobiography, and will (hopefully) provide interesting reading when it is discussed in the next section of this book.

In 1905, Wallace published a two-volume autobiography, *My Life*. Recalling himself as a young man, Wallace wrote that he had been raised in a family with a conventional Low Church philosophy. The family attended Church as a matter of form, but with no real conviction. He became convinced by the Unitarian argument that the miracles recorded in the New Testament were invented by over-enthusiastic followers of the early Church and were not historical fact. In retrospect he described himself at that time as having been 'agnostic'. Later, he was to become a spiritualist. Rather than considering an interest in spiritual matters 'unscientific', Wallace held that it was those who refused to investigate spiritualism because it did not conform with their pre-conceived ideas who were 'unscientific'. Spiritualism, which involved making contact with the dead, either through direct trance or by means of the Ouija board, was extremely popular in the second half of the 19th century. In addition to the receiving of messages at these meetings, on occasions objects, some quite large, were claimed to have moved. Most meetings were conducted in a darkened room, which added to the mystique. Wallace spent years investigating spiritualism, attending many meetings, and became convinced of its truth. He recorded his experiences and conclusions in *Miracles and Modern Spiritualism*, which he published in 1896. Wallace experienced no difficulty in combining his scientific and spiritual interests. By contrast with Wallace, Darwin set out on his voyage holding orthodox Christian beliefs, but ended his life an agnostic. From such different positions and perspectives, the professional amateur, the amateur professional, the believer, the non-believer, two paths met and upon that meeting point was raised the edifice known as 'evolution by natural selection'.

As a spiritualist, Wallace believed that all souls were equal in the sight of God and that all had an inalienable right to equality of opportunity, be they aristocrat or labourer, 'civilized' or 'savage', male or female. Wallace believed that no person deserved praise or blame for ideas that came to him, since they had come from another source, but only for actions taken as a result of those ideas. Darwin and Wallace both became interested in the origin of species at about the same time; both completed the working out of their theory at about the same time. Darwin, however, had put far more effort into the publicizing of his theory than had Wallace and this may have been a factor in Wallace's deference towards Darwin. Wallace left the theory in Darwin's hands, turning his own energies and attention to numerous other matters, as already mentioned. Wallace also believed that promulgation of ideas should be free, uninfluenced by praise or blame, reward or punishment.

Wallace came to believe that the evolution of physical forms had been influenced by outside causes (environment). By contrast, mental (spiritual/philosophical/artistic) attributes came from within and they affected the environment – the exact reverse. He believed that humans had mental/spiritual characteristics which separated them from the rest of the animal kingdom and which could not have come about as the result of natural selection (Wallace 1889b). Neither Darwin, nor any of his followers since, has ever explained the beginning of life on Earth, how inanimate matter became living, able to reproduce itself, and die, for death is as much a mystery as is life! Life was as inexplicable to Wallace as it was to anybody else; like other deists, he accepted the 'self', the non-material aspect of our being, as 'spiritual' and 'eternal', although neither of these terms can easily be defined.

Wallace rejected the Malthusian doctrine that the poor and weak members of society should be left to struggle and make their own way – hence his work as a social reformer. He was aware of the development in Germany before the First World War of what later became known as 'Social Darwinism'. He rejected eugenics and the idea that dominant (successful) races were entitled to suppress (or eliminate) less successful races. This was something which Darwin had predicted to be inevitable and implied was even desirable.

In 1910, at the age of 87, Wallace published the last of his works on nature: *The World of Life*. It was a scholarly work, covering a wide range of subjects on the topic of evolution. He included, not only his own observations, but some of those of other people. I have reproduced two such anecdotes, which I found particularly interesting, in Appendix I. Wallace was quite outspoken in relation to those aspects of evolution upon which he and Darwin had disagreed. He recorded that Darwin had been 'quite distressed' that he, Wallace, had rejected Darwin's conclusion that Man's highest qualities and powers had developed from those of lower animals by natural or sexual selection (p. 315). More importantly, Wallace rejected the idea that macro-evolution was the result of natural selection. Darwin had denied macro-evolution. Others, such as Huxley, Weismann and Wallace, could see no other explanation for the establishment of new features, distinct orders and classes. (Lamarck's book concentrated on macro-evolution, micro-evolution (gradual change/adaptation) being assumed rather than argued.) Wallace had concluded that much of creation could only have come into being as the result of the workings of a far Higher Mind. The most powerful statement of his changed attitude and belief was made by the subtitle for the book: *A Manifestation of Creative Power, Directive Mind and Ultimate Purpose.* 'Directive Mind', 'Ultimate Purpose' – the very opposite of the tenets of Darwinism! How different might things have been if Wallace had published his paper before Darwin published any of his work, if he had become the accepted voice of evolutionary theory in the 19th and 20th centuries? The debate about a possible conspiracy to suppress Wallace's work in favour of Darwin's is the subject of the next two chapters.

Wallace followed in the footsteps of Paley by citing striking examples as evidence for planning in nature, not just evolution by chance. His first example was that of the feathers of birds. Having briefly drawn attention to the changes necessary in the bony structure and musculature of birds to support the action of the wings, Wallace described the intricate structure of the feathers themselves, with their hooked barbs and barbules, horny plates which grow obliquely outwards towards the tip of the barb. These produce an air-tight structure when in flight but can be fluffed up when the bird needs extra warmth or to dry its feathers. Each feather is composed of hundreds of thousands of tiny parts, all of which adjust in relation to each other during flight to allow upward/downward movement, etc., yet the matter of which feathers are composed is dead. There is no circulation in any part of a fully grown feather. While each feather is a replica of its counterpart on the other side of the bird's body, no two feathers on the one side are exactly the same, since each must fit the exact requirements of its particular position. Wallace saw in the wonderful structure of the individual feather and the complete wing, evidence of an organizing Mind.

Wallace wrote of the process of insect metamorphosis, which he found to be one of the most marvelous occurrences in the whole organic world. Much research had by then been carried out on this phenomenon and its processes were well understood. Rudimentary structures, such as the wings, were already present in the larva, but never developed. The internal structure of the larva - muscles, intestines, nerves, respiratory tubes, etc. - along with these rudimentary structures, were gradually dissolved into a creamy pulp from which the imago developed, the information required for this transformation having remained dormant until required. In all the books on the subject of evolution that I have read, I have never found one which made even an attempt to explain how this complex process had been developed by the operation of natural selection.

Bird feathers and insect metamorphosis were discussed in a chapter entitled *Proofs of an Organising Mind.*

Wallace rejected the Christian concept of a God who was infinite, eternal and omnipotent (pp. 392-394). Wallace did not believe that there was a great chasm between humans and God, filled only by a hierarchy of angels who had little to do except act as attendants and occasional messengers. Rather he envisaged a whole host of spiritual beings of infinite variety, from the highest grade of power down to the lowest level of consciousness, or almost unconsciousness, such as would be manifested in "cell-souls". The Infinite Being would have determined the broad outline of the Universe, but then the initial great properties and forces, such as ether, light, gravity, etc., would have been brought into being by other, not quite so exalted Beings, and so on down to the creation of matter, atoms, minerals, etc., simple living and more complex cells/beings, such as ourselves (p. 395):

> At successive stages of development of the life-world, more and perhaps higher intelligences might be required to direct the main lines of variation in definite directions in accordance with the general design to be worked out ... Some such conception as this — of delegated powers to beings of a very high, and to others of a very low grade of life and intellect — seems to me less grossly improbable than that the infinite Deity not only designed the whole of the cosmos, but that himself alone is the consciously acting power in every cell of every living thing that is or ever has been upon the earth.

The image may be likened to that of a global corporation or conglomerate, with each department vying for its share of available resources, while being aware that the needs of the other departments must also be met if the whole enterprise is to be successful. Each contributes to the working of the whole. Wallace would have developed this belief system through his contact with spiritualists, who were under no illusions that the spirit(s) whom they contacted were 'God' or an 'angel'. Mostly they were departed relatives. Sometimes they were 'spirit guides', developed souls who took on the role of guide/teacher, even though they had no direct relationship with the person/people with whom they were communicating. Spiritualists believe in a multitude of heavenly hierarchies, as, of course, do the Hindus.

Chapter 15

From Inception to Deception

This chapter examines the controversy surrounding the claim made by Brackman (1980), Brooks (1984) and Davies (2008) that a small group of people conspired to use their positions of power and influence in Victorian society to suppress the work of Alfred Wallace and to promote Darwin as, if not the sole, then the primary developer of the theory of evolution by natural selection. It concludes by presenting a new scenario which, it is claimed, better accommodates the known facts.

The truth will never be known 'for sure'. Too many letters between the leading characters in this drama, Darwin, Hooker, Lyell, Gray and Wallace, are missing (lost or destroyed) for that ever to happen. Much is conjecture and speculation and, much like the fossil evidence, there are times when one has to 'assume' a missing piece to fill in the gap to make sense of the whole.

Not everyone will wish to spend the time trying to navigate these uncertain waters. I am here giving a brief synopsis of the events and those who wish to do so may then skip the rest of this chapter and the next.

While in Ternate, Wallace sent Darwin a paper outlining his latest thoughts on evolution, which Darwin claimed to have received mid-June, 1858. On 1st July, 1858, it was read, along with a section from Darwin's *Foundations* essay of 1844 and a 'letter' from Darwin to Asa Gray, before a meeting of the Linnean Society.

Brackman, Brooks and Davies all claimed that Darwin had received Wallace's paper earlier than admitted and used the time to rewrite part of his 'big book', *Natural Selection,* upon which he had been working for two years. Darwin's supporters reject this claim.

I have accepted the claims of Brackman, Brooks and Davies and am further suggesting that the 'letter' read before the Linnean Society was not the 'short' one Darwin wrote to Gray, but a summary of Darwin's ideas that Hooker had written before the meeting when he realized that Darwin's actual letter was of little account compared with Wallace's full paper.

If that is enough, move to Chapter 17. If you like mysteries, read on!

The first cracks in the armour of defence surrounding Darwin's name and his work began to appear at the time of the centenary in 1959 of the publishing of *The Origin.* It was shown in Chapter 9 how this had led Eiseley to conclude that Darwin had made unacknowledged use of the work of Edward Blyth. As more of Darwin's original journals and notebooks were published and became available for general study, for example the facsimile copy of the journal kept by Darwin on his voyage (G. Darwin 1979) and Darwin's notebooks from 1836-1844 (Barrett et al. 1987), it became increasingly clear that Darwin's thinking had undergone a radical change in the years after certain work by Alfred Wallace was published.

This work included, not only Wallace's two major papers to be discussed below, but other papers written while on the Archipelago, (Wallace 1854, 1855a, 1855b, 1855c, 1856a, 1856b, 1856c, 1857a, 1857b, 1858a, 1858b, 1858c), as well as letters sent to his agent, Stevens, which were read at society meetings. Some of Darwin's letters home during his voyage had similarly been read at meetings and helped to establish Darwin's name before he had returned home. Hooker (Huxley 1918: 144) was quite annoyed when he learned that his letters were also being read: "I do extremely dislike having my letters shown to those I do not know ...". However, his father, Sir William Hooker, had little choice but to comply when Prince Albert asked to read them!

In the 1980s, two books (Brackman 1980; Brooks 1984) were published which not only accused Darwin of using Wallace's work without acknowledgement, but also accused him of deliberately holding back a paper of Wallace's so that he (Darwin) could make amendments to his own work and falsely claim that he had developed the theory of evolution by natural selection before Wallace. The debate was re-ignited by the publication of a book (Davies 2008) which accused Darwin of perpetrating (p. xix): "... a deliberate and iniquitous case of intellectual theft, deceit and lies". He (p. xix) continued by claiming that Darwin: "... committed one of the greatest thefts of intellectual property in the history of science".

Before proceeding, I would point out that all of Darwin's surviving correspondence is now available on www.darwinproject.ac.uk/. However, this work is not about Darwin, or any other individual theorist. It is about how people evaluated evidence, as it related to evolution in general and human evolution in particular, how their pre-conceived ideas may have prejudiced their conclusions. Until recent times, such information was disseminated via books and articles. I will continue to reference Darwin's correspondence as it appeared when published by his son, Sir Francis Darwin, and others, such as Leonard Huxley, who published both Joseph Hooker's and Thomas Huxley's *Life and Letters*. While archival material was also available for the serious researcher, most readers, be they academic or general, were dependent upon published material to draw their conclusions and, in the final analysis, it is acceptance or rejection by the wider community which determines the success or failure of any theory or opinion. It is the material which was available to the authors under consideration at the time they formed their opinions which is relevant here, not the information which is available to us today via the internet.

In February, 1855, while in Sarawak, in the Malay Archipelago, Wallace wrote a paper entitled *On the Law which has Regulated the Introduction of New Species*, which was published in September of that year in *Annals and Magazine of Natural History*, which journal published the majority of his papers (see Appensix II). The 'Law' which Wallace proposed was that "Every species has come into existence coincident both in time and space with a pre-existing closely allied species". In other words, God had not created one creature here, another there, any resemblance between them being solely due to the whim of the Creator. Wallace was supporting the notion of evolution, as put forward by the author of *Vestiges*, which book he had read when it was first published in 1844. He supported his hypothesis by reference to geological changes, citing Lyell and the fossil record to show affinities (close relationships) or analogies (distant relationships) with later species. Wallace demonstrated how his Law, based on the assumption of evolution, could account for rudimentary organs, inexplicable under the doctrine of Special Creation. Wallace also suggested that if one species merely became modified into another 'new' species, then progression would be simple. However, if different populations of a species varied in more than one way, then it would be possible for diverse 'new' species to replace the original form, increasing the diversity of life forms. Wallace was not arguing for the fact of evolution – that was already well established – but his 'Law' clearly assumed a form of 'natural selection', although he did not use this term.

In November, 1855, two months after the publication of the paper, Sir Charles Lyell started to keep his own notebooks on the 'species question'. His first entry, dated 28th November, 1855, began with 'Wallace' and referred to Wallace's Law paper. In April 1856, Lyell visited Darwin at Down and his entry for 16th April contained reference to the theories of both Darwin and Wallace, concluding "The reason why Mr. Wallace's introduction of species, most allied to those immediately preceding in Time ... seems explained by the Natural Selection theory" (Brooks pp. 259-260). Wallace's theory that new species always - and only - appeared in environments in which a closely-related species had previously existed differed from Darwin's 1844 Essay in that *Foundations* had assumed isolation, not proximity, as the primary requisite.

It appears that this was the first time Darwin had confided his ideas to Lyell, who immediately urged him to publish (F. Darwin 1887, vol.1: 84; vol.2: 67). Records from the Lyell archives show that Lyell followed up his visit with a letter to Darwin of 1st May, 1856, urging him to publish. In his reply of 3rd May, Darwin agreed to reflect on the idea, saying "I rather hate the idea of writing for priority, yet I certainly should be vexed if anyone were to publish my doctrines before me" (Brackman p.260). This sentence is interesting for two reasons. First, it shows that, as early as May 1856, Lyell had warned Darwin that, if he did not publish his ideas, Wallace would forestall him. Second, it is interesting that, even at that early time, Darwin was referring to his ideas as his 'doctrine'! Darwin had already read Wallace's paper and his notes indicate that he had dismissed the article as containing "nothing of great interest" (Davies p. 1). This was notwithstanding the fact that, in Darwin's copy of the *Annals*, Wallace's paper was heavily marked with lines, or double lines, in the margins, underlining of text and marginal notes, although Brooks believed that some of these notations were added on re-reading. Davies believed that Darwin had, in fact, not merely read, but studied, Wallace's paper before Lyell drew it to his attention; he simply did not own up to the fact. It is, of course, possible that the annotations were made after Lyell's visit, although it is difficult to believe that Darwin would not have read and noted a paper on a subject in which he, himself, had such an interest.

Whatever the facts, it was following Lyell's visit that Darwin started to write. His diary entry for 14th May, 1856, states "Began by Lyell's advice writing species sketch". Darwin found it too difficult to condense his ideas and by mid-June had abandoned the proposed paper in preference for a book. In November, 1856, he confessed to Lyell (F. Darwin 1887: vol.2: 85):

I am working very steadily on my big book; I have found it quite impossible to publish any preliminary essay or sketch; but am doing my work as completely as my present materials allow without waiting to perfect them. And this much acceleration I owe to you.

On 20th July, 1857, Darwin, who was notoriously secretive about his work, even with close friends, wrote to the American botanist, Asa Gray (1810-1888) giving a brief outline of his hypothesis (F. Darwin, vol.1: 78-79):

> ... Nineteen years (!) ago it occurred to me that whilst otherwise employed on Nat. Hist., I might perhaps do good if I noted any sort of facts bearing on the question of the origin of species, and this I have since been doing. Either species have been independently created, or they have descended from other species, like varieties from one species. I think it can be shown to be probable that man gets his most distinct varieties by preserving such as arise best worth keeping and destroying others, but I should fill a quire if I were to go on. To be brief, I assume that species arise like our domestic varieties with much extinction; and then test this hypothesis by comparison with as many general and pretty well-established propositions as I can find made out, - in geographical distribution, geological history, affinities, &. &. And it seems to me that, supposing that such hypothesis were to explain such general propositions, we ought, in accordance with the common way of following all sciences, to admit it till some better hypothesis be found out. For to my mind to say that species were created so and so is no scientific explanation, only a reverent way of saying it is so and so. But it is nonsensical trying to show how I try to proceed, in the compass of a note. But as an honest man, I must tell you that I have come to the heterodox conclusion, that there are no such things as independently created species — that species are only strongly defined varieties. I know this will make you despise me ...

(Darwin then wrote a few sentences about distribution, especially in relation to climatic and geological changes, Gray's area of special interest, before continuing) ...

> I must say one more word in justification (for I feel sure that your tendency will be to despise me and my crotchets), that all my notions about *how* species change are derived from long-continued study of the works of (and converse with) agriculturists and horticulturists; and I believe I see my way pretty clearly on the means used by nature to change her species and *adapt* them to the wondrous and exquisitely beautiful contingencies to which every living being is exposed ... *(italics in original)*.

The remainder of the letter was not reproduced by Sir Francis, nor was Gray's reply, although this information is now available on line. Sir Francis gave the date of 20th July, 1856, for this letter, but it is clear from Gray's reply of August 1857 (www.darwinproject.ac.uk/entry2129) that the revised date of 1857 is correct.

Wallace was continuing his study of birds, butterflies, beetles and other interesting fauna, such as the orang-utan and, of course, humans. Living for so many years among so many diverse peoples, Wallace became

155

increasingly interested in the evolution of the various races of humans, his conclusions in respect to these then being applied to animal species in the formulation of his theory (McKinney 1966).

Alone (i.e. the only white man) on Aru, Wallace had no one with whom to discuss his ideas. Letters to and from his faithful friend, Bates, took months to exchange. Replying on 22nd December, 1857, to Darwin's first known letter to him of 1st May, 1857, Wallace had confided that he had been disappointed that his 'Law' paper had not excited discussion or even elicited opposition (Wallace 1905/1969: 355).

Wallace's first known letter to Darwin was dated 10th October, 1856, but the content of this letter is not known, since it is missing. However, in his reply to that letter (1st May, 1857) Darwin wrote (F. Darwin vol. 2: 95):

> I am much obliged for your letter of 10th October, from Celebes, received a few days ago; in a laborious undertaking, sympathy is valuable and real encouragement. By your letter and even still more by your paper in the Annals, a year or more ago, I can plainly see that we have thought much alike and to a certain extent have come to similar conclusions … I agree to the truth of almost every word of your paper; and I dare say that you will agree with me that it is very rare to find oneself agreeing pretty closely with any theoretical paper; for it is lamentable how each man draws his own different conclusions from the very same facts. This summer will make the 20th year (!) since I opened my first notebook, on the question how and in what way do species and varieties differ from each other. I am now preparing my work for publication, but I find the subject so very large that though I have written many chapters, I do not suppose I shall go to press for two years …

> I have acted already in accordance with your advice of keeping domestic varieties, and those appearing in a state of nature, distinct; but I have sometimes doubted the wisdom of this, and therefore am glad to be backed by your opinion.

The words, 'in a laborious undertaking, sympathy is valuable and real encouragement' would indicate that Wallace was aware of some, 'large work,' that Darwin was undertaking. The further statement, 'I have acted already in accordance with your advice of keeping domestic varieties, and those appearing in a state of nature, distinct,' indicates that by October, 1856, these two men were already discussing the issue of domestication *v* natural change (evolution). We therefore know for certain that Darwin and Wallace were exchanging views by the second half of 1856. It is likely that the correspondence commenced after the September publication of the Sarawak (Law) paper in 1855. Darwin was writing to numerous people; why would he not write to Wallace who was working in an as yet unexplored part of the world - unexplored, that is, from the naturalist point of view? It is highly unlikely that Darwin gave details of his work on his first letter. An exchange of views on change under domestication and in the wild would surely not have taken place at least until a second letter? I consider October/November, 1855, as the most likely time for this correspondence to have commenced.

Davies (p. 105) noted that Darwin's reply to this letter, dated 1st May, 1857, stated that he had received Wallace's letter "a few days ago". Davies (2008) claimed that this was an attempt at deception which had taken place because Darwin had used the intervening four months to incorporate Wallace's ideas into his own work. In support of this claim, Davies (pp. 107-108) pointed out that on 31st March, 1857, Darwin included, for the first time, in the chapter he was writing on 'Extinctions' reference to the principle of divergence (which had not been included in his 1844 *Foundations* essay). Shortly afterwards, on 12th April, 1857, Darwin (F. Darwin vol.2: 90-91) wrote to Hooker suggesting that species were but strongly marked varieties. This marked a departure from the view that he had expressed in *Foundations* that species appeared as new entities after extinctions brought about as the result of geological change. This new position was to form the basis of all of Darwin's arguments in *The Origin*.

In January, 1858, Wallace wrote to H. W. Bates apologizing for a six month delay in replying to a letter because he had been on a seven months' voyage. Later that year, he told his sister, Fanny, in a letter dated 10th December, 1858, "... I am now going out of reach of letters for six months..." (Marchant 1916: 52). Wallace was often out of touch with mail services and it is possible that his letter had not been dispatched until some time after it was penned. It is difficult to understand what benefit Darwin would have gained by any deception. He would not even have considered it deception, because that was what Darwin did. From his home at Down, Darwin created his own 'world wide web'. He rarely wrote letters simply for social reasons; he wrote to gain information. He received responses to his many requests from all over the world. He was a victim of his own success, for so much information did he receive that he had difficulty in assimilating, sorting and using it – hence the long delay in the production of his book! Darwin did not write to Wallace for social reasons; he wrote to gain information for his book. I have absolutely no doubt that he cogitated upon what Wallace wrote and incorporated anything he considered worthwhile into his theory. Academia today will not tolerate plagiarism. Everything has to be referenced *ad nauseum*. It was different during the 19th century. Darwin was criticized for not citing any references in *The Origin* but the criticism was not enough to destroy him, or his work, as it would today. The very fact that Darwin considered it acceptable to publish an 'abstract' of his planned 'big book' without any referencing, and that the book was so well received, shows just how much things have changed.

Davies' point that there is no record of Darwin ever having mentioned Wallace's name in any of his voluminous correspondence to his many friends, although he mentioned the names of many others, is worthy of note, as is the fact that it was during the four months during which Davies claimed Darwin was incorporating Wallace's ideas into his own work that he first mentioned some of these ideas in his correspondence with Hooker.

In his (presumed) next letter to Wallace, dated 22nd December, 1857, Darwin told Wallace that he should not think that no notice had been taken of his first paper since both Blyth and Lyell had especially called his attention to it (Wallace 1905/1969: 355; F. Darwin 1887, vol.2: 108). Darwin told Wallace that, while he agreed with his conclusions, he believed he went much further than Wallace with his own theory. It is possible that Darwin enclosed with this letter a letter to him from Lyell, referring to both their work and its similarity (see below). All of Wallace's letters to Darwin written while in the Archipelago have been destroyed. The earliest surviving letter is one which Wallace wrote to Darwin after he returned to England in 1862. We only know of Wallace's comment about his disappointment that little notice had been taken of his 'Law paper' because one small piece has survived. The comment was written on the back of a segment on black panthers, information which Darwin had requested, hence its having been cut out and kept.

Wallace continued to send articles to journals for publication. After the Sarawak paper, he published in *Annals* (1857) the last of three articles on the orang-utan, suggesting either man had evolved from an ape-like species or, possibly that apes had evolved from a more man-like one. He also published (Wallace 1856b, 1857a, 1857b) articles in which he compared the foot structure of birds from the Archipelago with those he had seen in South America, noting that, despite superficial variances, especially in size, there was a basic similarity of structure between those species which caught their prey on the wing and those which scavenged for their prey on the ground, even though they now inhabited lands half a world apart. The progression of Wallace's thinking is clearly shown in his published works.

While on the Aru Islands in February, 1858, Wallace suffered another bout of malaria and was confined to bed for several days. It was during this period of enforced rest that Wallace worked out the final details of his theory on natural selection (Wallace 1870/1973, 1905/1969). A few weeks previously, (4th January), Wallace had written to Bates that he had prepared the plan and written a portion of a work embracing the 'whole subject' [of the origin of species] (Wallace 1905/1969: 358). The subsequent article is reprinted here as Appendix III.

Quite why Wallace made the fateful decision to send his completed paper to Darwin, with the request that he show it to Lyell, rather than sending it directly to *Annals*, which journal regularly published his work, will probably forever remain uncertain. Whatever his reason, Wallace's fateful paper, *On the Tendency of Varieties to Depart Indefinitely from the Original Type*, was dispatched from Ternate on 9th March, 1858. When it was delivered to Darwin is a matter of debate. Both the original document, and the envelope in which it was delivered, are missing. Brackman and Davies both believed that it was delivered to Down House on 6th June, 1858, the same day that a letter sent by Wallace to his friend Bates, for onward forwarding by his brother, Frederick, was delivered in London. Brooks believed delivery to have been earlier, as will be discussed below.

Brackman stressed that he was not the first person to suggest that Darwin had received Wallace's paper earlier than generally believed. He drew attention to the fact that John Brooks had given a summary of "his forthcoming work" in the *American Philosophical Society's 1968 Yearbook*, claiming that the paper had been received by Darwin on 18th May, 1858 (pp. 18-19). Presumably Brackman had decided that twelve years was sufficient time for Brooks to complete his "forthcoming work" and published his own account.

Four years later, Brooks (1984) published. He had obtained, and reproduced facsimile, copies of hand-written mailing documents showing that the shipment of letter mail dispatched from Ternate on 9th March, 1858, was received in London on Friday, 14th May, 1858, at 10.25 p.m. It should have been delivered to Down House either Saturday, 15th May or Monday, 17th May, 1858. Brooks thus held that the letter arrived three weeks earlier than suggested by Brackman (1980). His documented evidence is difficult to dispute.

After receiving Wallace's proposed paper, Darwin wrote to Lyell a letter simply dated "18th" (F. Darwin vol.2: 116-117):

> Some year or two ago you recommended me to read a paper by Wallace in the "Annals", which had interested you and, as I was writing to him, I knew this would please him much, so I told him. He has to-day sent me the enclosed, and asked me to forward it to you. It seems to me well worth reading. Your words have come true with a vengeance — that I should be forestalled. You said this, when I explained to you here very briefly my views of "Natural Selection" depending on the struggle for existence. I never saw a more striking

coincidence; if Wallace had my MS sketch written out in 1842 he could not have made a better short abstract! Even his terms now stand as heads of my chapters. Please return me the MS, which he does not say he wishes me to publish, but I shall of course, at once write and offer to send it to any journal. So all my originality, whatever it may amount to, will be smashed, though my book, if it will ever have any value, will not be deteriorated, as all the labour consists in the application of the theory.

Darwin asked Lyell to send a copy of his letter, and of Lyell's reply, to Hooker, for his advice.

All three of the authors whose work is being considered, agreed that this letter was sent to Lyell on 18th June, 1858, as claimed by Darwin. Brackman and Davies believed that it had been written after Darwin had spent two weeks amending his work. Brooks felt that the poignancy of the letter indicated that it had been written very soon after Darwin had received Wallace's letter, which he (Brooks) claimed had been no later than 17th May, 1858. He suggested that, having decided not to send the letter immediately, Darwin retained the letter until he had finished the 'corrections' to his manuscript, dispatching it on 18th June, the co-incidence in the date being purely fortuitous. This explanation is not entirely satisfactory. It is here suggested that Darwin both wrote and sent the letter on 18th May, 1858, after he had received Wallace's paper, and that both Lyell and Hooker were aware of the situation in which Darwin found himself for a full month before the events took place which led to the joint reading of Wallace and Darwin's work before the Linnean Society on the evening of 1st July, 1858.

Brackman, Brooks and Davies all believed that receipt of Wallace's paper had resulted in Darwin reworking his manuscript to include Wallace's ideas on divergence. They accepted that Lyell had been deceived into thinking that the paper had arrived mid-June. Kohn (1981), responding to Brackman's account, refuted this accusation, preferring to trust Darwin than the speed of the postal service. Brooks' subsequent reproduction of the postal records cast doubt on this objection. Brooks made no mention of either Brackman or Kohn anywhere in his book.

More than a year earlier, Darwin had noted in his diary on 31st March, 1857, that he had completed Chapter 6 of his major work on 'Natural Selection' (Brooks p. 230). Darwin's pocket diary contained an entry for 12th June, 1858, stating that he had that day completed correcting Chapter 6 (Brackman p. 19; Brooks p. 230).

Brooks supported his claim by the examination of the eleven chapters of Darwin's proposed major work which still survive, including Chapter 6. The section on 'Distribution', which followed 'Extinction', had been rewritten.

The original folio page 26 was missing. In its place were forty-one new pages. These were written on different paper. Darwin numbered the inserted pages using an * for the first addition, i.e., 26, 26*, and then letters, i.e. 26a, 26b, and so on. The inserted pages were numbered up to 26nn. Pages 51-76 were also written on the same alternative paper and their later time was confirmed by a footnote on page 53 which Darwin had dated June 1858. In all, more than sixty pages had been (re)written.

If the date proposed by Brackman and Davies is correct, then Darwin finished sixty pages of amendments within a week. If Brooks is correct, it took Darwin more than three weeks, a more likely amount of time.

Darwin wrote a second letter to Lyell (F. Darwin 1887: 116-117), simply dated, 'Friday', presumed to have been sent on Friday, 25th June, but here believed to have been sent on Friday, 21st May, which showed that he was quite distraught. He claimed there was nothing in Wallace's paper which was not contained in his 1844 sketch, which he had shown to Hooker, which point he was to re-iterate in his Introduction to *The Origin*. Although there was a remarkable similarity between the two pieces of work, there was one major difference. It was not that Wallace now emphasized the struggle for life in which all wild creatures were constantly engaged - that concept had been brought to the attention of interested persons in 1852 by Herbert Spencer (albeit anonymously) in a well-written and widely-read article. What differed was the main thesis of Wallace's paper - that it was not appropriate to extrapolate from change under domestication to change in wild nature because the circumstances were so different. Wild animals were engaged in a constant hunt for food and were in constant danger of being killed. Domestic animals were fed a constant supply of food, never had to exert themselves to obtain it and were never threatened with death. Random changes which occurred under domestication could be preserved, if they took the fancy of the breeder, whether they were advantageous to the animal (or plant) or not. In the wild, only advantageous changes had any chance of becoming fixed, and those but rarely.

In this letter Darwin told Lyell he had a copy of a letter he had sent to Asa Gray "about a year previously", giving a short sketch of his views (F. Darwin vol.2: 117):

> ... so that I could most truly say and prove that I take nothing from Wallace. I should be *extremely* glad *now* to publish a sketch of my general views in about a dozen pages or so; but I cannot persuade myself that I can do so honourably ... But I cannot tell whether to publish now would not be base and paltry. This was my first impression, and I should have certainly acted on it had it not been for your letter ... *(Italics in original)*

Clearly Lyell had replied to Darwin's first letter of "18th" but this letter is missing. Darwin underlined the word 'extremely' once and the word 'now' twice. Omitted by Sir Francis (F. Darwin vol.1, p. 117) was the paragraph (Brooks p. 264):

> I should not have sent off your letter without further reflection, for I am at present quite upset, but write now to get subject for time out of mind that I confess it never did occur to me, as it might, that Wallace could have made any use of your letter.

This sentence is difficult to understand but may indicate that Darwin, at some time, impulsively sent Wallace a letter he had received from Lyell, presumably referring to their common interest and possibly to Darwin's proposed book. Brooks (p. 201) mentioned:

> ... a puzzling, undated entry in one of Wallace's notebooks. Under the heading, "Sketch of Mr. Darwin's 'Natural Selection,'" is a list ... of fourteen chapters.

Darwin might have enclosed this letter with the one he sent to Wallace on 22nd December, 1857, in which he told Wallace that Lyell and Blyth had both mentioned the Sarawak 'Law' paper to him or he might have sent it even earlier, possibly with what I believe to be the missing letter of early 1857. Quite what information would have been contained in a letter written by Lyell, which would have been of use to Wallace, it is difficult to say. However, it should be noted that Lyell's published work had been as much of an inspiration to Wallace as it had been to Darwin and something Lyell had written could have been useful to both men. It does, however, seem unlikely that the letter from Lyell would have contained a list of Darwin's proposed fourteen chapters! Surely this information must have been sent by Darwin himself? I am absolutely convinced that there are *at least* two missing letters exchanged between these two men.

Returning to the events of May, 1858, I suggest that, upon seeing Wallace's letter and proposed paper, Lyell, no doubt believing that Darwin's book was further advanced than it actually was, urged Darwin to make haste with his amendments with a view to forwarding his manuscript to a publisher. I also suggest that Hooker knew of the situation. When Leonard Huxley published his biography of Hooker, he stated (L. Huxley 1918, vol.2: 465) that it was Hooker's recollection that it was to him that Darwin first confided the receipt of Wallace's unexpected communication. At that time, Joseph Hooker was Darwin's closest friend and the only person with whom Darwin had shared any of his ideas, before that day when Lyell had visited Down House and drawn Darwin's attention to Wallace's work. That Darwin should have confided in Hooker, before forwarding the package to Lyell, seems not unreasonable. So much correspondence is known to be missing, that it is difficult to be certain exactly what transpired.

The urging of Lyell, and also possibly of Hooker, it is suggested, was the catalyst which spurred Darwin into the flurry of activity which resulted in him completing the revisions, a large amount of work, in what was (for him) a small amount of time. One can imagine the consternation of Lyell and Hooker when it became apparent that, despite Darwin's 'Herculean' effort, his manuscript was far from ready for submission.

Mid June there was a sudden flurry of activity. For a month, Darwin had been working steadily at his manuscript. No attempt that we are aware of had been made to arrange any form of publication for either Darwin's or Wallace's work, neither by its submission to a journal nor by the reading of any material at a monthly meeting of any of the Societies. Suddenly, the matter became urgent. I will give my explanation for the urgency after outlining what actually happened.

The death of the Vice President of the Linnean Society had caused their June meeting to be postponed until 1st July, 1858, and this presented the opportunity for Thomas Huxley, another close friend of Darwin, to ask the President to allow Darwin's and Wallace's work to be read at that meeting instead of the papers previously announced (Brackman p. 63). This occurred, despite a further complication: Darwin's household was in turmoil. Members had been struck down by both diphtheria and scarlet fever. Darwin's youngest son, Charles, died. The day of the Linnean Society meeting was the day of his funeral.

It was Hooker and Lyell who made the final decision on what was to be read and it was they who, on Darwin's behalf, presented three pieces to the meeting: an extract from Darwin's 1844 *Essay*, a 'letter' to Asa Gray, claimed to have been written 'October' 1857, later amended to 5th September, 1857, and Wallace's Ternate paper.

Partridge's (2015a) account portrays meetings as rather informal affairs (by today's standards) with members sometimes bringing their papers with them. He pointed out that the date of receipt of papers was often listed in the Society's records as being the same date as that of the meeting. He stated that formal agendas were not published ahead of the meeting, rather decisions on whose paper would be read, and in what order, were taken on the night. The night of 1st July, 1858, was a busy one and not all papers were able to be read. Two were held over for the next meeting. Lyell and Hooker had urged the importance of the Darwin/Wallace papers and they were read first.

Little of the correspondence between Darwin, Hooker and Lyell for most of May/June, 1858, remains, but two letters addressed to Hooker dated Tuesday, 29th June, 1858, have survived (F. Darwin vol.1: p. 119-120). In the first of these two letters, Darwin wrote:

> ... I have received your letters. I cannot think now on the subject but soon will. But I can see that you have acted with more kindness, and so has Lyell, even than I could have expected from you both, most kind as you are.
>
> I can easily get my letter to Asa Gray copied, but it is too short ...

How many letters Darwin had received from Hooker, and their content, is unknown, since they are now missing, but it is clear that Hooker had written to Darwin more than once 'on the subject'. It is difficult to understand how Darwin could have referred to the undated letter to Asa Gray, read at the Linnean meeting on 1st July, 1858, as 'too short' since it ran to more than five printed pages (F. Darwin vol.1: 120-125), even though Sir Francis omitted a section on the variation of large genera, which he did not consider relevant. I suggest that the letter Darwin was referring to may have been the one he sent to Gray on 20th July, 1857, quoted above, which had, indeed, been sent "about a year previously", or, more likely, one written in October, 1857, now missing. I will give my reasons for believing this after I have finished recounting what happened.

Later that day, Tuesday, 29th June, Darwin wrote a second letter to Hooker, which he simply dated 'Tuesday night' (F. Darwin 1887: 119-120):

> I have just read your letter, and see you want the papers at once. I am quite prostrated and can do nothing, but I send Wallace, and the abstract of my letter to Asa Gray, which gives most imperfectly only the means of change, and does not touch upon reasons for believing that species change. I dare say it is all too late. I hardly care about it ... I send my sketch of 1844 solely that you may see by your own handwriting that you did read it. I really cannot bear to look at it. Do not waste much time. It is miserable in me to care at all about priority ...
>
> I would make a similar, but shorter and more accurate sketch for the 'Linnean Journal'.
>
> I will do anything. God bless you my dear kind friend.
>
> I can write no more. I send this by my servant to Kew.

This reproduction of the letter is not completely accurate. Sir Francis showed no emphasis on any of the words. In fact, the word 'only' was underlined twice and the words 'the means of change & does not touch' were underlined once.. The word 'solely' was also doubly underlined, making quite clear that Darwin did not expect anything to be read from his 1844 *Essay* at the meeting.

Darwin was clearly distraught, not just in regard to Wallace's paper, but in regard to the death of his son from scarlet fever, which had occurred on 28th June, 1858, the previous day. On the one hand, he said he hardly cared, but on the other, he wrote that he would, 'do anything'. How could Hooker not have been moved to compassion for his dear friend? This letter tells us that Hooker must have told Darwin that he did not recall the 1844 sketch. Darwin enclosed this sketch solely to remind Hooker that he had, in fact, read this piece. Hooker spent ten days at Down with Darwin in November, 1846, and it was during that visit that Darwin confided his ideas to Hooker and Hooker took Darwin's *Essay* away with him (Davies p. 62). It seems strange that Hooker appears not to have remembered this, but twelve years had passed and, no doubt, they had discussed much else in the meantime.

Quite how late in the day Darwin wrote this letter we will never know. It was June, the days were at their longest. Twilight would have fallen late. That Darwin wrote 'night' rather than 'evening' would seem to indicate that it was already dark and the fact that Darwin did not wait until the morning alerts us to quite how emphatic Hooker had been in his request for the material to be sent 'immediately'.

The papers were with Hooker the day before the meeting. It may well have been Hooker's first opportunity to study Wallace's paper. In this latest paper, Wallace had built up his argument by referring to the *Struggle for Existence, the Law of Population of Species, Adaptation to Conditions of Existence, Increase of Useful Variations and the Survival of Superior Variations*. Competition for available resources would occur, not only between individuals, but between species and varieties of species. Wallace suggested that should conditions change, then a variety might find itself better placed to survive than the original species. A new variety might have some slightly increased power of preserving its existence and would inevitably in time acquire a superiority in numbers. In addition Wallace stressed the difference between survival of variation in the wild and under domestication. Re-iterating the argument already presented in his 1855 Sarawak 'Law' paper, Wallace argued that if one variety could thus become established as a species, why should this species not give rise to one or several more new varieties, which might themselves, if changing circumstances permitted, out compete the original species? If one species merely changed its form, or the original form became extinct (shortly) thereafter, there would be no increase in the number of species. The survival of both the original and changed form, with both being able to give rise to further mutant species before eventually becoming extinct (as most ancient forms seemed to have done), gave potential for an increase in the number of species. This was the principle of divergence.

Unlike Darwin, who was in the process of amassing large numbers of examples to illustrate his points, Wallace outlined his theory in general terms only, making very few references to specific animals, and none to plants. Nevertheless, he had covered the principal points. I came to the conclusion that Hooker was concerned that the Gray letter was not sufficient and made the fateful decision to 'rewrite' the letter.

The supposed letter of 5th September, 1857 (see Appendix IV), which accompanied the summary of Darwin's ideas, commenced in a rather strange way (F. Darwin 1887, vol.1: 120):

> My Dear Gray, - I forget the exact words which I used in my former letter, but I dare say I said that I thought you would utterly despise me when I told you what views I had arrived at ... Permit me to say that, before I had ever corresponded with you, Hooker had shown me several of your letters (not of a private nature), and these gave me the warmest feelings of respect for you ...

It was Hooker who had first introduced Darwin to Gray during the latter's visit to London in 1839, the two men meeting again in 1851 at a luncheon hosted by Hooker at Kew (Gray 1894: 117, 380). Hooker regularly corresponded with Gray. However, he would not have known what correspondence had passed between Darwin and Gray. If it was indeed Hooker who wrote the cover letter, this may account for the rather strange manner in which the letter commenced.

The second paragraph was also interesting. It commenced by thanking Gray for his last letter and saying that he (Darwin) agreed with every word of it. After a few general comments, and a reference to the futility of Lamarckian concepts, it continued (p. 121):

> ... I will enclose (copied, so as to save you trouble in reading) the briefest abstract of my notions on the means by which Nature makes her species ...

The paragraph ended strangely with a request for Gray not to mention Darwin's doctrine to anyone (p. 122):

> The reason is, if anyone like the author of 'Vestiges', were to hear of them, he might easily work them in, and then I should have to quote from a work perhaps despised by naturalists, and this would greatly injure my chance of my views being received by those alone whose opinions I value.

Chamber's *Vestiges*, you will remember, was published in 1844. This could explain Hooker's decision to include passages from the 1844 *Essay* which clearly showed that Darwin had been working on his theory for a long time, since before *Vestiges* was published. The request would have been understandable in 1844 but was rather out of place in 1857! I think the overworked and hurried Hooker became temporarily muddled! The letter concluded with six paragraphs outlining Darwin's theory. Sir Francis published the letter and the enclosure as one continuous document, although the archival material held in Cambridge and the Gray Herbarium is, in each case, two separate documents.

Three days after the meeting, on Sunday, 4th July, Darwin wrote to Gray asking if he still had, "my little sketch of my notions of 'natural selection' & would see whether it or my letter bears any date ... I am sure it was written in September, October or November of last year". Presumably this letter was posted Monday morning. There is no record of Gray's reply. That same morning saw the delivery of a letter from Hooker to Darwin reporting on the meeting of 1st July. On that Monday, 5th July, Darwin wrote to Hooker, thanking him for his note telling him that "all had gone prosperously at the Linnean meeting" (F. Darwin vol.1: 126). He continued: "I do not at all understand whether my letter to A. Gray is to be printed; I suppose not, only your note ...". The confusion regarding whether or not the Gray letter was to be printed would have occurred if Hooker had returned it. Darwin would have expected the Gray letter to be retained by the Secretary for the preparation of the *Proceedings*.

It is here suggested that the letter which had been left with the Secretary was Hooker's 'reworked' version; Darwin received back what he had sent, which puzzled him.   Darwin requested the return of his 'interleafed folio' – his copy of the 1844 *Essay* – which he would have expected to receive back, since he had not anticipated that any part of that 'old' document would be presented to the meeting. It had – and it had been retained by the secretary in preparation for the publishing of the *Proceedings*.

The reply to Darwin's request to Gray for the date of his letter could have provided vital support for Darwin's case. It is inconceivable that Gray did not send one – not only because of Darwin's request but because he would also have offered his condolences on the death of Darwin's child. It would be understandable for the personal portion of such a letter to have been destroyed, but it is hard to understand why the portion confirming the September date would not have been kept, if confirm the date it did.

Despite having written, "September, October or November" to Gray and having led Lyell and Hooker to understand, "October", in his *Autobiography* Darwin (F. Darwin 1929: 58) claimed that this letter had been written on 5th September, 1857. This is the date under which it was published by Sir

Francis, who added a footnote (F. Darwin 1887: 120ff):

> The date is given as October in the 'Linnean Journal'. The extracts were printed from a duplicate undated copy in my father's possession, on which he had written "This was sent to Asa Gray 8 or 9 months ago, I think October 1857".

Notwithstanding Sir Francis' comment that October was the date given in the Linnean Journal, the date given in the Journal was 5th September, although, 'October,' was retained for the letter of transmission, read to the meeting by Lyell by way of introduction and printed in the *Proceedings*. No explanation was offered for the changed date.

The words 'duplicate undated copy' are also interesting. The copy in Sir Francis' possession when he was preparing *Life and Letters* for publication was not the copy Darwin had kept in his files and of which he sent an abstract to Hooker. It was not the copied abstract which Darwin had sent to Hooker. It was a duplicate copy, which must have been prepared after the meeting, which had been in the possession of Darwin and which was now in the possession of his son.

Most importantly it was undated, as must have been the letter presented to the Linnaen Society, otherwise why would Darwin have written to Gray 4th July asking Gray to tell him the date?

Darwin never wrote explicitly about the happenings at the meeting. He thanked both Hooker and Lyell (F. Darwin vol.1: 126, 129) for their kindness, but was clearly uncomfortable about *something* that had happened because, in his first letter to Hooker after the meeting, dated 5th July, he wrote (F. Darwin vol.1: 127):

> Lastly, you said you would write to Wallace; I certainly should much like this, as it would quite exonerate me.

'Exonerate.' That is a strong word and one not lightly used. How can a person be 'exonerated' if there is no fault from which to be cleared? Most probably, the 'exoneration' related to the reading of material by Darwin. Wallace's paper should have been read on its own. That Hooker should have suggested he write to Wallace is indication that Hooker, himself, felt that there was something that he needed to explain. On 13th July, 1858, Darwin wrote again to Hooker (F. Darwin vol.1: 128):

> Your letter to Wallace seems to me perfect, quite clear, and most courteous. I do not think it can possibly be improved, and I have to-day forwarded it with a letter of my own. I always thought it very possible that I might be forestalled, but I fancied that I had a grand enough soul not to care; but I found myself mistaken and punished. I had, however, quite resigned myself, and had written half a letter to Wallace to

> give up all priority to him, and should certainly not have changed had it not
> been for Lyell's and your quite extraordinary kindness. I assure you I feel it and
> shall not forget it. I am more than satisfied at what took place at the Linnean
> Society. I had thought that your letter and mine to Asa Gray were to be only an
> appendix to Wallace's paper.

Darwin's letter confirms that the reading of the extract from the 1844 *Essay* was a late decision and that that decision was taken by Hooker and Lyell. This is the second time, when writing to Hooker, that Darwin referred to "your letter". Darwin seems to have been aware of another letter which Hooker had written. What was it and where is it? I believe it quite possible that Hooker had written to Gray mentioning Darwin's work and that Hooker had offered this letter to be read in support of Darwin's. At the time Darwin wrote his letter of thanks to Hooker he was not aware of the material which had been presented to the meeting.

Further doubt is cast upon the series of events having happened, as we have been led to believe they happened, by the letter Darwin wrote to Gray on 11th August, 1858 (F. Darwin vol.1: 135):

> Your note of the 27th July has just reached me in the Isle of Wight. It is a real
> and great pleasure to me to write to you about my notions; and even if it were
> not so, I should be a most ungrateful dog, after all the invaluable assistance
> you have rendered me, if I did not do anything which you asked.

> I have discussed in my long MS, the later changes of climate and the effect on
> migration, and I will give you an abstract of an abstract (which latter I am
> preparing of my whole work for the Linnean Society). I cannot give you facts ...
> I may just mention, in order that you may believe that I have some foundation
> for my views that Hooker has read my MS, and though he at first demurred to
> my main point, he had since told me that further reflection and new facts have
> made him a convert.

The remainder of the letter is about changes during the glacial period, Asa Gray's area of special interest. There is no acknowledgement of Gray having responded to his request regarding the date of previous correspondence, which should have been mentioned in the first sentence, according to protocol. There was no acknowledgment of condolences, no mention of Natural Selection. Indeed, Darwin writes as guardedly as ever about his 'notions', not as to one who had been his special *confidant*: "I cannot give you facts". An explanation is offered in the next chapter.

When Darwin and Hooker wrote to Wallace advising him of their action in publishing his paper, Wallace was pleased. He wrote to his mother on 6th October, 1858, saying how gratified he was and expressing his belief that he was now assured of their assistance when he returned home (Brooks p. 201). In July, 1859, Wallace sent another manuscript to Darwin, which

Darwin read to the Linnean Society on his behalf on 3rd November, 1859. Wallace was clearly still happy at this time to forward his work to Darwin, who had received this second paper on 7th August, 1859, just as he was completing the manuscript of *The Origin*. Darwin wrote to Wallace on 9th August, 1859, (F. Darwin, vol.2: 161-162):

> I received your letter and memoir on the 7th, and will forward it to-morrow to the Linnean Society ... Had I read it some months ago I should have profited by it for my forthcoming volume. But my two chapters on this subject are in type, and though not yet corrected, I am so wearied out and weak in health that I am fully resolved not to add one word, and merely improve style. So that you will see that my views are nearly the same as yours, and you may rely on it that not one word shall be altered owing to my having read your ideas. Are you aware that Mr. W. Earl published several years ago the view of distribution of animals in the Malay Archipelago in relation to the depth of the sea between the islands? I was much struck with this, and have been in the habit of noting all facts on distribution in the Archipelago and elsewhere in this relation.

Darwin mentions Wallace's soon-to-be-published work in *The Origin*, in one of only two references to Wallace made within the text (Darwin 1859/1998: 299):

> ... there is also a relation ... between the depth of the sea separating an island from the neighbouring mainland, and the presence in both of the same mammiferous species or of allied species in a more or less modified condition. Mr. Windsor Earl has made some striking observations on this head in regard to the great Malay Archipelago, which is traversed near Celebes by a space of deep ocean; and this space separates two widely distinct mammalian faunas ... we shall soon have much light thrown on the natural history of this archipelago by the admirable zeal and researches of Mr. Wallace.

Three brief mentions were made of Wallace in *The Origin*. The first occurred in the Introduction, in which Darwin outlined his long involvement with the theory of evolution and the circumstances leading to the joint reading of his and Wallace's ideas before the Linnean Society. The next did not occur until Chapter XI (Darwin 1859/1998: 269):

> This view of the relation of species in one region to those in another does not differ much ... from that lately advanced in an ingenious paper by Wallace, in which he concludes that 'every species has come into existence coincident both in space and time with a pre-existing closely allied species'. And I now know from correspondence, that this co-incidence he attributes to generation with modification.

The 'ingenious paper" to which Darwin referred was Wallace's 'Law' paper, published in 1855, while he was in Sarawak. The third reference to Wallace in *The Origin* was the one mentioned above and occurred in Chapter XII. Notwithstanding the prior work of Mr. Earl, the deep channel dividing the Malay Archipelago is still known as the Wallace Line.

Far from assisting Wallace, Lyell, Hooker and Darwin's good friend, Thomas Huxley, set about establishing Darwin as the true originator of the theory of evolution by natural selection. It is clear that Darwin himself was part of the 'Darwin Movement' by the time he composed his book. From start to finish, in *The Origin* Darwin spoke continually of 'my theory'. Within the body of the book, there was no mention whatsoever of Wallace's Ternate paper.

Darwin wrote to Wallace in January, 1859, telling him that he had stopped work on his 'big book' and was now engaged in producing an 'abstract' for early publication, it being nearly complete. In this letter, he told Wallace that he "had absolutely nothing to do in leading Lyell and Hooker to what they thought was a fair course of action" (F. Darwin vol.2: 145). This statement indicates that, six months after the meeting, Darwin was still feeling uncomfortable about something that had occurred. Again, one has to ask: If there was something which Hooker and Lyell considered to be a fair course of action, does that imply that other people (including Darwin?) might consider that that action was unfair?

In a further letter to Wallace, dated 6th April, 1859, Darwin spoke of the 'abstract' that he was then writing (Brooks p. 214):

> The first part of my MS is in Murray's hands ... There is no Preface, but a short Introduction, which must be read by everyone who reads my book. The second paragraph in the Introduction I have had copied verbatim from my foul copy, and you will, I hope, think that I have fairly noticed your papers in the Linnean transactions. You must remember that I am now publishing only an Abstract, and I give no references. I shall of course allude to your paper on Distribution, and I have added that I know from correspondence that your explanation of your law is the same as that which I offer.

This is an accurate account of what finally happened. Clearly Darwin had planned early what reference he would make to Wallace's ideas, that the paper read to the Linnean Society would only be mentioned in the *Introduction*. Within the text, only the paper on 'Distribution', written in Sarawak and published in 1855, would be mentioned, not the 1858 paper, written in Ternate, on 'Generation with Modification'. This letter was not reproduced in *Life and Letters of Charles Darwin*, although letters to Gray and Murray of 4th and 5th April, 1859, respectively, were (F. Darwin vol.2: 154-155).

McCalman (2009: 317-329) rejected the claims of Brackman and Brooks. He accepted the June delivery of Wallace's paper at Down, attributing the delay to the uncertainties of the postal service. McCalman denied that Darwin could have amended his writing on the grounds that the Darwin household was in a state of crisis due to it experiencing an outbreak of scarlet fever. However, Darwin's son, Charles, did not contract the disease until 23rd June (McCalman 2009: 320), after the time suggested by Brooks for the amending of sixty pages of his manuscript. McCalman (pp. 325-327) agreed that arrangements were made with all possible haste for the reading of the Darwin/Wallace papers.

Van Wyhe and Rookmaaker (2012) also rejected any suggestion that Darwin had received Wallace's letter any earlier than the historically accepted date of 18th June, 1858. They reached this conclusion based on the claim by Davies that Darwin's letter to Wallace, dated 22nd December, 1857, was delivered to Wallace, not on the late February steamer, but on the steamer which reached Ternate on 9th March, 1858. This, they claimed, would not have allowed Wallace to respond to Darwin's letter with his own dispatched by return of mail, which would have seen it delivered to London at the same time as Wallace's letter to Bates, i.e. 6th June, 1858. Rather, the letter to Darwin would have been dispatched two weeks later and received mid-June, as Darwin claimed.

Davies (2012) responded by denying Van Wyhe and Rookmaaker's claim that the mail steamer would have been docked at Ternate for a short time, possibly as little as one hour, thereby making it impossible for Wallace to have responded to Darwin's comment about Lyell's interest in Wallace's work by suggesting that Darwin show his draft paper to Lyell. Davies believed that the ship would have remained docked long enough for Wallace to have read Darwin's letter and written a reply, or even added a note on the back of the envelope/packet. The research of Van Wyhe and Rookmaaker led them to discover that the usual transit time from Ternate to Surabaya (Java) was about fourteen days and that this particular mail ship arrived back at Surabaya on Tuesday, 20th April, 1858. This would suggest it left Ternate on 6th April, a whole day after its arrival on 5th April. There would have been ample time for persons receiving correspondence to pen any replies considered to be 'urgent', obviating the necessity of their waiting for the next mail steamer. This would make good commercial sense. There is some disagreement between authors as to the frequency of mail deliveries. Some say two weeks, some say four. I tend to believe that mail ships left Surabaya fortnightly but that they had two different routes. The initial part of the journey was the same, but then deviated so that the outlying islands received mail alternate deliveries, i.e., every four weeks. If this was the case, then to assume that a mail ship, which called at a port only once every four weeks, would dock, off-load its cargo and leave, all within an hour, without allowing time for residents to read their mail and respond to anything perceived to be urgent, defies logic. Documented evidence of shipping schedules given by both Brooks and Davies showed that the usual stop-over time was 1½-2 days. If Darwin had indeed sent Wallace one of Lyell's earlier letters, this alone could account for Wallace's request that he show the paper to Lyell, irrespective of anything written by Darwin in his letter.

The more one considers the events of 1st July, 1858, the more strange they become.

Although Darwin subscribed to numerous journals, he led a reclusive life, rarely leaving Down House in Kent. Unlike many other people, he did not own a town house, nor come to London for the Season. He did attend some meetings, including those of the Linnean Society on at least one occasion, but he was not a regular attendee or contributor of articles to journals. In 1842, he had written a monograph on coral reefs, which had been very well received, and in 1851-1852 had produced a two volume work on *Cirripedia*, which is still highly regarded today. However, his first 'paper' on orchids was not published until 1862 and that on climbing plants not until 1865. In this he differed greatly from Wallace, who was a regular contributor of articles, but to the *Annals and Magazine of Natural History*, not to the Linnean Society. There was no precedent for the reading of papers from either of these men at Linnean meetings.

Since Lyell and Hooker were Darwin's closest friends, it might have been supposed that they would have been at Down House that day, supporting their friend by attending the funeral of his son. Instead, they were the first people to arrive for the meeting, this being shown by the fact that they were the first to sign in (Partridge 2015a). This meant that they were the first to register their intent to present a paper at that evening's meeting and they were, in fact, the first people to take the floor. And what did they present? Part of a private letter written to someone overseas the previous year and an extract from an unpublished essay, written fourteen years earlier, the two items being presented on the day of the funeral of the author's child! Hardly what was to be expected in a presentation to a scientific society! These were followed by the reading of a paper written by someone half a world away, who had never previously made any contribution to that Society, a paper which could easily have waited until the next meeting, which would have been in September. The papers held over from the postponed meeting were scheduled to be read. Some had to be held over for a second time to make way for Darwin's presentation. Why the haste? What was the urgency?

There was a clear implication made by Brackman, Brooks and Davies that Darwin, Lyell and Hooker had been involved in some form of conspiracy with the intent of promoting the ideas of Darwin above and before those of Wallace. This was based on their findings in relation to the date of receipt of Wallace's paper, the changes made by Darwin to Chapter 6 of his 'big book' and their arranging, with the help of Huxley, for the presentation of the material with such haste.

I agree with their conclusions and have added a further matter for consideration – that the 'Asa Gray letter' read to the meeting was not the one Darwin had written, but one composed by Hooker. In this case, there is no accusation of 'malicious aforethought'. It is concluded that Hooker acted in haste, motivated solely by a desire to help his good friend, believing that the letter he wrote did, in fact, not contain anything not already written somewhere by Darwin. Indeed, most of what he wrote was based on the 1844 *Essay*. I have concluded that there was no intent to deceive when the material was read to the Society, not on Darwin's part because I believe the letter he intended to be read was one that he had written to Asa Gray, not on the part of Hooker who genuinely believed that he was but putting Darwin's thoughts into clearer form, and not on the part of Lyell, who was not directly involved. However, all three men did allow the situation to stand. Darwin's obvious distress is seen as evidence that he was very unhappy with the situation as it unfolded but he allowed it to continue. What was said in their joint letters to Wallace we shall never know, because they are missing, but they seemed to satisfy Wallace, at least at the time. By the time Wallace returned to England in 1862, *On the Origin of Species* had been published and Darwin's reputation as the originator of the concept of evolution by natural selection well established.

Summary

There are certain facts of this case which are beyond dispute.

1. Darwin agreed with the suggestion made to him by Lyell and Hooker that his ideas be presented to the Linnean Society on his behalf by the reading to the Society of a letter which he had previously written to Asa Gray.

2. This letter, the 1844 sketch and Wallace's paper were delivered, by hand, to Joseph Hooker late 29th June, 1858.

3. Something took place during the meeting of the Linnean Society on the evening of 1st July, 1858, which Darwin had not expected which caused him distress and for which he felt the need to be exonerated.

4. After the meeting, Darwin was very evasive in all his correspondence when referring to this event and endeavoured, as far as possible, to distance himself from it.

What event could possibly have taken place at this gentlemen's meeting to cause Darwin such embarrassment and shame? Darwin was not even present at the meeting, so any complicity on his part must have occurred either before or after ther meeting. Surviving correspondence rules out any prior knowledge of, or involvement with, anything untoward which happened. Whatever happened, happened without Darwin's knowledge or consent.

What was it?

1. Darwin had agreed before the meeting that something he had written should be read in an effort to gain some recognition for his work, if not to give him priority then at least to give him parity.

2. It was not something to which he had previously agreed that was troubling Darwin. It was something done by Lyell and/or Hooker.

3. Inasmuch as Lyell does not seem to have been involved in any 'explanation' to Wallace, I have concluded that it was not the reading of the extract from the 1844 Essay, which seems to have

175

been Lyell's suggestion. It was something for which Hooker alone was responsible.

4. It is here suggested that the only plausible explanation is that the letter read at the meeting was not the one Darwin had written to Gray, not the one Darwin had sent to Hooker on 29th June, but a substitute one, one written by Hooker.

This matter is complicated. After I had completed my thesis, I re-examined this matter and identified a number of additional points which I believe support my hypothesis. The following chapter is a brief summary of relevant points already made in this chapter – and a little bit more!

Abstract Thinking

In the previous chapter, it was noted that Darwin destroyed all of the letters he received from Wallace while Wallace was in the Malaya Archipelago with the exception of one piece containing information regarding panthers. The earliest surviving letter is dated 7th April, 1862, after Wallace returned to England. Between then and Darwin's death twenty years later in 1882, the two men exchanged 146 letters (Marchant 1916).

Also in the previous chapter, mention was made of other missing letters. When Mrs. Jane Gray published *Letters of Asa Gray* in 1894, she made a point of emphasizing her surprise at finding no trace of the correspondence which had passed between her husband and Charles Darwin between mid 1858 and 1862. The only letter in her possession was a short one from Darwin, dated 5th January 1860, thanking Gray for offering to assist in the publication of an American edition of *The Origin*, which had accidentally been placed in the file containing the publisher's correspondence. The correspondence forwarded to her from Down by Sir Francis also pre- or post-dated those years. There was nothing from that crucial period of three-and-a-half years. It is clear that Mrs. Gray was not in possession of the 5th September letter, or the 'abstract', of which she surely must have been aware? Her husband actively supported the propagation of Darwin's theory in America, in 1876 publishing *Darwiniana: Essays and Views pertaining to Darwinism*. It is inconceivable that this husband and wife team did not discuss Darwin, his theory and Gray's part in the publication of Darwin's ideas. I can only assume that Gray stored the 5th September material in a safe place where it was later found. Darwin's 1844 *Essay* was found in a cupboard some time after his death and the original of the eleven chapters of 'Natural Selection' which Darwin completed before abandoning it to write *The Origin* was not found until during the Second World War (Brooks p. 231). The large houses in which the gentile lived contained attics, cellars, box rooms and plenty of storage cupboards, in which material could, and did, lie undisturbed for decades.

Lyell's correspondence for that period is also missing (Brackman p.351). The correspondence between Darwin, Gray, Lyell, and Hooker for this same more than forty month period is also missing. Wallace left the Archipelago late 1861 and returned to England in the Spring of 1862. It would seem that the four men jointly decided to destroy whatever 'evidence' of whatever 'misdemeanor' they had committed before Wallace's return. In addition, at least five years' of letters from Wallace to Darwin between 1856 and early 1862 are also missing. Darwin is believed to have written some 14,000 letters during his lifetime, of which over 7,000 have survived.

A letter of 11th August (darwinproject.ac.uk/2321) has been dated by archivists to 1858 on the grounds that Darwin was known to have visited the Isle of Wight in August of that year. The letter acknowledges receipt of a 'note' from Gray of 27th July, but no letter from Gray of that date has survived. The assumption would be that Gray's letter was written in response to Darwin's letter of 4th July, 1858, in which he asked Gray to check the date of the letter he (Darwin) had written, he thought September, October or November of the previous year, outlining his notions of "natural selection".

Darwin's letter of 4th July, 1858, had started "I have not answered your note of May 21 for I have had death & illness & misery amongst my children ..." That was all that Darwin told Gray about the outbreak of disease which had smitten Down House and resulted in the death of one of his children. (Gray had written to Darwin on the 21st June, not May.) The letter continued to discuss botanical matters (an essential ingredient in all letters which passed between these two men) particularly in relation to glaciation, Gray's speciality. The letter finished with Darwin outlining the events of 1st July and asking Gray if he had by any chance Darwin's little sketch in which he had outlined his notions of "natural selection and, if so, if he would be kind enough to let Darwin know the date?"

Gray's reply has never been found, which is very strange. If it contained the all-important date of 5th September, 1857, surely it would have been kept? There is a letter from Gray dated "August 1857". The date was given in square brackets, indicating that the date was not given on the letter but had been deduced by archivists. This was necessary because the first part of the letter was missing. Archivists dated it to "between 20th July and 5th September" because it appeared to be the reply to Darwin's letter of 20th July (see page 155) and would have been written before Darwin's letter of 5th September. Inter-continental mail took two weeks. If Gray replied immediately to Darwin's letter, the earliest his reply could have been delivered to Down would have been early August. It may, or may not, have been forwarded to Darwin holidaying in the Isle of Wight. However, August would have been the holiday season in Boston, as well as in England, and it is quite possible that Gray did not reply until the end of August, or even September, or that Darwin did not receive the letter until he returned from holiday. The letter, as it has come down to us, is short and gives the impression that the first portion is missing, although this is not stated in the footnotes.

Gray's August response to Darwin's 20th July letter was positive:

… when we see that every plant man takes in hand developes [sic] into varieties with readiness, when favourably circumstanced, we cannot avoid suspecting they may do the same thing — i.e. sport in some way in the wild state also — and that there is some law, some power inherent in plants generally prompting them to originate varieties — which is just what you want to come to and I suppose this is your starting point.

Here you begin with good, tangible facts; and I am greatly interested to see what is to be made out of them. First, can you get at the law of variation? Or throw any … [section missing]

I believe that, thus encouraged, Darwin took his time to compose in reply a letter which gave further details of his theory, as well as the name of his proposed book - *Natural Selection*. I believe he posted this in October, and, because it was an important letter, he kept a copy - unfortunately undated.

This letter from Gray becomes important for three reasons. Firstly, it tells us that Darwin's 'notions' were well received. Secondly, at the end of the letter, Gray requested: "Kindly post the enclosed." England had introduced the "Penny Post". Her postage rates were the lowest in the world. It became common for 'ex pats' to send packets of letters home for onward posting. Wallace sent such packets to the brother of his good friend, Bates, in London. What is noteworthy is not that Gray sent such items to England for onward transmission, but that he sent them to Darwin, not his close friend, Hooker. Darwin and Gray had met a couple of times during visits by the Gray family to England and clearly their two year correspondence had drawn them closer together. The third noteworthy point is the further evidence of their growing friendship, made clear from the next, and final, sentence of Gray's letter: *"I write in greatest haste, and am, with the highest regard Ever yours faithfully ...".*

The words shown here in italics were underlined. They were clearly a further endorsement of Darwin's work and would surely have pleased Darwin greatly.

Darwin's reply of 11th August 1857, I believe, not 1858 as shown in archive, was not as effusive, although he did start by saying that it gave him great pleasure to write to Gray about his notions. He added that he would be an, 'ungrateful dog,' if, after all the assistance Gray had given him, he did not do anything which Gray asked - a reference, presumably, to forwarding Gray's mail. Darwin told Gray that he had discussed climate changes in his long M.S. (manuscript) – i.e. the book he was writing, but would give Gray an "abstract of abstract" of his whole work, which he was preparing for the Linnean Society, the word 'abstract' being underlined in each case.. Then followed a portion about climate change/glaciation and flora. He ended the letter by referring again to the 'abstract' which he expected to be published 'next winter', by which I presume he meant winter 1858, not winter 1857, which would have been 'this winter'. That was when he then thought his book would be ready for publishing, which of course, it was not.

I agree that these comments could be interpreted as referring to the abstract of his book, *On the Origin of Species*, which he was planning to start writing when he returned from the Isle of Wight in August 1858, and which was published December 1859. However, the rest of the letter, which made no reference to the death of his child, the Linnean meeting, Gray's response to his request for the date of his letter, makes it highly unlikely that this letter was written in August, 1858, as suggested.

You will have noticed that 'abstract' was quite a favourite word of Darwin's. He was planning to write an article for the *Linnean Journal* at the end of the following year, which would be an 'abstract' of his book. (Writing articles/book reviews to publicize upcoming works is a tactic still used by writers today!) What he wrote to Gray was an 'abstract' of that future 'abstract'. He did conclude by saying that "I shall not give abstract on facts in regard to crossing, for they are too many to abstract –" which may be a piece of humour to indicate that he, himself, was aware of his overuse of the word.

What next? According to the orthodox version, the next letter which passed between these two correspondents was the one of 5th September, the one which started: "I forget the exact words I used ..." and which went on to remind Gray that, "before I had ever corresponded with you, Hooker had shown me several of your letters ... and these gave me the warmest feelings of respect for you". That Darwin should not remember the 'exact' words he used is understandable, because he did not keep copies of routine letters, and no one would expect him to. If the orthodox version is correct, the 'forgotten' letter was the one written on 20th July, six weeks previously, the one in which he had told Gray that it was now nineteen (!) years since he had started keeping notes and in which he, for the first time, shared some of his 'notions' with Gray. If my scenario is correct, the letter had been written but three weeks previously from the Isle of Wight. I find it highly unlikely that Darwin's memory was quite so faulty that he would not have remembered either of these two letters, at least in outline.

Here is what I believe happened.

Darwin's 20th July, 1857, letter would have arrived at the Gray residence early August. August was holiday month in Europe, and, no doubt, also in America. It is unlikely that Gray would have rushed to reply until the end of the month, at the earliest. I believe it was Gray who wrote to Darwin that September, not the other way around.

I believe that Darwin, in reply to this letter from Gray of August/September, 1857, did, in fact, write to Gray in October of that year, as he remembered and as he led Lyell and Hooker to believe he had. Although his 4th July, 1858, letter to Gray offered three possible months, September, October and November, the letter of transmission read to the Society stated "An abstract of a private letter addressed to Professor Gray, of Boston, U.S. in October 1857 ...". In this letter I believe Darwin gave further details of his theory, as it then stood, of 'Natural Selection', which was to be the title of his book.

I believe Gray replied again late October/early November. This reply is also missing, but this time we know for certain that Gray had responded to a letter from Darwin, because on 29th November, 1857, Darwin wrote to Gray: "I thank you for your impressions on my views". Whatever views Gray expressed, they were not entirely favourable, because Darwin continued:

> I had not thought of your objection of my using the term, "natural selection," as an agent; I use it much as a geologist does the word Denudation, for an agent, expressing the result of several combined actions. I will take care to explain, not merely by inference, what I mean by the term; for I must use it, otherwise I should incessantly have to expand it into some such (here miserably expressed) formula as the following, "the tendency to the preservation (owing

181

to the severe struggle for life to which all organic beings at some time or generation are exposed) of any the slightest variation in any part, which is of the slightest use or favourable to the life of the individual which has thus varied; together with the tendency to its inheritance". Any variation which was of no use whatever to the individual, would not be preserved by this process of "natural selection". But I will not weary you by going on, as I do not suppose I could make my meaning clearer without large expansion.

In the 6-point summary of 5th September submitted to the meeting, the term 'natural selection' occurs twice. Point 1 discussed change under domestication. Point 2 started 'Now suppose there were a being who did not judge by mere external appearances ...', a concept taken straight from the 1844 *Foundations* essay. Point 3 states "I think it can be shown that there is such an unerring power at work in Natural Selection (the title of my book) ...". Point 4 addresses environmental changes, the struggle for existence, occasional slight variation, better chance of survival of those fortunate beings, whose offspring inherit the beneficial variation " – natural selection accumulating those slight variations ...". I find it difficult to conceive of any letter which Gray could possibly have written, based on those two references to natural selection, the second fully and carefully explained, which would have resulted in Darwin writing the response that he did.

I do not believe the letter of 29th November addresses any letter Gray wrote in response to the 5th September letter. I believe Gray's response was to a letter Darwin wrote in October, 1857, in which he tried to outline his views, as they then stood, not very coherently, it would seem!

It was not common practice for people to keep copies of their hand-written personal letters to friends and Darwin was no different from anybody else in this respect. He only kept copies of important letters. Davies (p. 126) commented that it was fortuitous that Darwin kept a copy of his 5th September letter, but made no suggestion that any portion of it was not genuine.

Whenever Darwin referred to the document which he sent to Hooker on the evening of 29th June, 1858, he referred to it as a 'letter'. He never referred to it as an 'outline of my theory', or a 'summary of my notions', but as a letter. That Darwin should keep a copy of the theoretical part of his letter is very understandable. What is not so understandable is why he should bother to keep a copy of the personal part of his letter, the one that started: "I forget the exact words ...". Why would he bother to copy out and keep that piece? The letter in the archives at Cambridge was written in Darwin's own hand, the attachment by an assistant, this being the same as the documents held in the Gray Herbarium archives. So Darwin did write both copies of that letter. The question is, when?

I have already given my conclusion that, when Hooker received the Gray letter that evening, he realized that it was inadequate. If it had failed to impress Gray, why should it impress the meeting? I believe Hooker drew up those six points based on the 1844 *Essay* and the abstract from the genuine October letter to Gray, which he had in his possession that evening. He may, or may not, have included ideas from Wallace's paper. Without knowing the contents of the October letter, we will never be sure.

It is not known who composed the letter of transmission. The original was discarded, along with other material, after it was finished with by the printers. What we do know, from the *Proceedings*, is that it was dated 30th June. Whether it was delivered 30th June, or presented on arrival 1st July, is a matter of debate (Partridge 2015b). What is important is what was written. The introduction explaining "the accompanying papers" described the letter thus: "An abstract of a private letter addressed to Professor Gray, of Boston, U.S. in October 1857, by Mr. Darwin ...". What is particularly interesting is that, a few pages later, after the section from the *Essay* had been printed, item 2, the letter, was introduced thus: "Abstract of a Letter from C. Darwin, Esq., to Prof. Asa Gray, Boston, U.S., dated Down, September 5th, 1857". The change of date had obviously been agreed upon before the *Proceedings* went to print. Why the change of date? No letter from Gray responding to Darwin's request for information has survived. Based on my claimed scenario, that Hooker wrote the 'abstract' and Darwin wrote the copies of the letters and had copies made of the 'abstract' after the event, I offer the following explanation.

It had previously been agreed (by means of correspondence now missing) that Darwin's October letter would be part of Darwin's presentation. I say 'Darwin's presentation' because I am sure that the original intention would have been for Darwin to attend the meeting himself, to present Wallace's paper, along with an abstract of his long held theory, supported by the evidence of his correspondence with Gray. If it had not been for the outbreak of scarlet fever, hopefully two weeks would have been enough time for Darwin to pen an 'abstract'. When scarlet fever, and later diphtheria, did break out, it was clear that Darwin would not be able to attend the meeting. Apart from any natural desire to be close by his stricken family, Down House would have been under quarantine and Darwin would not have been able to attend a social event. The situation was saved when Lyell and Hooker offered to be there on his behalf. I believe this was the 'kindness' for which Darwin thanked them in his first letter to Hooker on Tuesday, 29th June. It is clear that it was Lyell's idea something be included from the 1844 *Essay* because of the comment Darwin made in his second letter that day that he was sending the document 'solely' (underscored twice) that Hooker could see for himself that he had indeed read it. Darwin did not wish anything to be read from that *Essay*.

I suspect that the composition of the letter of transmission was Lyell's contribution, partly because of its misleading nature. It gave the impression that the paper to be read was a joint paper. It also stated that the first item to be read would consist of "Extracts from a MS work on species ...", the word 'extracts' being in the plural, twice. In fact, but one long extract was read. This could indicate that Lyell composed the letter after it had been agreed that Hooker would choose some extracts to read and that Lyell was unaware of Hooker's final decision. The introduction also contained the words "...neither of them having published his views ..." which statement completely ignored Wallace's Law paper of 1855. The sentence continued by advising the listener that Mr. Darwin had been continually "urged by us to do so"! Partridge (2015a) also concluded that it was Lyell who had composed the letter of transmission, unaware of what Hooker was preparing to present, although Partridge did not question the genuineness of any of the material. It also stated that "both authors having now unreservedly placed their papers in our hands ...". Darwin wrote the date "1839" on the *Essay*, although we know from other of his records that he did, in fact, write the first draft in 1842. This seems to have been the full extent of Darwin's duplicity. I believe that the letter of October 1857 to Gray did exist.

When the conspirators realized that they needed Darwin to send a copy of the revised letter to Gray, and make one for his own files, it was worked out that, if they dated this spurious letter 5th September, it would have had time to reach Gray, and for Gray to have replied before Darwin sent his genuine October letter. The genuine October letter would fit neatly into the sequence. The decision to destroy correspondence was taken much later. The first part of the August letter from Gray was probably destroyed later, when they were covering their tracks, because its date did not fit the engineered scenario.

Hooker's wife helped by copying out the 'abstract'. On 13th July, 1858, Darwin wrote a letter to Hooker, at the end of which he asked Hooker to thank Mrs. Hooker for having made a copy of his 'messy MS'. There are two questions here. Why did he refer to it as 'messy' and why did he refer to it as 'MS'? Up until that time, Darwin had always referred to the document he was planning to send to Hooker, and which he did eventually send to Hooker, as a 'letter' or 'abstract' or 'abstract of abstract'. Now, it has become a 'MS' – manuscript. Quite so! Answering the second question was easy. What about the first? The only explanation I can offer is that Darwin had never been satisfied with the 1844 *Essay*, which he may have considered 'messy'. He never published it. He did not want it read at the meeting. He did not use it as the 'foundation' for his book, rather starting afresh. Since most of the material in the 6-point 'abstract' read to the meeting had been taken from his 'messy manuscript', I can only believe that that is how he thought of it, and how he referred to it when writing to Hooker.

Something else also occurred to me. Hooker would have had a very busy time, reading Wallace's paper, probably at least twice, re-reading Darwin's *Essay*, choosing which section to use, then deciding to make his 6-point abstract, which probably needed several drafts, as these things always do. He would have known that the Society would expect a letter – that had already been agreed. The genuine one could no longer be used. A new one had to be written. Did he ask his wife to help out by composing it? The letter reads very badly if one assumes that it was, indeed, written by Darwin. It reads but little less badly if one assumes it was written by Hooker – definitely below his standard. However, if one imagines Hooker's wife, poor woman, who would have had absolutely no idea how often Darwin and Gray corresponded, or what they wrote about, trying to come up with something, suddenly the letter makes a lot more sense!

The copy of the Gray letter which Darwin had forwarded to Hooker was returned by Hooker immediately after the meeting, leaving Darwin to wonder if it was going to be published. No, it wasn't! What was published was Hooker's version. I do not accept that when Darwin wrote to Hooker on 5th July: "... I do not in the least understand whether my letter to A. Gray is to be printed; I suppose not, only your note" and again on 13th July: "... I had thought that your letter and mine to Asa Gray were to be only an appendix to Wallace's paper" that he was referring to the letter of transmission as has been suggested. He would never have supposed that this introductory document would be "only an appendix to Wallace's paper". I believe that Hooker had offered, as supporting evidence, some letter he had written to Gray about Darwin's work, which, in the event, he did not use.

The euphoria that everything had gone so well at the meeting would soon have evaporated. It would have been clear to all concerned that Darwin needed to make copies of the material, which I believe he did, both for his own records and those of Gray.

If the 'letter' Darwin sent comprised solely the 6-point summary, what did Hooker return to Darwin immediately after the meeting? The summary was forwarded to the printers. If Darwin sent both components, letter and summary, and the summary had been forwarded to the printers, the personal letter being returned to Darwin, that would have been nothing other than Darwin would have expected. Why was he puzzled? Why did he not understand what had happened at the meeting, what was to be published in the *Proceedings*?

Darwin wrote to Hooker from the Isle of Wight on 21st July, 1858. The first sentence is uncontroversial: "I received only yesterday the proof-sheets, which I now return." These were the proof sheets of the *Proceedings*, which were in order. It is the next paragraph which needs an explanation: "I am disgusted with my bad writing. I could not improve it without rewriting all, which would not be fair or worth while, as I have begun on a better abstract for the Linnean Society. My excuse is that it was never intended for publication. I have made only a few corrections in the style; but I cannot make it decent, but I hope moderately intelligible. I suppose someone will correct the revise. (Shall I?)".

Now, clearly, by 'it' Darwin was not referring to the 6-point abstract, the proofs of which he had just approved and which he was returning by that post. Darwin had previously mentioned his intention of drawing up an abstract of his ideas for the Linnean Society, but had proved himself unable to do so. I believe the document ('it') which Darwin took with him to the Isle of Wight was the genuine October 1857 letter to Gray, which contained by far the most comprehensive expression of his views, as they had been the previous autumn, after two-and-a-half years working on his big book. What other handwritten written document, suitable for being amended into an article for submission to the Linnean Society, could there possibly be?

In his final paragraph, Darwin wrote: "I am very glad at what you say about my Abstract, but you may rely on it that I will condense to the utmost". In a footnote, Sir Francis explained that by 'Abstract' Darwin meant what was to become the *'Origin of Species'*. According to Darwin's personal diary, he began writing his "Abstract of Species book" 20th July, the day before his letter to Hooker. On 15th September, 1858, he recorded that he "Recommenced Abstract". Darwin always insisted that his 400 page book was but an *Abstract*. On p. 152 of *Life and Letters,* Sir Francis reproduced a copy of Darwin's originally proposed front page, showing that he had intended to entitle his book *"An Abstract of an Essay on the Origin of Species and Varieties through Natural Selection".*

Talking of abstracts (which Darwin constantly was!), look again at the letter Darwin wrote that fateful Tuesday evening. Earlier that day, Darwin had written to Hooker that he could easily get his letter to Asa Gray copied. Then he wrote (that night) "I send Wallace, and the abstract of my letter to Asa Gray ...". There is one problem with this scenario. Sir Francis did not accurately record what Darwin had written. What Darwin actually wrote was "I send my abstract of abstract of letter ...". Possibly Sir Francis thought that the second 'abstract' was an error but we noted earlier that Darwin had previously used that same phrase, reducing the chance that this was, in fact, an error. If it was an abstract of an abstract, then it was not the complete letter; part of it had been copied, probably by someone within his household. The local schoolmaster, Ebenezer Norman, usually helped with copying, but it is unlikely that he would have appreciated contact with a household under quarantine. Indeed, such contact would have been highly irresponsible on the part of both parties.

I believe Darwin asked someone to copy out the most relevant part of his letter. He may even deliberately have asked the copier to omit certain sentences of which Gray had been particularly critical, as critical he had been, as has been shown when discussing Darwin's November letter to Gray. In the haste, it was forgotten to date the letter/abstract.

The subject of haste, brings up another point which must be discussed. Why the haste? Why did not the three men make their comments about Wallace's work, post it back to him and leave him to submit it to the Journal? Why publish Wallace's work at all, let alone publish it with indecent haste, as indecent it most certainly was, considering the death of little Charles?

If Brooks was correct, as I believe he was, then there seems to have been no undue panic, time-wise, when Wallace's letter arrived mid-May. Darwin spent nearly a whole month amending Chapter 6, which he had completed the previous year. There seems to have been no idea of their presenting Wallace's paper to the Linnean Society at the June meeting – it was not listed. Yet when the June meeting was postponed for two weeks, there was a sudden scramble to have Wallace's paper read – along, of course, with something of Darwin's. Why?

The letter which accompanied Wallace's paper is missing but we do know something about it. In it, Wallace asked Darwin to show his paper to Lyell. He did not ask Darwin to publish. Also, he wrote in haste. Wallace later recorded how he had pondered over the problem of speciation during his bout of fever, had new thoughts, which he scribbled down that evening, then re-wrote over the next two evenings so that he might send them to

Darwin by the next mail, which was due to leave shortly. Now, did Wallace really want to wait several months for his paper to reach England and then be posted back before he sent it in for publication? Is it not possible that Wallace, in his hurry, simply forgot to ask Darwin to send it on to the *Journal* for publication if he and Lyell thought it had merit?

If this was the case, then is it not possible that, precisely one month later, the next mail boat carried with it a letter from Wallace to Darwin making that very suggestion? If that were the case, this second letter would have arrived exactly one month later - mid June. There would have been no question of the three men involved not submitting the paper, as asked. The problem was - how could they arrange for something of Darwin's to be published at the same time?

Darwin had shown himself to be totally incapable of producing a short, coherent account of his work. But he had a letter! Letters were often read at meetings, usually, of course, *from* abroad, not *to*, but it was worth a try! I am sure the original plan would have been for Darwin personally to present the material which would have made the presentation more powerful. Remember, in the scientific world, be it academic or commercial, credit goes to the person who publishes or who applies for a patent. Saying "I thought of that years ago, but did nothing about it" would be unlikely to result in any comment, other than "More fool you!". Darwin's shelved 1844 manuscript was useful as a back-up, but would not have been acceptable unsupported. Then disaster struck in the form of disease at Down, but Darwin's friends came to the rescue, offering to act as his representative.

Did Lyell lie? The letter of transmission is generally accepted as having been misleading, if not outright mendacious. The opening title gave the impression that the paper which was to be presented was a joint effort by Darwin and Wallace, although the following outline of the three components of the presentation were accurate. The introduction continued: "... both authors having now unreservedly placed their papers in our hands ...". Darwin may have done so, somewhat unwillingly, because of force of circumstance. But what about Wallace?

According to the succession of events as has historically been assumed, it would seem that this was a blatant misrepresentation of the truth. But what if Wallace had sent a second letter? What if in that letter Wallace had asked them to arrange publication of his paper, as they thought appropriate? Then this statement would not have been a lie.

We know that Hooker wrote a letter of explanation to Wallace, which Darwin enclosed with one he had written, dated 13th July, 1858. We know that these letters were received by Wallace, because he told his mother in his next letter home to her. What we do not know is what was said, because the letters are missing. I will hazard a guess that Hooker explained that Darwin had been preparing an abstract to present to the Linnean Society, based on the material he had been gathering for his book (about which Wallace knew), some ideas having been shared with Gray in a letter the previous year. After receiving Wallace's paper, it had seemed appropriate that Darwin should present Wallace's paper to the Society, along with his abstract, and (part?) of the letter containing his ideas which he had sent to Gray.

Unfortunately, due to illness, Darwin was not able to attend the meeting, so Lyell and Hooker had thought it best to present the material jointly on their behalf. (This was the 'exoneration' which Darwin so much appreciated.) Wallace could even have been left with the impression that the presentation was already arranged and that he had been very fortunate in being able to have his material presented at the same time, rather than published a month or two later – after Darwin! As we know from the above cited 1857 'abstract of abstract' letter, Darwin had, indeed, been planning such an abstract for the Linnean Society. He just had not gotten around to writing it!

What is harder to understand is why these two letters should be missing, since they would have been destroyed by Wallace, not Darwin or Hooker. The simplest explanation is that, when he was packing to return home, Wallace lightened his load by discarding unwanted material. These two personal letters would have had no botanical information in them and may have been considered unnecessary to keep. After all, *The Origin* had been published two years earlier.

One puzzle remains. Why did Wallace send the paper to Darwin rather than to his publishers? I believe that may be worked out from the remaining evidence.

The first surviving letter from Darwin to Wallace is dated 1st May, 1857, and commences by thanking Wallace for his letter of October, 1856. Darwin then told Wallace about the book he was writing and how long he had been working on the project. He continued:

> I have acted already in accordance with your advice in keeping domestic varieties, and those appearing in a state of nature, distinct; but I have sometimes doubted the wisdom of this, and therefore I am glad to be backed by your opinion.

When he published *The Origin*, Darwin did deal separately with change under domestication, which he covered first, the latter part of the book dealing with change in nature. However, there was a connecting link: his claim that what humans could achieve in a short time by deliberate selection under domestication, Nature could achieve over a longer period of time by 'natural selection'.

Now look again at the Ternate paper (see Appendix III) written by Wallace which was read to the Linnean Society. The second part is a point by point rebuttal of this idea. Domestic animals were shielded, fed regularly and protected from adverse environments. The weak or 'unfavoured' were not killed. Chance changes, even ones that would be detrimental in wild nature, could be preserved and multiplied. It was, Wallace concluded, not appropriate to extrapolate from change under controlled conditions, as with domestication, to change in nature.

How could Wallace have written a point by point refutation of Darwin's theory, which relied on extrapolation from Domestication to Nature, when Darwin's theory was not published until twenty-two months later? Surely the answer must be because Darwin had written to Wallace suggesting the idea.

We know that Darwin voluntarily took Gray into his confidence, telling him about his book and a few of his basic ideas. It has been suggested that Darwin felt Gray, being overseas, would be less likely to leak any of Darwin's ideas to anyone else. I am suggesting that Darwin felt the same way about Wallace. Wallace was even further away and was even more interested in the same ideas as Darwin than was Gray, whose main interest was in the newly launched concept of a previous Ice Age.

I am not suggesting that Darwin and Wallace conducted an extensive correspondence – Wallace's itinerary would have precluded that. I am suggesting that there is at least one letter from Darwin to Wallace missing and that in that letter Darwin outlined his idea that an extrapolation could legitimately be made from domestication to nature. Wallace may even have sent Darwin a letter indicating agreement. Whether he did or not, during that bout of fever, when he was thinking over the whole issue, he realized that Darwin was wrong. When he recovered he put his thoughts on paper to send to Darwin. He did not just address the domestication issue. After days of churning the whole thing over in his mind, he cleared his thinking by writing down all his current thoughts, reiterating parts from his Law paper and adding other parts which had become clear to him over the following three years. He concluded with the domestication issue. It was a lengthy missal, so long that Darwin thought it was a paper, which, he was most relieved to note, Wallace had not asked him to forward for publication! He would, of course, if asked!

And that is precisely what I think did happen. Darwin absorbed what he wanted, worked steadily on changes to his book (those sixty pages already discussed!) without undue haste. Then a bombshell dropped! Reflecting, Wallace, too, had decided that what he had written could, in fact, be a paper. By the next mail, exactly one month later, Darwin received another letter from Wallace and that was the one that caused all the panic!

I reconsidered whether it was possible that it was only at that time that Darwin first told Lyell and Hooker what had happened, that the '18th' letter had, in fact, been sent in June. However, if that had been the case, why was there such an unseemly haste to 'publish' Wallace's work when Darwin clearly stated in his letter that Wallace had not asked him to do so? The 'rush to publish' is the great mystery.

Darwin simply had to go public with his ideas and the delayed Linnean meeting provided the perfect opportunity. The three men were not complete rogues! They knew Darwin could not 'publish' alone. Wallace's work had to be presented, and it was, but after Darwin's.

One final point. Wallace suggested Darwin show his work to Lyell. Why? How did he know that Lyell was even aware of Darwin's book, let alone assisting him in any way he could? Answer: because Darwin had told him so!

We know that Wallace was not aware of the drama which surrounded the arrival of his paper until *Life and Letters* was published in 1887. Throughout his life, Wallace never said a bad word about Darwin. He had believed Darwin was far further advanced in his thinking and this was the position which he continued to take: "It was really Darwin's theory, not mine". The record of his demeanor at the 50th anniversary celebrations (Marchant 1916) are evidence of his continued deference to Darwin. Perhaps he destroyed the letters, when he came to realize that a cover up had occurred, as realize he must, when Mrs. Gray published Asa's *Letters*, if he had not realized before.

If everything which happened at the Linnean meeting was open and above board, why were three-and-a-half years' worth of letters exchanged between Darwin, Gray, Hooker and Lyell destroyed? (Darwin seems to have kept a few, although I suspect some of the ones allocated to this time in Darwinproject are incorrectly dated, possibly to fill in the gap.) Why did Darwin feel the need to be exonerated?

Three items were presented at that meeting – part of Darwin's 1844 *Essay*, the Gray letter and Wallace's paper. There is no question of their trying to suppress Wallace's paper. They owed Wallace an explanation of why his paper had been presented verbally rather than submitted to a journal, but that was all. There has never been any dispute regarding Darwin's writing of the 1844 *Essay*. They made no pretense that it was anything other than what it was. That only leaves the letter to Gray. If that was genuine, what was there to explain? That they had read material written by Darwin ahead of Wallace's paper? Possibly. That was certainly not usual practice. But why was three-and-a-half years' worth of correspondence between Darwin, Hooker, Lyell and Gray subsequently destroyed – and why did Darwin destroy all of Wallace's letters to him from the Archipelago, written over a period of at least five years?

It was not Darwin who did anything questionable. It was Hooker. The Hookers were an old established family, personal friends of Prince Albert, if not of Queen Victoria. Remember those letters home, written by the young Joseph Hooker on his travels, which Sir William read to Prince Albert? If Hooker's actions became known, it would be a scandal involving, not merely him, but his family! With hindsight, one can suggest that the best cover up would have been to say that, in the confusion reigning at Down that night, the wrong letter had inadvertently been handed to the servant – not the October, 1857, letter intended but one which Darwin was in the process of writing – hence it not having yet been dated. In the confusion after the event, I believe Hooker panicked and suggested Darwin make duplicate copies of the material, both for himself and for Gray. Remember, the close friendship which had existed for many, many years, between the Hooker and Gray families? Hooker would have explained to Gray and Gray would have done the 'right' thing. He knew of Darwin's long-standing interest in the subject and that Darwin had being writing a book for the past two years. He would, no doubt, have considered that preserving some degree of equivalence, if not priority, was but Darwin's due. His subsequent loyalty to Darwin, the help he gave in promoting Darwin's ideas in America over the next two decades, shows that he was fully behind Darwin.

I do not doubt that Darwin wrote those letters preserved in the archives. I merely question when he wrote them. Perhaps their water-marks, if they have any, would provide more information?

Note: I have, for some time, been in email contact with Professor Derek Partridge of Exeter University, England, who is currently working on a fresh account of Wallace's fascinationg life during his years on the Archipelago and of his friendship with the Rajh, Sir James Brooke. Professor Partridge initially rejected my hypothesis and does not accept it unreservedly, although admits that I have made a far better case than he ever thought I could.

I would like to express my sincere appreciation for his help. His lengthy responses, pointing out any and every inadequacy in my argument that he could find, were of invaluable assistance. They forced me to revisit the scenario again and again, checking every letter, every word, every date. He was always patient and his responses were always prompt and relevant. I hope these two chapters do justice to our combined effort

Chapter 17

With a Little Bit of Help from his Friends

Darwin was fortunate to be surrounded by a circle of friends who were prepared to offer him their help, support and encouragement, even though they received very little in return. Not that they needed help. Sir Charles Lyell, John Henslow, Sir Joseph Hooker and Thomas Huxley, among others, were authorities in their own right in their own fields and all achieved fame and acclaim by their own efforts. Yet without them, and many others upon whom Darwin relied (mostly by means of correspondence), not only would Darwin never have accumulated the large store of facts with which he bolstered his arguments, it is unlikely that he would ever have completed the manuscript of his theory to the point of publication, nor would his book, once published, have received the acclaim that it did.

Lyell's influence upon Darwin has already been covered (Chapter 10). This chapter will look at the role played by Henslow, Hooker, Huxley and Spencer in the establishment of evolution as scientific theory and natural selection as its most likely process.

John Henslow (1796–1861)

At the time Charles Darwin entered Cambridge University, John Henslow was Professor of Botany, having previously held the Chair of Mineralogy. He was keen to reform the teaching of science subjects at University at a time when university teaching was mainly classical, theological or medical. Although botany did not become a degree course in its own right until shortly before Henslow's death, it was his approach to the subject which paved the way for reform. Henslow endeavoured, as far as possible, to replace, or at least supplement, theoretical lecturing with practical experience. He organised field trips, some merely for a day, others for longer periods, to study geological formations and associated vegetation. Once a week he held 'open house' at his home where informal discussion took place between lecturer and pupils, and anybody else who was interested in attending. Charles Darwin was one of those interested people. Darwin's interest in entomology found a ready place within the circle of Henslow's botanical interests and the two men formed a friendship which was to last the rest of Henslow's life.

It was Henslow who recommended Darwin to Captain Fitzroy as naturalist on *HMS Beagle*. It was Henslow who gave Darwin the first volume of Lyell's *Principles of Geology* to read on his voyage, its principle of gradual change being crucial to Darwin in the development of his theory. It was Henslow who received the boxes of samples Darwin sent back during the course of his voyage It was Henslow to whom Darwin turned for help on his return when he realised that no one was interested in his specimens, not even the museums, which were overburdened with specimens sent to them by voyagers from all over the world. In retrospect, it was fortunate that Darwin was forced to do so much of this work, but without Henslow's help, Darwin's boxes may well have rotted in some cellar.

Darwin continued to rely on his good friend for help and advice. On 27th June, 1855, Darwin wrote to Henslow, asking whether he considered certain plants to be species or varieties, and why? He asked Henslow to collect seed heads from certain plants so that he could count the number of seeds and lastly, but by no means least (Barlow 1967: 175):

> ... busy as you are, can you forgive these several requests? ... I fear I am not a little unreasonable ... I want to know whether you would & this is the most troublesome job, (though I think it sounds more troublesome than it is) sometime, say in winter (or whenever you have the most leisure)[added] read over the names in the London Catalogue of Plants (& I wd send my copy) pencil in Hand, & mark with cross, all those species, which you believe to be (really) [added] species, but which are close species; - taking some definition for a "close species", as a form, which even to a (good) [added] Botanist is a little troublesome to distinguish, or which you conceive possible, though not probable, that further research will prove only to be varieties. I am really anxious for this, but I cannot explain my motive, otherwise it might unconsciously cause you to influence the result. I do not think it would take up (much) [added] more time, than (going) [del.] reading slowly over the names.

Self-deprecation was a technique Darwin used frequently and successfully, not only in correspondence, but also in *The Origin*. By pointing out problems and then suggesting that they were not quite as serious as at first supposed, Darwin disarmed opposition and brought people round to his way of thinking (Hull 1973). From his letters alone, it is difficult to assess just how truthfully he wrote, but his biographers were unanimous in claiming Darwin to have been very diffident, even humble, in his personal behaviour (for example, see Darwin 1887/1969; Desmond and Moore 1991), so it may be assumed that he wrote as he spoke. Darwin was very appreciative of Henslow's help and continued to consult Henslow regarding botanical questions until Henslow's death in 1861.

Sir Joseph Hooker (1817–1911)

The Hookers were an old, well-respected English family, with a recorded genealogy stretching back some four hundred years. Sir Joseph was born in Suffolk but by the time he was five, the family had moved to Glasgow when his father, Sir William Hooker, took up the position of Professor of Botany at the University of Glasgow. By the age of seven, he was accompanying his father to the University where he 'sat in' on lectures (Huxley 1918).

After leaving school, Hooker took up the study of medicine, qualifying just in time to take the position of Assistant Surgeon/Naturalist on board *HMS Erebus*, which was to sail to the Antarctic. He was thus on board the ship which was to discover that, unlike the Arctic, the Antarctic comprised a large mass of land. He was to make two voyages south, the first to Tasmania and the Antarctic, the second to New Zealand and the Cape of South Africa.

While Hooker was on his first voyage, his father was appointed Director of the Gardens at Kew. When Hooker returned to Britain at the end of 1843, he had more opportunity than ever before to immerse himself in practical botany. At this time Henslow forwarded to Hooker Darwin's collection of flora from the Galápagos Islands, which had not until that time been examined. Thus started the association between Hooker and Darwin, which was to ripen into a life-long friendship.

Unlike Darwin, Hooker loved being at sea, and remained in the service of the Admiralty for many more years, albeit as a naturalist, having shed the unwanted position of Assistant Surgeon. After the voyage to New Zealand, Hooker made two voyages east, to India and the Himalayas, returning to England in 1851, at which time he married Henslow's daughter, Frances, to whom he had been engaged for three years. His first task on his return was writing up his work. He was able to retire from the Royal Navy in 1853 when a position became vacant at Kew as assistant to his father. On his father's death ten years later, he took over the position of Director.

Two of the people upon whom Darwin relied so much for help and information were botanists. One of the main things which separated Darwin's book on evolution from any forerunner was the inclusion of so much evidence from botany, as well as from zoology. Plants were known to vary far more readily than animals, both in nature and under domestication, being able to reproduce by roots and shoots, by cuttings and corms; they were not dependent on sexual reproduction, as were most of the animal kingdom. On the whole, sexual reproduction tended to stabilise species, since productive mating could only take place between pairs which were very similar. The boundary between species and variety was far harder to determine with plants than animals and this problem interested Hooker as well as Darwin. Hooker was the only person to whom Darwin confided his views in the early stages of their formation by showing him his 1844 manuscript. It was a further ten years before Lyell was made privy to Darwin's secret and that only after Lyell had drawn Darwin's attention to Wallace's paper. Hooker does not appear to have had a full understanding of Darwin's theory until *The Origins* was published. On 21st November, 1859, Hooker wrote to Darwin (Huxley 1918: 510):

> ... to thank you for your glorious book. What a mass of close reasoning on curious facts and fresh phenomena; it is capitally written and will be very successful. I say this on the strength of two or three plunges into as many chapters, for I have not yet attempted to read it ... How different the book reads from the MS.

And in a further, undated, letter written early December, 1859 (Huxley 1918: 510):

> ... I have not yet got half through the book, not from want of will, but of time — for it is the very hardest book to read to full profit that I have ever tried; it is so cram full of matter and reasoning. I am all the more glad that you have published in this form, for the 3 vols, unprefaced by this, would have choked any Naturalist of the XIX century and certainly have softened my brain in the operation of assimilating their contents. I am perfectly tired of marvelling at the wonderful amount of facts you have brought to bear, and your skill in marshalling them ... Somehow it reads very different from the MS and I often fancy that I must have been very stupid not to have more fully followed it in MS.

Hooker described the book as 'clear' but 'the very hardest book to read' due to all the facts assembled.

By the time Darwin finally published his theory, Hooker was an acclaimed botanist who had published books identifying many new species and who held an eminent position at the Royal Botanical Gardens at Kew, of which his father was Director. Although Hooker's review of *The Origin* in *The Gardeners Chronicle* (31st December, 1859) was anonymous, as were most reviews, its strong endorsement of Darwin's views was no doubt influential:

> ... we have risen from a perusal of Mr. Darwin's book much impressed with its importance ... It is a book teeming with deep thoughts on numberless simple and complex phenomena of life; that its premises in almost all cases appear to be correct; that its reasoning is apparently close and sound, its style clear ... It is also a perfectly ingenuous book ... whatever may be thought of Mr. Darwin's ultimate conclusions, it cannot be denied that it would be difficult in the whole range of the literature to find a book so exclusively devoted to the development of theoretical inquiries, which at the same time is throughout so full of conscientious care, so fair in argument, and so considerate in tone.

The above favourable, review did not indicate unqualified support. In a letter to Darwin, dated 20th January, 1860, Hooker wrote (Huxley 1918, vol.1: 511):

> I finished Geolog. Evidences Chapters yesterday ... You certainly make a hobby of Nat. Selection and probably ride it too hard ... that is a necessity in your case. If improvement of the creation by variation doctrine is conceivable, it will be by unburdening your theory of Natural Selection, which at first sight seems overstrained, i.e. to account for *too much*. *(Italics in original)*

Hooker's glowing review had been written before he had finished reading Darwin's book! (If Hooker was prepared to write a review of a book he had not read, would he have been above rewriting a letter before presenting it to the Linnean Society? Probably not!) Notwithstanding Hooker having perused the *Foundations* essay, having been Darwin's closest confidant and having been present at the joint presentation of the Darwin/Wallace papers, this letter clearly shows that Hooker had not fully appreciated Darwin's theory. In May of 1860, Hooker wrote to Henslow regarding a recent meeting (Huxley 1918, vol.1: 512-513):

> Sedgwick's address last Monday was temperate enough for his usual mode of attack ... I got up, as Sedgwick had alluded to me, and stuck up for Darwin as well as I could ... I do not disguise my own opinion that Darwin has pressed his hypothesis too far.

Darwin's friends rallied to his defence motivated as much by their high regard for him as a person as for his hypothesis, which they clearly did not totally support. Although Hooker was yet to be appointed Director of the Royal Botanical Gardens and inherit the family title, at the time of the publication of *The Origin*, he was an established authority in his own right, an elite member of the informal academic circle which governed much of the thinking then current on the subject of evolution.

Thomas Huxley (1825-1895)

Born in Ealing, West London, in 1825, of an old, but not wealthy family, Huxley had one thing in common with both Darwin and Hooker: all three launched their careers by means of at least one long sea voyage of discovery. For Huxley, the ship was *HMS Rattlesnake*, and its voyage to Australia and southern New Guinea to map the north-eastern coast of Australia was to be his only one. Nevertheless, the voyage was to have a profound impact on his life.

The young Huxley was an avid reader. His father being a school teacher, many of the books he read were of an educational nature, such as Hutton's *Geology* (Huxley 1900). Huxley enjoyed investigating how things worked and was interested in anatomy and physiology for this reason.

The marriage of both his older sisters to doctors enabled him to become assistant to a London doctor, whose practice was among the poor in the dock regions of London. Years later he would comment that the slaves he saw on his travels seemed happier with their lot than the poor of England, and these early memories were a driving force in his later campaign to bring education to working men - and women.

Huxley was awarded a scholarship to study medicine at Charing Cross Hospital, where he completed his medical studies. He took an interest in botany. He entered a competition at the age of 16 and won second prize. Having determined to enter (Huxley 1900: 17):

> I set to work in earnest ... I worked really hard from eight or nine in the morning until twelve at night ... A great part of the time I worked til sunrise.

This was typical of the (manic) enthusiasm and dedication which Huxley brought to all his endeavours and which were to ensure him a place of high esteem both among his colleagues and the general public.

On completing his medical studies, Huxley applied to the medical service of the Royal Navy for a position and was eventually appointed to *HMS Rattlesnake* as assistant surgeon. There was an understanding that he would be able to carry out scientific work as time permitted. Like Darwin, he determined to make "one grand collection of specimens" which would be deposited at the British Museum, or some other public place. His enjoyment of the voyage was marred by recurrent bouts of severe depression, possibly schizophrenia (McCalman 2009).

To Liddle (1991) we owe knowledge of an interstinig event in which Thomas Huxley was a participant.

On 1st October, 1849, *HMS Rattlesnake* anchored at the most northerly tip of Australia, Cape York. There the young Huxley was to participate in the saving of the life of a young Scottish girl who had been captured by the Kauraregas tribe of Torres Strait Islanders some five years previously. In 1837 the Crawford family had arrived in Sydney from Scotland. In 1843, their 15 year-old daughter, Barbara eloped with William Thomson. The pair married at Moreton Bay (near Brisbane) and it was there they met a formerly ship-wrecked sailor who told them that his wrecked ship, with tons of whale oil on board, was still stranded on a reef in Torres Strait The three set out to retrieve the valuable cargo, sailing north in a small boat, with three other crew members.

They never found the wreck, but their own boat foundered. Barbara saw her companions drown as they tried to swim to shore, while she clung to the wreck. The next day she was spotted by some Kaurarega warriors, notorious head-hunters, but these men were not interested in beheading a young female, whom they took back with them as a 'trophy'. Their chief declared Barbara to be the returned spirit of his daughter, who had recently died and she was welcomed into the tribe, while always remaining a prisoner. She was never allowed to accompany other tribe members on visits to other islands and was hidden when a 'white' ship was in the area, less she be spotted and rescued. (There is no suggestion that she ever had any sort of sexual relationship with any tribe member.)

October 1849 found Barbara in poor health. She had developed ophthalmia and was nearly blind in her right eye. Perhaps this contributed to her fall on the rocks, which had resulted in a severe infection in her knee and her nose, both of which had been lacerated. A message was received telling the tribe of the arrival of the *Rattlesnake*, and two other boats, one of which was the mail ship. Her original rescuer, Tomogugu, wanted to take her with him to greet the boats, a suggestion opposed by other tribe members. As her rescuer, Tomogugu still technically owned her. Eventually all the tribe members set out together, not to fight off the 'invaders' but to greet them and to be the recipients of the gifts the white people had brought with them - clothes, biscuits, and - most prized of all - metal tools, axes, spear heads and so forth.

The Islanders could not sleep on the mainland, so they camped on a small island just off-shore and there Barbara had to stay, separated from her potential rescuers by a strip of crocodile invested water. Salvation came when gun shots were heard announcing that the white people were going into the bush to shoot birds, an opportunity not to be missed! All the Islanders scrambled into their canoes and, with members of the mainland Aboriginal tribe, were soon seen heading off into the bush. Barbara was left behind with the few considered too old to hunt – and Tomogugu. As soon as the hunting party were safely out of sight, Tomogugu bundled the 'left-behinds' into his canoe and headed for the mainland. He assisted Barbara, who had trouble walking with her infected knee, half-a-mile along the coast to the beach where the sailors were ashore. No reason was given for this action but one may surmise that Tomogugu realized that Barbara needed 'white-man's medicine' and saved her life for the second time.

Barbara was taken aboard *Rattlesnake*, much to the chagrin of the returning Islanders, where she would have been treated by *Rattlesnake's* Assistant Surgeon, Thomas Huxley. They left the Straits at the beginning of December, arriving in Sydney 3rd February, 1850, where Barbara was re-united with her family. Presumably Barbara made a full recovery under Huxley's care, because the next year she re-married and lived in Sydney until she died in 1912 at the age of 84.

From Sydney, Huxley wrote to his sister that he had already sent a paper to the Linnean Society on *Physalia* (Portuguese Man-of-War jelly-fish) and was in the process of writing two more papers on *Diphydæ* and *Physophoridæ*. These papers were well received and were to be the launching point of his career. They show Huxley's interest in evolution, modification, diversification and common descent.

Huxley returned home at the end of 1850 and wrote up his papers, as a result of which he was elected Fellow of the Royal Society at the young age of 25. He was later to be Secretary, then President, and in 1852, the Society honoured him with the award of the Royal Medal in Physiology.

Notwithstanding these honours, Huxley found it impossible to earn a living as a scientist, and seriously considered returning to Sydney 'to set up shop as a trade' as this seemed to be his only hope of being in a position to marry Miss Henrietta (Nettie) Heathorn, to whom he had become engaged while in Sydney, both being aged 22. The Heathorn family returned to England in 1855, and finally, after a six year engagement, the not so young lovers were eventually married, and were to remain so for forty years.

It is not possible here to detail all the positions held by Huxley. One can scarcely turn a page of Huxley's biography (L. Huxley 1900) without learning of another position, another endeavour, in which he had become involved. Huxley had a passion for improving education, particularly by increasing the science content. He campaigned for universal education and believed society had a responsibility to provide a good technical education for those persons not suited to a professional career. Huxley lectured to the Working Men's Club, as well as women's groups. He was a member of a number of commissions, not only relating to education, but also to the fisheries, he being concerned about over-fishing. His growing family and his modest pay ensured that he was financially 'straitened', at least during the time his family were growing up. When his brother died, Huxley was forced to sell his Royal Society Medal for £50 (the value of the gold) to settle his brother's debts.

Huxley became rector of Aberdeen University and received honorary degrees from Oxford, Cambridge and Trinity College, Dublin, quite an achievement for someone who had never attended University. Huxley continued to contribute scientific papers and after his retirement collected his most important essays together into ten volumes. The subjects ranged across Hume, Descartes, Jewish/Christian Tradition, Science and Education, Darwinia, Man's Place in Nature, biology, ethics and much more (Huxley 1893-1917/1968). Huxley was a fierce opponent of the Church, which he did not feel represented what may once have been the teachings of the Nazarene. However, he was not an atheist, inventing for himself the term 'agnostic'. Rather surprisingly, Huxley argued for the retention of Bible Study in schools. Without training in morals and ethics, he feared children might grow up to be 'prigs'. In keeping with his principles, Huxley declined a knighthood, but accepted a position in the Privy Council, as this was a working position which brought him into contact with the most influential people in the country.

Huxley was a true friend to Darwin, but, bearing in mind his wide range of interests, especially in the areas of education and social reform, and his propensity for publishing papers, one cannot help but wonder, if circumstances had been different, whether he might not have found an even more congenial friend in Wallace?

The 'X-Club'

Huxley and his close friends founded the 'X-Club', although there were nine, not ten, members. They met monthly to dine before attending the meetings of the Royal Society. They socialized at other times; their wives formed their own club: 'The Yves'. The two groups combined for family outings and holidays, yet the purpose of the X-Club was far from benign. Huxley had been frustrated at his first attempt to see Darwin awarded the prestigious Copley Medal, by what he and his friends perceived as the

'Closed Club' of the Upper Classes, strongly influenced by the Church, which was by no means unreservedly supportive of Darwin's work. 'Oxbridge' graduates influenced religious, academic and political life in Victorian England. The purpose of the X-Club was to undermine this influence (McCalman 2009: 355-357):

> The Xers made friendship a machine of war ... They were a meritocratic 'conspiracy' ... together they were unstoppable. They nominated each other for awards, refereed each other for jobs, published each other's work, sponsored each other's lecture tours, awarded each other grants, and circulated each other's achievements ... The Xers also founded or dominated major scientific journals. Huxley made the Reader 'an organ' of the Darwinists from 1863, and in November 1869 co-founded Nature, which became one of the most prestigious research journals in the world ... They set out to gain access to every major establishment of national power — government, parliament, universities, schools, the Admiralty, the arts, the Church ... [They] allied themselves with liberal Anglican clergymen ... to promote science in schools, appoint liberal clergymen to universities, and resist the inquisitional attacks of Church Tories.

Huxley embraced the concept of evolution but never accepted that natural selection accounted for more than micro-evolution. He enthusiastically supported Darwin, not because he thought that Darwin's theory was correct, but because it allowed the premise that God was not necessarily the (sole) cause of life as we know it, indeed that there may be no need to acknowledge the existence of God at all.

People naturally congregate with others of like mind, talk on topics of mutual interest and possibly discuss methods and means whereby their ideas may be promoted. This is not only natural, it is desirable, since open debate is healthy. However, a conspiracy occurs when a small group of people plot to further their aims in a covert and underhand manner, involving deception. There was a clear accusation of conspiracy made by McCalman that Huxley and his friends had their own agenda and that Darwin's theory provided a suitable vehicle by which that agenda might be advanced. If I have understood McCalman correctly, this is not only what happened, it was the prime purpose of the X-Club.

Man's Place in Nature

Perhaps the greatest service Huxley gave Darwin was the publishing in 1863 of a small book entitled *Evidence as to Man's Place in Nature* (Huxley 1863/1959). The purpose of this book was to put before the general public, as well as the scientific community, such evidence as could be gleaned from comparative anatomy regarding the place of human beings within Nature.

Huxley compared the anatomy of apes and humans, with particular reference to the gorilla and the chimpanzee, concluding the gorilla most closely resembled humans. However, he was at pains to point out that there was not a single bone in the body of the gorilla which could be mistaken for the corresponding bone in a human and that there was now no intermediate link bridging the gap between humans and apes. He concluded that humans might have originated by gradual modification from a man-like ape or "as a ramification of the same primitive stock as those apes" (p. 125). Like Cuvier before him, Huxley's 'conclusions' appear to have been based more on a pre-determined agenda which he was resolved to follow than on the evidence produced by his own work.

Huxley also made a comparative study of the skulls of the Engis and Neanderthal fossils from the valleys of Meuse (Belgium) and Neander (Germany) respectively. Scientific study of these remains was still in its earliest stages. Huxley's views on the Neanderthals helped to shape European thought on this (at that time) earliest of human beings. He concluded that both these skulls were within the range of modern humans (pp. 181-183).

Huxley (1893-1917/1968) saw three problems with Darwin's theory: saltation, hybridisation and speciation.

Saltation

Huxley was not happy that Darwin had specifically excluded saltation from his theory. His later writings placed less emphasis on this aspect, possibly because Huxley realised that it was subsumed under hybridisation and speciation. Nevertheless, he always remained aware of 'discontinuities' which, in his opinion, could not be explained by Darwinian gradualism.

Hybridisation

Like all naturalists of his time, Huxley was puzzled by the ability of apparently closely related species, such as the horse and the donkey, to produce healthy offspring which were nevertheless infertile. Huxley acknowledged that external conditions could have had a greater influence on the ability to breed than had yet been realised and accepted, conditionally, Darwin's idea that the correct breeding conditions for mules may not yet have been found. He was not entirely happy with this explanation, but if this were the only difficulty with Darwin's theory, then he would not consider that difficulty insurmountable.

Speciation

A cornerstone of Darwin's theory was that varieties eventually became species, yet some five thousand years of recorded domestication, and the production of countless varieties, had yet to produce one instance of two varieties, known to be descended from a common progenitor, which had become two completely separate species, i.e. were not inter-fertile. Bearing in mind the vast number of varieties of sexually reproducing animals which had been produced under domestication, at least one should be classifiable as a separate species if Darwin's explanation was the correct one. Until this happened, Huxley maintained that natural selection was an hypothesis, not a theory.

Herbert Spencer (1820-1903)

Herbert Spencer was not one of Darwin's close circle of friends, although he was a great friend of Huxley and was considered one of the greatest philosophical minds of the nineteenth century. The son of a school teacher, he himself never went to school or university (Spencer 1861/1966).

Spencer's first paid employment was as a worker on the London-Birmingham railway. This sparked his interest in geology and fossils. Reading Lyell's *Principles*, Spencer was unconvinced by Lyell's criticism of Lamarck's theory of evolution, becoming rather a convert to the idea.

In 1852 Spencer published a lengthy paper, *A Theory of Population, deduced from the General Law of Animal Fertility*, which concluded with a discussion of the effect of over-population. Spencer compared the easy life of Islanders, whose climate ensured the production of sufficient foods for their needs, with the hard life of Europeans, who for centuries had had to work and strive to produce sufficient for their growing population during harsh winter conditions. He claimed this had resulted in Europeans, not only working hard physically, but mentally, using and developing their minds as they explored scientific ways, such as chemistry and mathematics, to increase food production. (His argument failed to explain why the Islander population had not expanded to the limit of its food supply.) Spencer argued that the constant exercise of skill, intelligence and self-control would lead to a higher form of humanity. He even argued that "Difficulty in getting a living is alike the incentive to a higher education of children" and concluded that "it unavoidably follows, that those left behind to continue the race are those in whom the power of self-preservation is the greatest".

Spencer's conversion to the concept of gradual evolution of species rather than their coming into existence by means of millions of acts of special creation had been publicly expressed in an article, *The Development Hypothesis*, published in *The Leader* on 20th March, 1852. Spencer argued that it was unreasonable for opponents of gradual evolution to object that supporters could not explain precisely how this happened on the grounds that they could not explain how special creation happened either. He challenged supporters of special creation to explain how a new species was formed:

> Is it thrown down from the clouds? Or must we hold to the notion that it struggles up out of the ground? Do its limbs and viscera rush together from all the points of the compass? Or must we receive the old Hebrew idea, that God takes clay and moulds a new creature?

We may never have witnessed a species mutate but we had never witnessed a species being specially created either. Spencer argued that even to show how a new species might *conceivably* have come into existence under the *Development Hypothesis* was an improvement on the doctrine of Special Creation which was unable to suggest any conceivable method at all. Spencer argued that developmental theorists were able to show organisms changed when subjected to modifying influences.

Spencer then argued that it was unreasonable to suppose it impossible for humans to have developed from a single-celled organism over a period of many millions of years when a fully grown multi-celled human being regularly developed from a single-celled organism over a period of a mere twenty years.

In 1857, Spencer published an Essay on the subject of evolution, *Progress: Its Law and Cause*, but, like Lamarck and others before him, failed to suggest a mechanism by which evolution might have taken place. Like Lamarck, he did not feel it was incumbent upon him to do so since he was not arguing the details, but the concept. Suggesting a possible method was to be Darwin's and Wallace's great contribution the following year. However, Spencer did use the word 'evolution' in place of the more generally accepted 'epigenesis'.

Spencer dealt more fully with the subject of evolution in his *First Principles of a New System of Philosophy*, published in 1862.

In 1864, Spencer published *Principles of Biology*. It was in this book that he first used the term 'survival of the fittest': "This survival of the fittest, implies multiplication of the fittest" and "This survival of the fittest, which I have here sought to express in mechanical terms, is that which Mr. Darwin has called 'natural selection', or the preservation of favoured Races in the struggle for life". Later a collection of Spencer's *Essays* was published in book form, *The Man versus the State,* (1916) in which Spencer wrote: "Thus by survival of the fittest, the militant type of society becomes characterized by profound confidence in the governing power, joined with a loyalty causing submission to it in all matters whatever". Initially, Darwin resisted the term but, eventually, was forced to concede its use, entitling Chapter 4 of the 5th Edition of *The Origin* "Natural Selection, or the Survival of the Fittest". Spencer, correctly, attributed a more 'militant' undertone to this phrase than that present in the more gentle 'natural selection'. It was the 'survival of the fittest' implication of Darwinism which underpinned the Social Darwinism of the twentieth century, with such disastrous results.

Some, for example Ruse (2006: 19), attribute to Spencer the independent discovery of natural selection, which would raise to three the number of people who had made the same discovery by 1857: Spencer, Darwin and Wallace. Spencer's work, appearing as it did so closely before and after the publication of *The Origin,* was of great assistance to Darwin in establishing the concept of 'natural selection by the survival of the fittest' in the minds of the Victorian public. However, Spencer by no means accepted Darwin's theory unreservedly. Indeed, his doubts seem to have increased as time passed.

In 1893, a long article by Spencer was published in the *Contemporary Review* under the heading *The Inadequacy of Natural Selection.* Spencer contended that the process of evolution by natural selection was applicable only to features which were of 'life or death' importance. While fossil evidence indicated that human jaws had decreased in size over the millennia, Spencer asserted that there was no reason to suppose that a person with a slightly smaller jaw ever had greater life/reproductive potential than another person with a slightly larger jaw and that there was, therefore, no scientific support for this type of argument.

Spencer argued that the long front legs of the giraffe had come about in synchronicity with changes in other parts of the giraffe's body. The bone, muscle, tendon, etc., of the back and hind legs of the giraffe would have had to adjust to accommodate the longer forelegs to allow the giraffe to run, even if with a 'grotesque gallop'. Somewhat surprisingly, Spencer did not point out that the giraffe was the only mammal whose forelegs were so long that they had to be splayed outwards in order to allow their possessor to drink. Common sense would seem to suggest that the giraffe's forelegs should have stopped growing, or its neck should have continued to grow, to allow the animal to drink, rather than that the shoulder girdle should adapt in this unique way. (I have yet to see this point addressed in the literature.) In answer to people who claimed that co-operative parts changed together, Spencer pointed out that this would apply just as much in reverse, yet the blind cave crabs of Kentucky had lost their eyes but not the stalks carrying them.

Spencer made reference, as had Darwin, to the unusual case of a male quagga (a South African animal related to the zebra) owned by the Earl of Morton, which was mated with a black mare, which, not surprisingly, gave birth to a foal "with decided indications of her mixed origins" (p. 448). The mare was sold to Sir Gore Ouseley, who mated her with a black Arabian horse, but the first two foals of this later union were both bay in colour, like the quagga, and had dark stripes along the ridge of the back and across both the fore and back legs. A second case occurred with a pig. A black-and-white sow had been mated with a wild boar of chestnut colour and in some of her litter the chestnut colour of the boar was strongly marked. The sow was subsequently mated with a black-and-white domestic boar, but some of the litter were of chestnut colour.

These two cases proved to Spencer's satisfaction that Weismann (see Chapter 20) was wrong in alleging that somatic cells were completely independent of, and separate from, reproductive cells (p. 454). Spencer held that cells were interlinked by threads of protoplasm, which allowed information to be transmitted throughout the organism and thus characteristics acquired during the life time of an individual, even through a previous mating, could be passed on to the next generation. Spencer insisted that his comments were not to be taken as being contra Darwin. Rather they supported Darwin, who fully recognized and often insisted on the inheritance of acquired characteristics as an integral part of his theory. An alternative explanation was offered by Romanes (1893), that some sperm from the earlier mating survived in the female body and was responsible for the later mating.

The two people considered of greatest intellectual stature during the second half of the nineteenth century, Huxley and Spencer, both accepted evolution, but neither was completely satisfied with Darwin's theory of how it had come about. Huxley was concerned because natural selection did not account for major changes (speciation), while Spencer was concerned because it did not account for minor features, which would not have influenced survival or procreation.

Spencer (1879) was also interested in the evolution of societies. He believed that the more people lived together in small spaces (i.e. towns), the greater was the amount of (mental) energy circulating among them. Spencer believed that increased mental stimulus was the reason that town dwellers ('civilised' people) had increased intelligence. That having happened, it was inevitable that, as the human population continued to grow, the superior races would replace the inferior.

Spencer portrayed the path to civilisation as a progression from cruelty and callousness to kindness and caring. He believed there was no need to impose laws upon humans in regard to their behaviour because 'good behaviour' gave pleasure, while 'bad behaviour' resulted in pain. Humans would automatically do what was good, because it was also pleasurable. The torture inflicted upon unfortunate victims, especially during Tudor times and under the Inquisition, is evidence to the contrary. This was yet another example of prejudice (in favour of one's own opinion, people, way of life) over-riding truth.

The Oxford Debate

Darwin's friends did not only support him in print, they supported him verbally as well.

Debating was a very popular Victorian pastime, especially on a Saturday afternoon. Some addressed serious issues, others were more light-hearted and debated topics such as "It is better to have loved and lost than never to have loved at all".

There was an established format. One person proposed a motion, presenting a prepared speech. Another person would then present the opposing view, also by way of a prepared speech. The third person to speak was the seconder of the motion. He would not only reinforce the most important points made by the Proposer, his main task was to attack the points made by the person who had opposed the motion. Since he had no way of knowing exactly what these points might be, his speech was not prepared in advance. Speaking "off the cuff" (which was where they sometimes wrote reminder notes to themselves while listening to the previous speakers) the Seconder's role was extremely important. The audience was by now aware of the main issues and the second speaker's points were often more easy to assimilate and evaluate. The final speaker, whose main task was to demolish the defence presented by the Seconder of the motion, had the advantage of being able to address points raised by all three previous speakers and, again, his reply was 'off the cuff' and often very influential.

In 1860 a famous, or possibly infamous, debate took place at Oxford, during which Thomas Huxley supposedly routed Bishop Samuel Wilberforce by his brilliant defence of Darwin's theory. The popular account of this event held that Wilberforce (proposer of the motion contra Darwin) had asked Huxley whether he traced his ancestry back to the ape through his grandfather or his grandmother? This may seem innocuous by today's standards but, in Victorian times, to make any comment which could be seen as denigrating the female sex was absolutely 'out of order'. Apparently there was uproar. Some women were reported to have fainted! (It was a hot afternoon and the room was overcrowded.) In his biography, Sir Francis (1902: 238-239) gave a summary of the events which took place, based on notes taken by W. H. Freemantle at the time:

> The Bishop of Oxford attacked Darwin, at first playfully but at last in grim earnest ... The Bishop was declaring with rhetorical exaggeration that there was hardly any actual evidence on Darwin's side ... But he passed on to banter: 'I should like to ask Professor Huxley, who is sitting by me, and is about to tear me to pieces when I have sat down, as to his belief in being descended from an ape. Is it on his grandfather's or his grandmother's side that the ape ancestry comes in?' And then taking a graver tone, he asserted in a solemn peroration that Darwin's views were contrary to the revelations of God in the Scriptures. Professor Huxley was unwilling to respond but was called for ... [made several points] ... Lastly as to the descent from a monkey, he said: 'I should feel it no shame to have risen from such an origin. But I should find it a shame to have sprung from one who prostituted the gifts of culture and of eloquence to the service of prejudice and of falsehood.'

According to legend, it was Huxley's response which won the day, but Gould (2002) held that such reports had been misremembered, misrepresented and misreported.

The well-known American writer, Stephen J. Gould, researched the affair for one of his monthly articles in *Natural History Magazine*. According to Gould (1991: 385-401), there was no official record of the debate, since no stenographer was present. Recollections of what occurred varied. Gould based his conclusions on two accounts which he believed to be reasonably reliable. The first was a letter written by Balfour Stewart, Fellow of the Royal Society of the Kew Observatory (p. 389):

> ... the Bishop of Oxford and Huxley fell to blows ... I think the Bishop had the best of it ... The Bishop said that he had been informed that Prof. Huxley had said he didn't care whether his grandfather was an ape [sic for punctuation] now he (the bishop) would not like to go to the zoological Gardens and find his father's father or his mother's mother in some antiquated ape. To which Prof. Huxley replied that he would rather have for his grandfather an honest ape low in the scale of being than a man of exalted intellect and high attainments who used his power to pervert the truth.

The second source quoted by Gould was the memory of Canon Farrar, as communicated to Leonard Huxley (Thomas Huxley's son) nearly 40 years later (p. 396):

> ... he (Bishop Wilberforce) rhetorically invoked the help of feeling ... and said (I swear to the sense and form of the sentence, if not to the words), "If anyone were to be willing to trace his descent through an ape as his grandfather, would he be willing to trace the descent similarly on the side of his grandmother?" It was (you see) the arousing of antipathy about degrading women to the Quadrumana. It was not to the point, but it was to the purpose ... Your father's reply ... showed that there was a vulgarity as a folly in the Bishop's words; and the impression distinctly was ... that the Bishop had forgotten to behave as a gentleman ... the victory of your father was ... the fact that he had got a victory in respect of manners and good breeding ... You must remember that the whole audience was made up of gentlefolk, who were not prepared to endorse anything vulgar.

Gould further cites Hooker (Huxley 1918, vol.1: 596) as having told Darwin that "Huxley answered admirably and turned the tables, but he could not throw his voice over so large an assembly, nor command the audience; and he did not allude to Sam's weak point nor put the matter in a form or way that carried the audience". Two further speakers took the floor before Hooker was invited to speak on evolution from the botanical point of view. According to Hooker, it was he who gave the lively rebuff to Bishop Wilberforce now so frequently attributed to Huxley (Huxley 1918, Vol.1: 526-527; Gould 1991: 393):

> My blood boiled ... I swore to myself that I would smite the Amalakite. Sam, hip and thigh ... There and then I smashed him amid rounds of applause ... I hit him in the wind and then proceeded to demonstrate in a few words (1) that he could never have read your book and (2) that he was absolutely ignorant of the rudiments of Bot (botanical) science .. Sam was shut up ... had not one word to say, and the meeting was dissolved forthwith.

Hooker's account receives support from Cannon Farrar's memories given to Leonard Huxley (Gould 1991: 396):

> The speech which really left its mark scientifically on the meeting was the short one of Hooker ... no one really contributed any valuable point to the opposite side except Hooker ... but that your father had scored a victory over Bishop Wilberforce in the question of good manners.

There is no contemporary account of Huxley's famous whispered aside in which he is reported to have said "The Lord hath delivered him into my hands". This appeared for the first time in 1892, more than thirty years after the event, in a short biography of Darwin published by Sir Francis, to which Huxley contributed a letter, and was repeated when Leonard Huxley published his biography of his father in 1900. Bearing in mind the biblical

211

tone of Hooker's self-reported remarks (Huxley was a 'non-believer' after all), it seems possible that if the remark was ever made, it might have been made by Hooker.

After the four speakers had presented their positions, debate would have been thrown open to the floor. Members of the audience would have been invited to ask questions of the speaker. Some would have taken the opportunity to make short speeches of their own. Sir Francis' account mentioned others who spoke from the floor, among them Admiral (formerly Captain) Fitzroy who stated that he had often participated in discussions with old comrades from the *Beagle* which entertained views contrary to the first chapter of *Genesis*. Admiral Fitzroy was a religious man and it is interesting, not only that he attended the debate, but that he spoke up on behalf of his old 'ship-mate'.

I have concluded that there is probably truth in all the accounts. Both Huxley and Hooker vigorously defended Darwin and whose argument was the most effective would have been a matter of opinion. I tend to agree with Gould that the most passionate defence of Darwin may well have come from the unrehearsed speech by Hooker, if for no other reason than that his more informal participation in the debate would have allowed him more latitude. Be that as it may, Darwin's friends were speaking up on his behalf whenever and wherever they could.

Chapter 18

Darwin's Protegé

George Romanes (1848-1894)

We pass now to a consideration of the work of the next generation of evolutionary theorists, since George Romanes was but a boy when Darwin's book was published. Darwin had been as interested in the evolution of the mind and the emotions as he was in that of the body and his theories were the impetus for the serious study of human and animal psychology. It was this aspect of evolutionary theory which originally interested Romanes, his first three works being *Animal Intelligence* (1882), *Mental Evolution in Animals* (1884) and *Mental Evolution in Man* (1888). After reading a contribution which Romanes had made to *Nature*, Darwin invited Romanes to Down House. It was after this meeting with Darwin, and having formed a close friendship with him, that Romanes turned his attention to physical evolution, which he did as much to help the aging Darwin as he did to further his own interests.

Romanes assisted Darwin with some of his experiments undertaken with a view to supporting Darwin's theory of *pangenesis*. A comprehensive representation of the correspondence which passed between them, as well as accounts of Romanes' visits to Down, was given by Romanes' wife, Ethel, in her *Life and Letters of George John Romanes* (E. Romanes 1896) published after his death. The first of Romanes' three volumes on Darwin's theory, *Darwin and After Darwin*, was completed before his death, the remaining two being edited from notes and published posthumously (1892-1897).

His close friendship with Darwin, and his capabilities as a lecturer, made Romanes a well-respected authority on Darwinian theory in the years following Darwin's death. He had worked with Darwin in a more practical way than had Lyell, Hooker or Huxley. Romanes believed the "idea of natural selection is unquestionably the most important that has ever been conceived by the mind of man" (vol.2: 256-257). It is thus to Romanes we can trace the beginning of the movement which was to see Darwin and his theory elevated almost to a 'cult' status, with Darwin as its guru.

Romanes' books were written as a consequence of his years of lecturing to fill what he saw as a need to clarify Darwin's position, which he believed was already becoming distorted. In particular, he took issue with Wallace who, he claimed, attributed all evolutionary change to the process of natural selection, something which Romanes insisted Darwin never did (Romanes 1892-1897, vol.2: 1; Darwin 1859/1998: 7). Romanes noted that since Darwin's death, the Wallacean view of natural selection as the sole cause of organic evolution was receiving more and more support, thanks to the theory of inheritance put forward by Weismann (1893), for which Romanes coined the term "Neo-Darwinism", to distinguish it from Darwinism and Lamarckism. In fact, it is clear that as Wallace grew older and became more and more involved with Spiritualism, Wallace acknowledged more and more the role of 'the other side' in all things taking place on this Earth. His earliest papers may have given the impression that he attributed all physical evolution to the process of natural selection, but he always admitted that natural selection could not account for humanity's artistic and philosophical attributes. However, after he returned to England, Wallace left natural selection in Darwin's capable hands and turned his own attention to many other matters, as was shown in Chapter 14. Since Wallace's later books, including his autobiography, were written after Romanes' death, Romanes may well not have been aware of Wallace's changing attitude.

Lamarckism

Romanes did appear to have a good understanding of Lamarck's thesis. He understood that, for Lamarck, evolution revolved around the notion of an inherent force in nature, striving for growth, adaptation and diversification and that the twin theories of use/disuse and inheritance of acquired characteristics must be evaluated within that framework.

Romanes (vol.2: 255) acknowledged that the theory of the 'inheritance of acquired characteristics' should more correctly be associated with the names of Erasmus Darwin and Herbert Spencer than Lamarck, but decided to follow established custom and refer to this hypothesis as 'Lamarckian'. This decision was the one taken by many writers on the subject of evolution which effectively repressed and 'silenced' the true nature of

Lamarck's work and opinion. Romanes should have added Darwin's name to that short list, since Darwin's exposition of *pangenesis* was his explanation of how characteristics acquired by individuals could be passed on to their offspring. Indeed, much of the experimental work carried out by Romanes on Darwin's behalf was aimed at its confirmation (E. Romanes 1896: 223):

> Although I spent more time and trouble than I like to acknowledge (even to myself) in trying to prove Pangenesis between '73 and '80, I never obtained any positive results, and did not care to publish negative. Therefore there are no published papers of mine on the subject, although I may fairly believe that no other human being has tried so many experiments upon it.

Despite his brilliance and his dedication, Romanes allowed his admiration and friendship for Darwin to interfere with the reporting of his results, something which is very tempting to any scientist when the results do not turn out in the way expected or desired. There is a temptation to say to oneself "I must have made a mistake in the setting up of my experiment. I will carry out another one and this time the results will probably (hopefully) be different". Publishing work which disproves one's own theory (or that of a good friend) is not an easy thing to do.

Both theories, Lamarckism and Darwinism, relied heavily on the concept of 'use and disuse', but Darwin's theory, Romanes claimed, was concerned only with the transmission of *congenital* characteristics, not with characteristics acquired during the individual's lifetime. This is clearly inconsistent with Darwin's theory of *pangenesis* and it is difficult to understand how someone who had worked so closely with Darwin on this very point could so misrepresent Darwin's views. One explanation is that work was being undertaken aimed at elucidating quite how characteristics were passed from one generation to the next. Romanes' intent was to disprove Weismann's theory of the separation of germ-plasm (reproductive cells) from somatic (body) cells (vol.2: 240). If Weismann was right, then the ability for change which had occurred in one generation to be passed on to the next would be proven to be impossible and that would be a severe body-blow for the whole concept of 'natural', as distinct from 'theistic,' evolution.

Romanes believed not enough attention had been paid to the difference between selective and adaptive characteristics. A colour affording protection, and thus a greater chance of survival, was an example of selective evolution. The reflex action of removing one's foot from some source of irritation, or of a dog shaking water from its coat, could never have been of life-saving importance. These were examples of adaptive evolution, more readily explained by use inheritance. Continual repetition in a certain way to specific stimuli would eventually result in the inheritance of the appropriate reflex.

Certain adaptations, such as better eyesight or longer limbs, might evolve in different members of the same species and gradually spread until all members had both adaptations. This would be an example of cumulative blending of adaptations. Completely different were the cases of adaptations having no selective value by themselves but which were useful when occurring with other adaptations, such as the lowering of the larynx and the development of the hyoid bone, neither of which had any adaptive value on its own, but which together gave humanity its unique ability for complex speech. Adaptive value in such cases could only have been present if two or more adaptations occurred simultaneously in a co-ordinated manner in the same individual, since they were unlikely to have occurred together by chance in more than one. Romanes claimed that Wallace, and other Neo-Darwinists, did not recognize the problem of co-ordinated adaptations, holding that natural selection alone was sufficient to account for all evolutionary variation.

It was Romanes' opinion that too much attention had been paid to possible causes of isolation, such as geography, and not enough to type. Geographical isolation was indiscriminate. Pure chance dictated which animals were where when an earthquake caused an influx of water to isolate two pieces of land which had formerly been one. Discriminative isolation meant that some factor had separated out a particular subsection for isolation, such as happened under domestication. In nature, visiting insects could discriminate between various plants. In the higher animals, choice of partner could also be a discriminating factor, possibly effecting secondary sexual characteristics. Since in nature all available females would normally be mated, such evolution of secondary sexual characteristics, if it occurred, would affect only males.

Natural selection was a discriminating factor. The more fit survived, the less fit died out. The effects of discriminative selection were always cumulative. However, due to interbreeding, as the change(s) spread throughout the interbreeding population, they may diverge from their original type, but only one new variety/species would result. On its own, natural selection could only lead to monotypic variation, one parent type being replaced by one daughter type. This was a restatement of the problem of divergence which had troubled Darwin for so long. One population split into two following a geological disturbance might lead to the establishment of two different varieties/species, but, for such a complete break, the geological disturbance would need to be substantial. (How many earthquakes were so severe that they divided a population *completely* in two to the extent that there was no further interbreeding between the two new populations? There would have had to have been innumerable such catastrophic earthquakes to account for the immense number of different varieties of animals – and birds would simply fly over them! And what about fish? How did they become subdivided?)

While physiological isolation (infertility) was recognized by Romanes as essential for establishing a new variety or species, how this came about was unknown. Change needed to occur in the reproductive cells as well as in the soma (body) cells. If change took place in the reproductive cells of a plant or animal (in which a favourable variation had occurred) such that that organism was unable to procreate with any of its fellows, then that variation would not survive. If no change took place, then cross-breeding would swamp the variation out of existence. Romanes postulated that infertility between the old and new variety/species would at first be "... well-nigh imperceptible, and have to proceed to increases stage by stage" (vol.3: 43). How the 'well-nigh imperceptible' infertility could increase stage by stage is unclear, since the infertile ones would be eliminated and only the fertile remain. It will be remembered that Lamarck had suggested that all animals in a similar environment would alter in the same way and that this altered form would, somehow, be the one that eventually was inherited. This was Lamarck's best effort at explaining the origin of variation, something with which Romanes also struggled. Weismann's theory, if true, and it was becoming increasing well-known and accepted during the 1890s when Romanes was writing, would effectively deliver a death-blow to any theory of evolution which relied on change being passed from one generation to the next. Romanes efforts to oppose Weismann, to prevent him from obliterating Darwinism, led Romanes into uncomfortable waters.

It was difficult to explain the beginning of structures, especially of those which were useless to the animal until such time as they were at least somewhat developed and thus came under the influence of natural selection. Romanes' solution was to deny that all adaptations had to be useful from the first in relation to their final function. For example, the early change occurring in a forelimb, which would eventually become a wing, would not have had anything to do with potential flight, but might have been "for increased locomotion of other kinds" (vol.3: 355). This explanation is unlikely to have impressed Darwin's critics!

There was one case which had been advanced by the Duke of Argyle (vol.3: 364), which Romanes agreed was very difficult to explain by natural selection – that of the electric organ in the tail of the skate. While a number of creatures were known to possess the ability to discharge an electric current, that of the skate in question *(Rala radiata)* was so mild that it was only discovered because a Professor Sanderson heard electrical interference when he touched a skate in its tank while talking on the telephone. The mild electrical discharge from the skate was too weak to be felt by the hand, and certainly too weak to do any damage to either predator or prey. To make the problem even more severe was the fact that electric organs in animals required the transformation of muscle tissue into a specialized type of nervous tissue, which was far more expensive to maintain from a metabolic point of view than normal muscle tissue, to say nothing of the 'wasting' effect this development had on the propulsive ability of the much thinned tail. Why would this, or any other species of fish or eel, undertake the metabolically expensive and time-consuming task of evolving electric cells at a point when the cells were incapable of actually discharging any electric impulse at all, let alone continue to select for this feature during the time the discharge was too weak to be felt? Romanes concluded (vol.3: 371):

> The structure of the electric organ is far too elaborate, far too specialized, and far too obviously directed to a particular end, to admit of our conceivably supposing it due to any accidental correlation with structural changes going on elsewhere.

For Romanes, this was the most difficult case yet encountered for the defenders of Darwinism, and had there been many other cases like it, he would have been forced to admit that the theory of natural selection would have to be discarded. Since it was a solitary case, Romanes was convinced that a solution would eventually be found under the general law of selection.

In *The Origin* (p. 146) Darwin had stated that his theory would "absolutely break down" if any complex organ could be shown to exist which could not have been formed by "numerous, successive, slight alterations". Implied, but not explicitly stated in this sentence, is that these "numerous successive, slight alterations" would be preserved by natural selection because they were advantageous. It is difficult to explain the advantage of an electrical charge which is too weak to be felt. Nor can it justifiably be claimed that the skate was an isolated instance. The same problem of early development applies to all venoms, be they that of a bee, a snake or a skate. Animals are not alone in possessing venom – the stinging nettle and some cacti, for example, also have the capacity to envenomate. Many plants are poisonous. All must have passed through a stage when these mechanisms were too weak to be effective and therefore 'selected'.

Darwin (p. 146) had been aware of a difficulty with electric organs:

> The electric organs of fishes offer another case of special difficulty. It is impossible to conceive by what steps these wondrous organs have been produced but ... their intimate structure closely resembles that of common muscle ... Rays have an organ closely analogous to the electric apparatus and yet do not ... discharge any electricity, we must own that we are far too ignorant to argue that non transition of any kind is possible.

> The electric organs offer another and even more serious difficulty; for they occur in only about a dozen fishes, of which several are widely remote in their affinities.

Darwin appears to have been more concerned with the problem of parallel evolution than he was with how evolution itself could be accounted for by natural selection.

Romanes also puzzled over the origin of the instinct of the *Sphex* insect to sting a caterpillar precisely on nine separate nerve ganglia, to paralyze, and thus preserve it as food for its young. How many chance times would be needed for the *Sphex* directly to envenomate one ganglion so that it became ingrained as an instinct? Romanes (E. Romanes 1896: 221) suggested 100,000 as a working hypothesis. The number of repeated 'chance hits' to establish an instinct accurate enough to envenomate *nine* ganglia were so astronomical that they could not be calculated – 100,000 x 100,000 x 100,000 x ... : "unity to one thousand million billion trillions". This was not enough to shake his faith in Darwin's theory (E. Romanes 1896: 222):

> Of course, I do not rely on calculations of this kind for giving anything like accurate results (mathematics in biology always seems to me like a scalpel in a carpenter's shop), but it makes no difference how far one cuts down such figures as these. Therefore, if Lamarck won't satisfy such facts, neither do I think that Darwin *minus* Lamarck can do so. We must wait for the next man.
> (Italics in original)

Romanes' three volume treatise concluded with the following words (vol.3: 142):

> Having unanimously agreed that organic evolution is a fact and that natural selection is a cause or a factor in the process, the primary question in debate is whether natural selection is the only cause, or whether it has been assisted by the co-operation of other causes. The school of Weismann maintain that it is the only cause; and therefore deem it worse than useless to search for further causes. With this doctrine Wallace in effect agrees, excepting as regards the particular case of the human mind. The school of Darwin, on the other hand — to which I myself claim to belong — believe that natural selection has been to a considerable extent supplemented by other factors.

This, of course, was the conclusion to which Darwin had come in his final years (see Chapter 13).

Romanes acknowledged that the greatest obstacle yet to be overcome in relation to Darwin's 'true theory' was the vexed question of the possible inheritance of acquired characteristics. It was his hope that the next ten years would finally bring a solution to this outstanding question (vol.3: 142):

> ... thus that within the limits of an ordinary lifetime the theory of organic evolution will have been founded and completed in all its parts, to stand for ever in the world of men as at once the greatest achievement in the history of science and the most splendid monument in the nineteenth century.

Alas, Romanes died well short of his expected allotted time span, at the age of 46, his life cut short by a tumour of the brain. He never lived to see the fulfilment of his dream of the final establishment of Darwin's 'true theory'.

Chapter 19

Reactions to Darwin's Theory

D espite anonymous authorship being accepted practice in the Victorian era, and despite the anonymity providing fertile grounds for speculation as to authorship, it was definitely to the advantage of *The Origin* over *Vestiges* that its author's name was known. Charles Darwin had built upon the good reputation of his family name, not only by his voyage, but also by the quality of his subsequently published work, especially that on barnacles, which was acknowledged as being of high calibre. He was to continue to build upon this reputation with his future scientific papers on subjects as diverse as orchids and worms.

The opinion of a person so well-respected carried considerable weight with a British public, already becoming more and more convinced of the fact of evolution by the discovery of more and more fossils. Huxley (1893/1917: 352) estimated that between 30,000 and 40,000 fossil species had been discovered. Despite being a strong supporter of Darwin, Huxley nevertheless admitted that if all of Darwin's contributions to the subject of evolution were to be omitted, the reality of evolution would stand on the fossil evidence alone (L. Huxley 1900, vol.2: 12).

During the second half of the nineteenth century, Richard Owen took over from Cuvier as the authority on fossil animals and he acknowledged an evolutionary process, attributing it to the 'active and anticipating intelligence' of the 'Great Cause' (Owen 1890: 450). Owen accepted Lamarck's concept of continuous 'spontaneous generation' as more logically consistent than Darwin's theory of life having been created once only and remained a critic of Darwin's concept of natural selection.

Hull (1973) collected together sixteen of the most important reviews of Darwin's work, including those by Sedgwick, Owen, Jenkin, Mivart and

Agassiz. Some were for Darwin, some against, others tried to give both sides equal weight, leaving the reader to decide. Many were lengthy, well-considered critical analyses. The review by Fleeming Jenkin was the one which Darwin found most troublesome, because it was long and well-argued; that by Mivart was to Darwin the most annoying, because it supported the concept of evolution, but decried Darwin's processes of thought (F. Darwin 1887/1969; Darwin 1903). Reviews were discussed in the correspondence between Darwin, Huxley, Henslow and Hooker, and between these men and their individual correspondents. Naturally, there was a considerable degree of overlap between the writers in regard to their criticisms. The following is a synopsis of the main points (Darwin 1887/1969; Darwin and Seward 1903; Barlow 1967; Hooker 1918; Hull 1973; Huxley 1900; Peckham 1959; Vorzimmer 1970).

Variation

While it was admitted that possibly a greater degree of individual variation existed in nature than was immediately obvious, it was nevertheless questioned whether small degrees of variation were sufficient to make the difference between life and death for the individuals concerned. Such degrees of difference needed to occur not once, but many times, for changes to accumulate such that they would warrant identification as distinct species. Darwin was sensitive to this criticism. In later editions of *The Origin*, phrases such as 'very small' were reduced to 'small'. However, Darwin's critics were not satisfied, inasmuch as these small, cumulative changes did no more than postulate longer legs, stronger wings, etc., which was hardly a novel idea. They did not explain the origin of anything.

Speciation

Following from the above, it was pointed out that thousands of years of domestication had failed to result in any new animal (sexually reproducing) species. Pigeons were still pigeons, cows were still cows and, most importantly, dogs were still dogs. No other creature had undergone such a metamorphosis as had the dog, yet, not only were dogs clearly dogs in human eyes, dogs themselves were never deceived. Selective breeding was able to produce structural divergence as great as those of species, but not physiological divergence.

Darwin had used variation under domestication as the main platform for his theory. If speciation could not occur under domestication, could it occur at all? Darwin argued that his critics were asking too much to expect such a speciation to occur during an observable length of time.

Although he acknowledged that Darwin's hypothesis would be 'utterly shattered' if it were to be demonstrated that it was impossible to breed forms mutually sterile from descendants of the same stock, Huxley saw no reason why such mutually sterile forms should not be produced and confidently expected that they would be. Placing the burden of proof at some time in the future made Darwin's theory impossible to falsify.

Saltation

Leading on from the problem of speciation was that of saltation, which Darwin had denied. That 'sports' (unexpected variations) did occur, especially among domesticated plants, was a fact. Huxley accepted that such 'sports' probably did occur in nature. Nevertheless, a 'sport', such as a flower with extra petals, still had produced no new *feature*. Petals were still petals.

While Darwin claimed that all life had descended from one, or a very small number, of forms, his theory failed to explain the origin of any feature - despite the title of his book!

Intermittent forms

The problem of intermittent forms fell into two categories, their survival during selection and their survival during fossilization. For example, how did a reptile survive for the thousands, possibly millions, of generations necessary for its forelimb to convert into the wing of a bird, during which time it would be possessed of neither a fully operational forelimb, nor of a wing? Of great concern was the lack of intermittent forms in the fossil record. Darwin attempted to counter this criticism by appealing to the imperfection of the fossil record. In this he was not entirely successful. His critics pointed out that it was now acknowledged that certain geological formations, such as chalk cliffs, had taken untold ages to form. However, fossil evidence showed very little change in life forms between those existing during the earliest sedimentation and those of later deposits. If it took so long for even slight changes to take place, how long would it take for the establishment of a new, perfected form? There was no evidence of the evolution of these later forms, which just appeared.

Darwin had suggested that forms new to one area may have arrived from another, but as more and more fossils were found, there was no evidence of the previous homes of these incoming forms. Darwin also suggested that their former homes were now beneath the sea, an hypothesis completely untestable (at that time), which was accepted by his supporters but considered a lame excuse by his detractors.

Connecting links

Since evolution was ongoing, connecting links (intermittent forms) should be found between varieties and species, etc. Whereas supporters of Darwin were prepared to see links between, for example, fish and birds in the flying fish or the penguin, and between mammals and reptiles in the platypus, others saw these forms as completely separate. The platypus may be an egg laying mammal, but did this connect it to egg laying fish, egg laying reptiles or egg laying birds? Some fish and reptiles gave birth to live young. Which should be considered connecting features and which lines of demarcation? Such 'links' as appeared to exist did not satisfy the full requirements of 'connections' between all families, genera, species, etc.

Simple organisms

It was asked 'If evolution is ubiquitous, why are there still simple organisms?' Darwin claimed that, since there was still a place in nature for simple organisms, not all would have evolved. It was queried why simple organisms should have evolved at all. Agassiz pointed out that earlier fossil forms may have been different from later ones, but they were not necessarily less complex. Indeed, in some instances, more complex forms preceded simpler ones.

Personification of Nature

It was claimed that Darwin had substituted 'Nature' for 'God' and had spoken as if Nature planned and carried out evolution with intent. There was a discrepancy between the impression given in some parts of his book and the overall impression that evolution was without guidance. While Darwin endeavoured to convince people that he held the latter view, he had difficulty himself with completely letting go of the former. In a letter to Hooker, dated 12th July, 1879, he wrote (Darwin and Seward 1903: 321):

> My theory is a simple muddle. I cannot look at the universe as the result of blind chance, yet I can see no evidence of beneficent 'design', or of design of any kind, in the detail.

Since Darwin was unclear in his own mind, it is not surprising that some of his readers were also unclear.

Spontaneous generation

Darwin rejected Lamarck's concept of spontaneous generation, i.e. the generation of life by 'natural/chemical' means. In the very last sentence of *The Origin*, Darwin wrote of life and its several powers "having been originally breathed into a few forms or into one" (Darwin 1859/1998: 369).

Pictet argued that there were only two alternatives: either life had arisen spontaneously (internally) or it had been 'breathed' into matter by some other force (externally). Since Darwin had rejected the first alternative, his final conclusion was a genuine statement of his position, not an attempt to forestall criticism by the Church, as some people believed.

Mivart and Pictet argued that a belief that generation was ongoing, either spontaneously or through the action of the Creative Force, was less an appeal to the miraculous or supernatural than was a belief that generation of life had happened only once. Others, such as Owen, also argued for 'continuously operating creative laws'.

It was argued that Darwin and his supporters had no right to divide people into two groups, Creationists (belief in God and planned creation according to *Genesis*) and Evolutionists (atheists denying any plan to creation), when many believed evolution was planned and guided by God. This latter group could, and should, be further divided into those who believed that God had set His whole plan in motion from the very beginning and those who believed that God was continually active in His creation.

Unconscious selection

Mivart agreed with Pictet and in 1871 published his own book, *On the Genesis of Species*, suggesting that species evolved through saltation, driven by an innate internal force. Mivart was a Catholic and endeavoured to show that evolution was compatible with biblical teachings. He believed two forces were at work, one active (generative), the other passive (selective) and claimed that Darwin also postulated two such forces. Natural Selection (passive) could only operate on changes which had appeared spontaneously, i.e. had been brought about by some mechanism within the plant or animal itself (active).

With successive editions of *The Origin*, Darwin came to rely increasingly on what he called 'unconscious selection' where whole populations altered at the same time in the same way due to their being acted upon by changed external forces or conditions (passive change). (This clearly resembled Lamarck's approach.)

Blending

Darwin proposed 'unconscious selection' in response to the problem of 'blending'. Since it was assumed that the characteristics of the two parents were blended, and therefore halved, at each union, it seemed obvious that any new characteristic passed on by inheritance would soon be swamped by the old characteristic at subsequent matings. This posed a big problem for Darwin, one he never completely solved. Suggesting that a whole population would change concurrently due to experiencing the same external changes was an attempt to overcome this problem. Indeed, Darwin argued that blending would actually speed up the process of spreading new characteristics. Blending thereby became a means for maintaining species as well as promoting change.

Covering both sides

Attention has already been drawn to Darwin's tendency to cover both sides of an argument. This was noted by his critics. For example, in one of the most comprehensive reviews of *The Origin* published in 1867, Fleeming Jenkin took exception to Darwin citing the same forces as barriers to the spread of variation that he also used to explain distribution (Hull 1973: 342):

> Darwin calls in alternately winds, tides, birds, beasts, all animated nature, as diffusers of species, and then a good many of the same agencies as impenetrable barriers. There are some impenetrable barriers between the Galapagos Islands, but not between New Zealand and South America ... However an animal may have been produced, it must have been produced somewhere, and it must either have spread very widely, or not have spread, and Darwin can give good reasons for both results.

So good was Darwin at coming up with explanations, that he earned himself the nickname 'Wriggler', an epithet which seems to have amused, rather than displeased, him (Darwin and Seward 1903).

Limitation of Variability

From the third edition of *The Origin* (1861) onwards, Darwin included attempts to address points raised by his critics, to such an extent that by the 5th edition, about half his book had been altered. Darwin rarely yielded a point completely. For example, in response to Harvey's comment that Darwin had *assumed* variability to act indefinitely, Darwin responded (Vorzimmer 1970: 139-140):

> That varieties more or less different from the parent stock occasionally arise, few will deny; but that the process of variation should be thus indefinitely prolonged is an assumption, the truth of which must be judged of by how far the hypothesis accords with and explains the general phenomena of nature.
>
> On the other hand, the ordinary belief that the amount of possible variation is a strictly limited quantity is likewise a simple assumption.

His opponents rejected this view. Domestic breeders of both plants and animals had found that in the initial stages of breeding to develop a feature, rapid progress could be made. After a few years, it became harder to gain any further ground in that particular direction and attention was then frequently diverted to the production of a new variety. Darwin's critics claimed that the evidence supported limitation to variability and that Darwin's claims were merely hypothetical or conjectural.

Hybridization

One of the issues which Huxley believed prevented Darwin's hypothesis from being accepted as a theory was that of hybridization. Darwin considered he had 'shown' in *The Origin* that domestic dogs were descended from two or more species of dog. While Huxley himself concluded that small domestic dogs were probably descended from jackals and large ones from wolves, nevertheless he declined to accept the matter as proven until fertile hybrids between jackals and wolves were produced, something which he was confident would shortly be achieved.

Several species combining to produce one, even if that species had several varieties, was hardly the direction by which Darwin was arguing that *diversification* had been produced under natural selection. Quite the reverse.

Darwin partially withdrew hybridization as a factor in natural selection when he suggested that, while he was sure several of our domestic animals were the result of interbreeding between separate wild species, he conceded that there was probably something in domestication which made the reproductive system more plastic, eliminating the natural sterility of species.

Moral attributes

By 'moral' was meant not only such attributes as altruism, but the ability to philosophize, to construct mathematics and to participate in artistic endeavours such as music, painting and poetry, none of which seemed capable of being explained by natural selection.

Unsound reasoning

As already mentioned, Darwin found Mivart's critique the hardest to answer. Mivart himself supported evolutionary theory, following closely the debate surrounding *The Origin*. It was not until after the publication of the 5th edition of *The Origin* and the publication of *The Descent of Man* in 1871, that he finally entered the public debate. He published a scathing attack on Darwin's process of reasoning. With every edition, Darwin was modifying his thesis. Points which Darwin had previously stated 'must be', he now found only 'may be', or admitted were entirely wrong. Darwin appealed to a 'belief in the general principle of evolution', considered things 'possible', 'probable', that we had 'every reason to believe', 'no doubt', 'almost certainly', 'can hardly be a doubt', etc.

Nor was Mivart the only person to object to Darwin's method of dismissing difficulties by declaring them not to be insuperable. Hopkins, in a lengthy review published June, 1860, drew attention to Darwin's manner of asserting that "if this be true ... then it would not be an insuperable difficulty to my theory if ..." when arguing for the existence of former forms, no evidence of whose existence had been found (Hull 1973: 263-264):

> We had not dreamt that because the objections to a theory could not be proved to be absolutely insuperable, we were called upon to accept it as true. We had fancied that the laws of reasoning in such matters ... were still in force.

Time

Fossil evidence, such as chalk cliffs, did indeed support the notion that change took place very slowly - if at all. To this length of time had to be added great stretches unaccounted for by 'missing' geological strata and associated fossils. Lord Kelvin had calculated the rate at which it had to be assumed that the Earth would have cooled and concluded that the Earth could not be older than 40 million years. This was far shorter than needed for the Darwinian hypothesis of evolution. Radioactivity is now known to help counter loss of heat from the Earth into space, but Darwin refused to accept the best scientific estimate that was available at that time and this did not help his reputation for basing his hypothesis upon unproven data.

Different languages - different paradigms

Darwin and his supporters seemed at times to be speaking a different language from that of Darwin's critics, a not uncommon phenomenon during a time of paradigm change. Darwin's critics were asking questions about the origin of features, such as the first feather, the first nerve sensitive to light, indeed the first nerve, or the first bone. In reply, Darwin gave suggestions regarding *adaptations*, the thicker fur on the fox or rabbit in colder climes, the longer legs of the deer, the longer neck of the giraffe, the more magnificent horns of the stag. Darwin did not seem to understand that, if he wanted his theory to account for more than variation, then he must provide *extra* information before extrapolating to families, orders, to four or five original types, or even to one.

As the Victorian era drew towards its close in Britain, the religious ideology still dominated the secular within the general community. Christianity was still largely unchallenged. However, within the scientific community, increasing emphasis was being placed upon the secular position. Owen (1890) was one of the last evolutionary theorists to write from a religious perspective, although he used the term "Great Cause" rather than "God", thereby distancing himself from the established Church. Before the publication of *The Origin,* the religious perspective was still dominant, although under threat. By the end of the century, scientists were viewing evolution predominantly from a secular perspective.

The fact of evolution was largely unquestioned.

The theme that one either accepted evolution by natural selection, without the direction or assistance of 'God', or one believed in creation by God in accordance with the account in *Genesis,* was already being orchestrated by Darwin's supporters. This was not a position taken by Darwin himself. Many people claimed to believe both in evolution and 'God'. Nevertheless, the dichotomy, evolutionism or Creationism, was one which was to become increasingly evident during the 20th century and by the beginning of the 21st century was to be the main area of debate.

Darwin's theory of natural selection had caused what would later become known as a 'paradigm shift' (Kuhn 1962/1970). Popper (1972) held that one piece of contradictory evidence was sufficient to undermine a theory, the 'falsification' with which Popper's name is now indelibly associated. Romanes' suggestion that the theory should be upheld, despite contradictory evidence, was an attempt to put in place a 'protective belt' to 'save the theory' (Lakatos 1970), something which many scientists felt should be allowed while further evidence was being gathered. The number of 'protective belts' around Darwin's original theory have been steadily accumulating. The Kuhn/Lakatos theory predicts that at some point the number of 'protective belts' will become so large and cumbersome that the paradigm will be abandoned and another paradigm, which better explains the inconsistencies, will take its place.

Part III

AFTERMATH

Before 1950

Application of Darwin's

Theory and its Ramifications

Chapter 20

Germ Plasm and Immortality

August Weismann (1834–1914)

The name of Weismann came up with increasing frequency as consideration of the 19th century drew to a close and this was appropriate because the name and work of Weismann was being increasingly cited during that time. Weismann was a firm believer in 'selection' and worked out a far more comprehensive 'germ plasm' theory than that offered by Darwin's theory of *pangenesis* to explain how change came about and made itself available for selection. These were the subject of lectures which he gave over a period of twenty years at the University of Freiburg in Bresnau. He published these 36 lectures in 1904 and it is for this reason that consideration of his work has been held back until now. It took time for his theory to take its final shape and it is that final shape, as published in 1904, which will be considered here.

It is to be regretted that the work of August Weismann is little known today except for his refutation of Darwin's doctrine of *pangenesis* by having cut off the tails of twenty-two generations of mice in order to show that characteristics acquired during an individual's lifetime were not passed on to the next generation. In fact, few scientists (other than Darwin) had ever believed that the results of trauma were inherited in any way and the docking of the tails of dogs and sheep over many generations was seen as evidence enough by most people that this property was not inherited. Had Darwin restricted his doctrine to changes which occurred spontaneously in response to changes in the environment, his theory might have achieved wider acceptance. It was Darwin's insistence that *all* change, however caused, would be registered by the gemmules, and therefore become available for inheritance, which made his theory problematic.

Over the course of the 19th century, microscopes had advanced and their use had led to a greatly increased understanding of the structure of cells, in particular the difference between meiosis and mitosis. Mitosis is the process of division which takes place when body (somatic) cells subdivide and replicate themselves. Meiosis is the process of cell division which takes place in the reproductive cells. Spermatozoa had been seen under the microscope as early as 1677 but the witnessing of the fertilization of an egg by sperm had not occurred until 1875 (Dobzhansky 1964). However, by the time Weismann was giving his lectures during the 1890s, the fact of this process was well established.

Like Lamarck, Weismann was most interested in the invertebrates, particularly butterflies, bees, cockroaches, crabs, Daphnids and the unicellular organisms. Anything which could be magnified was grist for Weismann's microscope. The presence of chromosomes was well established by this time, but their exact role was not, and Wesimann's germ plasm theory was an attempt to explain the role, not only of the chromosomes, but also of the small particles of which they could be seen to be composed.

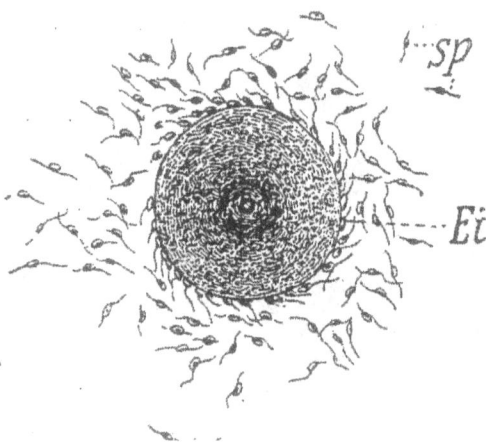

Fig. 20.1: Illustration from Weismann (1904, vol.1: 272) of the ovum of brown seawrack (Fucus platycarpus) surrounded by swarming sperm cells.

Weismann was aware of the work of Tschermak, de Vries and Currens and of their rediscovery of Mendel's earlier paper. While Weismann was of the opinion that Mendel's Law did account for a number of phenomena, he felt that there were some anomalies and that further work needed to be done before it could receive full evaluation.

The process of division of the nucleus of the cell was also known, if not completely understood, by the turn of the century.

Weismann had correctly identified three levels of participation in the reproductive process, although his terminology was to become obsolete.

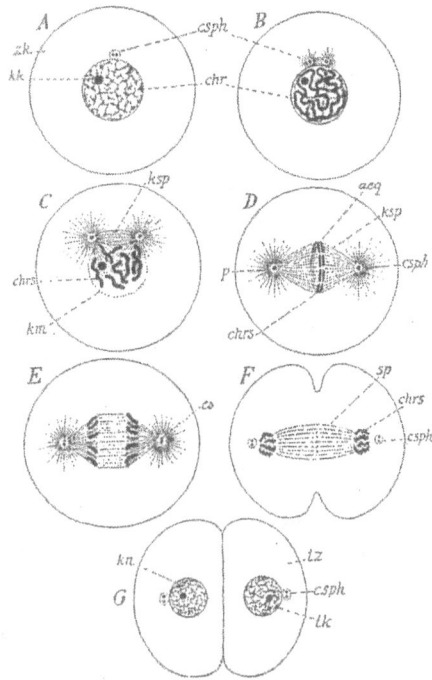

Fig. 20.2: Nuclear Division as illustrated
by Weismann (1904, vol.1: 288)

Ids

That there were different numbers of chromosomes in different species of plants and animals had been established, although accurate counting was still half a century or more away. Weismann believed that each individual chromosome contained all the information necessary for the formation of a new individual. Division of chromosomes appeared to be random and the chance of a group of chromosomes from the other parent containing the precise missing information were so small that Weismann could only assume that all the necessary information was contained within each chromosome. While some small chromosomes appeared to be single entities, others, with a banded appearance, he believed to be a combination of several different pieces of chromosome. These pieces he called *ids*. Since he believed that it was the *id* which was the smallest component containing all the information necessary for the formation of a new individual, he referred to *ids*, not chromosomes, throughout the remainder of his lectures.

Determinants

The particles which could be seen beneath the microscope forming part of the *ids*, Weismann believed to be responsible for the development of particular cells and parts of the body. In support of his theory, he cited not only small peculiarities which he had seen passed on through the generations in human families, but peculiarities passed on by varieties within the same species, such as the small, but consistent, differences in butterfly markings in different localities.

Because these particles determined the nature of the body part for which they were responsible, Weismann called them *determinants*. It was known that these *determinants* were capable of reproducing themselves, this having been observed during cell division. Weismann concluded that they must, therefore, be living and not only capable of reproducing themselves but of taking in nutrients and growing.

Biophors

Because *determinants* were capable of taking in nutrients, growing and reproducing, they must themselves be complex organisms, made up of simpler parts, rather as a molecule was made up of atoms. The smallest living particle in germ plasm could not be seen, but its existence could be deduced. These particles must be larger than any chemical substance since they, themselves, were made up of groups of molecules. Being living, they must also have the properties of movement and sensibility. These sub-microscopic particles, being the 'bearers of life', Weismann termed *biophors*.

Weismann concluded that *determinants* were responsible for those parts which were capable of varying *en bloc*, such as blood cells, the cells of the liver and other body parts, the tissues of the nose or the ear, which must vary in unison if that body part was to alter. This became known as 'hard' inheritance. However, he concluded that there had to be another agent capable of making changes to smaller constituent parts. As examples, Weismann mentioned the specialized odiferous scales which occurred on the wings of some butterflies, which were adaptations of the normal wing scales which gave the butterfly its colour. He also drew attention to the fact that some grasshoppers 'fiddle' by rubbing 'teeth' on one vein of a wing against the 'teeth' of a particular part of the other wing. These 'teeth' were adaptations of hairs which were found on all the other veins of the wing. *Determinants*, he decided, determined the form of the wing, but something else was responsible for the individual scale or hair on the wing, or the vein of the wing, and this, he assumed, was the *biophor*. Individual variation was the result of 'soft' inheritance.

Amphimixis

The process of the 'pooling' of *determinants* and their *biophors*

Weismann termed *amphimixis*. The *ids*, with their *determinants* and *biophors*, had been handed down through the generations. There was a natural limit to the number of *ids* which any germ nucleus could contain. At each *amixis* (fertilization) half of the *ids* failed to be passed on to the next generation. A subsequent mating, even with the same partner, would almost certainly result in a different set of *ids* being passed on. At *amixis*, characteristics making up the new individual, for example blue or brown eyes, were controlled by the majority of *ids* present.

Weismann's theory answered at one and the same time the problem of the tendency for species to be continued, almost unchanged, and their tendency to vary enough to be acted upon by natural selection. *Biophors* could vary only in a plus/minus (larger/smaller) direction. If N represented the normal (most numerous) form, L represented a larger form and S represented a smaller form, at *amixis* there would be six possible outcomes: LL, LN, LS, NN, NS and SS. Four of these conjunctions would tend to preserve the species in its present state: LN, LS, NN and NS. NN would bring about no change. LN and SN would retard a potential for change and LS would average out to N. Thus there were only two possibilities for establishing variants: LL and SS. These altered *biophors* would continue their trend towards variation in subsequent generations, if further LL or SS combinations occurred. 'Unsuccessful' *biophors* were not eliminated but continued to exist in the germ plasm pool. Weismann's theory accounted for both stability and gradual change.

Spontaneous generation

In his final lecture, Chapter XXXVI, Weismann tackled the difficult subject of spontaneous generation. He dismissed the suggestion that life had been 'seeded' on Earth from other parts of the Universe on the grounds that no living matter, however small, could survive the extremely low temperatures such a journey would entail. Weismann concluded that life on this Earth had originated on this Earth.

While rejecting old ideas of the spontaneous generation of complex organisms, Weismann also rejected Pasteur's experiments as insufficient on the grounds that they were concerned only with microscopic cellular organisms. Having concluded that the smallest living particles, the *biophors*, were submicroscopic, he maintained that it would never be possible to witness their generation, yet generate they had out of inorganic material at some point in time.

No experiment would ever be able to ascertain whether *biophors* could be generated anew today because they could never be seen. In any event, any experiment which proved that spontaneous generation had not occurred, only proved that it had not occurred under the conditions of that particular experiment. Since conditions on Earth many billions of years ago might have been quite different from those on Earth today, it would not be possible to replicate all possible conditions, even if *biophors* were capable of being detected, which they were not. Either *biophors* generated spontaneously, or they generated as the result of the action of some outside 'vital force'. It was not possible to prove that a 'vital force' was involved, any more than it was possible to prove that it was not. Within science, it would only be necessary to admit the action of a 'vital force' when all possibilities of spontaneous generation without the action of a 'vital force' had been eliminated. That would never happen.

Germinal selection

There were two major hurdles which Darwin's theory of natural selection struggled to overcome. First, there was the problem of just how much difference was necessary for a change to be acted upon by natural selection? Second, there was the problem of vestigial organs. As his example, Weismann chose the whale. Many and wonderful were the changes which had been wrought in the body of this large land mammal as it adapted to a marine environment but most could be explained as adaptation. This explanation faltered with the whale's residual pelvis. Darwin's process of natural selection, Weismann believed, could only account for shrinkage through disuse of the hind limbs of the whale as far as their disappearance within its body. After that, there would have been no selective pressure upon these bones, yet they had clearly continued to shrink.

Weismann attributed such shrinkage to changes in the germ-plasm, that is, to the shrinkage of the relevant *determinants* and/or *biophors*. Weismann argued that there was no reason to suppose that nutrition at the intra-

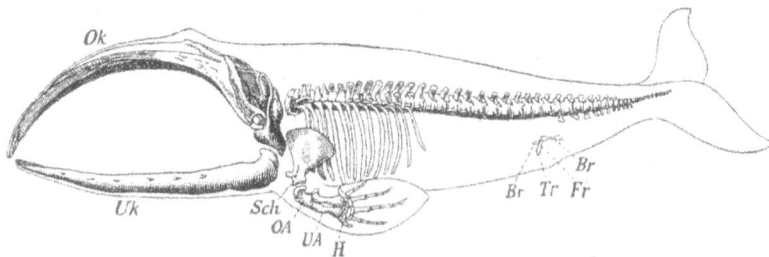

Fig.20.3: Residual pelvic bones of a whale
(Weismann 1904, vol.2: 313)

cellular level was any more evenly distributed than at the inter-cellular or inter-individual level. Some *determinants/biophors* would assimilate more nutrition than others. These would grow larger and stronger than their fellows and the parts of the body for which they were responsible would likewise grow larger and stronger. Other nearby *determinants/biophors* would be comparatively deprived of nutrition and would grow smaller and weaker. The initial change would be insignificant, but once *determinants/biophors* started along the path of receiving more or less than their fair share of nourishment, they would most likely continue on that path. In this way, insignificant changes would gradually become significant ones, which would be acted upon by natural selection.

Weismann thus proposed a two-tiered process, that at the level of the germ-plasm and that at the level of the soma. The concept of the separation of the germ-plasm from the soma was becoming increasingly popular during the 1890s. Like Darwin, Wesimann concretized a known concept into a practical theory, with which his name became associated.

While the development of characteristics, such as the possession of venomous stings/bites which would not be effective until they had reached a certain stage of development, and the residual presence of vestigial organs past the point at which they could have been subject to selection, were one side of the coin, the other side was the over-development of organs/structures which fossil evidence suggested had preceded the extinction of a number of species. The coiling of the shells of *Gryphea* continued until it must have been almost impossible for them to open, other animals developed enormous horns and spines before becoming extinct (Haldane 1932). If small amounts of favourable change were enough to ensure that those favoured would out-compete those less favoured, then small amounts of unfavourable or disadvantageous change should have been enough to ensure that the unfortunate recipients of these changes would be eliminated before their disadvantageous characteristic could spread.

Weismann held that each *id* contained all the *determinants* and *biophors* necessary for the development of the individual. There were more *ids* in the germ-plasm than were needed. These had been inherited from parents and grandparents, who, in turn, had inherited their *ids* from their ancestors, in an unbroken line back into the mists of time. The concept is similar to that held in relation to mitochondrial DNA (mtDNA) today. Indeed, it must be true of nuclear DNA as well. The germ-plasm was essentially immortal, since it traced its ancestry back to single cell organisms, which reproduced by simple cell division and were not subject to death by aging, only by trauma.

Problems

By a process of deduction, Weismann had arrived at a concept of heredity which closely resembled modern theory inasmuch as the germ-plasm was a three-tiered structure. *Ids, determinants* and *biophors* may be compared with chromosomes, genes and DNA. *Biophors* could only grow or shrink, they could not change their nature. How, then, did a single-celled organism evolve into something more complex? How did mitosis give way to meiosis for the production of new individuals? How did qualitative changes take place?

Weismann's theory needed to be able to account for qualitative changes. The fossil record gave a hint of the vast number of different forms which had previously occupied the surface of this planet. More forms would no doubt evolve. Many possible forms may never have eventuated. Thus Weismann concluded (Weismann 1904, vol.2: 390):

> ... an incredible wealth of animal and plant species was potentially contained in these simplest and lowest *'Biophorids"* ... an indefinitely greater wealth than has actually arisen ... it will hardly be disputed that potentially the first *Biophorids* contained an absolutely inexhaustible wealth of forms of life.

This is difficult to imagine, but were it not for mutation/duplication of DNA, that would be the same conclusion to which geneticists of today would have had to come in relation to genes.

How far was Weismann's vision from that of the perfect world of Plato, of the concept that all forms which could be created had been created, for to think anything else would be to think God's world as incomplete and, therefore, less than perfect!

Chapter 21

The Birth of Genetics

The 20th century was to bring the greatest changes the world had ever seen. It somehow seems appropriate that it was the year 1900 which saw the rediscovery of the work of Gregor Mendel, an important event which gave rise to modern genetics, a peaceful and innocuous event, but one which, combined with Darwinism, was to give rise to Social Darwinism with its disastrous consequences. However, that was in the future and the birth of what was to become known as genetics was an exciting event, full of promise.

While those, such as Lyell, whose work during the 19th century was focused mainly on geological evolution, had been able to call on increasingly firm scientific evidence for support, Darwin had been forced to argue his case on the basis of what he thought was 'probable', or merely even 'possible'. Fossil evidence may have put the fact of change beyond dispute for all but a small minority of people, but how change had come about was still much disputed. Now, at last, scientific facts were emerging which were beginning to show how genetic information was passed from one generation to the next. For the first half of the 20th century, evolutionary theorists devoted their time to trying to meld Darwin's theory of evolution by natural selection with this new knowledge, the emergent theory becoming known as 'The Modern Synthesis', 'The Evolutionary Synthesis' or 'The Modern Evolutionary Synthesis', or simply 'Neo-Darwinism'. It was not yet clear exactly how change came about, but, nevertheless, for the first time, there was a path to follow, a path which it was confidently expected would lead them to their Holy Grail.

While great interest in the subject of evolution had been aroused by the works of Chambers and Darwin, mature reflection had presented many problems. As the century drew towards its close there had been a falling away of support for Darwin's theory, the problem of blending, in particular, seeming insurmountable. The concept of discrete particulate inheritance, provided by Mendel's theory, supported by work in both field and laboratory, offered a solution to the conundrum. Mathematics were quickly applied, although the results were not always quite what supporters of evolutionary theory might have wished!

Gregor Mendel (1822-1884)

Mendel was a German monk and his life-style within the monastery allowed him plenty of time for gardening, reading and reflection. He was as interested as any other naturalist in the causes of variation and set out to investigate the matter for himself with the plants available to him in his monastery garden. Mendel carefully selected peas *(Pisum)* as the plant of choice for his experiments because they were self-fertilizing. He could thus be sure that the hybrids would be completely protected from any foreign pollen (Mendel 1865/1966). Artificial fertilization could be achieved by carefully opening the not quite fully developed bud, extracting the stamen and dusting the stigma with the desired pollen.

It was the fact that these plants were normally self-fertilizing that had given rise to clearly observable characteristics, such as colour, size and shape of the seeds and pods, which Mendel selected to study with such meticulous care. These conditions were quite rare in nature. For several years, Mendel manually, and with extreme care, cross-fertilized his peas, noting the number which manifested each characteristic, generation after generation. As a result of his careful observations, he came to realize the fact, not only that some characteristics appeared to be dominant over others, but that this domination was numerically consistent. He came to know that when a plant with a less common characteristic was pollinated by another plant with that same less common characteristic, that less common characteristic was produced. When plants exhibiting the dominant characteristic, but of mixed lineage, were crossed, the majority of new plants would exhibit the dominant characteristic, but some would manifest the less common, showing that somehow the less common characteristic had been retained and was available for inheritance. He realized that some characteristics were dominant over others but that the recessive characteristics were not necessarily eliminated. They were preserved and could reappear in a future generation, if the same recessive gene was available for inheritance by having been preserved also in the genetic material of the partner. He used 'A' and 'a' to represent the dominant and recessive characteristics of one parent and 'B ', 'b' those of the other. (Stern and Sherwood 1966: 20):

> ... when two kinds of differing traits are combined in hybrids, the progeny develop according to the expression Ab + Ab + aB + ab + 2ABb + 2aBb + 2AaB + 2Aab + 4AaBb

His paper was full of possible A a B b representations of the various experiments he had carried out. Although the concept of the double helix format was not proposed until 1953 (Watson and Crick), Mendel did realize that an equal contribution was made by each parent and henceforth any suggestion that one parent supplied the inheritable material while the other parent supplied the nutrition was abandoned forever.

Unfortunately, although his work was published in 1865, it excited very little attention. It was not rediscovered until the year 1900, when Tschermak announced his rediscovery of Mendel's work. Hugo de Vries and Carl Correns had been about to publish similar work of their own, which they did the following year (Stern and Sherwood pp. 107-132).

Still unanswered was the question of how much change was needed for selection to occur, whether evolution was as gradual as Darwin had proposed or whether mutation could result in noticeable changes occurring within one generation, i.e. by saltation, which would make the advantageous change more readily selectable. Early critics of Darwinism had objected that natural selection was not, of itself, creative. Laboratory experiments during the first half of the 20th century showed that mutant genes were almost invariably recessive and they were invariably disadvantageous. However, there was always the possibility that an advantageous mutant genetic variation *could* happen, but, being recessive, it would not be expressed in the phenotype; it would not be available for selection. It was now suggested that the 'undercover' spread of recessive mutant gene combinations could be a source of variation. It appeared that almost all heritable variation owed its origin to recessive genes. Any mutation would be recessive when it first occurred and it would be necessary for at least two members of a breeding population to carry the same mutation for it to be inherited by the next generation. Only when a mating took place between two parents, both of whom carried the recessive gene, did these more rarely seen (possibly new) characteristics manifest. However, if change occurred in response to changed environment, and all members of the population experienced the same environment, it was not seen as unreasonable that such parallel change would occur in numbers of individuals. Some degree of isolation for this mutant population was still seen as being necessary for the recessive gene to become dominant, but a possible mechanism for a new variant characteristic to become established now offered itself. The problem of 'blending out' which had so troubled Darwin had been solved!

Ford (1932) acknowledged that in laboratory experiments, mutations were almost invariably recessive, were almost invariably associated with decreased vitality or viability and were often lethal when homologous (when two individuals carrying the recessive gene were mated and the offspring was double recessive). Fruit flies were often used for these experiments – they were cheap and bred rapidly. The usual method of inducing variation was to subject them to radiation, often intense. Interbreeding of these flies would sometimes produce mutants and further interbreeding could establish this variation.

Although Ford used the word 'almost', neither he, nor anyone else, suggested an alternative cause for heritable variation and it would seem that this word was merely scientific caution. Despite the fact that recessive genes produced during laboratory research were unlikely to establish themselves in nature (because they were disadvantageous or lethal) Ford claimed that laboratory results could legitimately be used to interpret natural phenomenon.

The names of Ronald Fisher, Edmund Ford, Theodosius Dobzhansky, Ernst Mayr and George Gaylord Simpson are those most associated today with the attempt to synthesize newly discovered genetic information with Darwin's theory of evolution.

William Bateson (1861-1926)

William Bateson was the first evolutionary theorist unreservedly to accept and incorporate Mendelian genetics into the theory of evolution. In 1909, the centenary of Darwin's birth and fiftieth anniversary of the publications of *The Origin*, Bateson published an extensive work entitled *Mendel's Principles of Heredity* (Bateson 1909a). He not only gave a thorough explanation of Mendel's own experiments, but attempted to apply mathematical principles to determine the ratios for any number of pairs of dominant/recessive factors. Bateson recognized that Mendel's theory of particulate inheritance obviated the problem of swamping due to blending. Because the factors, be they dominant or recessive, segregated, no obliteration took place. The recessive factor could continue unexpressed for an unlimited number of generations. Bateson also saw that genetic variation was a phenomenon of individuals – each new character was formed in the germ-cell of some particular individual at some point in time. This position continued to be recognized but continued to cause problems. Muller (1949/1963): 431) pointed out that:

> It would have to be postulated that several such identical mutants had arisen at once, and, presumably by virtue of some feature conferred upon them by the same mutation, succeeded in finding one another effectively enough and over a long enough period to establish a permanent line.

Without isolation, Muller believed, this could not happen. The mathematics of genetics made the spread of a genetic variation occurring in one individual, even with the help of natural selection, very unlikely. The advent of genetics had done nothing to change the point of view of earlier theorists, such as Romanes and Darwin, regarding the important role played by isolation.

Also published in 1909, as part of the anniversary celebrations, was a compendium of essays by thirty different authors, entitled *Darwin and Modern Science* (Seward 1909). Bateson contributed an abbreviated version of his book, in which he pointed out that, although the characters of living things were dependent upon definite elements or factors, which were treated as units, nevertheless these units did not always act separately. They sometimes interacted with each other, producing varied effects (Bateson 1909b). An understanding of the interactive nature of genetic material was to prove essential to the growing understanding of the mechanism of genetics (Dobzhansky 1937; Fisher 1929/1948; Ford 1931; Goldschmidt 1938, 1940; Haldane 1932; Huxley 1942b; Muller 1949/1963; Simpson 1944; Stern 1949/1963). Weismann's contribution to the 'anniversary' book was a restatement of his earlier work, untouched by Mendelian theory. Notwithstanding his having been one of the three people credited with having rediscovered Mendel's work, de Vries' contribution on *Variation* was written entirely from a Darwinian perspective, with no mention of Mendel. Sedgwick wrote on the importance Darwin had placed on embryology. Frazer's interesting essay, *Some Primitive Theories on the Origin of Man*, stimulated much of the later work of Sir Arthur Keith, which will be discussed later. There were chapters on Darwin and geographical distribution, religious thought, sociology, in fact, upon just about every aspect of evolution upon which Darwin had written, but, apart from Bateson, there was virtually no reference to Mendelian genetics. The book had, after all, been published in honour of Darwin and a synthesis between Darwinian theory and Mendelian genetics was not achieved without difficulty.

Neo-Lamarkism

One of the most surprising findings of laboratory and field research during this time was the differing effect of environment upon the development of some organisms of the same genotype. For example, it was possible to cause birds such as the scarlet tanager *(Piranaga erytherometas)* and the bobolink *(Dolichonyx anyvivorous)* to retain their breeding plumage throughout the whole year by means of fattening food, dim illumination and reduced activity. An increase in temperature could cause pupæ of a variety of butterfly from central Europe to produce butterflies which resembled varieties from Syria or southern Italy, while treatment with cold of certain central European butterflies resulted in butterflies resembling varieties from Scandinavia (Dobzhansky 1937/1951; Peake and Fleur 1927). This showed not only that some 'species' were, in fact, polymorphic varieties, but that environment did affect morphology. Although this phenomenon was not mutation, *per se*, nevertheless its results were highly pertinent to the course of evolution, since one form might be more successful than another in a changing environment (Simpson 1944). These findings lent support to both Lamarck and Darwin in their belief that organisms were affected by their environment but Neo-Darwinists denied that new mutations adaptive to new conditions were produced in this way. Neo-Lamarckists believed new mutations could occur as an adaptive response to changes in the environment but they did not accept Lamarck's thesis of an inherent progressive evolution.

Julian Huxley (1942b) believed that polymorphism would be to the long-term advantage of an organism since it could come into play if conditions changed. Simpson referred to high variability in a population as a sort of 'bank' in which mutations were held on deposit, available when needed, their immediate availability making possible more rapid evolution. Dobzhansky also considered that no single genotype, however plastic, could function with maximal efficiency in all environments and that natural selection had, therefore, preserved a variety of genotypes, more or less specialized, to render organisms efficient within a range of environments. All these statements contain an element of teleological thinking which is hardly in keeping with the doctrine of purposeless evolution.

Certain characteristics were known to be inherited together, such as deafness and eye colour in cats. Mutual mutation of genes was seen as the possible explanation for the evolution of disadvantageous characteristics, such as over burdensome horns borne by some stags or extreme coiling of some snail shells. If the disadvantageous character was associated in some way with an advantageous character, and if the advantage outweighed the disadvantage, then both might continue to evolve together. Huxley (1942b) argued that just because no natural explanation for a process had yet been discovered did not mean that one did not exist. There are no living ammonites. We cannot be aware of all the circumstances of their existence, making it dangerous to claim that extreme characteristics, such as over-coiling of their shells, had no good (natural) explanation.

Mutation

Gene mutations were acknowledged to be rare in nature. It was originally estimated that they occurred in one in every fifty thousand to one hundred thousand individuals, but this included mutations in body (somatic) cells and these mutations would not have contributed to evolution. For germ cells, mutation rates of one in 1,000 million individuals, or even higher, were postulated (Huxley 1942b). For animals, such as humans and elephants, with a comparatively slow rate of reproduction, such a low mutation rate posed a problem.

Slow mutation rates had also been a problem for scientists in their laboratory work until it was found that they could be increased by the use of radiation. Haldane (1932) suggested an increase of one hundred and fifty times, while Ford (1931) suggested an increase of *fifteen to twenty thousand* percent. Despite such a large measure of artificial interference, results were extrapolated across to Darwinian theory, i.e. that of *natural* selection.

Some genes were more liable to mutation than others, opening up the possibility that a mutation might occur in more than one individual, thus considerably increasing its chance of survival (Dobzhanzky 1937; Simpson 1950). It became known that the phenotypic effect of a gene could be both local and general (Muller 1949/1963; Simpson 1944; Stern 1949/1963) and that a gene could act in one combination as dominant and in another as a recessive (Goldschmidt 1938), that some genes might control rate of growth of certain tissues which could have profound effects on the entire organism, especially if the tissues affected were hormonal or had their major influence early in the embryonic stage (Goldschmidt 1938; 1940).

The effect of genes could be altered by temperature, food or chemical reactions (Goldschmidt 1938). During development, factors such as temperature and moisture could affect the embryo (Ford 1931; Goldschmidt 1938). However, a given gene, or combination of genes, would always give an identical result under identical conditions (Ford 1931). One gene could affect more than one character (pleiotropism) and one character could be affected by many genes, the entire process of branching and anastomoses constituting a complicated biochemical network.

The blending of discrete populations, both genetically and phenotypically, was the reverse of the process required to establish differentiation/speciation. For differentiation to occur, not only was the production of new (recessive) genes essential, it was necessary for these to occur/be maintained in a small isolated population, so that they could become dominant and, therefore, phenotypically expressed and available for selection.

During the first two decades of the 20th century, at the time when Mendelian genetics were becoming established and incorporated within Darwinian theory, there was no organized resistance to evolutionary theory, which was accepted throughout the scientific community. However, for the first time, theorists were seeking 'discontinuity' to explain evolution rather than the 'continuity' which had been such a dominant feature of Darwin's original theory. Discrete particulate inheritance, with the possibility of genetic material being carried forward by an individual even if not expressed in the phenotype, was embraced as the solution to the problem of blending. Science, including mathematics, could be applied to the theory of evolution for the first time.

Chapter 22

The Spread of Change

The 19th century had been a century of great change. It had witnessed the greatest changes which had occurred in the history of humanity - arguably, in the history of the world. The Industrial Revolution had been underway for some time but it was the 19th century which saw the origin of mechanized transport, principally the railway system, which was increasingly snaking its way across Britain and the rest of the Western world. For the first time, ordinary people were able to leave their home town and visit another place, miles away, and return back home the same day! Although slower, the canal system was also a miracle of engineering. It could only exist thanks to the invention of the system of lochs.

But it was not only physical communication which was transforming Victorian society. Radio waves had been discovered; there was the telegraph and the telephone allowing (almost) instant communication with persons in distant lands. (Just stop to think for a moment quite what a momentous change that was!) Electricity was transforming people's daily lives. Photography was invented - or should that be discovered? What was more, thanks to the X-Ray, it was possible to photograph the inside of a person's body! We may take these things for granted today but they were truly revolutionary in their time. Towards the end of the century, the automobile was born and, somewhat late in time, also the bicycle. Flight had not yet been achieved but the concept was being worked upon and would soon 'take off'. Wallace published a book entitled *This Wonderful Century* (1898) in which he recalled the innovations which had occurred during the 19th century and it may fairly be said that for the first half of the 20th, innovations were but advancements on these already discovered principles. Then came the age of computers and electronics and inventions took another leap forward.

However, it was not only upon the material plane that change took place during the 19th century. There was probably a greater shift in patterns of thought during that time than any other in the history of humanity, including the last century.

When the century had dawned, Christianity, as portrayed by the Church, was dominant. There may have been some individuals who questioned the existence of God, but few expressed these views publicly. The miraculous birth of Jesus, the miracles he performed, his resurrection, these were accepted as fact. Equally accepted was the Old Testament account of the creation of the world, which had been accomplished in just six days. This creation had taken place but a few thousand years previously, the history of the persons of the Old Testament being accepted as fact to such an extent that it had even been possible to estimate the year of creation by working backwards through the genealogies given in the Bible!

As scientific knowledge produced more and more real-life 'miracles', so belief in the miracles of the Bible waned. While most people continued to believe in God and attend Church, the stories of both the Old and New Testaments were increasingly questioned, especially that of the Flood and the saving of all animal life by Noah in his Ark. Allegorical interpretations replaced literal.

The French Revolution, at the end of the 18th century, had rocked all European monarchies. There was a growing 'anti-establishment' feeling in the air, which affected attitudes towards both Church and State. The Humanist movement, as we know it today, became established. Change was in the air and 'Evolution' was the catch-cry, whether spoken or implied. Darwin, with his theory of evolution by natural selection, was but part of a far bigger movement which was sweeping across Europe - without which his theory would never have been accepted. It is time now to look at some of these other changes, some of these other evolutions of thought.

Karl Marx (1818-1883)

Karl Marx read Darwin's *The Origin* in 1860, soon after its publication, and was struck by how Darwin's views on the evolution of species mirrored his own views on the evolution of human societies. Marx equated the struggle of workers to improve their position in society with that of a species struggling to establish itself within its environment. He applauded the fact that, in his opinion, Darwin's work had finally disposed of religious teleology.

Karl Marx was quite a character. He was constantly in financial

difficulties and eagerly sought public attention for himself and his views as a means of increasing his income. Becoming engaged at the age of 18 to a young lady (Jenny von Westphalen) four years older than he, Marx was unable to marry for seven years, until he finally found work which promised to pay a salary adequate for marriage. Initially relying on his wife's mother for financial support, including the inheritance she left her daughter, he was at times supported by money raised by his friends and admirers, a habit which he was to retain his entire life.

A radical student, Marx had embraced Hegelian views, frequently publishing radical pamphlets and newspapers, often at his own expense, which failed to provide him with an adequate living. The man who sought to teach the civilized nations of Europe how to run their financial affairs was himself incapable of running his own household! Marx' main assistance came from his friend, Engels, who continued to support him throughout his life, and Marx' daughter after Marx' death. Marx justified this arrangement by claiming that he and Engels were a partnership: Engels provided the finance, he did the work. At times, Engels also did the work. When Marx was commissioned to write a series of articles for the *New York Tribune*, not being confident of his English, he asked Engels to write them for him. Later, he wrote the articles himself in German and Engels merely translated. Eventually, Marx did the work himself (McLellan 1973).

In 1844, Marx put together the kernel of his views in a collection of essays which became known as the Paris essays. This was the same year that Darwin wrote his *Foundations* Essay – and the same year, incidentally, as Chambers had published *Vestiges*. In 1848, Marx and his family moved to London, that place being tolerant of dissenting views. In preparation for the revolution which Marx was sure was about to happen, he instructed his disciples not only whom to kill, but how, naming the shop in London which sold the sharpest axes! Concerned citizens failed to have any legal move made against Marx, Scotland Yard asserting that there was nothing illegal in *talking* about killing, even members of the Royal family, just so long as the member they were *talking* about was not the Sovereign herself! How times have changed!

Marx frequented the Reading Room of the British Museum where he read and absorbed an immense amount of literature on the subject of economics. He became an obsessional reader and, like Darwin, had difficulty in condensing his thoughts into readable form. In 1858, Marx prepared an abstract, *Outlines of a Critique of Political Economy*, of what he planned to be his major work, a strategy also adopted by Darwin, and commenced in the same year. However, Darwin's 'abstract' became his major work, while Marx was able to produce two of his three planned volumes (Marx 1867/1954; 1885/1957), the third volume of *Capital* being published by Engles some years after Marx' death.

Dialectical materialism

It is not intended here to enter into a discussion of Marx' teaching, but merely to draw attention to the fact that Marx, together with Engels and Lenin, were proponents of a political paradigm which greatly influenced thinking in the latter part of the nineteenth century and throughout the twentieth. Marx' philosophy was materialist and traditional materialism had seen matter as passive, merely responding in predetermined ways to any form of energy which acted upon it. Since matter always acted in response to predetermined laws, there was no need or place for intervention by a force (God) in the operation of the created Universe. However, an outside force (God) was still needed to set immobile matter into motion in the first place. In *The Origin,* Darwin had made reference to a First Cause. Like others, he was unable completely to eliminate God from his theory. Even this briefest of introductions to materialism will have shown how closely the two theories embraced each other.

Engels, in accordance with the science then emerging, saw matter as being dynamic. According to his doctrine of emergent evolution, the transformation of energy into matter and matter into energy, eliminated the need for a First Cause. The Universe required neither a ruler nor a Creator. The interaction between matter and energy, between material goods and their means of production, between wealth and work, as opposing forces, became known under Lenin as Dialectical Materialism. This triad of teaching (by Marx, Engels and Lenin) was creating its own atheistic paradigm at the same time that Darwinism was working towards a similar end. It was perhaps inevitable that the supporters of the one paradigm would tend to draw support from the other.

Ernst Hækel (1834–1919)

Ernst Hækel was born in the same year that Darwin wrote his *Foundation* essay, Chambers published *Vestiges* and Marx his *Paris* essays. Quite a year! Like Marx, Hækel was greatly influenced by the evolutionary thought of his time, but his interpretation of Darwin's theory was diametrically opposed to that of Marx. Hækel was neither agnostic nor an

atheist. His family belonged to the Free Evangelical Church, of which he was to remain a nominal member for most of his life (Gasman 1971). As a child he was much drawn to botany, keeping an herbarium, but after qualifying in medicine, he took up the post of lecturer in zoology at Jena University, at which institution he was to stay all his working life. His *Evolution of Man* (1874/1906), published when he was barely 30 years of age, was a comprehensive two volume work dealing with the development of animals and humans from conception, through their life as embryos, to maturity. He considered the fundamental law of evolution to be ontology – the recapitulation in the embryo of the developmental progress of its ancestors through time. Hækel's writing was both comprehensive and orderly in a way which Darwin's failed to be.

As a zoologist, Hækel was well aware that Darwin's emphasis on the individual's struggle for existence was but half the story of evolution. Extra speed might be necessary for the individual predator/prey to survive, but most animals lived in some type of community and it was the strongest communities which led the evolutionary way. Two packs of wolves living in the same forest did not support each other, they competed.

The Monist League

Hækel grew up during the time when Germany was recovering from the devastation inflicted upon her by the Napoleonic dynasty and he rejoiced at the reunification of Germany under Bismarck. Napoleon, Marx and Hækel had one thing in common – their opposition to Christianity. Hækel's love of nature lead him to see Nature as one great entity. The whole Universe was alive, all atoms possessed souls, which were eternal. God was everywhere. There was no difference between matter and spirit (no dialectic), no difference between the souls of men, the souls of animals, the souls of plants, the soul of the Earth. The Christian distinction between matter and spirit, between the human soul and that of animals (if, indeed, the animals were seen to possess souls at all), Hækel called 'Dualism', referring to his own philosophy as 'Monism'. Hækel founded the German Monist League which became very influential, not only in Germany, but throughout Europe.

Marx' interpretation had been based on Darwin's theory of *pangenesis* in which all members of an evolving population contributed equally to change. Hækel's Monism was hierarchal. Individual groups of social animals owed their survival to a strong leader. There was never equality in nature.

253

Subordinate members of a pack or troop must obey their leader and this was also true of human society. The State must rule over the individual. If necessary, the individual must be prepared to sacrifice himself to/for the State. Strength and survival came through struggle.

Secure as he was in his belief that all living things had not only a right, but a duty, to strive to out compete their rivals, Hækel was an unashamed proponent of territorial imperialism. The Monist League joined together with other like-minded organizations to form the Pan-German League with the intention of expanding German territory. Germany had been slow to annex territory overseas, unlike England, France and Spain. Hækel urged his fellow Germans to correct this situation, since Germany would need more territory to accommodate its rising (superior) population. It would be a loss for humanity if Germany were to neglect to ensure the establishment of a healthy and biologically fit German community in all parts of the world.

The essence of natural philosophy had been present throughout Europe for at least a hundred years, but under Hækel it assumed a more organized form. Hækel urged people to spend time in the countryside, rather than in Church, where they could feel themselves at one with nature. The sun was revered as being the source of all life on Earth. Great emphasis was placed upon healthy living, especially for the young. Good food, plenty of exercise, fresh air and sunshine would promote strength and fitness both of body and mind. Alcohol and illicit sex were seen as weakening, degrading influences and therefore condemned. As a lecturer, Hækel was extremely interested in the education of the young and his position as a respected Professor enabled him to campaign for a restructuring of the German education system, in such a way that nature, not the past exploits of human beings, would be the principal subject of study. Children would be taught not only to serve the State, but to do so by being both mentally and physically fit. Hækel was central to the formation of Germany's Youth Movement. With his interest in healthy living and exercise spreading throughout Europe, it is no surprise that the Olympic Games were resurrected in 1896.

Hækel promoted the concept of *Volkism*. Even in English, the word 'folk' has an 'earthy' connotation. It relates not to humans as a whole, but to groups of humans occupying specific localities, villages, towns, or places environmentally defined, such as 'country' or 'mountain'. In nature, no area was shared by members of two groups of the same species. One would dominate, be 'selected' or 'chosen, the other eliminated. In Germany, there was only one 'chosen' people, and it was not the Jews! Hækel campaigned for a 'pure' German people and was openly anti-Semitic. He also espoused the same principles of eugenics that did Darwin's cousin, Sir Francis Galton. The sick, weak and feeble should not be allowed to breed. Indeed, they should not be kept alive. 'Civilized' society was working against nature by protecting those whom nature would not have allowed to survive.

Since the German House of Hanover occupied the British throne, Hækel had hoped that Great Britain would ally herself with Germany in establishing world domination. When Britain failed to comply, Hækel became staunchly anti-British, welcoming the war as a means of liberating the world from "the insufferable despotism of Great Britain" (Gasman 1971: 135). He held the English responsible for the war, which had been started by their "egoism" and their quest for world domination!

Hækel had spent his youth during a time in which Germany was rediscovering itself as a nation, a time of great hope and optimism. He ended his life witnessing Germany's defeat after World War I, having steadfastly opposed any notion of surrender until the bitter end. His dream of World Domination was not dead. The flame had already been kindled in the heart of Adolph Hitler, whose Youth Movement followed that of Hækel's. The Nazi symbol depicted the rays of the sun in acknowledgement of the role of the sun as the progenitor and supporter of all life.

Marx' interpretation of Darwin's theory was instrumental in the establishment of Communism in both Russia and China, with the loss of countless lives. Hækel's interpretation was instrumental in the outbreak of two World Wars, also with the loss of countless lives. How surprised would Darwin have been by these applications of his theory? He concluded *The Origin* by pointing out that behind the apparent peace of nature there was being waged an eternal war, through which, by the means of "famine and death, the most exalted object which we are capable of perceiving, namely the production of the higher animals, directly follows" (p. 369). He cannot be held entirely blameless, even if he never foresaw the precise outcome of his work.

Chapter 23

Change of Mind, Change of Heart

Victorian England was a strange mix of stability and change.

The great Age of Exploration had drawn to a close. The nations of Europe had between them discovered the world. They had broken through the boundaries which had held them for so long – the Atlantic Ocean in the west, the Mediterranean Sea in the south, the Himalayan Mountains in the east and were even starting to explore the Arctic regions in the north. The Mediterranean Sea was no longer thought to be in the 'centre of the Earth'. Africa had been circumnavigated; its immense size was now known and so were its people; the 'mysterious East' was now a valued trading partner; America had been found – and lost – at least as far as the English were concerned, but Australia and New Zealand had been found – and claimed – and served as a replacement.

There was no land left to find. The Victorian attention turned avidly to the treasurers of the newly discovered lands – not just gold and diamonds, spices and cotton, but to its incredible flora and fauna, samples of which were filling the newly established museums and zoological gardens. Comparison between novel specimens and familiar ones from home led to a new interest in exactly what there was at home and the Victorians studied their native flora and fauna with an enthusiasm never before shown. Darwin's teenage interest in beetles was part of this 'awakening'; his theory of evolution but one of several which the changing attitudes of the times allowed to take hold and flourish. Following the Battle of Waterloo in 1815, Britain experienced nearly a century of peace, both at home and in her colonies – known as *Pax Britannica*. Even her monarch, Queen Victoria, was stable, reigning longer than any monarch had done before her.

Material stability allowed a spirit of change to take hold in Victorian thought on many levels and it is now time to consider two of these, one philosophical, one psychological, one home grown, the other imported from the Continent, but enthusiastically embraced.

First to be considered will be the change in thinking relating to psychology which spread from Europe. The next chapter will consider a profound change in philosophical thought, nurtured from within.

Sigmund Freud (1856-1939)

Sigmund Freud was three years old when Darwin's *On the Origin of Species* was published in 1859. Darwin's work was soon translated into German and was well received. Freud's schooling took place during the time of 'The Great Debate', which was won by the evolutionists. The first of Darwin's books which Freud owned was *The Descent of Man*, and although he subsequently purchased others, it was *The Descent of Man* which was to have the greatest influence on Freud's own work (Ritvo 1990). It will be remembered that Darwin had not only been interested in physical evolution, he had been greatly interested also in the evolution of the mind, of instincts and emotions. It was these areas which fascinated the young Freud – not just their presence in humans and how they influenced human life, but why. How and why had our emotions and thought patterns evolved? Why do humans behave the way they do? Do we control our emotions, our thoughts, or do they control us? Although Freud chose medicine as his profession, it was the mind, rather than the body, which always interested him (Masson 1984, Ritvo 1990).

In 1990, Ritvo wrote a book entitled *Darwin's Influence on Freud* in which he explored the influence which Darwin's theories had had on Freud, particularly in relation to the inheritance of acquired characteristics. Freud was a strong believer in the inheritance of acquired characteristics and carefully studied Darwin's accounts of the evolution of emotions in animals and children. Although Freud made a number of mentions of Darwin's theories in his work, he never mentioned Lamarck, having correctly associated this theory with Darwin (p. 23). He believed that repeated patterns of thought/emotions were passed on to future generations and human beings born today still carried in their psychology, as well as their physiology, reminders of earlier stages of evolution. The concept of 'recapitulation' was being explored, particularly on the Continent, by scientists such as Hækel. It was thought that each human being, during the course of its evolution in the womb, recapitulated the evolution of human beings from the single cell, through the fish to the air breathing mammal, finally to the human form. It is now realized that this 'recapitulation' is not as precise as once thought, but the general idea still has appeal.

Following Darwin's theory in regard to human evolution as outlined in *The Descent of Man*, Freud saw in the mind and emotions of the present day child a recapitulation of the former adult, with primitive emotions and sexual needs. These needs could not be fully expressed or satisfied in the immature body, leading children to fantasize, guilt about which, Freud came to believe, was the basic cause of neurosis.

Darwin had believed that humans originally lived in small groups, or hordes, ruled over by a single dominant male, who may have physically castrated his sons in order to maintain his position of leadership and the right to mate with all the females, including his own daughters. This 'acquired memory' was reinforced sufficiently often over repeated generations for it to become an 'inherited characteristic'. Freud, like Darwin, found in this 'inherited memory' the basis for the hostility felt by children, especially sons, towards their fathers. Eventually a son, presumed to be non-castrated following the intercession of the mother, would rebel against the father, killing him and usurping his position, this giving rise in his descendants to the Œdipus complex (Kline 1984, Ritvo 1990, Bocock 2002). These 'inherited memories' are not to be confused with Jung's *Collective Unconscious* in which resides a number of *Archtypes* (Jung 1939).

Freud and Humanism

The role that modern Humanism played in demolishing the last bastion of the Christian 'trinity of virtues', poverty, obedience and chastity, will be discussed in the following chapter. The rise of Humanism and the rise of Freudian psychoanalysis took place concurrently, making it difficult to determine which should be discussed first!

I decided to put Freud first because the claims of Humanists in the first half of the twentieth century were based largely on his writings. Freud came to claim that sexual longings occurred very early in life and, being unfulfilled or suppressed, could manifest in symptoms associated with neurosis or psychosis. Freud's whole system of psychoanalysis came to be based on the (presumed) suppressed sexual urges of infancy and early childhood, urges which found form in fantasies which were falsely remembered as real events. At the beginning of his career, Freud had believed his patients when they recounted childhood sexual abuse but later came to believe that he had been naïve in accepting his patients' accounts (Masson 1984).

Jeffrey Masson (1941-)

Jeffrey Masson received his Ph.D in Sanskrit, only later completing his clinical training as a psychoanalyst. His fluency with languages led to him being appointed Projects Director of the Sigmund Freud Archives, entrusted with editing and translating copies of Freud's letters to his good friend, William Fleiss. Unfortunately, Freud had destroyed the letters he had received from Fleiss. Some of Freud's letters to Fleiss had already been published but Masson came to see that there had been a deliberate omission of certain correspondence which reflected none too well upon Freud and Fleiss. Also omitted were articles and letters which gave the history of Freud's early adoption of the *Seduction Theory*, and its abandonment, largely, it appeared, due to the influence of Fleiss.

In 1981, Masson gave a preliminary account of his findings, which resulted in his being sacked from his position as Projects Director (Masson 1985). No criticism was raised in regard to the accuracy of the material he cited, objection being raised merely to the fact that he cited it at all. In 1984 Masson published a book entitled *The Assault on Truth*, which appeared in hard back and was republished in paperback the following year (1985), with an additional *Afterword*. This saga is of interest here as Freud's theories played an important role, along with those of Darwin, in establishing the Humanist Movement in its current form, which Movement, in its turn, is currently sustaining Darwinism.

According to Masson, the story of Freud's life-long interest in psychoanalysis and its relation to sexual matters started in Paris in 1860, when forensic pathologist, Dr. Ambrose Tardieu (1818-1879) published a paper entitled *A Medico-legal Study of Cruelty and Brutal Treatment inflicted on Children*. Perpetrators claimed they were merely chastising their children in the interests of good parenting, or insisted that injuries sustained by children were the result of accidents. Tardieu showed these claims to be false by detailing thirty-two cases which he had examined on behalf of the Court, which involved excessive cruelty and torture, only one of which included sexual abuse (Masson 1984: 18-19):

> ... from the most tender age, those defenceless unfortunate children should have to experience, every day and even every hour, the most severe cruelty, be subjected to the most direct privations, that their lives, hardly begun, should be nothing but a long agony, that severe corporal punishment, tortures before which even our imagination recoils in horror, should consume their bodies and extinguish the first rays of their reason, shorten their lives and, finally, the most unbelievable thing of all, that the executioners of these children should more often than not be the very people who gave them life — that is one of the most terrifying problems that can trouble the heart of a man.

It did not trouble the heart of the Parisians – they ignored Tardieu and they ignored his book. Nineteen years later, the year of his death, Tardieu published a second book on injuries in which he lamented the fact that his previous article had not awakened the indignation and interest he had expected.

Masson gave an abbreviated account of the one case included by Tardieu which had involved sexual abuse, to which Tardieu had devoted thirteen pages. In December, 1859, the case of a seventeen year old girl, Adelina Defert, had come before the court at Reims. Her parents had kept her in a coffin like box, 1.86 cms. long, 70 cms. wide and 48 cms. high, which box was fastened by a heavy chain and lock. There was a small air hole to allow her to breathe. The box was lined with straw, thistles and nettles and, because the straw was never changed, was teeming with insects. Rags, soaked in pus, served as blankets. At times (it is not clear how often), her father took her out of this coffin, tied her to a wooden bench and beat her. He took live coals from the fire and rolled them along her legs and back, rekindling the coals in the fire as they cooled. Her mother washed her wounds with a sponge dipped in nitric acid. The neighbours could hear the girl screaming for hours and one assumes that they finally reported the matter to the authorities, but not, apparently, until her torment had been in progress for about eight or nine years. The 'sexual' assault occurred when she was tied to the kitchen table and a piece of wood from an elder tree was thrust into her vagina. The examining doctor noted that she was not able to urinate or defecate without experiencing indescribable torture.

Tardieu's second attempt to bring these matters to the attention of the public was more successful. His book cited many instances of the sexual abuse of children, mostly girls, sometimes involving the use of turnips or potatoes, leading one to assume that the men were acting out their anger at their own impotence. After Tardieu's death, his work was taken up by Paul Brouardel and it was then that Freud entered the picture

Before completing his medical studies in Vienna, Freud travelled to Paris where he spent four months studying under Charcot, who, along with Breuer, was the acknowledged authority on the use of mesmerism (hypnosis) for assisting in the psycho-analysis and treatment of neuroses. In collaboration with Magnan, Charcot had published in 1882 a long paper (over 300 pages) on sexual perversion, in which they had claimed that many sexual assaults were committed by 'lucid lunatics', i.e. people who appeared perfectly normal, even intellectual, but who nevertheless had perverted sexual appetites. Brouardel had pointed out that many convicted rapists were 'excellent family men' and Magnan had referred to these people as 'dégénérés supérieurs'. Later, one of the objections raised against Freud's theory was that acts were alleged to have been committed by people 'above reproach'.

While in Paris, Freud also attended lectures by Brouardel, Tardieu's successor in the post of forensic medicine. Brouardel allowed medical students to attend autopsies three days a week. Freud recalled that, while he missed many of the other lectures, he rarely missed one given by Brouardel. How many of the autopsies which Freud attended involved a child subjected to abuse cannot be known, but he later wrote (Masson 1984: 32-33):

> While I was living in Paris in 1885 as a pupil of Charcot, what chiefly attracted me … were the demonstrations and addresses given by Brouardel. He used to show us from post-mortem material at the morgue how much there was which deserved to be known by doctors but of which science preferred to take no notice.

Sexual attacks on young children were generally carried out by a member of the household, be that person a family member or a member of staff. The father was frequently involved. The perpetrators were not sensitive as to which orifice they used, buccal and rectal penetration occurring as well as vaginal. Freud believed that neither Charcot nor Breuer had pushed far enough back into their patients' histories when they attributed hysterical reactions, particularly in girls, to sexual advances in early puberty. The tentative holding of a hand or the brushing of a knee could not explain such severe reactions as vomiting. Such neurotic reactions were far more common in females than males.

Attacks carried out on young children were usually quick and violent, since they needed to be accomplished before another adult came on the scene. Sexual assault involving an older child (between 8 and 12) was different. Here the child needed to be 'seduced'. The perpetrator, who, being a member of the household, had generally established some form of physical contact with the child, might start with a pat, a stroke, a cuddle, and gradually proceed to more advanced forms of petting, although not always to intercourse. The child might be both puzzled and frightened by what was happening, although the adult would take pains to reassure the child.

Freud wrote a lengthy paper on what he believed to be the long-term effects of sexual abuse, which he presented before the Society for Psychiatry and Neurology, Vienna, in April 1896.

Like Tardieu before him, Freud was proud of what he believed was an original contribution to psycho-analysis, one which would revolutionize the understanding and treatment of neuroses and obsession. Like Tardieu's before him, his paper was ignored. It was usual for the Society's *Journal* to record, not only the title of papers presented at meetings, with the name of the presenter, but also a brief abstract. In the case of Freud's paper, his name and the title of the paper were given, but there was no abstract. Freud wrote to Fleiss that his lecture met with "an icy reception", that it had been described as being like "a scientific fairy tale": "The word has been given out to abandon me, and a void is forming around me" (Masson 1984: 9-10). Eventually Freud would deny his insight and join his colleagues in asserting that his patients had invented these fantasies to cover their own childhood sexual longings, often directed towards a parent, and their own childhood practice of masturbation, transforming sexual fantasy into reality to relieve themselves of guilt.

Kline (1984) acknowledged that Freud's essential insight had been gained from the French school of psychotherapy and was to become, albeit changed, the central core of psychoanalysis. This statement is of particular interest since the book in which it appeared was published the same year as Masson's *Assault on Truth*. It implied that even before the publication of Masson's work, Freudian practitioners were aware of Freud's original work and how it had been changed, but they were not prepared to discuss that change. This is reminiscent of the references to Lamarck's ideas not having been as they were generally portrayed, which also lacked extrapolation.

William Fleiss (1858-1928)

Freud wrote extensively to his friend, Fleiss, about his theory and it was while translating these letters that Masson came to understand what lay behind Freud's abandonment of his *Seduction Theory*.

Dr. Fleiss had postulated that, because both were subject to bleeding, there was a connection between the nose and uterus. He believed that neurasthenia and dysmenorrhœa both resulted from covert sexual practices. To prevent relapses after standard treatment, Fleiss proposed to operate on the nose and Freud suggested one of his patients, Emma Eckstein, as a trial patient. After the operation, Fleiss returned to Berlin. Eckstein's recovery was difficult, with persistent swelling of the nose, haemorrhaging, a purulent discharge and pain requiring constant morphine. One day two bowlfuls of pus were collected. Another doctor was called in, who pulled half-a-metre of iodoform gauze from Emma's nostril, resulting in a further severe hæmorrhage: "There was bleeding as though from a carotid artery; within half a minute she would have bled to death" (Masson 1984: 71) had not the wound be repacked. Freud hastened to assure Fleiss that no one could hold him to blame: "The tearing off of the iodoform gauze remains one of those accidents that happen to the most fortunate and circumspect of surgeons".

Freud continued regular visits to his patient. Gradually, as recorded in his letters to Fleiss, Emma's complaints came to be seen as psychosomatic, an unconscious ruse used by Emma to bring Freud to her side. They were the result of 'longing', a form of 'wish fulfilment'.

On 21st September, 1897, Freud wrote to Fleiss that he no longer believed in his *neurotica* (Masson 1984: 108-109):

> In all cases, the *father*, not excluding my own, had to be accused of being perverse … Now I have no idea of where I stand … It seems once again arguable that only later experiences give the impetus to fantasies … It is strange that no feelings of shame appeared, for which, after all, there could well be occasion … The expectation of eternal fame was so beautiful, as was that of … lifting the children above the severe worries which robbed me of my youth … *(Italics in original)*.

This letter, which contains more than a hint that Freud himself had been the subject of abuse as a child, marked the beginning of Freud's open rejection of his own theory.

William Fleiss' son, Robert, also became a psychoanalyst, with particular interest in the psychology of the perpetrator, many of whom he had found to be respectable family men, operating in society quite normally, except for this, their specific perversion. Robert Fleiss called these persons 'ambulatory psychotics'. Masson (1984: 142) quoted R. Fleiss:

> … the child of such a patient becomes the object of defused aggression (maltreated and beaten almost within an inch of his life), and of a perverse sexuality that hardly knows an incest barrier (i.e. is seduced in the most bizarre ways by the parents and, at his or her instigation, by others).

In his book, Robert Fleiss admits to having himself been an abused child (Masson 1984). It would seem that Freud's best friend and confidant was also a perpetrator! Is it any wonder that he did his best to persuade Freud that the children were either deliberate liars or, at best, deluded? Darwin's theory that the dominant ruling male in early human societies had not only castrated at least some of his sons (and/or other males in the group?) but also had sexual relations with his own daughters, and that these group memories were inherited as a form of acquired characteristic, could explain why so many children 'remembered' such acts being perpetrated against themselves.

The whole scenario was further complicated by the fact that the relationship between Freud and Fleiss was sexual (confirmed by Freud's daughter, Anna) (Masson 1984). At that time homo-sexuality was forbidden, which would have placed an extra strain upon this illicit relationship.

Freud was influenced by Darwin's ideas. Freud accepted the belief that humans were but animals in fancy clothing, whose natural sexual instincts were being suppressed by society, resulting in neuroses. Freud's claim that society would be better if sexual needs were more openly acknowledged and allowed more freedom of expression became a fundamental tenet of Humanism during the first half of the twentieth century. Although Humanism has now somewhat relaxed its open espousal of sexual freedom in response to public concerns regarding child abuse and child pornography, nevertheless Humanists are active in promoting Darwinism, since it supports the Humanist position in relation to atheism, and draws upon Freudian theory in as far as it relates to the right to gratify 'animal needs' and materialistic pleasures over (supposed) spiritual needs, which theory was itself based upon Darwinian ideas (which are not necessarily the same as Darwin's ideas).

It is difficult to understand how Freud, after his experience in Paris, could have so abandoned the children he had sought to help for what appears to be no other reason than to salve his conscience, and that of his friend (lover) Fleiss. The account given above of the Emma Eckstein affair is but brief. Further details of the mismanagement of this case by these two medical men were given by Masson (1984), who also reprinted Freud's original paper in full (see Appendix V). How different things might have been for women in the first half of the 20th century if Freud's original theory had been accepted, instead of his second, with its resultant lengthy, expensive, and often ineffective, psychoanalysis. In his autobiography, Freud did not mention Fleiss.

Chapter 24

The Rise and Rise
of Humanism

Although the philosophy of humanism was not to bear full fruit until the twentieth century, its roots reach back to the rediscovery of the ideologies of the Ancient Greeks during the fourteenth and fifteenth centuries. To understand the humanism which was becoming established in the second half of the nineteenth century, and which melded with Darwin's theory of evolution to become the dominant paradigm of the twentieth century, it is necessary briefly to summarize the rise and establishment of humanism over the previous half millennia.

The Reformation

Prior to the Enlightenment, the ideal Christian life was one of simplicity and humility. Those devoting their lives to the service of the Church took vows of poverty, chastity and obedience. If there was one characteristic which bound together the communion of saints, apart from their belief in Christ, it was poverty. The founders of religious orders, such as St. Francis and St. Benedict, owned nothing but their simple (sack-cloth) clothing, their staff and their begging bowl. Nor was such an attitude to simple living unique to Christianity. The Buddha only found Enlightenment after renouncing his worldly wealth and princely position, and the custom of holy men in India relying on charity to survive persists to this day.

Not only was Christ himself of humble birth, but his work as teacher and healer was unpaid. Christ and his disciples were dependent upon charity and the gospels abound with exhortations to give alms to the poor. Indeed, those unfortunate enough to have been born to wealth had very little chance of entering the kingdom of heaven except by virtue of their charity. The parable of the rich man who failed to give succour to the beggar, Lazarus, only to see Lazarus in the bosom of Abraham while he suffered the fires of hell (Luke 16: 19-31) was very potent. It was the custom in pre-Reformation times for charity to be dispensed indiscriminately, without regard to the deservedness of the recipient, since it was the act of *giving* which was important, not that of receiving. Such an attitude was represented by Vincent de Paul, who exhorted his *Sisters of Charity* to "have as their end first their own perfection and then the relief of the sick" (Todd 1987: 245).

The Reformation was to bring a change of attitude in favour of personal responsibility. No longer was it desirable, or even acceptable, to rely on the authority of the Church for one's faith and one's salvation. The Scriptures were to be available to all in their vulgar tongue, to be studied, understood and accepted as an act of personal faith and commitment. The Catholic Church, led by the Jesuits and the Dominicans, rejected the move towards individual Faith, campaigning instead for the absolute authority of the Pope, and the Faith as promulgated by those educated in spiritual matters, the bishops and the clergy. The acceptance of this view led to the establishment of the Inquisition in Catholic Europe, under which anyone shown to question the teaching of the Church could be required to recant or suffer the consequences.

The establishment of the Anglican Church in England was part of this movement of reform. While still recommending the guidance of the bishops and clergy in matters ecclesiastical, it did not frown upon personal study of the scriptures. Indeed, the very establishment of the Church of England was founded upon a rejection of the authority of the Pope, and the translation of the Bible and Prayer Book into English was an important part of Anglican autonomy.

In Europe, those who protested against the dictatorial authority of the Pope and the Church of Rome in matters spiritual became known as Protestants. Led by Calvin and Luther, they insisted on personal responsibility in matters of faith, through individual study of the Bible, religious instruction in the home accompanied by daily worship led by the head of the household (Bullock 1985). However, both Luther and Calvin persisted in a belief in the essential sinfulness of Man, a result of which was the "hell fire and damnation" teaching which so appalled Erasmus Darwin

Many of these Protestants preferred not to concentrate on the sinfulness of Man following the 'Fall', but on the intrinsic worth of humans, who had been created in the image of God. Pride was no longer seen as the most deadly of sins. Instead, a proper sense of self-worth was encouraged as being appropriate for a child of God and a potential inheritor of the

Kingdom of Heaven. People holding these views upheld the intrinsic worth of human beings, rather than denigrating them and seeing them essentially as evil, came to be known as 'Humanists' - the philosophy of Humanism had been born.

While many Protestants were also Humanists, and many Humanists were Protestants, the two ideologies were not mutually inclusive. Protestants and Humanists shared a belief in the efficacy of education. For Protestants, education was the means by which the Scriptures could be studied and understood by the individual; it was a means for furthering true Christian belief. For Humanists, education was the means by which the poor could be enabled to better their position. Social conditions which held back such change needed to be challenged. Social change was the objective of Humanists and education was to be its chief tool (Todd 1987, Elton 1990). Many Catholics supported the Humanist belief in personal responsibility and education as the means of enabling the poor to better their situation (Bullock 1985, Todd 1987).

The rise of Protestantism and Humanism coincided with the start of the Industrial Revolution. Wealth was no longer handed down from generation to generation through the stewardship of the higher classes but was able to be earned and accumulated by any (or all). The accumulation of wealth for personal, rather than social, use became not only acceptable, but desirable. The distinction was made for the first time between the 'deserving' poor and the 'undeserving' poor.

While the Catholic Church continued to maintain the virtue of indiscriminate generosity, Protestants came to see misplaced benevolence as depriving the truly needy of their due. Giving alms to the undeserving became, at best, undesirable, and at worst, actually sinful. In places, such as Norwich, Ypres and Lyons, private almsgiving was prohibited, it being seen as encouraging slothfulness and drunkenness. Instead, a 'poor tax' was levied to be distributed among the deserving poor. Where possible, able bodied men were to be found work. Those not prepared to work were not to receive the dole. The relief of the poor fell upon the local parish, each of which was responsible for the care of it's own residents. Workhouses were established which ensured that no person was denied a roof over their head or basic food in their belly. However, as their name implied, these refuges were designed to be tough, to ensure that only the truly needy would come to them for relief. Orphanages were run after the manner of child workhouses to discourage mothers from abandoning their children to their care, except under extreme duress.

Education was not to be limited to academic subjects; it was to be extended to cover the acquisition of a trade, by which all able-bodied men were to support themselves and their families. In this way, poverty was to be overcome and charity restricted to those who were unable to work. This became known as the 'Protestant Work Ethic'.

Desiderius Erasmus (1460-1536)

Understandably, a renewed interest in the culture, learning and society of ancient Greece and Rome started in Italy, from where it gradually spread across Europe. The years 1350-1600 A.D. are those associated with the Renaissance, its establishment in England taking place in large measure due to the influence of Desiderius Erasmus. Indeed, so deep was the influence of Erasmus, both in England and across Europe, that early Humanism is sometimes referred to as Erasmian Humanism, or Erasmianism (Todd 1987; Matheson 1990).

Erasmus was born in the Low Countries and brought up by the Brethren of the Common Life, an Order of laymen devoted to following the example of Christ by living a simple life (Bullock 1985). The Order rejected the cult of relics, miracles, the veneration of saints and the ostentation of Catholicism. Erasmus traveled widely across Europe. In Italy he was impressed by the teachings of Cicero but rebuked the 'Ciceronians' for taking these teachings out of context. Cicero was Roman and the Roman society in which he had lived was different from the European society of the fifteenth century. It was important for all teachings to be taken in context and this applied as much to the teachings of the Bible as it did to the works of the Greeks and Romans. In order that these may be correctly understood, it was important that they be studied in their original language, be that Latin, Greek or Hebrew.

Erasmus lived for some years in England, teaching Greek at Cambridge University and it was there, in 1516, that he published his Greek New Testament. It was due to his influence that humanities came to be included in the University curriculum in addition to the traditional classical subjects (Bullock 1985). Erasmus' writings were widely read. In 1520, the Oxford bookseller, John Corne, sold more of Erasmus' works than those of Aristotle. For Erasmus, education was the means to a better understanding of Christ's teachings and to the leading of a more Christ-like life. Society, based upon Christ's teachings, would benefit and changes would follow, but it was not Erasmus' primary aim to change society, and certainly not to lift people out of poverty, since it was the simple life that he was advocating (Tuck 1990).

A supporter of Erasmus was Thomas Moore (1478-1535), whose *Utopia* was published in 1516, portraying a mythical island inhabited by people living in an ideal society. *Utopia* was a republic, as had been ancient Greece and early Rome, which were considered by all humanistic philosophers to be superior to later, cruel and decadent, Imperialistic Rome.

On the island of *Utopia*, land was held in common, working hours were restricted and the laws were passed by a consensus of the people. However, these benefits were offset by the fact that the citizens of *Utopia* were expected to submit to these laws and were under constant supervision. The regime was very restrictive and the subdued population seemed to find its greatest source of pleasure in sexual activity (Elton 1990). Those not prepared to accept the rules of the society were to be expunged, either by exile or execution. Whether this mythical population lived in an island paradise or on an island prison, it was up to the reader to decide.

Theistic humanism

Erasmus' dream that education would lead to a greater study of the Bible and, therefore, to a greater appreciation and practice of Christianity, was not to be realized. An educated study of the Bible led many people to question the story of the Creation, of Adam and Eve and of Noah's Ark, all of which were seen to be highly improbable, and reduced in the minds of many people to the status of myth or allegory. While accepting the life and teachings of Christ, as told in the New Testament, many people questioned the accuracy of the Old Testament. Others rejected the Bible, and Christianity, completely. This did not mean, however, that they necessarily rejected the concept of God.

Humanism, which rejected Christianity and the authority of the Church, but which still adhered to a belief in some supernatural Divine Force, became increasingly popular and was espoused by many philosophers: Matthew Arnold, Francis Bacon, Jeremy Bentham, Samuel Taylor Coleridge, Thomas Cromwell, Erasmus Darwin, René Descartes, Denis Diderot, Emile Durkheim, George Eliot, Thomas Huxley, Emanuel Kant, John Milton, Isaac Newton, John Jacques Rousseau, Herbert Spencer, Voltaire, Alfred Wallace and many more (Bullock 1985; Todd 1987; Elton 1990; Tuck 1990). Essential to Humanism, be it Christian or Theistic, was a belief in the possibility of personal improvement, both spiritual and material, by personal effort assisted by education.

269

Since it was possible for poor people to improve their situation, and to accumulate some degree of wealth by their efforts, it was axiomatic that some degree of change in society was not only possible, but inevitable. It increasingly became the prime objective of humanists to bring about social change, not by an increase in Christian faith, but by the redistribution of wealth and authority. Increasingly, the authority of both the Church and the Sovereign were questioned. While many Christian Humanists were still prepared to accept the establishments of Church and Crown, their absolute authority was being undermined. Both the Christian and Theistic Humanists began demanding consensus of opinion, which was increasingly extended to include all men, although the concept of the vote for women was somewhat late to take hold.

Humanism reversed

By the nineteenth century, humanism was well established in its two forms, Christian and Theistic. It had, in effect, reversed two of Christianity's three maxims: poverty, chastity and obedience.

Poverty was no longer seen as inevitable, or desirable. Rather, accummulation of wealth became a sign of God's favour.

Obedience, the corollary of humility, advocated that it was one's duty to serve and, in the case of the Church and the Sovereign, to serve without question. With the rejection of the Old Testament came the rejection of the Ten Commandments and the concept that the Church and the Sovereign were God's representatives on Earth, to be obeyed absolutely. Humans came to be seen as being subject only to human law, which could only be imposed by common consent, but which, once imposed, was to be obeyed by all. Those who declined to accept such rules, agreed by the majority, were to be severely dealt with. After the second Revolution in France in 1792, Robespierre and St. Just, following the teachings of Rousseau, proclaimed that Christianity was to be replaced by the worship of a Supreme Being. Those who did not accept the new order were liable to arbitrary imprisonment and death, as recommended by Rousseau. This led to the death of some 40,000 citizens between 1792 and 1794, including both Robespierre and St. Just (Bullock 1985).

Theistic Humanists were no more tolerant of dissenting opinion than were the Christians of the Inquisition.

Secular humanism

Atheism has been known since Biblical times: "The fool hath said in his heart 'There is no God'" (Psalm 14: 1). By the end of the eighteenth century, 'the fool' had become 'the philosopher' and he was no longer keeping his thoughts hidden in his heart, but proclaiming them to all who would listen. Atheism was gaining such ground that Paley had felt constrained to address his book, *Natural Philosophy*, against atheism (see Chapter 8).

By the nineteenth century, there was a growing resistance to the concept of the miraculous, at least as far as saints were concerned. Miracles attributed to Christ might still be accepted, although not always, but the Catholic claims of many miracles worked by saints, or upon those who prayed to them, was coming under increasing suspicion. The 'supernatural' was being seen as 'unscientific' and for some people the term 'supernatural' included the concept of God. If explaining the origin of the Universe without reference to a Creative Force may have been difficult, explaining the myriad of incredible life-forms inhabiting this Earth without reference to either design or a designer was even harder. Darwin's theory of Natural Selection was the first theory to offer a possible explanation for how evolution may actually have occurred and it did so with only a passing reference to a Creative Force. Setting aside the First Cause, which was a matter for the philosopher, it was clear that the practical aspect of evolution by means of natural selection could be utilized by the atheist as much as by the theist. As mentioned in Chapter 17, Huxley and the X-Men were prime proponents of agnostic (Secular) Humanism, based upon Darwin's theory of natural selection.

One of the reasons for the rapid acceptance of Darwin's theory was its malleability to so many different perspectives. The believers in the literal truth of the *Genesis* account of Creation, now known as Creationists, rejected Darwin's theory, as they rejected all theories of evolution, but most other groups, be they Christian, theist, atheist or agnostic, found Natural Selection compatible with their views. Most Christians were quite prepared to accept the possibility that the Old Testament may be flawed, having been 'contaminated' over the centuries by human error. It was pointed out that the *Genesis* account was evolutionary in a manner very similar to that first proposed by Buffon: the Heavens, then Earth, the seas and dry land, plants, fish, land animals and birds, and finally humans. There were problems with this account. On the first 'day', God created light but the Sun and Moon were not created until the fourth 'day', leaving one to wonder how there could have been 'evening and morning' of the first, second and third days. Clearly, the term 'day' was not to be taken as 24 hours, but as a period of time (Age) which might be very extended.

Some theists believed in a Creative Force that had set the Universe in motion. It could allow the Universe to evolve in a pre-ordained manner, after natural law. Other theists believed that the Creative Force made 'adjustments' at times, thus overcoming the problem of speciation. It was as though the Universe, or at least the Earth, was operating on 'automatic pilot' but the pilot (God) could take over the controls whenever he wished to alter direction. Atheists, who had no problem in accepting that the Universe was self-created, had no problem with speciation. The precise details might not yet be clear, but that was a minor issue.

Atheism was far more dependent upon Darwinism than Darwinism was upon atheism.

Christian Humanists who embrace the ideal Christ-like life of simplicity and poverty espoused by Erasmus would be hard to find today, although Mother Teresa could no doubt be held up as a twentieth century example. However, Christian Humanism which embraces personal study of the Scriptures, a personal relationship with God not necessarily mediated by the priesthood, personal development through education, work and the accumulation of personal assets, is very active, no more so than in the United States where the concepts of religion and wealth have been welded together as nowhere else on Earth, today or at any period in the past. Evangelical Christians consider personal wealth a blessing from God. Organizations, such as Christian Aid Abroad and World Vision, place as much importance on schooling and the learning of a trade as they do on the teaching of the Gospels.

Secular Humanism is firmly established today on both sides of the Atlantic – and in both hemispheres. Unlike theistic humanists, secular humanists organized themselves into a body for the purpose of furthering their views in a politically active manner. In 1933, they issued a Humanist Manifesto, reproduced in full by Lamont (1949/1965: 285-289). The preamble claimed that Humanism was the 'religion for today' and the first precept stated "Religious humanists regard the universe as self-existing and not created". The manifesto went on to assert that humans had emerged as a result of a continuous process, that there is no mind (thought process) independent of the body, and therefore no survival after death, that modern science made unacceptable any supernatural involvement in human values, that distinction between the sacred and the secular could no longer be maintained but that all things, such as work, art, science, philosophy, love, friendship, recreation, which made life intellectually satisfying or physically pleasurable, were equally 'sacred'. They further held that the religion of today (i.e. humanism) must work towards joy in living by fostering creativity and encouraging a sense of achievement, by the individual and society: "The goal of humanism is a free and universal society in which people voluntarily and intelligently co-operate for the common good" (p. 288). Lamont summed up the philosophy of humanism as "mankind's interests upon this earth are the first and the last word" (p. 19). Since Lamont wrote those words, many Humanists have become actively involved in the Conservation movement. It would no longer be accurate to claim that they put the interest of humans above all else, although active conservation tends to highlight those life forms which humans find most interesting and rewarding.

In 1973, the *Manifesto* was updated (LaHaye 1980) and included the propositions that "The only happiness humans will ever have is that which they experience during this lifetime", "we strive for the good life here and now", "moral values derive their source from human experience" and "individuals should be permitted to express their sexual proclivities and pursue their life-styles as they desire". Freedom and dignity would be achieved by allowing each individual to experience the full range of civil liberties. The insistence upon personal sexual freedom disposed of the last of the three great pillars of Christian living, that of chastity.

The American Humanist Association was founded in 1941. The British Humanist Association was not formed until 1963, when it was launched under the Presidency of Sir Julian Huxley at a large dinner held in the British House of Commons. Lamont (1949/1965) listed the names of many famous people from Humanist ranks, including Luther Burbank, Pierre and Marie Curie, Karl Marx, Albert Einstein and Sigmund Freud, all of whom either rose to prominence in their chosen fields or created their chosen field. Others, such as Sir Julian, who at one time served as Director General of UNESCO, attained office thanks to the Humanists' policy of supporting each others' career advancement. The seed sown by Sir Julian's 'great' grandfather, Thomas Huxley, and the X-Men flowered and bore rich fruit as the 20th century progressed.

Darwin's version of evolution was the only one compatible with the humanist belief system and secular humanists adopted Darwin's *The Origin of Species* as their 'Bible'. Sir Julian Huxley's '*Modern Evolutionary Synthesis*' was a synthesis of evolutionary theory with genetics, but it was also a synthesis of atheistic views with evolutionary theory. The interweaving of humanistic philosophy with evolutionary theory is exemplified by comments, such as those made by Simpson (1950: 229-230):

> There was no anticipation of man's coming. He responds to no plan and fulfils no supernal purpose. He stands alone in the universe, a unique product of a long, unconscious, impersonal, material process, with unique understanding

and potentialities. These he owes to no one but himself, and it is to himself that he is responsible.

From Huxley to Dawkins, atheism has increasingly become a necessary component of evolutionary theory, even to the extent of being incorporated by law into the teaching of evolution in American State schools, which event will be discussed in the final part of this book.

After the horrors of World War II, an attempt was made to establish a new interpretation of humanism via a series of books, of which Theodosius Dobzhansky's *The Biology of Ultimate Concern* (1967) was the second to be published. Dobzhansky's name appeared as one of the editors of the series, along with Eric Fromm, Fred Hoyle and more than a dozen others. Dobzhansky's work in relation to evolution will be considered later, but here it is of interest to note that an evolutionary geneticist of Dobzhansky's standing should have become so intimately involved with this new expression of Humanism.

The new approach was explained by Ruth Anshen in the preface (Dobzhansky 1967: vii-xvii). 'Scientific' Religious Humanism was seen to have failed. Anshen drew attention to Aristotle's claim that "the soul is the meaning of the body" and claimed that it was ultimately impossible to make a true statement about the physiological dynamics of the human body without taking into consideration the spirit which formed the flesh. While nature was "a miracle defying understanding", yet underlying apparent change there was a constancy which, for humans, was their essence, their nature. The role of New Humanism was "to reveal the true nature of man and the task for which he was born" (p. xi). It was time for science once more to accept metaphysics, else a complete category of human experience would forever remain excluded from scientific investigation purely on *a priori* grounds.

Humanism was the common metaphysical faith behind the religious/philosophical systems of both East and West (p. xii). Anschen acknowledged that there was a division within the ranks of Humanists and that (pp. xv-xvi):

> ... the idea we form of humanism will have wholly different implications according to whether we hold or do not hold that there is in the nature of man a constant, an essence, something which breathes an air outside of time and a personality whose profoundest needs transcend time and space, and even the self...The authors of this series work to substitute for the inhuman system currently confronting us a new form of civilisation which would outline and represent humanism both sacred and secular.

Hitler's thinking was quasi-religious, but that of Stalin was purely atheistic. At the time the New Humanist movement was launched, the horrors of Stalin's communism outweighed the horrors of Hitler's Social Darwinism in the minds of many, especially of the Americans. Stalin's regime is believed to have been responsible for the deaths of between 50 and 100 million people, up to ten times that for which Hitler is believed to have been responsible. The establishment of the Chinese atheist Communist regime was equally brutal. As a Russian, who had left his homeland shortly after the Russian revolution to pursue his career in America, not as a refugee but as a scientist, Dobzhansky was dismayed at the turn of events in Russia. However, he and his New Humanist colleagues were unable to turn the tide of atheistic Humanism, a tide which gathered even more force with the support of Richard Dawkins, whose influence will be considered in Part V of this book.

The Core Principles of Secular Humanism may be found on the internet (http//upconnect.net/slsoc/manussa/coreprin.htm). There is no longer any claim that Humanism is a religion, possibly because the American Constitution bans the teaching of religion in any form in its State schools. Humanity and its environment are seen as the only relevant spheres for consideration, any form of Divine agency is denied. Humans owe no duty to anyone other than themselves (each other), and evidence for persistence of life after death is denied. Clause 3 requires that beliefs be founded on reason and human experience. This nevertheless (presumably) excludes any human experience of psychic phenomena, such as clairvoyance, mediumship and near-death experiences. Police forces are becoming increasingly willing to consult with psychics with proven ability to assist them in solving murder cases, as evidenced by television programmes, such as *Psychic Detective* and *Sensing Murder.* However, these results are not considered 'evidence' by Humanists. It is noted that while many scientists will write of the relativity of time and space, of the concept that time and space may be 'curved' rather than 'straight', most scientists still resist the idea that it is possible in any way to make contact with the past or the future. Clause 7 of the Principles, which requires that children not be indoctrinated into any religious or political belief, presumably does not include indoctrination into a belief in Humanistic principles? Gone from the Humanist Principles is an overall permission for any form of sexual activity in which the human being may wish to indulge. 'Sexual violence and misconduct' are acknowledged to exist, presumably as a result of growing community concern about sexual abuse of both women and children and child pornography, although adult pornography is still permissible. Increased importance is placed on human rights, although these do not include drunkenness or the use of narcotics and mind altering drugs. Nevertheless, the increasing community acceptance of drunkenness and some degree of drug use has undoubtedly been fostered by humanistic ideas permeating today's society.

In the preamble, particular mention is made of the role of Darwinism in establishing Humanism as a viable philosophy. If natural selection was shown to be insufficient to account for evolution, this would be a serious blow for the Humanist movement as it exists today.

The Humanism of Erasmus was very simple. It had two basic precepts: each human being was responsible for his/her own beliefs and actions which should not – could not – be dictated to them by some other authority, such as the Church or the State and that human beings were not 'fallen' or still falling all the way to Hell, but were worthy beings, created by God, the pinnacle of God's creation on Earth. Today, Humanists still hold that human beings are responsible for their own beliefs and actions, which should not be dictated to them by any other authority; they still hold that humans are worthy beings, but, rather than being the pinnacle of God's creation of this Earth, they have now become the pinnacle of existence anywhere in the Universe, since all angelic or divine beings have been eliminated. The early Humanist was still responsible to God; today's Humanist is responsible to no one, other than his fellow humans and then only as deemed fit or necessary.

Chapter 25

Darwinism in Practice

Neither philosophical nor political ideologies are of much use unless put into practice. Freud's ideas took firm hold during the first half of the 20th century, not just among the medical practitioners who embraced Freudian psychoanalysis for many of their patients, but the general public, whose vocabulary became enriched with terms such as 'anally retentive' and who became aware of the importance of 'correct' potty-training for their toddlers! Darwin's ideas regarding the inheritability of acquired characteristics were also put to practical use - in both sensible and strange ways.

This chapter will commence with a brief consideration of three changes which took place in Darwin's own home country. Then consideration will be given to the work of two men from Europe, the one German, whose work was suppressed and forgotten, the other Russian, whose work was elevated and practiced, with disastrous consequences - at least for Russia.

Social Reform in England

Darwin had written of the importance of educating young women of marriageable age so that their improved mental abilities might be passed on to their children by inheritance but it was Huxley and Wallace who actively campaigned to bring about changes to the education system such that not only females, but the working classes, would all equally be educated. For the first millennium A.D., literary skills were mainly confined to the monasteries and the Courts. Alfred the Great, in the 9th century, was the first English King able to read and write, succeeding to the throne above his older brothers because he possessed this skill. For the next thousand years, it was mostly the sons of the wealthy who were educated, although some daughters also received basic education in the 'Three R's'. The first Public

School for girls was established in 1697 by the Worshipful Company of Haberdashers. By the 18th century, some working class children were receiving basic education at Parish Schools but it was not considered that they had the innate intelligence to benefit from more advanced study.

Many years ago, I was shown a large medical book, published in the 19th century, which stated that females should not receive education, or, indeed, indulge in any strenuous mental activity after puberty, because such stimulus would draw blood to the brain from the reproductive organs and make them sterile! Ladies should confine themselves to light piano playing and embroidery! Quite why educating males did not make men sterile was not explained. I cannot remember the title of the book or the name of its author, but I believe it to have been published in the second half of the 19th century – i.e. during the time in which Darwin's ideas were taking hold.

It was against a background of such thinking that Huxley, Wallace, and others, strove to have education, not only made available to all, but made *mandatory*. All children were *compelled* to attend school. It was now believed that educating the entire population would result in the mental abilities of the entire population being rapidly raised as the improvement achieved in one generation would be passed on to the next. In Germany, Kindergartens were not gardens *for* children, they were gardens *of* children. This decision that the State should provide, free of charge, education for all its citizens had enormous consequences, not least being the cost to the tax payer. That the practice continues to this day indicates that the benefit to the community is held to outweigh the cost.

I do not believe that Darwin's book was the sole cause of these changes. Rather I believe that Darwin's book had the success that it did because a change in thinking was already spreading across Europe and Darwin, after so much procrastination, was fortunate enough to ride the crest of the wave. However, I do believe that Darwinism and Humanism combined to bring about the next change.

It was the teaching of the Church that humans were born with a propensity for evil. John Locke (1632-1704) had argued that children were born neither good nor evil, but were a 'blank slate' *(tabula rasa)*. This concept may have been the subject of philosophical argument, but it had not impressed itself upon the Courts. That changed. It was now accepted that each individual grew and changed during the course of his/her life according to the circumstances (environment) in which they found themself. Children, especially female children, were now seen as 'innocent'. If a child behaved badly, it was the fault of the parents, of society. Special Courts were instituted to hear cases involving children and special Detention Centres (not prisons) were built for juveniles. I cannot speak for other parts of the world, but in Australia the names of children appearing before the Children's Court are not released. 'Under age' convictions are not considered at adult trials. Even the names of persons accused of violating a child in some way are suppressed less the release of their name identify the child.

Today, lawyers for the Defence can argue that adverse family/social circumstances during the childhood of an offender caused that offender to become delinquent and to ask that the Court take such factors into consideration when sentencing in the expectation of their client receiving more lenient treatment. These arguments carry greater weight in the Children's Court than the adult Court, but even there it may be asked that they be taken into account, as well as other adverse circumstances which may have arisen during adult life.

The third change which took place was probably the most profound. The State granted unto itself the right to remove a child from its parents if the State considered that the parents were not suitable or were failing the child in some way. A whole system of Welfare Services was founded upon the belief that, if external circumstances could be changed, the person could be changed. Parenting may previously have been a shared responsibility in some households/communities, unmarried mothers may have been pressured into giving up their children for adoption, but never before had the State assumed the right and the duty forcibly to remove a child from its parents against the parents' wishes. Countless thousands of social workers the world over now have jobs that were not even conceived before the time of Darwin.

Evolutionary thinking inspired revolutionary thought.

Russian Darwinism

As outlined in Chapter 22, the Darwinism which had inspired Marx and his colleagues was adopted by the post-revolutionary Soviet Government under Stalin sixty years later, after the First World War.

Contrary to the Social Darwinism (cream rises to the top) which was to be embraced by Hitler, Stalin embraced the view that change (improvement/progress) occurred, not in the favoured few who had reached the acme of their evolutionary possibilities, but in the lower echelons of society, who were in the process of changing/evolving through struggle, undertaken in response to the external conditions in which the 'mass' of people found themselves. It was supported by Trofim Denisovich Lysenko,

President of the Lenin All-Union Academy of Agricultural Sciences (L.A.A.A.S.). As a result, Russian agricultural policy was based on acceptance of the doctrine of the inheritance of acquired characteristics in accordance with Darwin's theory of *pangenesis* (Medvedev 1969; Adams 1980; Dobzhansky 1970).

Stalin was more inclined to support, and raise to positions of power, peasants who offered practical solutions to farming (agrobiological) problems, as Lysenko claimed to do, than to support the laboratory work of 'educated scientists', whose status still retained something reminiscent of the bourgeoisie (Joravsky 1970). Lysenko was seen to be giving the Russian people grain while Mendelian geneticists in Western Europe were studying the colour of the eyes of fruit flies.

While the doctrine of evolution had been welcomed in post-revolutionary Russia as an integral part of the 'new world view', not all academics supported Lysenko's position. There was bitter rivalry between academic institutions in the interpretation of Darwinism, according to Dobzhansky (1970) who, it will be remembered had fled from Russia to America to escape Stalin's regime.

Not only were the Soviet hierarchy convinced that their agricultural policies were in accordance with Darwinian theory, they considered that the Western (especially German) interpretation of Darwinism was a bourgeois plot to "justify the fact that, in the capitalist society, the great majority of people, in a period of overproduction of material goods, live poorly" (Medvedev 1969). The Weismann/Mendel theory of genetics seemed to provide a scientific basis for eugenics (the elimination/sterilization of the 'non-desirable'), the politics of race, serving as a useful tool for Hitler's elitist theories. The engineering of hybrids was a plot by the capitalist firms to produce seeds available to the ordinary farmer only by purchase, at least in the first instance. Some people today view the genetic engineering of food crops in the same way! Medvedev, himself an opponent of Lysenko, appeared to consider Lysenko's claim that his work was rooted in Darwinian theory to be inaccurate and misguided, that Lysenko was, in fact, embracing the ideas of Lamarck. At no point did Medvedev refer to Darwin's theory of *pangenesis*, of which he appeared to be unaware. Joravsky (1970) asserted that Lysenko claimed his theories were his own, not taken from Lamarck. He made no mention of Darwin at all in relation to Lysenko's work. Either he was not aware of any claim by Lysenko that his work was 'Darwinian' or he dismissed it as irrelevant.

Lysenko's great discovery was 'Vernalization'. Lysenko believed that the number of hours of daylight a crop received was the crucial factor in its growth. Higher latitudes had a greater number of daylight hours during the summer months, but winter sown seed could not survive the low winter temperatures and reduction in light. According to Medvedev, Lysenko based his 'discovery' on the single planting, in one pot, of two seeds of two varieties of winter wheat. The seeds were planted in March, 1935, kept in a hothouse, but at a cool temperature, until the end of April. One pair of plants died without 'heading' in late autumn. Of the other pair, one plant also died but the surviving plant 'headed' and its seeds were collected for a second planting, which took place in September, 1935. A third generation was planted March, 1936, and Lysenko reported his success upon sowing the fourth generation in September 1936. Lysenko had been inspired to undertake his small experiment by his father who, in 1929, had soaked some seed to germinate it, stored it over winter under a bank of ice, sown it in the spring and obtained a good yield.

The first major trials of 'vernalization' followed the loss of more than thirty million acres of winter wheat during the very severe winters of 1927-1928 and 1928-1929. The winter wheat that Lysenko used naturally germinated in late autumn, suspended its development during the winter frosts, recommencing its growth cycle in spring. If the winter was very severe, a whole crop could be lost. Lysenko instituted his method of soaking the seed for several days under controlled conditions, the germinated seeds being stored by freezing over winter. Lysenko believed that the crops would gradually become adapted, generation by generation, to the changed conditions, but his project had to be abandoned because the labour and expense involved outweighed the advantage gained by the small increase in yield in anything other than a very severe winter. Nevertheless, between 1926 and 1970, Russia registered a greater percentage increase in wheat yield than did America, although Russia did start with a lower yield per acre (Levin and Lewontin 1985). Joravsky supported the claim that Lysenko's system did bring some measure of success, so Medvedev's condemnation of Lysenko's work may not be totally justified. Levin and Lewontin mentioned more than thirty scientists whose work, mostly undertaken during the 1920s-1940s, at least gave credence to some measure of inheritance of acquired characteristics.

Lysenko had been impressed by Darwin's teaching regarding the detrimental effect of inbreeding and the advantages to be gained by regular cross-breeding. Lysenko applied the principle of cross-breeding even to self-fertilizing wheat varieties. He required that the collective farmers removed the anthers from the spikes of their wheat using tweezers so that their crops would be fertilized by wind-born pollen from their neighbours' farms. Lysenko believed that populations of plants and animals were, to some extent, self-controlled, that predation and lack of food were not the only factors involved. He believed that if too many young were born, or too many plant seeds fertilized in a given area, some would die in a 'self-sacrificing' way. Nor were Lysenko's interests confined to wheat. He also experimented with cattle in regard to the butterfat content of milk. Lysenko believed that he could predict the outcome of cross-fertilization through knowledge of the parents' characteristics. For example, if a large bull mated with a small cow, the zygote would sense that, if it developed after the manner of its sire, it might be too big to pass through the birth canal, at least with any ease. It would, therefore, 'choose' to develop after the type of its mother. Therefore, it would be better if a small bull from a line of high butterfat cattle mated with a large cow from a low fat herd, rather than the other way. The offspring could be encouraged to attain the larger size by doubling or tripling the food consumption of the gestating cow, since this would stimulate the growth of the fœtus (although not enough to impede birth), the stimulus to growth continuing in the calf during its post-birth development.

The Soviet Minister of Agriculture banned all genetic research in animal husbandry and the liquidation of all research projects not 'in the spirit of Lysenko'. Only professors teaching Lysenkoism were permitted to graduate students. If this sounds extreme, it must be remembered that Lysenko firmly believed his methods to be in accordance with the teachings of Darwin, whose theories had so impressed Marx. At that time, some States in America had passed laws banning the teaching in State-funded schools of pro-evolutionary teaching (see next chapter) and America has now passed laws banning the teaching in State schools of anti-evolutionary subject matter (see Part V). Government interference with what may or may not be taught in schools and universities is not unique to any particular regime.

The 'five-year agricultural programmes' instituted by the Russian government were based, at least in part, on the belief that this was the length of time needed for a new crop to become established. The failure of Darwinian theory within this context was to have a profound, and detrimental, effect on the Russian economy and would be instrumental in precipitating the downfall of communism within Russia.

Darwinism in Germany

To understand not only Lysenko, but the adherence of so many of his Russian colleagues to a belief in the inheritance of acquired characteristics, it is necessary to consider the work of the Austrian, Paul Kammerer, who was the most influential scientist working in this field during the first quarter of the twentieth century. In 1971, Arthur Koestler published an interesting and informative book about Kammerer's work, entitled *The Case of the Midwife Toad*, which is well worth reading.

For the first thirty years of his life, Paul Kammerer (1880-1926) may well have considered that he had been born 'under a lucky star' (Koestler 1971). Vienna during the last two decades of the 19th century was a prosperous and happy place, and Kammerer's family was prosperous, even by Viennese standards. They were a musical family and Kammerer first studied music at the Vienna Academy before studying zoology at the University, where he obtained his Ph.D. He was an accomplished pianist and composer and his employer, Professor Przibram, was later to say that much of the antagonism towards Kammerer's work was due to his being first a musician and second a scientist.

As a child, Kammerer developed an abiding interest in the lizards and frogs which he found in the grounds of his family home. He developed a reputation for being 'a wizard with lizards'. An article Kammerer wrote on the care of animals in captivity motivated Hans Przibram to employ the young Kammerer in the Institute for Experimental Biology which opened in 1904. The Institute was very modern for its time, being equipped with an early form of air-conditioning which made it possible to keep temperatures in the laboratory constant, as well as to control humidity. Failure of other scientists to replicate Kammerer's work may have been due in part to the lack of equal facilities, but it was also due, according to Przibram, to Kammerer's devotion to, and affinity with, his experimental animals. He regarded reptiles and amphibians as sensitive creatures and refused to buy animals from dealers, considering them 'spoilt', over or under fed and often unwilling to mate. He collected all his laboratory animals himself from the wild. Another factor which impeded other scientists from replicating Kammerer's work was their inability to keep these animals alive in captivity at all. The breeding of even one generation under standard conditions proved almost impossible, let alone the breeding of several generations under abnormal conditions, which was the hub of Kammerer's work.

In 1924, Kammerer published an account of his work, intended for the general reader, and still well worth reading today. The book was entitled *Inheritance of Acquired Characteristics* and the final chapter, *Darwinism and Socialism*, was devoted to an outline of Kammerer's belief that the Neo-Darwinism so popular in Germany, Britain and America and other non-socialists countries, was 'aristocratic', not 'democratic' (p. 63):

When he was invited to join the team at the newly formed Institute of Experimental Biology in 1904, Kammerer was aware of the new theory of Mendelian inheritance, which he accepted. His initial experiments were intended to throw light on the problem of atavism (reappearance of former characteristics). The male Midwife Toad *(Alytes obstetricans)* winds strings of eggs around the upper part of his hind legs, so that they develop to the tadpole stage out of water. Kammerer was able to hatch some of the eggs in water, which could be described as atavism – a reversal to a previous habit. Subjecting some eggs, not to immersion in water, but to relative aridity and darkness, caused the tadpoles to remain inside the eggs (which became 'gigantic') until the tadpoles had grown their hind legs (p. 52):

> These eggs and tadpoles produce dwarf-like toads which now, from generation to generation, produce eggs that are proportionately more limited in number, but are larger and larger, and more rich in yoke. If the environment continues to be warm, rather dry, and quite shady, tadpoles emerge from these eggs with completely developed hind legs. If restored to normal conditions, tadpoles are produced with just the beginnings of the hind legs.

This was just one of the experiments that convinced Kammerer that, by changing the environment, it was not only possible to cause atavism, but also to produce a novel condition, in this case the eggs continuing to develop out of water to an unprecedented degree.

Fig 25.1; The Midwife Toad (Alytes obstetricans), Male with tighly packed eggs around the thighs.
(fKrammerer 1924; 55)

While the ability of the organisms to respond to changing environmental conditions was, to some extent at least, what Kammerer had been expecting, what he had not expected was that the new characteristic would be passed on to subsequent generations. The eggs/tadpoles of 'water' and 'air' developed toads which differed in several ways, not just in the development of hind legs. 'Water' egg tadpoles, over several generations, developed three gill arches, instead of the usual one. The eggs became smaller (poorer in yolk) but the gelatinous coating became thicker. Eggs of 'abnormal' Midwife Toads (those which did not take care of their eggs any more but simply deposited them in water) produced specimens in which the instinct to attach the eggs to their thighs was lacking, even though they had themselves passed their period of development on land. Kammerer 'controlled' for his experiments by subjecting some of the toads to reverse conditions (p. 60):

> The most important variation in the case of the Midwife Toad is the voluntary relinquishing of carrying the eggs and taking to the water at the mating period, even after the influence which brought about these changes of propagation has again been eliminated. The unassailable proof of genuine inheritance was brought about here by the aforementioned controlling tests and strengthened by the fact that, in crossing "abnormal" Midwife Toads and "normal" ones, the hybrids are subject to the Mendelian Rule.

Males of the third or fourth generation also tended to develop a rough, blackish nuptial pad. It was this which was to lead to Kammerer's downfall.

Kammerer also experimented with salamanders. The spotted Fire Salamander *(Salamandra maculosa)* naturally inhabits moist woods. The female gives birth in water and the fifty or so young live in water for several months, with clusters of gills for respiration and a finned tail for swimming. Kammerer removed the female from the water, forcing her to birth on land, where the young would have died had he not placed them in water (pp. 88-89):

> Death by drying up would also have been the fate of the next issue — usually born at intervals of six months — had not the mother salamander delivered larger larvæ which, within the womb of the mother, passed the period meant for development in the water. Generally, beginning with the fourth pregnancy, at the conclusion of the second year of experimentation, the young ones, born on land, are no longer in any danger of death by drying up. They are completely developed little salamanders breathing through lungs and, thanks to sturdy little legs and a cylindrical finless tail, they have the ability to move with ease upon solid ground ... instead of fifty progeny ... only six, four, or even two are born at one time, the salamander's womb allowing space for no more.

These changes took place, not in subsequent generations, but in subsequent pregnancies! Kammerer's work showed that, not only were acquired characteristics inherited, but they might be acquired in a very short space of time. Time had been a problem for evolutionary theorists ever since Darwin first proclaimed his theory, particularly in the case of large, slow breeding mammals. It might have been supposed that evolutionary theorists would have welcomed Kammerer's work, and many did, but a relentless campaign by the Englishman, William Bateson, discredited him and his work, not difficult to do bearing in mind the enmity which existed between Britain and Germany during and after the First World War. The South European Wall Lizard *(Laceria serpa)* lays elongated eggs, covered in a soft onion-like skin (p. 181). Kept at warmer temperatures, the female lizard laid eggs with a thicker shell, not so elongated, eventually laying perfectly round eggs with calcified shells, similar to the geckos *(Gecconida)* in warmer (tropical) climates. Young female lizards hatching from the hard-shelled eggs laid hard-shelled eggs, even if they were kept at an intermediate temperature, as did the parent lizard when returned to normal conditions. It appeared to Kammerer that once the calcium-secreting glands of the oviduct had been stimulated, the new pattern became fixed. Laying eggs with calcium rich shells in a cool climate is not detrimental to the embryo in the same way that is the laying of thin-skinned eggs in a warm climate.

Kammerer also referred to the well-known phenomenon of changes in egg-shell colour under hybridization. If a hen which usually laid white-shelled eggs was mated with a rooster from a line laying brown-shelled eggs, the eggs would be brownish, as would future eggs laid by that hen, even though she was then mated with a 'white egg' rooster. The same phenomenon was known to occur with finches and canaries. Many animal breeders refused to use females which had been mated with another 'line' believing that the female was from then on 'impure' in some way. (Some human societies, even today, have similar views regarding women who have, voluntarily or non-voluntarily, had sexual relations with an 'undesirable' male). Kammerer believed that some of the superfluous sperm penetrated (was absorbed into) the cells of the oviduct and some of the genetic material became integrated with the genetic material of the female bird. Since this material was only integrated with the cells of the oviduct, which produced the material for the shell, and not with the reproductive cells of the female bird, the male characteristics were not truly 'acquired' by the female and were not passed on to the next generation and was not the same phenomenon as the inheritance of an acquired characteristic.

The Case of the Midwife Toad

The ongoing saga of the acquired nuptial pads of the Midwife Toad had its beginnings with Bateson during the 1880s. Bateson, enthusiastically embracing Darwin's teachings, tried to find evidence for the inheritance of acquired characteristics, but was unsuccessful. He was to find the explanation for his failure in the theories of Weismann and Mendel (Bateson

1909a), which he then enthusiastically embraced. For the next twenty years, he rechannelled his energies into disproving the very theory he had once held so dear. Bateson had some misgivings when he heard of Kammerer's work. In September, 1910, Bateson visited Kammerer at the Institute of Vienna and at that time became openly hostile. In a letter to his wife, Bateson wrote (p. 54):

> ... there is no denying any longer the extraordinary interest of what he is doing. The Brauftschwielen [nuptial pads] cannot be produced. Somehow or other I have hit on a weak spot there ... But he has certainly done a very fine lot of things and he comes uncommonly near to showing that an acquired adaptation is transmitted. I don't like it, and shall not give in till no doubt remains.

In a paper published after the war, Kammerer took the opportunity to explain that the Midwife Toad, adapted by him to breeding in water, developed nuptial pads for a short time during the mating season. The mating season was still several weeks away at the time of Bateson's visit (pp. 58-59). Kammerer further explained that only a very few of the experimental eggs developed into breeding adults and he had been reluctant to kill a breeding male, during the mating season, simply to preserve such a specimen.

During the war, Kammerer had been conscripted to work in one of the Ministries. The Institute was unable to maintain its high standards and most of the animals died. One male Midwife Toad survived, developing the nuptial pads even though there was no female present. This specimen was killed and preserved as evidence. After the war, the economies of Germany and Austria collapsed. The Institute was in great difficulties. Kammerer was forced to try to support both himself and the Institute by undertaking lecture tours. In April 1923, the Cambridge Natural History Society sponsored a visit by him to England at which he displayed the specimen of the male Midwife Toad, with its nuptial pads. By that time, Bateson's belief that these pads neither existed nor were inherited had become a public source of contention. Bateson did not attend the meeting. The meeting was so successful that the lecture was repeated in London on 10th May. This time, Bateson did attend and, although he did not remove the specimen from its jar for examination, he did withdraw his charges against Kammerer and accepted his published results as genuine.

In September of that year, Bateson expressed a desire to see the specimen again and offered to defray expenses if Kammerer would bring the specimen to England a second time. The Institute, in the person of Przibram who owned the specimen, declined to subject the valuable specimen to further travel, but offered to accommodate Bateson at Przibram's house, should Bateson wish to come to Vienna. Bateson declined, and there the matter rested – until 1926.

After nearly a year's leave of absence, during which Kammerer lectured both at home and abroad, and wrote extensively, Kammerer left the Institute in October 1924 to continue his activities on an independent basis. He lectured in Russia, where his work was well received. Pavlov trained mice to respond to the sound of a bell which announced the arrival of food (Koestler 1971). The first generation of mice needed 300 trials, the fourth only five. It was Pavlov's hope that mice would eventually be bred that responded to the sound of the bell without any training, i.e. without the arrival of food. However, when attempts to replicate Pavlov's work proved negative, Pavlov withdrew his claims, blaming an assistant for faulty experimentation.

The purpose of Pavlov's experiments had been to show the inheritance of learning. Experiments by Harvard Professor, William MacDougall, showed that rats learned the escape route through a water maze more quickly with each generation, but Professor Agar of Melbourne subsequently showed that not only the experimental rats learned more quickly with each generation, but so did the controls! It appeared that merely being bred under laboratory conditions improved learning ability (Koestler 1971).

Pavlov's belief in the inheritance of acquired characteristics led him to invite Kammerer to oversee the building of a new biological research laboratory, to be affiliated with Pavlov's Institute (Koestler 1971; Adams 1980). Kammerer was due to take up his position 1st October, 1926 when he was to receive a Professorship and be in charge of a new facility. One week before he was due to commence his new duties, Kammerer was found dead from a bullet wound to the head, with a suicide note in his pocket.

In February, 1926, Przibram's Institute had been visited by Dr. Noble, the Curator of Reptiles at the American Museum of Natural History. A known opponent of 'Lamarckism', Noble examined the preserved specimen of the male Midwife Toad and declared that the nuptial pads had been faked by the injection of Indian ink. Przibram examined the specimen and concurred. His initial reaction was that someone working at the Institute had noticed that the dark colour of the pads had faded due to exposure of the jar to light, and had tried to help by recolouring the specimen. He changed his mind and concluded that the fraud had been committed with the intention of discrediting Kammerer. He believed he knew who the perpetrator was but had insufficient evidence to make a public accusation. Kammerer absolutely denied any involvement and his denial was obviously accepted by Pavlov, since the discovery of the fraud took place at the time Kammerer's contract was being negotiated.

Koestler tried to duplicate the fraud. He found Indian ink gave good results but they were only temporary. With the specimen preserved in alcohol, the ink 'ran'. With the specimen preserved in formaldehyde, the colour faded within two weeks. Mixing the Indian ink with another substance prior to injection gave equally unsatisfactory results. In glycerol, the Indian ink dissolved and 'ran'. Paraffin oil did not take up the ink. With gelatine, the ink did not fade but the patch coagulated, and looked very artificial. Koestler concluded that the fraud had been carried out using simple Indian ink shortly before Noble's arrival. Since Kammerer had not worked at the Institute for three years by then, clearly the fraud was not perpetrated by Kammerer. However, Koestler's experiments were not carried out until 1970. None were tried at the time of the accusation.

According to a news report of the time (Koester: 1971: 6): "Two days before his suicide, Dr. Kammerer visited the Soviet Legation in Vienna and with much zest gave instructions regarding the crating and transport of the scientific apparatus and machines which he had ordered ...". What caused such zest to be transformed into suicidal depression? One suggested reason was the refusal of Kammerer's current mistress to accompany him to Russia.

Leaving the dispatch of his equipment in the hands of the removalists, Kammerer paid a final visit to his favourite holiday resort, Puchberg, where he arrived in the evening of Wednesday, 22nd September, 1926. The following morning, he went for a walk along a narrow footpath leading from Puchberg to Humberg. He was found at 2 p.m. that day, in a sitting position, leaning against the Theresa Rock on the Schneeberg Pass. In addition to the suicide note in his pocket, Kammerer had posted four other suicide letters the day before. Or had he?

Although the gun was still in his right hand, the bullet had entered his left temple, just above the ear, exiting through the right temple, damaging the right eye, indicating an angle of shot which would surely have needed the abilities of a contortionist? All the suicide notes had been typed with only a signature, which could easily have been forged if the death was indeed a professional 'hit'. The note in Kammerer's pocket suggested that his body be donated to a laboratory for dissection (Koestler 1971: 1):

> I would actually prefer to render science at least this small service. Perhaps my worthy academic colleagues will discover in my brain a trace of the qualities they found absent from the manifestations of my mental activities while I was alive ...

This was a strange bequest from someone planning to blow his brains out! Although the full extent of Hitler's brutal exercise of power was yet to be felt, political assassinations were occurring with increasing frequency in Germany and Austria and it cannot be ruled out that the attempt to discredit Kammerer and thus prevent his being offered the contract by Pavlov having failed, and his defection to Russia being imminent, a resort was made to extreme measures.

Bateson died in February, 1926, just as the 'fraud' was being discovered. Had Bateson lived to know of this new Kammerer controversy, he may not have been as pleased as might have been expected. By 1924, he had come to the conclusion that Neo-Darwinism (the amalgamation of Darwinian theory and Mendelian genetics) was not a sufficient explanation for evolution and that "it was a mistake to have committed his life to Mendelism, that this was a blind alley which would not throw any light on the differentiation of species, nor on evolution in general" (p. 119).

The geneticist, Steve Jones (2002) recorded several instances in which changes occurring in the life time of plants or animals due to changed conditions appeared to be inherited by their offspring, even if the conditions were reversed in subsequent generations. Referring to such instances of *epigenetic variation*, Foder and Piattelli-Palmarini (2010: 67) stated that "this domain has raised perplexities ... caused by the fear that Lamarckism may be making a comeback".

Following upon his early work, (Steele 1979), Edward Steele and his colleagues (Steele et al. 1998) found that, in certain instances, there was evidence of soma to germ-line feedback associated with the immune system. They concluded that the Weisman barrier was not as impenetrable as had been assumed by the Neo-Darwinists. While their work with the immune system was highly specialized, they drew attention to some simple examples of what appeared to be acquired characteristics being inherited. Ostriches and warthogs both develop callouses where they sit or kneel and it is difficult to consider these as anything other than an acquired characteristic, yet these callouses are already well-formed in the embryo, implying that they are germ-line encoded.

While accepting that Lamarck believed in the inheritance of acquired characteristics, here disputed, Steele et al. also made repeated reference to Darwin's theory of *pangenesis*, which they felt may need to be revisited. Steele was inspired by the work of Arthur Koestler (Tynan 1994), who became his mentor. Thus through Koestler and Steele, the work of Kammerer is not totally forgotten.

Chapter 26

Plenty of Scope

At the same time that Stalin, in charge of the fledgling Communist regime in Russia, was actively promoting the ideas of Darwin as he understood them, even entrenching them in law, there was resistance in the United States of America, not so much to the idea of evolution itself, but to the idea that humans had not been specially created by God, but were merely another animal, descended from the apes. Some States were to enact laws forbidding the teaching of this pernicious doctrine within State schools, showing that the leaders of free and democratic societies can be just as entrenched in their thinking on certain issues as any other leaders, be they of Church or State - or, indeed, as any other human beings at all.

Until the time of the First World War, religion had not featured prominently in evolutionary debate. Possible religious implications were of concern to many individuals, but no denomination of the Christian Church had made acceptance/rejection of Darwin's theory of evolution the subject of doctrinal edict, nor had any government felt it necessary to impose/forbid the teaching of the theory in its schools. The majority of Church-going people increasingly accepted the Biblical account of Creation as allegorical. Those who rejected evolution and accepted the Biblical account of Creation as factual did not present an organized opposition. This was about to change.

In 1925, a famous trial took place in the United States of America, the effects of which reverberated, not only throughout America, but across Europe and other Western nations. Later known as the 'Monkey Trial', *Tennessee v. John Thomas Scopes*, saw a 25 year old school teacher convicted of teaching evolution contrary to a Bill which had been enacted by the Tennessee House of Representatives on 18th January, 1925. The Bill had passed by a vote of 71-5 and been signed into law on 21st March, 1925.

This law provided (Scopes and Presley 1967: 51):

> ... that it shall be unlawful for any teacher [at a State funded educational institution] to teach any theory that denies the story of the Divine Creation of man as taught in the Bible, and to teach instead that man has descended from a lower order of animals.

The law did not proscribe the teaching of evolution *per se*, only the teaching of evolution in relation to humans. Whatever their differences, all Christian denominations accepted that humans were in some essential way different from animals. Humans had been endowed with free will, an ability to choose between right and wrong, to make independent decisions about their behavior in a way not evident in members of other animal kingdoms. Humans were seen as a special creation, as having been made 'in the image of God', with unique potential and unique responsibilities. It was generally accepted that only humans had souls and only humans were capable of earning for themselves the right to 'eternal life', and a place in Heaven. This understanding was perceived to be under threat from the spread of Darwinism and its associated teaching, that human beings were descended from apes, or a form closely related to the ape line. Even if evolution were accepted as having occurred, and the law certainly did not deny this, Divine intervention must have occurred at some point, namely when 'man became a living soul' *(Genesis II: 7)*.

William Jennings Bryan (1860–1925)

The person whose campaigning brought about the introduction of this legislation, and similar legislation in other states, was William Jennings Bryan. He was one of the most eminent persons in America at the turn of the 20th century. He stood three times for the Presidency of the United States, the youngest person ever to do so. Although he was never successful, he did hold the position of Secretary of State under President Woodrow Wilson, whose term of office finished in January, 1917. In this position, Bryan would have been more aware than nearly any other citizen of America of the situation abroad, especially in Germany before, during and after the First World War. He would have known of the Monist League with its anti-Christian philosophy (Wallace 1910) and of the fledgling Nazi movement, already active by 1925 and of the growing interest in eugenics (Ginger 1958, Grabiner and Miller 1974, Shermer 2006).

Bryan was "an educated man who held seven doctorates of law and numerous other degrees" (Ginger p. 88). It is inconceivable that a man so well-educated and so politically astute should not have been aware of the communist interpretation of Darwinism, including atheism. Two differing interpretations of Darwin's doctrine of evolution by natural selection had led to devastating social upheaval in Europe and Bryan was determined to do all he could to stop what he saw as a pernicious doctrine from becoming established in America. In addition to being a politician, Bryan was also a lay preacher.

John Scopes (1900-1970)

In 1967, Scopes published his autobiography (Scopes and Presley 1967). Unless otherwise indicated, the following account of his life and his progression of thought is taken from there.

Born at the beginning of the 20th century, Scopes received his early education in an America for which the memory of the Civil War was still vivid. The Scopes family believed passionately in freedom, both physical and intellectual (p. 46):

> ... early 1920s, after Prohibition had been legislated and the country's morals were taken care of, science became the primary target of those self-anointed crusaders .... They equated evolution with agnosticism, which in turn they made synonymous with atheism ... intolerant Fundamentalists, who wanted to foist their beliefs onto everyone else. Here was the crux of the controversy as far as I was concerned. The Fundamentalists had an inalienable right to believe what they did, but when they insisted that others hold those beliefs too, they were violating other people's rights ... It was a specific example of the universal conflict of the narrow-minded and intolerant against the broad-minded and tolerant.

As a child, John had been encouraged by his father to read widely and to be an independent thinker. For a time the Scopes family had lived in Salem, the birth place of William Jennings Bryan. Bryan made regular visits to Salem and Scopes had heard him preach a number of times. Scopes grew up to be a 'civil libertarian' and when the American Civil Liberties Union sent out a plea for help, he answered the call.

While at University in Kentucky, Scopes had chosen his courses, not because of their subject matter, but for the quality of the lecturers. In his final year, he realized that his unsystematic following of personalities rather than a degree plan had precipitated a crisis – his subjects were so scattered that he had insufficient in any one area for a major, or even a minor! During his final year Scopes studied what was in effect first year law, some child psychology and some geology and this strange assortment, together with his previous studies, resulted in him being the only person to graduate from the University of Kentucky with a Bachelor of Arts while majoring in Law!

Armed with this strange degree, Scopes found career opportunities somewhat limited! Scopes was offered a temporary placement (September 1924 – May 1925) in Dayton, Tennessee, as sports coach and part-time teacher in algebra, physics and chemistry. Towards the end of the first term of 1925, the headmaster, who taught biology, fell sick and Scopes was asked to take some of his classes, which included revision for the senior boys in biology. There being much work to cover in only two sessions, Scopes omitted the short reference in the school text book which dealt with evolution, so he never, in fact, taught evolution to the boys (p. 60).

Scopes had been enjoying a Saturday afternoon game of tennis with some of the boys when he decided to buy a soda drink at the local drug store. Seated at the table by the soda fountain were a group of men engaged in deep discussion. They included Doc Robinson (the owner of the drug store), Mr. Brady (owner of the rival drug store), Mr. Hicks and Mr. Haggard, the town's lawyers and George Rappelyea, a businessman, who was to be the moving force in the upcoming events. (In passing I cannot resist mentioning that Mr. Hick's first name was Sue – so there really was 'A Boy named Sue'!) They were discussing an advertisement which had appeared that day in the *Chattanooga News*, placed by the American Civil Liberties Union (ACLU), which offered to pay the expenses of anyone willing to test the constitutionality of the law forbidding the teaching of evolution in public schools.

They asked Scopes if he had been teaching biology and whether it was possible to teach biology without teaching evolution. Scopes answered the first question in the affirmative, the second in the negative. They then asked Scopes if he would be willing to stand trial in a test case. "At the end of the term I had substituted in the classes of the principal while he was ill; I assumed that if anyone had broken the law it was more likely to have been Mr. Ferguson ... to tell the truth I wasn't sure I had taught evolution" (p. 60). Mr. Robinson walked over to the telephone and called the city desk of the *Chattanooga News* – and so the die was cast.

"If I had been the regular teacher at Rhea County Central High School, I wouldn't have let the law restrict my teaching the truth" (p. 53). Clearly this trial was not about truth as far as Scopes was concerned. Scopes had agreed to allow himself to be accused of something he had not done in the name of *freedom*, not truth. Several of his students were called as witnesses. One boy tried to run away, because he did not wish to testify. Scopes followed him to persuade him to take the stand: "I told him to go ahead and testify to what he had been told to say because he would be doing me a favour" (p. 105). Scopes, armed with his recently acquired legal knowledge, was well aware of the workings of the American legal system. He knew that he would not be required to take the stand. Had he been required to do so, he would either have had to perjure himself or admit that he never actually taught evolution (p. 134). Nevertheless, Scopes pressured the young boy into doing what he was reluctant to do himself, lie under oath. Scopes mentioned that one of the boys, Harry Shelton, was seventeen years old. These were not primary school children, who might not have understood the situation. They were boys who were nearly adult and who would have understood the meaning of taking the Oath and the requirement to tell the truth.

According to Scopes (pp. 47-49), William Jennings Bryan had descended upon the state of Kentucky in January 1922 "crying out for a return to Fundamentalism ... he stumped around the towns and cities of Kentucky with the fervour and energy of a political campaigner". Scopes was at Kentucky University at that time. The University opposed the introduction of the Bill banning the teaching of "evolution as it pertains to man". A tour of the University was held so that all of the legislators could see the operation of a free educational system (pp. 48-49): "They brought out the university's prettiest co-eds as special guides for the tour and to dazzle the lawmakers ... Organized opposition paid off". The Bill was narrowly defeated in both Houses.

But when Bryan's campaign hit Tennessee, there was no organized opposition and Bryan and his supporters were triumphant. The Bill, as cited above, was passed into Law.

The night before the trial began, Bryan addressed a dinner held in Dayton. Scopes recalled that Bryan said "If evolution wins, Christianity goes. Not suddenly of course, but gradually, for the two cannot stand together". Yet it became clear during the trial that Bryan was no Fundamentalist. He was deeply devout, but much to the chagrin of some of his supporters, admitted on the stand that creation may have taken place over an extended period of time, even millions of years.

Questioned in court about some of the myths of the Old Testament, Bryan found himself forced to defend Jonah being swallowed by a 'big fish' and living to tell the tale, to account for the date of the Tower of Babel and explain how long it had taken languages to spread therefrom, to explain from where Cain had found a wife and how the snake moved around before being cursed by God and forced to go upon its belly. As a speaker, Bryan had been able to choose his topic. As a witness, he was forced to answer difficult questions, a process which he found not merely difficult, but devastating. When the trial was over he was a 'broken man'. He died a few days later.

Bryan's adversary, and leader of Scopes' defence team, was Clarence Darrow, an outspoken atheist, who had earned his reputation as a leading defence lawyer by defending two teenage boys who had confessed to abducting a fourteen-year-old boy and clubbing him to death (Shermer 2006). Darrow had argued that the boys were not ultimately responsible for their actions because human volition was fiction: every act followed a cause and the cause in this case was 'environmental influence' (Shermer 2006). Support for the anti-evolution laws had, in part, been fanned by a fear that Darwinism would lead to a break down in ethical standards on the basis that humans were, after all, but beasts by nature, which position the aforesaid case illustrated. ACLU supported the employment of Darrow as leader of the defence team, but others tried to persuade Scopes to engage someone less controversial. Scopes quietly held his ground – he wanted Darrow (Ginger pp. 76-77).

Darrow's interest was not in Scopes' innocence or guilt. Although acting as Counsel for the Defence, Darrow urged the jury to convict! A conviction was necessary for an appeal to be lodged. An appeal would be heard in the Federal Court, which would have the authority to determine the legality of the disputed legislation, something which the local court at Dayton could not do.

No sooner had the jury been sworn in than they were asked to leave the Court, while legal arguments were heard regarding the breadth of argument that would be admitted. Was the case merely about whether or not Scopes had taught that Man was descended from a lower order of animals, was it about the validity of evolution in general, of evolution by natural selection in particular, about the (non) compatability of science and religion? The jury did not miss out on hearing the arguments. From their seats on the grass outside the Court House, the jury was able to hear all the arguments from inside the Court, thanks to the heat-wave conditions Dayton was enduring, which had resulted in every available window being fully opened. The front lawn was also the venue for the retirement of the jury while it considered its verdict, which it reached very quickly. In all, the jury was in the Court for only about one hour.

The legal wrangling had caused Bryan to take the stand, a position to which he, as a lawyer, was not accustomed. His testimony lasted four days. He expected to cross-examine Darrow the following day, but Judge Raulston decided that only Scopes' innocence or guilt on the charge of teaching evolution was at issue and struck Bryan's testimony from the record. Raulston could have made this decision on the first day of the trial, but he and Rappelyea were determined that the trial would 'put Dayton on the map', increasing business for Rappelyea and prestige for Raulston, whose re-election to the position of Judge was imminent. It was in the interests of both men that the trial should attract the maximum amount of attention, last long enough to maintain the interest of the journalists, but not so long that it would lose its place as front page news.

All the participants in the Scopes trial had had their own agendas. Originally from New York, Rappelyea was Dayton's largest employer and was determined to make the trial as profitable as possible. Anticipating at least 2,000 visitors to Dayton for the trial, Rappelyea and other local businessmen set about making preparations for their accommodation. An eighteen-room house, the largest in the County, which had been vacant for a decade, was in a dilapidated state with no lights and no plumbing. It was renovated in a manner suitable to house the lawyers and scientists "with Japanese seats and iron cots" (Ginger pp. 81-82). The Pullman (railway) Company was asked to sidetrack cars at Dayton to serve as temporary accommodation and the War Department was asked to supply tents (p. 73). The local hotel was repainted and had beds placed in corridors while the Dayton Progressive Club announced that visitors would be given a medal showing a monkey wearing a straw hat (p. 84). Those who could not be accommodated in Dayton were housed in Chattanooga, from where special trains ran each day to Dayton (Scopes and Presley p. 95). There were dance parties every night, amusements galore, hot-dog stands, lemonade peddlers, booths selling books on every relevant topic, Bible preaching and even circus performers who brought with them two chimpanzees. There were photographers everywhere and Judge Raulston was happy to pose for pictures, along with the lawyers. Scopes posed for many photographs and on one occasion rode through Dayton sitting on the gas tank of a low yellow racing car, arms linked with an attractive girl. Movie cameras and telegraph wires were rigged up in the court room, a novelty for the time (Ginger pp. 84-85).

Having gone to so much trouble to promote this trial and secure Dayton's place in history, Judge Raulston (and Rappelyea?) had no intention of allowing their moment of glory to be eclipsed by a later Federal trial. The jury having done what both the Judge and the Defence told them they were required to do, duly returned a verdict of 'Guilty'. Raulston asked the jury if they had fixed the fine, to which they replied that they would "leave it to the Court". The Act Provided for a fine of between $100 and $500 and Judge Raulston fixed the fine at $100. The American judicial system differs from that of many other countries in that it gives far greater power to juries in matters of sentencing. In Tennessee, a judge (magistrate) was not allowed to fix a fine of more than $50. Sums higher than that had to be handed down by the jury. As an experienced judge, Raulston must have known this.

On this technicality the verdict was later overturned, thus preventing any appeal being lodged. There was no second trial. Dayton retained its place in history.

The case against John Scopes was initiated and funded by the America Civil Liberties Union. ACLU was the political front of the Humanist movement, which, from the time of Huxley and his X-Men, had been working on both sides of the Atlantic towards the undermining of established Christianity, often using Darwinism to further its objectives.

Before the Scopes trial, Christian teaching had been focused mainly on the New Testament. The Old Testament was significant for two main reasons. Firstly, it contained the record of the Ten Commandments, upon which all ethical behavior was seen to be founded; secondly, it provided evidence of God's special relationship with his 'Chosen People', the Jews, with whom he had entered into a Covenant. This special relationship was seen to extend to the followers of the teachings of Jesus under the New Covenant. Apart from these two important factors, there was little concern with whether the tales of the Old Testament were fact, myth or allegory.

The Scopes trial focused attention on the literal interpretation of the Old Testament, especially that pertaining to the story of Creation as told in *Genesis*. Whereas before, the Fundamentalism of the Baptist Church related to whether or not complete immersion was necessary at the time of baptism, it now extended attention to acceptance of other biblical scenarios. They accepted that the world was created in six days, as written. They also accepted the age of the Earth as being little more than 6,000 years. This had been worked out during the 19th century by Bishop Usher on the basis of chronologies given in the first five books of the Bible, the Pentateuch. Increasing fossil evidence undermined this view, but the issue was not overly divisive. The age of the Earth, and the manner of its creation, were not matters of doctrine. They now became so for an increasing number of

Christian denominations, this belief becoming known as 'Creationism'.

Although the Court had considered evolution only in as far as it applied to the human race, its decision had the effect of suppressing the teaching of evolution in general. The ripple effects of this trial were to spread across the Western world, resulting in further legal action, which will be considered later. A belief in Creation according to the Bible came to be interpreted by the Creationist movement as a need totally to deny the fact of evolution. This was not the case at the time of the Scopes trial.

During the Scopes trial in 1925, Clarence Darrow had been ably supported by his 'second chair', Mr. Dudley Malone. Malone gave an impassioned speech, which Scopes (p. 154) reported in detail, saying that the Court room "went wild when Malone finished ... the judge futilely called for order. The Chattanooga policeman applauded too, pounding a table with a night stick that must have been loaded with lead; he split the table top". The central theme of Malone's argument had been that both points of view should be taught (pp. 152-154): "give the next generation all the facts ... all the theories ... let the children have their minds kept open ... Make the distinction between theology and science. Let them have both. Let them both be taught. Let them both live ... The truth is no coward. The truth does not need the law ... The truth is imperishable, eternal, and immortal, and needs no human agency to support it".

In a postscript to his account of the trial, Scopes summarized the feelings which he still had forty years after the trial. He wrote of the repulsiveness of any law restricting the constitutional freedom of teachers, of how such limitations would make robot factories out of schools.

Later in the century, this legislation would be reversed and at the beginning of the 21st century, legislation would be introduced banning non-Darwinian teaching in State funded educational institutions (including universities) (see Part V).

Chapter 27

Humans –

Ancient and Modern

While politicians and scientists were pondering the ramifications of evolutionary theory within the confines of their offices and laboratories, archæologists in England and France were quietly going about their work in the field, uncovering the history of human evolution through the study of fossil remains, which they were finding in increasing numbers. At this time, finds were being made in Europe and Asia. The finds of *Australopithecus* and *Homo erectus* in Africa came later. While the fact of evolution in general had been widely accepted, the course of human evolution was a matter of much debate.

The first ancient human remains to be found were those of the fossil which came to be known as *Gibraltar 1*, these having been located in a cave on that peninsular in 1848. Although they were recognized as a curiosity, their significance was not. It was not until eight years later, when another ancient human fossil was discovered in the Neander Valley, Germany, that the idea that there had once been a type of human walking this Earth which differed substantially from those alive today came to be accepted. Incidentally, and by the most fortuitous of co-incidences, 'Neander' means 'New Man' and thus this fossil, and others like it, received the most appropriate of names! The find was too late for inclusion in Darwin's book, but, as mentioned earlier, Huxley discussed these remains at length, and entered the argument which was then current, as to which ape these remains most closely resembled. While they were seen as a link between apes and humans, their essential humanity was not questioned. More about that later.

During the first half of the 20th century, the position became confused. Neanderthals were accepted, if not as direct ancestors of modern humans, then as a 'race' or 'variety', rather than as a distinct evolutionary line. The finding of a child's skull in Taung, South Africa, caused quite a stir. It being so young, it had not assumed its full adult shape, which led to confusion as to its correct evolutionary position. While more Neanderthal remains had been found in France, for example at La Chapelle-aux-Saints in 1908 and La Ferrassie in 1909, what appeared to be even earlier specimens had been found in parts of Germany, at Mauer in 1907 and Steinheim in 1933. Another find had been made in southern Africa, at Kabwe, Zambia, in 1933. Confusion was increased by the practice of allocating new names to new finds. The term *Homo* refers to the genus, not the species. Terms such as *Homo heidelbergensis*, *Homo rhodesiensis* and *Homo neanderthalensis* implied speciation, although the remains were generally accepted as belonging to the human genus.

Once the reality that there had at one time existed a form of people quite different from those alive today had taken hold in the public imagination, far more care was taken by miners and other excavators, not only with the preservation of any skeletal remains or artifacts they uncovered, but in the noting, and where possible preserving, of the provenance of such finds. In 1864 King had created the nomenclature *Homo neanderthalensis* to indicate that archæologists were dealing with a species separate from *Homo sapiens*. He felt that, the skull being so simian, "the thoughts and desires which once dwelt within it never soared beyond those of the brute" (King 1864: 88). At that time, only the one skullcap had been identified, but as more remains were uncovered, together with beautifully worked stone tools, it became increasingly accepted that Neanderthals were fully human. They used fire and buried their dead, often with artefacts and/or food, such as joints of meat (Keith 1915). The discovery of a jawbone in ancient gravel at Mauer, near Heidelberg, confirmed that there had once lived in Europe a people even more primitive than the Neanderthals. The term *Homo heidelbergensis* came to be used by some to refer to any pre-Neanderthal remains discovered in Europe, and sometimes also in Africa. None of the early writers, such as Boule, Sollas, Vallois or Keith, had any doubt that these ancient people were fully human.

The man from La Chapelle-aux-Saints, as well as the people at La Ferrassie, had all been buried with their feet facing west. The westward direction for burial continued into the Azilian (late Magdalenian ~12,000 ya) period and, indeed, into Celtic times. If the Neanderthal mind was already philosophizing about life and death, then "clearly ... Neanderthal man does not represent the human dawn" (Keith 1915: 117). Burials were often protected by slabs of stone, a practice which persists in one form or another until the present. Of particular interest were the cup-hole marks in some of these stones, often presumed to have been associated with grave offerings, or some other ritual, although at La Ferrassie the cup-holes were on the underside of the stones. Early (Neanderthal) man was seen to have been a social being, with males bringing flesh food to the family group and with ceremonies accompanying his last journey to the grave.

In France, Marcellin Boule attempted to reconstruct the skeletal remains of 'The Old Man from La Chapelle', these being the most complete. Unfortunately, Boule failed to make allowance for the fact that this man had suffered from degenerative arthritis which had distorted, not only his limbs, but also the angle at which he held his head – something commonly seen in older people. The result of his reconstruction was the portrayal of a person unable to stand fully erect. From this he concluded that the Neanderthals still retained many simian characteristics.

While conceding that Neanderthal brains were comparable with those of modern humans in regard to size, Boule claimed that the design of the convolutions of their brains indicated that they had only rudimentary intellectual faculties and language. Furthermore, the Mousterian stone tools associated with the Neanderthals were considered very primitive which argued against any superiority of the brain. In fact, the Mousterian tool kit was no more primitive than that of the

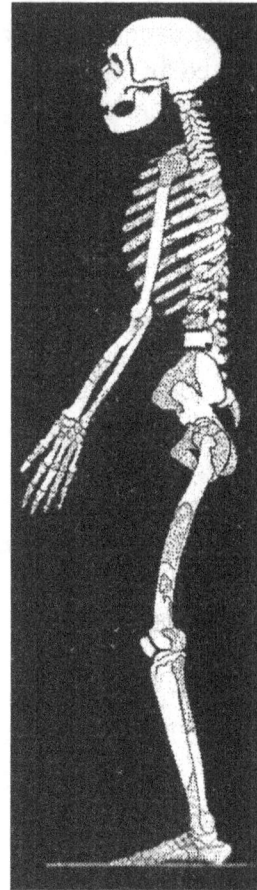

Fig.27.1: Boule's Reconstruction of the Neanderthal from La Chapelle aux Saints (Boule 1923: 225)

Australian Aborigines, especially the Tasmanians. Sollas (1924) believed that, were it not for the different stone from which they were made, it would be difficult to distinguish between Neanderthal and Tasmanian tools.

Boule acknowledged the rightful place of Neanderthals within the genus *Homo*, although he did believe that they were a distinct species. Nevertheless, he did not deny the possibility that some Neanderthal blood may have entered that of other human groups by way of hybridization, although such infusion would have had very little overall effect. (Interbreeding would not have been possible if the Neanderthals were a truly separate species, at least not beyond the first generation.)

Sollas was far more cautious than Boule in his interpretation of Neanderthal features. He disputed the consistent portrayal of the Neanderthal jaw as prognathous, pointing out that the skull remains from Krapina and Gibraltar were as orthognathous as many white men, while many Australian Aborigines were as prognathous as were some of the Neanderthals.

Keith (1948) mentioned that Vallois had claimed only five per cent of Neanderthals had lived over the age of forty, but gave no evidence in support of this claim. Since *Sinanthropus pekinensis (H. erectus)* lived even earlier than *Neanderthalensis*, he assumed that the former lived no more than twenty years (Keith 1948). These assumptions are still affecting interpretations today. They appear to have been based on nothing more scientific than the knowledge that, in the wild, chimpanzees may live for approximately twenty years. If chimpanzees live for approximately twenty years and humans for approximately seventy, then early humans must have lived for a length of time somewhere in between these two figures. That forty was chosen, rather than fifty, shows the inclination of the writer to prefer to consider the Neanderthals closer to simians than to modern humans.

Experts can only estimate the *biological* age of skeletal remains, not their *chronological* age (Hunter et al. 1996). For an infant or child, the chronological age may usually be estimated with a fair degree of accuracy, based on the development of teeth and ossification of the bone, whether or not the epiphyses have fused, etc. With age, the cranial sutures ossify and these provide a further biological means for making a chronological estimate. Once tooth and bone development are complete and sutures closed, accurate aging becomes increasingly difficult.

When human remains in the Spitalfields Cemetery in London needed to be relocated due to road works, forensic scientists took the opportunity to conduct a study to see how accurate were their estimates of age at death compared with the information given on the plates on the lids of the coffins. Results showed a tendency to over-age the young and under-age the old (Molleson and Cox 1993; Hunter et al. 1996). Nor were these errors minor. Even persons in their fifties were under-aged by as much as ten or fifteen years, those in their seventies, eighties and nineties sometimes by as much as twenty or twenty-five years, or more. There was clearly a reluctance on the part of the scientists to accept that people in the Middle Ages lived beyond seventy years, which many of them did. There were equally large over-estimates of age made in respect of some of the younger remains, with errors of twenty, thirty, even forty years being made.

Hunter et al. concluded that the theory that the majority of people died young in antiquity might not be true, since people age biologically at very different rates. By 'antiquity' Hunter et al. were referring to the last few hundred or thousand years. Nevertheless, their conclusions are equally true of the most ancient of remains.

Life expectancy during Mediæval times was low for town dwellers, due to overcrowding and poor hygiene. Indeed, life expectancy remained artificially low until the end of the 19th century. The connection between impure drinking water and disease was not made until the middle of that century and, even after it had been recognized, it took time to connect everybody to a supply of clean water such as people had enjoyed in times past. It also took time to ensure the proper removal of waste products – not just human waste, but rubbish in general, which attracted rats. Outbreaks of the plague and the Black Death had periodically decimated populations – as much as one third of Europe's population is believed to have died in one outbreak, leading to an 'outbreak' of prosperity among the survivors, who inherited a share of the deceased's wealth!

In addition to these 'natural' causes, there was also sickness and death brought on by ill-advised medical practice. As scientific fervor took hold among the educated elite, the medical fraternity started using metals as medicines: gold, silver, arsenic, antimony, mercury and many, many more poisonous substances replaced traditional herbal medicines in the apothecary's store. Nor was that all.

Disease was proclaimed to be caused by 'bad blood' and it became the surgeon's duty to remove as much of this bad blood as was necessary, or possible, from their long-suffering patient – and I mean 'long-suffering'! While some physicians were relatively modest, others bled their patients regularly, some even using their thumb to press upon the vein to squeeze out the last possible drop. Nor did the conscientious doctor minister only to the sick. Pregnant females were not forgotten! Many women died in childbirth, so medical attention during pregnancy was seen to be of the utmost importance to those who could afford it. Unfortunately, this attention resulted, at its worst, in about half of the female population dying in childbirth – not necessarily the first child, of course, but, even so, at one time half the female population was dead before their fortieth birthday! If women faced dangers on the home front, increasing numbers of men were facing danger at work. There was no such thing as 'Workplace Health and Safety' and it was not just in the mines that men were at risk. For example, many men died as the result of working in factories such as those making that miracle of modern science - the match! These were made from phosphorus, which was deadly poisonous and which resulted in many deaths before the problem was realized and rectified by the introduction of the 'safety match'.

It was during these times that centralized records started to be kept of births, deaths and marriages. Of course, parish records had been kept for many centuries but the keeping of this information by the State at a centralized location was new. A problem arises when people today extrapolate from this data to assume that the increasing longevity to which this restricted data bears witness is evidence, not only of an on-going trend which will continue into the future, but of a trend which has been happening since the time of the Neanderthals. I do not share the optimism of some today that the average age at death will rise beyond 100 by the end of this century - not unless something is done about the consumption of drugs and alcohol and poor diet so prevalent in many today. I do not share the retrospective pessimism which assumes that the life expectancy of ancient people was but half ours.

Today's increased longevity is as much due to improved hygiene and sanitation as it is to improved medicine. The Neanderthals lived in small groups, ate fresh food and breathed clean air. There is no reason to suppose that their life expectancy was any less than ours today, except as the result of direct trauma and trauma-induced infection.

All human beings alive today had ancestors walking this Earth in Pleistocene times, but not all people of Pleistocene times have descendants alive today. The question that increasingly occupied the minds of archæologists during the first part of the twentieth century was: to which of the human remains being unearthed could we claim relationship? How much could we learn from modern hunter/gatherer tribes about the possible life-styles of our ancestors and, more importantly, was it possible to glean evidence of direct descent from any Pleistocene people? Sollas believed that it was.

Sollas looked to the Tasmanians for the closest analogy to the life style of the Neanderthals, although he stressed that he did not postulate any direct blood link between the two. The Tasmanians had certain physical characteristics that distinguished them from other groups of present-day humans but did not isolate them as a separate species. If the way of life and

305

stone tool technology of the Neanderthals was similar to that of the Tasmanians, there was no reason to suppose that the Neanderthals were any less human than the Tasmanians.

Sollas commented that rafts, presumably used to reach Australia, had not been preserved in the archæological record. There was no way of knowing whether Neanderthals had used any form of water craft. The fact that none had been preserved was not evidence that none had ever existed.

The Tasmanians made a 'wine' by fermenting juice they tapped from a particular gum tree *(Eucalyptus resinifera)* and the Bushmen of South Africa made forms of mead by crushing the bodies of honey ants, from honeycomb or by infusing honeysuckle flowers or the fruits of various trees. We have no way of knowing whether the Neanderthals made any form of alcoholic beverage, although the fact than one, and possibly two, of the Neanderthal males from Shanidar suffered from diffuse idiopathic skeletal hyperostosis (DISH) may indicate that they did. More on that later.

Sollas also noted other cultural similarities between Australian aborigines and people in other parts of the world, both ancient and modern. The practice of implanting hand prints, both in outline (negative) and in full (positive) was also practiced in Africa, Arabia, Arizona, Babylonia, California, France, India, Mexico, Palestine, Peru, Phœnecia and Spain from Aurignacian times onwards. These hand prints appeared to show that the practice of amputating one or more finger joints was common all over the world, although the possibility that some digits may have been severed accidentally could not be discounted.

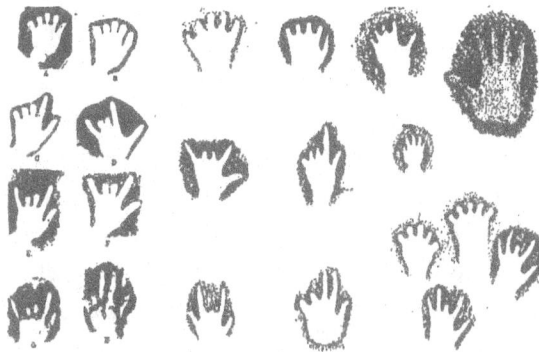

Fig.27.2: Hand prints
(Sollas 1924: 418)

Fraser (1909) published his famous book, *The Golden Bough*, to celebrate the centenary of Darwin's birth and the half-centenary of the publication of *The Origin.* In it he traced creation myths from around the world, which he found to be surprisingly consistent, many involving the taking of a bone/rib from a man to create the first woman. The Tahitian word for bone was *ivi*, pronounced 'eevee'. Rather surprisingly, Sollas did not cite any of these similarities in support of his theory of cultural continuity, possibly because he could not be sure that 'contamination' by ancient travellers had not occurred.

Most people accepted that the skull named *Pithecanthropus* found in Java by Dubois was a connecting link between the human and ape lines. Brain casts showed development of the area associated with speech, indicating that even at this early stage of human evolution, *Pithecanthropus* was capable of at least rudimentary speech. The discovery of *Sinanthropus pekinensis* at Chou K'ou Tien, near Peking in China, will be covered later. These discoveries came too late to be included by Boule in his major work but were recognised by evolutionary theorists of this time as having predated the Neanderthals, as had *Pithecanthropus*. Vallois, when he updated Boule's work (Boule and Vallois 1957), recognized *Sinanthropus pekinensis* as having been fully human since he had kindled fire, made hearths and had a stone and bone tool industry. Boule correctly identified the lower jaw bone found at Piltdown as that of an ape, which he believed had mistakenly been associated with the cranium found nearby. He was of the same opinion in relation to a skull found in Java in 1890, questioning whether the femur was necessarily from the same individual as the partial cranium, just because they were found in the same locality.

Inasmuch as *Sinanthropus pekinensis* also walked erect, it became clear that, if Boule's reconstruction of the La Chapelle-aux-Saints skeleton was correct and Neanderthals did not walk fully erect, then they must have been a degenerate species. This view became less and less popular, but was still supported by Vallois inasmuch as he did not edit or amend that part of Boule's original work. *Pithecanthropus* and *Sinanthropus pekinensis* both later became known as *Homo erectus.*

The discovery in 1921 at Broken Hill, Rhodesia, of what appeared to be the remains of an ancient human, which post-dated *Homo erectus*, was the first to be made in Africa and deepened the problem of human evolution, rather than providing any solution. The skull was seen as bearing a degree of similarity with that of the Neanderthals, but was believed to be of a later date because this person, named *Homo rhodesiensis*, walked upright. Boule believed that the Neanderthals, *H. rhodesiensis* and the Australian Aborigines had a common origin, the Australian Aborigines being their final representatives. Boule did not say that either *H. rhodesiensis* or the Australian Aborigines were descendants of the Neanderthals, merely that they had a common ancestor.

Although the precise place of ancient humans within the human genealogy was a matter for debate, all writers during the first half of the twentieth century agreed that the skeletal remains found at Cro-Magnon in 1868 were representative of modern humans. The four skeletons were described by Keith (1915) as tall and lanky, their average height of almost 6 ft. (180 cm.) being greater than the average height of Europeans even today and certainly more than that of Europeans during the Middle Ages. The tallest of the Cro-Magnon was estimated to have been 6 ft. 4 ins. (191 cm.). Keith suggested that they resembled people from the Punjab, or possibly Africa, since there were similarities with Negroid races, such as limb proportions. Sollas noted that their fingers were short compared with the overall size of the hand.

A number of skeletal remains were found in the *Grotte des enfants* at the Grimaldi Caves, north-west Italy. Some were of the Cro-Magnon type, but at a deeper level remains were found of two people of considerably shorter stature, about 5 ft. (152 cm.). Boule and Sollas both believed that these people were also Negroid, there being two races of modern humans in Europe during Aurignacian times, both with some Negroid affinity. Evidence for the existence of these smaller people was also seen in small hand prints in the caves at Gargas. On the floor of one of the chambers in the Tuc d'Audoubert Cave imprints had been preserved of small human feet. Notwithstanding the difference in size, Keith saw the Grimaldi people as members of the same race as the Cro-Magnon people and this is the opinion generally held today.

As late as 1948, Keith, still reliant upon relative dating techniques, assumed that the Cro-Magnon people had replaced the Neanderthals "quite suddenly, some 100,000 years ago". From where had the Cro-Magnon people come? This mystery seemed to be solved when an expedition during the early 1930s discovered the remains of ten people at Mount Carmel, Palestine, which appeared to be transitional between the Neanderthals and Cro-Magnon. This led Keith to suggest that "if all turns out as we anticipate we may claim that the Caucasians of S. W. Asia still occupy the original area of their evolution". The Garden of Eden, if not intact, was still in place!

Boule, conversely, claimed that the Neanderthals could not be the ancestors of *Homo sapiens* since both species were contemporary. It is difficult to see how one species could give rise to the other unless they were contemporary, at least for a short period of time.

By the Late Pleistocene, mammoth had generally been replaced in Europe by reindeer. Human dependence on reindeer was such that it led Boule and

Sollas to suggest that, as the ice retreated and the reindeer moved north, so, too, did people, eventually following the reindeer into the arctic regions and, possibly, across the Bering Strait and the Aleutian Islands into Canada and other parts of the Americas.

Sollas suggested there was a genealogical connection between the Eskimo (Inuit) and the Magdalenian people. The Eskimo made great use of seal intestines, stretching and drying them, then cutting them into strips which, when sewn together, made extremely lightweight and water-proof overalls. Other pieces were used to make windows. These uses serve as a timely reminder that the artefacts which survive in the archæological record need, by no means, be all that were produced.

The position within the field was confused. The Neanderthals were accepted, if not as the ancestors of modern humans, as a 'race' or 'variety', one which later became extinct. While the human race/species was seen to have come into existence with the appearance of *Homo erectus*, confusion was introduced by the practice of giving new names to new finds. This was done as much to establish the name of the *discoverer* of the fossil as it was to establish the name of the fossil! There was a degree of distinction associated with being the discoverer of a new 'species', although no scientific evidence of speciation was required. Small, but distinct, differences in skull shape were all that was necessary. This differentiation was quite inadequate. No forensic pathologist today has any difficulty identifying modern human remains as being either Negroid or European, Asian or Australian Aborigine, yet these people are all of the one species. The debate as to exactly when, and how, speciation occurred continues to this day.

Chapter 28

The Mathematics of
Evolution

W hile exciting discoveries were being made in the field, exciting discoveries were also being made in the laboratory. The four years of the First World War precipitated the largest change in the shortest period of time in the entire history of human social existence. Gone was the privileged position of the aristocracy, the gentleman naturalist/scientist. Indeed, gone was the 'gentleman' – and the 'lady' too, as more and more women took positions of paid employment in society. By the time of the outbreak of the Second World War, so emancipated had women become that, for the first time in history, they were conscripted into the armed forces.

I cannot help but feel glad that Alfred Russel Wallace died in 1913. He lived long enough to know of the foundational principles of modern genetics, of particulate inheritance which, it was thought, overcame the problem of blending which, at the end of the 19th century, threatened to consign the principle of natural selection to the pages of history. Darwinian (and, therefore, Wallacian) evolution enjoyed a resurgence in popularity as the microscope offered up the secrets of the reproductive cells. He died a 'gentleman' and was spared the trauma of witnessing the devastation of war.

The war over, the 'gentleman scientist' was replaced by the scientist in paid employment, which had the advantage of allowing laboratories which invested in the latest equipment, which brought together, several men all interested in the same objective, working together towards a common goal. The study of genetics attracted some of the brightest of minds, not because genetics offered to illuminate the dim reaches of humanity's past, but because it promised to provide a productive path forward for medical research. Certain conditions, such as hæmophilia and mongolism (as it was then called), were recognised as having an aberrant genetic basis. Understanding how genetic changes came about and, more importantly, how undesirable changes might be reversed, demanded the attention of some of the best minds in medical science and evolutionary theorists benefitted from this increased understanding.

Much of the research work done in laboratories during the first half of the 20th century was carried out on insects, because of their quick rate of reproduction, their large numbers and their comparative cheapness. Small mammals, such as rats and guinea pigs, were also used, as were amphibians, such as frogs. The most work was carried out on the various species and varieties of the fruit fly, *Drosophila*, which possessed particularly large cells in their salivary glands, observed easily under a microscope. However, these cells, being somatic (body) cells, only underwent mitosis, not meiosis, which occurs only in reproductive cells. Somatic cells could differ from each other, through mutation, but it was realized that such somatic mutations were not hereditary and could not be passed on by inheritance. Although drawing some conclusions as to occurrences during meiosis from mitosis, care had to be taken not to extrapolate from the one situation to the other too freely. In 1942, Sir Julian Huxley published a widely read and influential book entitled *Evolution: The Modern Synthesis* in which he summarized the then current thinking which sought to bring about a synthesis between genetics and Darwinism. Huxley downplayed the step between mitosis and meiosis (p. 132):

> The existence of mitosis, of however simple a nature, presupposes the need for accurate mechanical division of the hereditary substance; and this in turn would not be necessary unless the hereditary substance were differentiated into specialized parts, each with their appropriate functions. Thus the mitotic organism has reached a stage of particular inheritance, based on spatial differentiation of the germ-plasm ... Apparently, once the detailed differentiation of the germ-plasm into accurately-divisible chromosomes had been accomplished, it was comparatively simple to alter the timing of the various processes involved in one cell division, so as to produce meiosis; and this was fraught with such advantages that it was all but universally adopted.

No suggestion was offered as to why or how genes and chromosomes came to exist in single-celled organisms, interest being in the evolution of life forms, not their origin. Huxley appeared to blur over not only how, but why, meiosis was first established, possibly because it was difficult to explain why a single-celled organism, able to live and multiply unassisted, should 'choose' to become reliant on another organism.

Notwithstanding the fact that a single mutation resulting in the formation of a new recessive gene could not be expressed in the phenotype (i.e.

manifest in the body) and, therefore, could not be available for 'selection', it became accepted that *recessive* mutations (even if originally disadvantageous) could spread through the gene pool for many generations, forming different combinations, possibly even becoming dominant under certain conditions (Goldschmidt 1938). Eventually conditions might arise such that formerly disadvantageous mutations might manifest in an advantageous way, making them subject to natural selection. Nevertheless, it was recognized that, while the amount of genetic variation might increase within an interbreeding population, this increase in genetic variation would not, of itself, cause the formation of distinct groups, i.e., it would not cause speciation (Dobzhansky 1951; Goldschmidt 1940).

Darwin had earlier recognized the importance of isolation in the formation of new species. He was later to reduce the importance of geographical isolation by the introduction of sexual selection. As the twentieth century progressed, the importance placed on sexual selection was reduced, since it was acknowledged that the number of species in which a dominant male was responsible for most of the offspring was really quite small and in nature all females were mated, no matter how 'unfavoured'. (The much beloved merekats would appear to be the exception to this rule, but this was unknown at this time). Evolutionary theorists once again turned their attention to isolation as the means of speciation, be it geographical or reproductive.

Dobzhansky argued that geographical isolation and time were all that were required for a species to subdivide into distinct races/varieties. Of particular interest were a number of intergrading subspecies which formed a loop or overlapping circle, the terminal forms of which no longer interbred, even though they existed in the same localities. For example, in Britain there are two distinct species of gull: the Herring Gull and the Lesser Black-backed Gull. They do not interbreed. However, they are connected by a series of varieties around Europe, Siberia, Alaska and North America and back to Europe, each of which can interbreed with its adjacent variety. If variety 'a' can interbreed with variety 'b', which can interbreed with variety 'c', and so on, but variety 'f' cannot interbreed with variety 'a', are varieties 'a' and 'f' different species? How can they be, when there is a known gradation of interbreeding varieties between them? Yet inability to interbreed became accepted as the definition of a distinct species, a convenient definition such as the biologists of the previous two centuries had struggled in vain to operationalize. Mayr concluded that at some point various subgroups of the original population isolated themselves from each other as a result of preferences for different ecological niches and this isolation eventually resulted in speciation. The isolation was not necessarily geographical. Dogs today could be regarded as such intergrading subspecies. All dogs belong to the same species and are theoretically capable of interbreeding but there are physical differences which would make the mating of a St. Bernard and a Chihuahua impossible. If breeders were to decide, for some reason, that they only wanted to breed very large and very small dogs, and all medium sized dogs disappeared, two 'species' would result.

Conversely, geographical isolation could occur over extended periods of time, yet no new species be formed. The connection between the Pacific and Atlantic Oceans at Panama had been broken some two to three million years ago, but some of the species of fish and crustaceans were still the same on both sides of the Isthmus (Mayr 1942). Certain plant species found in eastern Asia and western North America, which had probably been separated for millions of years, were still morphologically indistinguishable. Geographical and biological isolating mechanisms were required to work in conjunction.

While all theorists were in agreement that geographical isolation was a prerequisite for speciation, just what constituted geographical isolation? A population of butterflies in a meadow? A population of deer in a wood? But might not butterflies be blown to a new meadow, where they might mix with others similarly displaced? Might not deer leave their usual grazing ground in search of food in hard times? Pacific Islands were evidence for long migrations. Even the Galápagos Islands were home to land turtles. Mayr argued that, far from enhancing the potential for evolution, islands were evolutionary traps for such species as settled them.

Dobzhansky (1937: 228) held that Romanes' contention that without isolation, evolution was impossible was subject to misinterpretation:

> The difference between individuals and groups may be due to a single gene or a single chromosome change. Such differences can never be swamped by crossing, since in the offspring of a hybrid segregation takes place, and the ancestral traits reappear unmodified. No isolation is needed to preserve the variation.

Recessive genes were both a blessing and a curse. On the one hand, they could not be 'swamped out' - they could be preserved through generations, unmodified. However, they did not manifest in their recessive state, so Dobzhansky's claim that "the ancestral traits reappear unmodified" must be seen as referring to a dominant gene which acted as a recessive in certain combinations but which was still preserved and able to resume its dominant position when circumstances changed. The situation was not that of a recessive gene becoming dominant for the first time, but of a previously dominant gene resuming its former status.

Ford (1931) and Simpson (1944) agreed that a small degree of isolation was all that was required for change to become possible within a population, even without mutation, due to chance increase or decrease in the proportion of various genes. However, no basically new types of organisms could arise by this means, nor could new species arise (Dobzhansky 1937; Simpson 1950). Dobzhansky and Simpson rejected Ford's claim that gradual change in the gene-complex alone would be sufficient to "lead to the establishment of genetic incompatibility between them [colonies]. This will be slight at first, but will end in a condition of partial and later of complete 'inter-specific sterility'" (Ford 1931: 90).

Numerous species of fish, many closely related, existed side by side in some of the great African lakes: 171 species of *cichlid* in Lake Nyasa, more than 300 species of *Gamimarid* crustaceans in Lake Baikal (Mayr 1942). It had been suggested that these species had evolved from a common ancestral species by adapting to different ecological niches, a process known as sympatric speciation. Authors such as Mayr (1942) and Muller (1949/1963) considered this was no longer an acceptable solution, since it was apparent that, even between fish occupying quite different parts of the lake, or living at different depths, there was no barrier which provided absolute isolation. Intermediate areas, with intermediate varieties, must exist and these would serve as a genetic bridge. They concluded it was more likely that the various species had previously existed in the rivers which were later, as a result of geological movement, to feed into the same lakes. How these fish/crustaceans, having evolved in separate rivers, came to be such closely allied species was not explained.

Goldschmidt (1940) questioned the whole concept of 'incipient' species. Any isolated group within a population would have the same chance as any other of proceeding towards speciation. The only advantage a so-called 'incipient' species would have had would have been that a few mutations had already been accumulated.

In 1943 a committee had been established in the U.S.A. by the National Research Council to consider "Common problems of Genetics, Paleontology and Systematics". The publication in 1949 of *Genetics, Paleontology and Evolution* was the record of this committee's work, aimed at bringing about 'a meeting of minds' in the fields of genetics and palæontology. In this aim they were only partly successful. Of the twenty-two papers included, sixteen were from the perspective of palæontology and/or systematics, which made only passing reference to genetics, and six primarily addressed genetics, with only minor references to palæontology and/or systematics.

On the other side of the Atlantic, attention was also being focused upon a possible synthesis between Darwinian evolutionary theory and modern genetics which was to culminate in the *Evolutionary Synthesis*, as summarized and expounded by Sir Julian Huxley.

As the first half of the twentieth century drew to a close, reproductive isolation induced by geographical isolation became established as the essential criterion for the establishment of new species (for example Ford 1949/1963; Goldschmidt 1938, 1940; Lack 1949/1963; Mayr 1942; Muller 1949/1963; Wright 1949/1963). Although Darwin had noted that large land masses gave rise to the greatest diversity of species and Mayr had concluded that isolated islands were evolutionary traps, nevertheless the mathematics of evolutionary reproduction led to the conclusion that a new genetic variation would have the best chance of becoming established if it occurred in a small population. But even the assumption of small populations could not provide a complete answer. Dobzhansky (1937: 284-285) summarized the problem:

> It is indeed difficult to conceive how isolation between two groups of individuals might arise through a single mutation. Mutations that change the sexual instincts, or the structure of the genitalia, or the physiology of the gametes, or some other properties of their carriers that are essential for reproduction, may occur. Such mutations may prevent crossbreeding of the modified and the ancestral types, but this is not yet sufficient to produce a workable isolating mechanism. For isolation encountered in nature has always two aspects: the crossing of individuals of Group A with those of Group B is made difficult or impossible, but individuals of A, as well as of B, are fully able to breed *inter se*. A mutation that would produce isolation must therefore not only prevent crossbreeding between the mutant and the original type, but must simultaneously ensure that internal crossability of the individuals carrying the mutations. Such a coincidence can hardly be imagined to be a common occurrence ... With mutation rates that are as low as those observed for most genes in the laboratory, the number of mutants produced in each generation would be so small that they could hardly find mates.

The rarity of mutations occurring without laboratory interference, most of which were either neutral or detrimental, even lethal, necessitated consideration of the mathematical possibilities and probabilities of the evolutionary consequences of genetic mutations.

Within any inter-breeding population, genetic uniformity would eventually be reached. One mathematical law formulated by Hardy in England and Weinberg in Germany, subsequently known as the 'Hardy/Weinberg Law (1908)', stated that if no new genetic material was introduced, gene frequencies would remain constant from generation to generation indefinitely, the equilibrium of AA, Aa, aA, aa in a randomly breeding population being reached in a single generation for a two–allele system.

While equilibrium might be the rule throughout a large breeding population, there would be fluctuations within small, local populations. This might result in a mutation becoming 'fixed' (universal) throughout a small population, or being eliminated altogether, both events being possible without the participation of natural selection. The maintenance of the genetic equilibrium was seen as a conservative, not a progressive, factor. Notwithstanding the Hardy/Weinberg Law, which he, himself, had cited, Dobzhansky (1937/1951: 131-132) went on to claim:

> If the heterozygote Aa (a being a mutant gene decreasing viability) is as viable as the ancestral homozygote, AA, the frequency of the gene a will be allowed to increase until the Aa individuals become so frequent in the population that their mating together is likely to take place, and the homozygous aa are produced. The aa being unfavourable, the aa individuals will be eliminated, and this will impose a check on the further spread of the mutant gene a in the population.

The proportion of Aa heterozygotes would remain the same, according to the Hardy/Weinberg Law, yet Dobzhansky assumed that "a will be allowed to increase" until the double recessive, aa, made its appearance. Dobzhansky assumed that the recessive a would be unfavourable, which was in accordance with laboratory findings that most mutations were neutral, unfavourable or lethal. Other theorists had to assume that some, at least, of the recessives were potentially beneficial, just waiting to be selected. Incipient favourability was the explanation of a mutant gene's ability to overcome the mathematics of genetics according to Hardy/Weinberg. If aa were to be lethal, their carrier would die, but a would continue in its heterozygous form, a point made strongly by Haldane (1938) when writing in opposition to eugenics.

Gene mutations of detectable magnitude were estimated to occur once in every 100,000 to 1,000,000 individuals (Ford 1931). An advantageous mutation would be very rare, occurring perhaps once in 10^9 (10,000,000,000) individuals. With the entire human population of Africa two million years ago not reaching anything like this number, these figures raised considerable questions as to how positive evolution could have reached the stage it had in slow breeding organisms. Evolution had to be measured in generations, which meant that slow breeding mammals had evolved at a rate many times more rapid than that of quickly breeding animals, and even more rapidly when compared with fast breeding insects.

The apparent slow rate of mutation would be lessened if more than one mutation occurred at the same time, as Haldane (1932) suggested would have been necessary for organs such as the human eye or hand. However, the simultaneous appearance of several gene mutations in one individual had never been observed (Simpson 1944: 54):

> Postulating a mutation rate of .00001 and supposing that the occurrence of each mutation doubled the chances of another mutation in the same cell — a greater departure from random incidence than is likely to occur — the probability that five simultaneous mutations would occur in any one individual would be about .000000000000000000001 (21 noughts). In an average population of 100,000,000 individuals with an average length of generation of only one day, such an event could be expected only once in about 274,000,000,000 years — a period about one hundred times as long as the age of the earth.

These calculations were based on the age of the earth as estimated at that time. The large populations with short generation spans postulated by Simpson were presumably fruit fly, generally used in laboratory experiments. A smaller population of larger animals reproducing annually would considerably inflate the figures. Larger mammals with smaller populations and longer reproductive spans, would have involved figures that were truly astronomical! Human populations may be large now, but during the time that humans were evolving, they were quite small. It was not unusual, among hunter-gatherer tribes, for between two-to-four years to elapse between infants, at least those that survived. If this rate of reproduction was similar to that of early humans, and their antecedants, that would place huimans as among the slowest reproducers, making their 'advanced' evolution even harder to accommodate within the assumed time frame.

Simultaneous mutation, providing several copies of a new gene, would seem to be necessary to allow manifestation of a new, recessive characteristic but (Dobzhansky 1937/1951: 40):

> Mutation changes one gene at a time; simultaneous mutation of masses of genes is unknown ... A sudden origin of a species by mutation, in one thrust, would demand a simultaneous mutation of numerous genes. Assuming that two species differ in only one hundred genes and taking the mutation rate of individual genes to be as high as 1:10,000, the probability of a sudden origin of a new species would be 1 to 10,000,[100] [a further one hundred noughts].

Fisher had thought that for a mutation to play any role in evolution, the gene would somehow have to change from being recessive to being dominant. Laboratory work showed that, while most mutant genes were recessive, some were not completely so, i.e. had some immediate influence,

317

helping to give rise to an appearance of 'blending'. According to Huxley (1942: 56):

> A mutation with partial dominance occurring once in 10^5 [100,000] individuals will, if selectively neutral, take a period of somewhat over 10^5 generations to establish itself in half the individuals of a species. If there were the faintest adverse selection against it, it would never increase at all. But if it conferred an advantage of only 1% ... then it would establish itself in half the individuals of the species in a period of only about 10^2 generations.

Mendelian genetics *alone* could not account for dissemination of a mutation throughout a population. For that, it was necessary to call upon positive selection, and mathematics was used to estimate its effect. The foremost proponent of this approach was Fisher whose book was so full of obtuse formulæ, incapable of being understood by anyone not well versed in statistics, that it is frequently mentioned but rarely discussed. Haldane claimed that he could write with authority on the subject of natural selection because he was one of only three people who understood its mathematical theory, the other two being Fisher and Wright.

Dobzhansky (pp. 178-179) was by no means convinced that mutation and selection accounted for evolution:

> The number of generations, and consequently the amount of time needed for change, may be so tremendous that the efficiency of selection alone as an evolutionary agent may be open to doubt ... Combined with mutation, the process of selection may be either enhanced in speed, or, vice versa, slowed down still further [because mutations are reversible] ... Whether the combined forces of mutation and selection are sufficient for a sustained progressive evolution is not immediately clear.

Dobzhansky made no suggestion that some superior creative force was involved but it will be remembered that he was one of the people who tried to established 'New Humanism' which would have allowed for both theistic and atheistic views.

Ford suggested that a non-adaptive gene would continue to affect only the same *proportion* of the population as at its first appearance, although if that population were to increase in size, the *number* of those genes would increase. Then, if the size of the population were to be reversed, the spread of non-adaptive genes would be checked, and those of semi-lethal character quickly eradicated.

Huxley (1942: 58) believed that repeated mutations and a large aa population were necessary for the establishment of an advantageous mutation:

> Even with a definite selection advantage such as 1% ... the chances are

strongly against a lone mutation surviving in the species ... Thus repeated mutation together with a considerable-sized population, are necessary for new mutations to have an evolutionary chance.

Here Huxley was agreeing with Darwin and Mayr that a large population was more conducive to evolution. This was contra Haldane (1932) who argued that mutations had the best chance of becoming established in small populations, that single mutations would almost certainly disappear unless the population was highly inbred. Dobzhansky, as a result of his mathematical calculations, concluded that the majority of mutations would be lost within a few generations irrespective of whether they were neutral, harmful or useful to the organism.

The large *versus* small population conundrum was resolved by Huxley thus: for advantageous mutations to occur, a large population was needed and for such an advantageous mutation to become *established*, it was necessary for the same mutation to occur numerous times. Only in a population of infinite size, mating entirely at random, would the proportions decreed by the Hardy/Weinberg Law operate precisely. Since all populations are, in fact, limited and mating is never completely random, gene frequencies in the smaller (sub)populations would be subject to chance variation and offer higher possibility for a particular gene to become fixed or deleted entirely.

Simpson (1953) estimated that it would take some 500,000 years, and often much longer, for one species to change into another. Mayr (1942) estimated that the time required for the formation of a new subspecies in the Holarctic region had been between 5,000 and 15,000 years, concluding that if such long periods of time were generally required for the production of *subspecies*, all geological time would not be sufficient to explain the present diversity of animal and plant life.

Notwithstanding the statistics cited, all of the above authors continued to support Darwin's theory of evolution by natural selection. Their own calculations told them of the astronomical amounts of time which would be needed for a beneficial mutation to occur and become established. However, evolution by natural selection had become the dominant paradigm and it was upheld, despite, rather than because of, the emerging evidence. Within the scientific community there was no alternative position. Funding for research was not available outside the established scientific community, leaving dissenters with no avenue for research within the established, accepted scientific community.

The work of a number of theorists of this time, including the

319

mathematician, Ford, and the geneticist, Dobzhansky, produced evidence of astronomical odds against the viability of natural selection being the sole, or even principal, means of evolution, yet they still persisted in their support of this doctrine. Gradually a shift started to occur. Some scientists started to express reservations about the process of natural selection itself and about the uses to which the theory was being put.

Eugenics

The social implications of evolution by natural selection had been recognized by Darwin. While Darwin spoke of superior races replacing inferior ones, his cousin, Francis Galton, pursued the possibility which presented for the improvement of the civilized races by encouraging reproduction among the more favoured *individuals* and discouraging it among the less. If necessary, a process of compulsory sterilization was suggested for individuals with gross physical and mental disabilities (Galton 1892/1962). William Bateson (1909a) considered that we [civilized societies] should prepare ourselves for the practical application of genetic science to human affairs, something which many people considered had already become 'urgent'. By the 1930s, compulsory sterilization laws had been passed in both the United States and Germany and were proposed for Britain. Haldane was vehemently opposed to eugenics and in 1938 published *Heredity and Politics* in which he pointed out the difficulty legislators had had in defining quite what conditions justified such drastic measures. Many instances of both physical and mental disability were congenital, not hereditary, or were the result of some later misfortune.

Haldane argued that even in cases known to involve an hereditary component, sterilization would not necessarily eliminate the problem, illustrating his argument with the disease hæmophilia.

At the time Haldane was writing, it was thought that hæmophilia was carried by the female but manifested in the male. This is now known not to be completely correct. The faulty genetic material is, indeed, carried on the X chromosome. If a female inherits this recessive gene, she has a good chance that her second X chromosome will be normal (dominant), thus preventing the development of hæmophilia. The male has no second X chromosome and will suffer hæmophilia if he inherits the one defective X chromosome. For a female to be affected, she must be 'double recessive'. This rarely happens. Severity varies, but affected males frequently die before reproducing, this being the most severe form of 'natural selection'.

Haldane gave family pedigrees showing how hæmophilia could be passed on, generation after generation, despite the death of double recessive carriers. He gave similar pedigrees for other, less severe conditions, such as colour blindness and brachydactylism, arguing that in order to eliminate these conditions, it would be necessary not only to sterilize the double recessive carrier, but also all the heterozygous carriers as well. At that time, the only way this could be achieved was by wholesale sterilization of all family members, whether they appeared affected or not, genetic testing as we know it today not then being available.

So long as there was even one heterozygous female, there was a chance that that female would give birth to another, or more than one, heterozygous female child. If one such female were fortunate enough to have only sons, then another, according to mathematical probability, would have only daughters, some of whom were mathematically likely to be heterozygous for the recessive gene. Haldane calculated that one normal X-chromosome gene mutated to the hæmophilia gene in every fifty thousand generations, thus replenishing those recessive genes lost by the death of the double recessive carriers.

Haldane argued that if it would be so difficult deliberately to eliminate a lethal recessive gene by eugenics, how much harder would it be to eliminate a merely 'undesirable' trait, one that allowed normal reproduction? If it was so difficult to eliminate an 'undesirable' recessive gene by the application of eugenics, how much more difficult would it have been for natural selection to have accomplished this feat? Haldane opposed eugenics. He wrote to that end and having accomplished his purpose, he appears to have been satisfied. He continued to support Darwin's theory as it was then being incorporated with genetic theory into the *Modern Evolutionary Synthesis.*

Macroevolution

By the time he emigrated to the United States in the late 1930s, Richard Goldschmidt (1878-1958) had earned himself the reputation of being Germany's foremost authority on evolution. He was welcomed by American academia and in 1939 was invited to give the Silliman Lectures at Yale University. Part of the requirement was that an expanded version of the lecture material be made available in book form. *The Material Basis of Evolution* (Goldschmidt 1940) was that book.

If Goldschmidt himself was welcomed by academia, his ideas were not – at least, not by his fellow evolutionary theorists, such as Ford, Fisher, Dobzhansky and Mayr (Brown 1999). Goldschmidt did not believe that the problem of evolution had been solved as far as its genetic basis was

Fig.28.1: Three species of Lymantria butterfly.
(Goldschmidt 1940: 145)

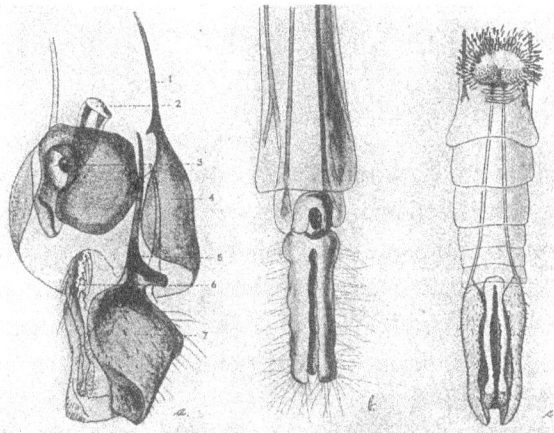

Fig.28.2: Genital armature of female Lymantria butterflies shown above.
(Goldschmidt 1940: 146)

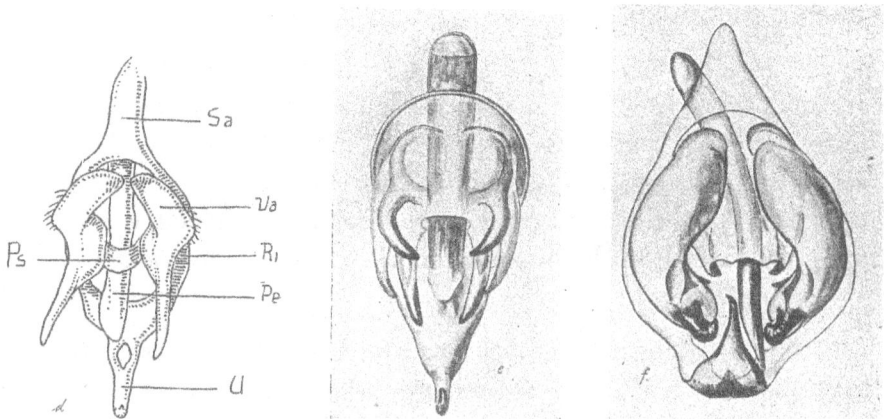

Fig. 28.3: Genital armature of male Lymantria butterflies shown above
(Goldschmidt 1940: 47)

concerned. He questioned the extrapolation of facts pertaining to microevolution (change observed within the span of a human lifetime) to macroevolution (changes occurring on a geological scale). He pointed out that, despite the fact that very few characteristics similar to those obtained in the laboratory by radiation had ever become established in wild *Drosophila*, yet the rate of *Drosophila* mutation in the laboratory under radiation was being made the basis for theoretical conclusions.

Goldschmidt's book was divided into two parts, the first dealing with microevolution, upon which he was in general agreement with his colleagues. He did dispute the existence of so-called 'incipient species'. At some point, he argued, 'incipient' subspecies and distinct species had to be divided by a 'bridgeless gap'. He illustrated his point by reference to three species of the butterfly *Lymantria dispar, Lymantria mathura* and *Lymantria monarchia*.

Although believed to be closely related, not only did their caterpillars and the butterflies differ in shape, size, colour and hairiness, their genitalia were different (see Figs. 28.1, 28.2 and 28.3). Furthermore, they had completely different egg-laying habits. How could micromutations account for differences in reproductive organs? At what point did one generation come to possess generative organs which differed from that of its parents, in however small a degree? Goldschmidt considered these species to be an example of a 'bridgeless gap', that it was not possible to account for their having descended from a common ancestor by the process of micromutation/microevolution. Goldschmidt cited the mathematical calculations of Wright, Fisher and Haldane, used by them to support the concept of the accumulation of micromutations as the basis of speciation, as evidence that the process of forming a subspecies, let alone a species, a genus or a family, was so drawn out that he doubted whether this method could ever be anything more than a theoretical possibility.

Goldschmidt (p. 193) ended the first part of his book with the conclusion and proposition:

> Subspecies are actually, therefore, neither incipient species nor models for the origin of species. They are more or less diversified blind alleys with the species. The decisive step in evolution, the first step towards macroevolution, the step from one species to another, requires another evolutionary method than that of sheer accumulation of micromutations.

The second half of Goldschmidt's book was devoted to macroevolution. He argued that gene mutations were insufficient to account for macroevolution, which needed to be considered in the terms of rearrangements of chromosomal patterns. Goldschmidt had come to the conclusion that evolutionary theory was being handicapped by its adherence to a belief in genes as discrete units, e.g. 'beads on a string'. Goldschmidt maintained that there was no such thing as a particular gene, proposing that chromosomes should be considered as if they were a very long molecular chain. He suggested that genetic material should be considered as letters of the alphabet, whose arrangement and rearrangement could either make complete sense (healthy condition), partial sense (viable but disadvantaged) or no sense (not viable). He had come to the conclusion that macroevolution could only come about when there was a complete rearrangement of a chromosomal/molecular chain. He was well aware that small genetic changes influencing hormones, etc., could result in large mutations, but still did not believe that these were enough to bridge the 'bridgeless gap'.

Goldschmidt was also concerned about chromosome numbers. He noted that related species or genera often had different numbers of chromosomes. It seemed as if, in some cases, one or more chromosomes had broken into two (which would require the acquisition of a new spindle fibre) or, conversely, some smaller chromosomes had united into a larger one, leaving the spare spindle fibre and its centromere unaccounted for. Goldschmidt pointed out these problems, but did not attempt to solve them. This task White (1937) had already attempted, with mixed success.

Michael White lectured in chromosome-cytology at University College, London, before moving to Melbourne in the 1950s. Like so many other lecturers, he wrote a text book, *The Chromosomes*, to assist his students and others interested in this subject (White 1937). The first edition was a slim pocket-book devoted to basic explanations of mitosis and meiosis. By the time the 6th edition was published in 1973, it had grown considerably.

White was more interested in chromosomes than he was in genes, carefully describing their role during the various stages of mitosis and meiosis: leptotene, zygotene, pachytene, diplotene, prophase, metaphase, anaphase and telophase. He explained how some plants were able to sustain chromosomal deviations, such as polyploidy, which was impossible for sexually reproducing organisms, and also how some Orders, such as insects, had complicated reproductive mechanisms involving several x and y chromosomes, which enabled them to produce 'inter sexes'. Mammalian reproduction was, by comparison, simple, involving but two sex chromosomes.

At meiosis, each chromosome (or chromatid as White called the haploid chromosome) was attached to a spindle fibre, which formed during metaphase. For successful reproduction, there had to be one fibre for each chromatid, no more, no less. Each chromatid had one centromere, no more, no less. The centromere might be towards the centre of the chromosome, in which case the chromatids would form a "V" shape when their centromeres attached to the spindle and migrated towards one pole. Other chromosomes had their centromeres more towards one end, in which case the chromatid formed a "J" shape. Other chromosomes had their centromere so close to one end that it had been thought that these centromeres were terminal. All chromosomes had some genetic material, however small the amount, on each side of their centromere. The position of the centromere was constant for each individual chromosome.

The nature of the 'double helix' was not understood until after Watson and Crick published their famous paper in 1953. Until that time, it was thought the two chromosomes came together, exchanged material in some 'random' way so that no two offspring produced by the same parents were ever *exactly* the same, and then separated again. The new, separated, chromosomes were drawn towards spindle fibres at each pole, to which they attached. There was always the correct number of spindle fibres for the chromosomes, never more nor less.

Mutations occurred when a piece of genetic material broke away from its chromosome, its ends being 'sticky' for a short time. It might reattach to its original chromosome but inverted, or change places with another piece of genetic material which had become loose from elsewhere. These changes had to take place before the 'sticky' ends sealed. Neither mitosis nor meiosis would be possible if there were extraneous 'sticky' ends which would cause the whole process to degenerate into a shambles.

Of particular interest were the so-called microchromosomes, whose role was not clear. These were found in the *centre* of the spindle. There may be none, or their number may be less than, equal to or greater than, the number of chromosomes and, therefore, of the number of centromeres and spindle fibres. Were they extra centromeres, which had somehow come into existence and which might be used at some future time to increase chromosome numbers, or were they discarded centromeres, the result of a reduction in chromosome numbers? It is perhaps unfortunate that White's illustration (reproduced here as Fig. 28.4) portrayed a nucleus with an equal number of centromeres and microchromosomes, since the text made clear that it was more usual for there to be considerably less microchromosomes than centromeres.

White placed great importance on the role of the centromere. When a chromosome was broken as the result of irradiation, the portion with the centromere attached to the spindle fibre, the remaining fragment being lost. Sometimes under irradiation, two chromosomes fused so as to form a compound chromosome. If the centromeres travelled to opposite poles, the chromatids were stretched and broken. If they travelled to the same pole,

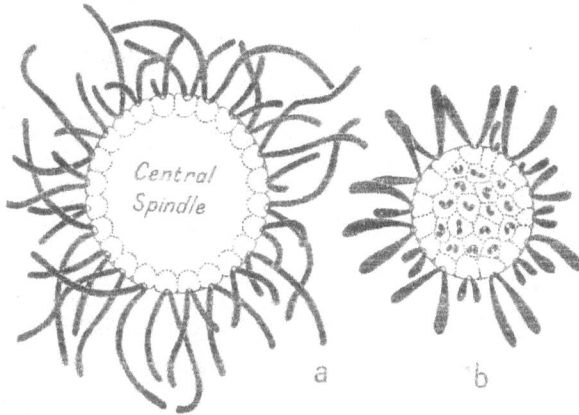

Fig.28.4: (a) Organism with 24 chromosomes attached
to spindle fibres.
(b) Organism with 16 attached chromosomes and 16 microchromosomes in
centre of spindle.
(White 1937: 16)

they might survive that phase of cell division, but at the next cell division, half would travel to opposite poles and become broken. After a few cell divisions, all would be broken, such chromosomes standing no chance of becoming permanent. 'Double' chromosomes had never been observed in natural cell division, only under irradiation. Successful cell reproduction (mitosis or meiosis) could only take place with chromosomes which contained but one centromere each.

White discussed at length the exchange of genetic material which took place at meiosis. Hybrids might be formed where the parent species had a different chromosome number or different chromosome sizes. Where the chromosome number of the parents of a hybrid species differed, one of the two haploid sets in the hybrid would contain more chromosomes than the other and the 'extra' chromosomes would necessarily be unable to pair at zygotene, making the zygote non-viable. Differing chromosome numbers clearly acted as an isolating mechanism, far more efficient than any form of 'sexual selection' envisaged by Darwin, but White was unable to explain how chromosome numbers had come to change in the first place.

Chromosomes and evolution

In the final chapter of his book, White considered the role chromosomes had/had not played in the process of evolution. White saw all gene mutations, such as inversions, translocations, etc., as being small structural alterations involving a mere rearrangement of genes on a small section of the chromosome, insufficient to cause macroevolution, which he concluded could only have come about as the result of changes in the chromosomes *per se*. Evolutionary change, such as polyploidy, was simple for non-sexually reproducing organisms, mostly plants, but was more difficult to explain in sexually reproducing organisms. Yet it was clear from the differences in chromosome numbers which existed throughout nature that some mechanism must exist whereby chromosome numbers could be altered (White 1937: 94-95):

> It used to be supposed that two or more chromosomes could merely fuse together to form a single one, and alternatively that one chromosome could break into a number of pieces, each of which would behave in future as a separate chromosome ... We now know, however, that each chromosome contains a single spindle attachment which is a self-perpetuating body; new spindle attachments only arise from pre-existing ones. Moreover, although spindle attachments divide longitudinally at mitosis, they do not appear to be transversely divisible. It is possible that in some cases V-shaped chromosomes have spindle attachments in the middle only separated by a very short interstitial region; in this case the two spindle attachments may be expected to function mechanically as a single unit ... breakage of the interstitial region will give two chromosomes with quasi-terminal spindle attachments ... Conversely two chromosomes with quasi-terminal spindle attachments may fuse together so as to give a V with two attachments in the middle. Any other kind of fusion will give a chromosome with two widely separated spindle attachments which will break at anaphase and any other kind of breakage will give rise to a fragment with no spindle attachment, which consequently cannot form an independent chromosome. Thus apart from these two special cases, it does not seem possible that 'fusion and fragmentation' have played any part in the evolution of new chromosome numbers.

White was suggesting that two centromeres from two separate chromosomes could combine so closely together that they *appeared* as if they were only one centromere but actually had a miniscule amount of interstitial material between them and that they could then separate again to form two separate centromeres/chromosomes. This had never been observed to happen, even under irradiation, but it was the only explanation which White could offer.

This hypothetical scenario was proposed by White in the case of *mitosis* (see line seven of above quotation), applying to somatic, not germinal, cells. It was known that some species of *Drosophila* had somatic material with different chromosome numbers from that of their reproductive organs. Since somatic cells reproduced by mitosis, rather than meiosis, it was thought that somatic cells could be more amenable to mutation – as happened with other non-sexually reproducing organisms. What White did not explain was how such changes in somatic cells could be inherited, i.e. how this scenario could be extrapolated to *meiosis*. White further did not address the problem of how a sexually-reproducing organism which had achieved a change in the number of its chromosomes could find a mate.

White concluded that the primary origin of new species lay in some accident in the chromosome set, completely unconnected with natural selection, although natural selection was an important factor in the establishment of subsequent morphological change.

Both Goldschmidt and White were clearly dissatisfied with natural selection's ability to explain change beyond the micro/variety level but both nevertheless continued to support natural selection as an important contributor to evolution. They were apparently content to await further developments rather than to attempt to propose any other solution. Goldschmidt was the last major theorist to raise the possibility of macro-evolution. The Neo-Darwinism which emerged in the second half of the 20th century, based on the *Evolutionary Synthesis*, postulated gradualism. It was accepted that only very small genetic mutations could be tolerated. Larger changes did not result in a viable zygote.

White continued to be a prominent contributor to genetic research into the 1970s and his later work will be considered in Part IV. No other theorist attempted to solve the problem of how change in chromosome numbers came about by natural selection.

IMPLICATIONS

Towards 2000

All roads lead to Darwin

Chapter 29

Evolutionary Synthesis

It was during the darkest days of World War II (1942) that Huxley published his book, *The Modern Synthesis*. It contained no new work but brought together that of previous theorists in such a way that the '*Evolutionary Synthesis*' soon became accepted as the dominant paradigm.

Many of the major proponents of the evolutionary synthesis continued to be active participants in evolutionary discussion well into the post-war years. Particularly prominent were Dobzhansky (1951, 1959, 1962, 1967, 1970) and Mayr (1963, 1972, 1976, 1977, 1982, 1991, 2001; Mayr and Provine 1980). They were ably supported by writers such as de Beer (1958), Haldane (1951, 1954), Levins and Lewontin (1985), Maynard-Smith (1958, 1982, 1984, 1987, 1989) and Simpson (1950, 1953). It was not just the number of books that increased, but their size. Mayr's (1982) *Growth of Evolutionary Thought* ran to 858 pages of text. New authors, such as Stringer (1985, 1988, 1994, 1996), Tattersall (1998, 1999, 2002), and Trinkaus (1983a, 1983b, 1985); Trinkaus and Smith (1985) made their contributions, but the evolutionary synthesis having been established, they concentrated on other issues, such as the status of Neanderthals. Richard Dawkins (1976, 1982, 1986,1995, 1996, 1998, 2003, 2006, 2009) wrote forcefully in a manner designed to engage the attention of the general reader, concentrating more and more on philosophical issues. The last of the major books (to date) on the evolutionary synthesis, *The Structure of Evolutionary Theory*, was written by Gould (2002) and exceeded 1100 pages.

Despite the plethora of words, little that was new was added to the debate. Writers sought to reinforce opinions already voiced by the addition of more and more examples. There was ongoing work on the genetics of quickly breeding laboratory populations, mostly insects and rats. Very early it had become apparent that the original concept of a gene as a bead on a string, transmitting information about characteristics from one generation to the next, was not valid and it was an understanding of how genes interacted with each other, as much as an increased understanding of genetic mutations, which had led to the concept of the evolutionary synthesis. Debate as to how exactly diversification had occurred (sympatric v allopatric) continued, but the examples given tended to be very repetitive: gill arches to jaws to ear bones; reduction in the number of horses' toes; swim bladders to lungs, although Mayr (1976: 99) pointed out that early fish had primitive lungs, which were converted into swim bladders, not the other way around.

Sympatric v allopatric speciation was not the only source of confusion in relation to diversification (Dobzhansky 1962: 5):

> To make Darwin's theory as shocking as possible the proposition "man and apes have descended from common ancestors" was garbled into "man has descended from the apes". This, of course, is obvious nonsense, since man's remote ancestors could not have descended from animals which are our contemporaries.

Why not? If the human line has continued to exist, why should not the apes?

Versions of this statement are common throughout the literature and it is they, not the so-called 'garbled' version, which are 'obvious nonsense'. It matters not whether the human line diverged directly from the ape line, or whether humans and apes both diverged from a common ancestor. The divergent form must have been contemporaneous with the original form. To postulate an hypothetical common ancestor, thus moving divergence back one step, may be psychologically satisfying, but is not scientifically necessary and only goes to prove the resistance there is in the human mind (as much today as in Victorian times) to the notion of the direct descent of humans from apes. No line can diverge from an extinct ancestor. The statement by Zimmer (2001: 324) that "The many whales with legs that palæontologists have now uncovered are probably ancient cousins of today's whales rather than direct ancestors" is a further example of similar thinking.

Where one form *changed* into another form, contemporaneous existence could have been for only a short period of time, but where *diversification* occurred, i.e. a genetic mutation led to the establishment of a distinct species, then there is no reason why the two species should not have continued to co-exist, and diversify independently, especially if the parent population was geographically and/or numerically large.

Goldschmidt was the last person to make a serious attempt to establish micro, macro, and even mega, evolution as separate events. The knowledge that the genetic alphabet contained a mere four letters and that these letters could be repeated in endless combinations, gave the proponents of the evolutionary synthesis all the means they believed they needed to account for evolution. Weismann, having proposed the complete separation of the reproductive genetic material from the somatic, had struggled to explain how evolutionary change could have occurred and had been forced to conclude that the first *biophors* had contained all the information necessary, not only for all forms of life which had ever manifested, but for those which might manifest in the future. Now that it was known that DNA could not only replicate itself, but rearrange itself by means of inversions, translocations, etc., the genetic 'alphabet' was seen as capable, not only of remodeling a pre-existing part, but of initiating something totally new, purely by chance. Thus macroevolution no longer needed explaining. In response to the self-imposed question: "Is a New Evolutionary Synthesis necessary?", Stebbins and Ayala (1981) answered "No" on the grounds that the only way change was possible was through mutation, which was a gradual process involving small changes. Small changes (microevolution), as postulated by the evolutionary synthesis, were all that was possible and the evolutionary synthesis was all that was needed to explain all evolutionary change.

Some authors, it must be said, dealt with this matter in a less than forthright manner. Haldane (1954: 1) opened his book, *The Biochemistry of Genetics,* by expressing the opinion that:

> Genetics is concerned with differences between similar organisms, and mainly with those differences which are not due to causes acting during the lifetime of the organisms concerned. This distinction works fairly well for higher organisms, but breaks down completely for unicellular organisms. If a cell can divide once an hour, but takes a day to adapt itself to ferment a type of sugar to which it is unaccustomed, a growing population can only adapt if the adaptation (and even the beginning of the adaptive process) is inherited.

Having made this statement, Haldane (1954) concentrated thereafter on higher organisms, despite the fact that by volume, unicellular life exceeds that of multicellular by several orders of magnitude. To admit that the one behaves differently from the other, and then to extrapolate from the one to the other, is not clear thinking.

Major changes, such as the appearance of the first feather, may have been cited at the beginning of a discussion, but later concrete examples examined in some (but not much!) detail, tended to concentrate on structures which could readily be imagined as having resulted from microevolution, such as the size of horses' teeth or the reduction in the number and size of their toes. Mayr (1982) delighted in pointing out that eyes cannot have been very difficult for natural selection to produce since they had appeared at least forty times in separate lines of evolution. Whether this made their appearance 1/40th as wonderful or 40 times as wonderful depended upon one's point of view!

Feathers were said to have first appeared as a form of insulation, but this 'explanation' merely moved the problem back one stage and invoked teleology, which was not permitted, since it presumed preplanning. It did nothing to explain why a feather should appear instead of (more) hair/fur. Cold blooded animals have neither fur nor feathers, nor fat for that matter. Indeed, since they draw their warmth directly from the sun, it is important for them to have no barrier between their blood supply and their source of heat. There was no reason for a cold-blooded reptile to develop fur/fat/feathers and yet no explanation is offered as to how the first warm-blooded creatures survived before these life-saving 'aids' appeared. Indeed, how did a cold-blood creature evolve into a warm-blooded one at all?

Surprisingly, no author questioned the first appearance of hair. To this day, there are many worms and grubs which are completely hairless. Why did a smooth skinned 'grub' first sprout a hair? Warmth or sensitivity to touch cannot be the *reason*, unless teleology is accepted, yet nowhere in the literature is this problem addressed. Did a single 'hair' develop or did many hairs appear all at once in that first hairy 'grub'? How did the hair grow and penetrate the skin without the assistance of a hair follicle? No hair follicle could have developed and been 'selected' by nature before there was any hair useful enough to be 'selected'. The presence of at least some hair/fur on land mammals is tacitly accepted yet its first appearance invites as many questions as does the appearance of the first feather.

Nowhere in the literature was a satisfactory explanation offered for the initial evolution of butterflies and moths. Countless 'grubs' reproduce in their 'worm' state, yet a few, such as butterflies, moths, cicadas, etc., transform into totally new forms. This physiologically expensive process is particularly hard to understand in the case of those butterflies which exist in the imago form for only a few hours, long enough to mate but not long enough to need to feed. Issues of polymorphism, of mimicry, were discussed, but not the process of transformation itself, which is far harder to imagine happening by the gradual process of natural selection.

Most evolutionary theorists concentrated on biota - flora and fauna.

Some (geneticists) addressed possible changes in DNA – its rearrangement, not its origin. Michael White alone seemed interested in changing chromosomes. Vilmus Csányi went further. His book, *General Theory of Evolution*, published in 1982, raised the issue of the evolution of chemical substances. Csányi claimed that, at the time of its creation, the newly forming Earth was not in a 'stable state'. It was an 'open system", being continually bombarded by energy from the Sun, which energy had to be stored or dissipated in some way. At some point, a maximum level was reached in the (then) existing system. It was the excess energy which 'mobilized' the basic elements, such as carbon, oxygen, hydrogen, nitrogen, to form molecules: "the system becomes reorganized and reaches a new stable state" (p. 18). This ceaseless bombardment of energy eventually precipitated the formation of more complex substances, such as amino acids, organic bases, sugars, etc. Csányi claimed that, whereas most evolutionary theorists stressed the importance of differentiation, it was the most commonly occurring 'patterns' which became established and eventually gave rise, not merely to more complex chemicals, but to nucleotides proteins, carbon compounds, DNA, and other cellular components. Replicative constancy was essential for these advances to take place.

Csányi was not content to consider only organic evolution. He followed the evolution of systems from molecular to cellular to organic to ecosystems to cognitive functions to behavioral, cultural and technological systems.

> Evolution is the process in the course of which an open system receiving an energy flow shifts from the state of equilibrium, in its structures replicative information arises that continuously increases, converging towards a maximum value until the whole system becomes one replicative unit (p. 93).

Csányi believed evolution at the more basic levels (molecular, cellular, organic) was now complete and that the only evolutionary change yet to be made would occur at levels from ecosystem through to technological. He expressed concern about humanity's interference with the ecosystem. Csányi's work was theoretical inasmuch as he provided no new evidence to support any claim that he made. His aim was, rather, to encourage evolutionary theorists to expand their horizon, to include within their field of vision the whole of creation, the inanimate as well as the animate.

If the second half of the twentieth century did little by way of adding new information to support the evolutionary synthesis, it compensated by way of adding stature to Darwin's image. The year 1959 marked the centenary of the publication of *On the Origin of Species* and it was to be expected that this anniversary would be marked by publications eulogizing Darwin's work. The eulogizing never stopped. By the turn of the century,

Darwin had been elevated almost into a cult figure (for example Dawkins 2006, Dennett 1995, Ruse 2006, Zimmer 2001). Anything which Darwin had said which was seen to have been vindicated by modern science was held up as a great insight, a piece of inspired wisdom, even if it was not original to Darwin, such as the concept of evolution itself. Dennett (1995: 21) wrote: "He [Darwin] would discover the single most important idea in the history of biology", placing Darwin's insights above those of Mendel. Anything that Darwin said which, with hindsight, appeared less than satisfactory, was not mentioned. Darwin's theory of *pangenesis*, based as it was on the acceptance of the inheritance of acquired characteristics, was either ignored, or attributed to an aberration of old age.

Zimmer (2001: 211) entitled his chapter on disease in the age of evolutionary medicine "Doctor Darwin", as if Darwin himself were in some way responsible for our understanding of the evolution of bacteria in response to modern medicines, such as antibiotics. A pride of lions was offered as an example of "Darwinian family life" (Zimmer 2001: 246). Hardly an aspect of life could not be identified in some way with the Darwinian revolution. Dennett (1995: 21) went further:

> If I were to give an award for the single best idea anyone has ever had, I'd give it to Darwin, ahead of Newton and Einstein and everyone else. In a single stroke, the idea of evolution by natural selection unifies the realm of life, meaning and purpose with the realm of space and time, cause and effect, mechanism and physical law.

By the end of the 20th century, the early 19th century paradigm of attributing everything to the province of some 'Divine Creator' or 'Supreme Being' had been completely reversed. Any mention of such a 'Being' had become anathema. The Darwinian revolution had surpassed the Galileo/Copernican revolution in its effect on how we viewed ourselves, this Earth and the Universe as a whole. *Darwinism* had become a pseudo religion, with Darwin its pseudo Pontiff. To criticize Darwin had become little short of blasphemy.

Chapter 30

A 'Noo' Theory
of Evolution

The two countries most involved in the development of archæology, France and England, each gave birth to a prominent philosopher during the nineteenth century, whose influence was to be felt, each in its own way, well into the twentieth.

Sir Bertrand Russell (1872-1970)

In England, it was Sir Bertrand Russell whose pessimistic views on the evolutionary future of humankind cast a deep shadow over the lives of those growing up in post war Britain. Born into an aristocratic family, his parents, Lord and Lady Amberley, both died before Sir Bertrand was seven years old and he was brought up on the estate of his grandmother, Lady Russell (Russell 1967), whose extensive library of books helped fill the lonely years of his childhood. At a young age he was versed in the philosophies of such people as Dante, Descartes, Huxley, Kent, Machiavelli, Swift and, of course, Darwin. At Cambridge he was introduced to Ibsen, Shaw, Whitman, Nietsche and others but it was his early readings which shaped his philosophy of life.

Russell was interested, not so much in where humans had come from, as in where they were headed. Staunchly anti-Christian, he embraced the humanist philosophy, believing that all humans, including children, should be afforded the maximum degree of freedom in which to express themselves and that they should be subjected to the minimum amount of authoritarian interference, even from parents/teachers. Russell believed it was lack of freedom and spontaneity which induced fear, resentment and anger, all of which could lead to violence and hence to war.

Russell also believed that eventually communism would replace capitalism (Russell 1961). Like Marx. Russell believed that humans adapted to external circumstances. Whereas Marx tended to look to macro changes in society, such as the redistribution of wealth, to increase human happiness, Russell, like Wallace and Spencer before him, was more concerned with micro changes, as in family or schooling.

Russell felt that children were born with a natural tendency to be co-operative and that recalcitrant, delinquent or criminal behavior developed in response to inappropriate and/or undesirable social conditions. This premise provided the basic philosophy for social work, where parental care/abuse, living conditions, schooling, (un)employment, nutrition, medical care and the (mis)use of drugs and alcohol were all seen to influence the child's evolving character. Under certain circumstances, for the first time in history, the State took upon itself the right to remove children from their birth parents and to place them in foster homes, the hope being, not only to increase their happiness, but to prevent the development of anti-social tendencies in response to an adverse external environment.

Russell compared the human mind to the Universe. For a number of centuries, scientists had been forced to expand their mental vision as the telescope expanded their range of physical vision. Not only was the human mind compelled to try to comprehend a space of immense magnitude, that space was filled (or, rather, loosely scattered) with stars, the numbers of which strained the abilities of mathematics to express. Furthermore, that space (and those numbers?) was believed to be constantly expanding. As with space, so with time. Time was expanding as it receded backwards, not in thousands, not even in millions, but in billions of years.

Russell believed that this expansion of the human consciousness was in accordance with the dictates of human evolution. He was concerned that human feelings were not evolving at a comparable pace. In the past, humans had been concerned with 'What is best for me, my family, my tribe?' Some people were expanding their field of feeling to include people of other nations. Unless these expanded feelings were developed and expressed in the majority of people, rather than being expressed only by the minority, then increased knowledge would be used to reap increased destruction, rather than increased happiness,

Russell's belief that the human mind was evolving, that it had indeed reached a critical point in its evolution which could take humanity to a whole new level, was shared (but in a completely different way) by France's great philosopher of the twentieth century, Teilhard de Chardin.

Teilhard de Chardin (1881-1955)

Pierre Teilhard de Chardin was at the same time a palæontologist and a Jesuit priest and, one might add, a free thinker, this latter quality rarely being seen developed to such a degree in someone so faithful to the Catholic Church. Born in 1881, the fourth of eleven children, he was sent as a boarder to a Jesuit college at the age of ten. It was during his school years that he became interested in geology and mineralogy, which interests later expanded to include the study of fossils and, inevitably, human evolution.

Before being ordained, Teilhard spent time teaching in Cairo, which served to enhance his interest in archæology. He was ordained in 1912, served as a stretcher-bearer in World War I and took his final vows in 1918. He was awarded a doctorate in geology in 1922 and, being forbidden to teach in Paris because of his unorthodox views, was sent to China, where he became actively involved in archæological excavations. While not directly involved in the discovery of 'Peking Man', he was involved with the interpretation of the finds (see next chapter). During the war, Teilhard was detained in China by the Japanese, not returning to Paris until 1946. By that time he had written many essays and books, but Rome refused him permission to publish. None of his major works were published during his lifetime.

Teilhard saw evolution as an ongoing process. Tielhard was not concerned with the possible future direction of physical human evolution, he was concerned with the evolution of the non-material aspect of our beings, our minds and, in particular, our capacity for reflection, since he saw this as the one feature which distinguished the human species from all others which had ever existed, or which might exist in the future.

It was Teilhard's aim to approach his subject, human evolution, from a position as free as possible from preconceived ideas, religious or secular. He made scarcely any mention in his books of any established religion, the major exception being the *Epilogue* to his major work, *The Phenomenon of Man*, which was published posthumously in 1955. The *Epilogue* had presumably been added in an attempt to placate Rome. In addition to avoiding any reference to religion, Teilhard avoided almost all reference to other theories of evolution. There were minimal references to Darwin and Lamarck.

Teilhard attempted to persuade by means of a step-by-step logical progression of thought. He contended that it was impossible for inert matter to become living. Life must be present in some latent manner for it to be able to manifest in inert (inorganic) matter to produce a living cell. While life was latent, as in inorganic or 'dead' matter, it was the natural order of things for degeneration to take place, for carcasses to rot, for vegetable matter to compost, for rocks to wear away and break down. Only in response to 'pressure' from 'life', Teilhard argued, would it have been possible for molecules to amass in such a way as to form the macromolecules which shaped the cells which hosted the most basic form of life. Teilhard thus supported the thinking of many Eastern peoples that life preceded matter, not matter life. How much his time spent in the Far East led to the solidification in his mind of these ideas, it is not possible to say. Certainly his line of thinking was not new, nor did he claim it to be.

Teilhard's thinking was different from that of either Lamarck or Darwin. He agreed with Lamarck that evolution took place in response to an internal 'pressure' from 'life'. However, while Lamarck restricted his theory to the evolution of organic matter, Teilhard believed that life existed throughout creation, being latent within inorganic matter. Once 'pressure' had built up to such an extent that life broke through the barrier holding it trapped within 'inanimate' matter, the dam had been breached. The life force was now free to expand, via the biosphere, across the whole surface of the Earth and into the atmosphere as well. While individual expressions of life (species) could disappear, new expressions were constantly being formed. Earth would never again be lifeless. Never again would there be the same build up of 'pressure'.

The concept of 'pressure' was central to Teilhard's thinking. Teilhard saw all progress as being brought about by pressure. The very shape of the Earth, round, was important. It 'trapped' everything associated with it, preventing anything from expanding indefinitely. Even gravity was important, because the round shape of the Earth and gravity combined to exert pressure on the very substance of the Earth itself, leading to the formation of compound elements and minerals beneath its surface. It was pressure that caused elements to combine together, to form simple molecules and macromolecules. The material of the universe became more and more concentrated which resulted in ever more organized forms of matter, a process which Teilhard referred to as 'complexification'.

The physical Earth, itself a sphere, was comprised of a number of spheres: the barysphere, lithosphere, hydrosphere, atmosphere and stratosphere. Eventually the pressure of 'complexification' caused the life energy to burst forth into a new sphere, the biosphere.

342

Every living cell had a *Within* (living or spiritual energy) and a *Without* (material form). The *Within* energy exerted its influence in two directions: *tangentially*, which caused each element to link with others of the same order, and *radially*, which caused greater complexity and centricity – i.e. towards forward progress. These two energies were balanced. As living matter increased in complexity, it increased both tangential and radial energy. Radial energy did not only exert an outward pressure, as its name might suggest, but exerted an inwardly directed pressure at the same time – since the one form of pressure cannot take place without the other. It was this inwardly directed pressure which would lead to the evolution of the 'inner person', the thinking, rational, philosophical, spiritual person that we are today. Teilhard's writings have a very 'Lamarkian' flavor inasmuch as Lamarck also believed that an inner force was driving living forms towards greater complexity. Both Teilhard and Lamarck held that this inner force was divinely driven and directed. Neither gave any explanation for the results of evolution other than that the pressure of this life force was irresistible and must constantly find new avenues for expression.

The question of whether increased size and complexity of the cranial cavity and tissue of the brain led to increased intelligence, reasoning powers and philosophical thought, or whether increased mental powers induced increased cranial capacity was debated during the second half of the 19th century by people such as Lyell. Equally, it was debated whether, if humans were descended (or ascended) from animals, all human emotions and other mental faculties must be present in some degree in the animal kingdom. If they were not present in some degree in the animal kingdom, then their presence in humans would indicate some sort of 'special creation' or intervention by the Creator, which was contrary to Darwin's theory. Actively seeking emotional responses in animals which could be said to be similar to those experienced by humans led to a change in attitude towards animals, who were no longer seen as automatons, acting purely out of instinct, but as feeling creatures with, at least among the higher animals, some ability for independent thought and decision making. Cruelty to animals became a crime.

Teilhard accepted that animals shared many emotions with humans and some evinced the appearance of some basic form of rational thought, which operated alongside instinct. However, he claimed that humans alone were capable of philosophical thought and artistic expression, which were closely related because they alone had penetrated the *'noosphere'*.

Teilhard theorized that the nervous system of the brain had evolved in response to radial pressure to accommodate incessant energy. Because the

343

hominid family, apes, *Australopithecus* and hominoids, were less specialized than many animals in areas such as muscle, claw, tooth, wing, etc., they became more specialized in the development of the nervous system and the brain. This line of thinking was very similar to that of Julian Huxley, whose writing he would not have read, being at that time interned in China.

At some point, at the beginning of the hominid line, radial pressure became so strong that it caused another 'dam' to burst and the hominids moved from instinct to reason, operating for the first time within the noosphere, Teilhard's term for the sphere of the reasoning mind. Human numbers increased as they spread across the globe. As they came more and more into contact with one another, they created a complex web of thoughts and ideas. Pressure built up, causing the mind to become more and more complex, more capable of rational and philosophical thought. In modern times, with millions of people living in very close proximity, radial pressure was constantly forcing further development of the mind. Although Teilhard believed that Neanderthals were a distinct species with no living descendants, he did believe that they were thinking beings, because they had evolved after the 'break through' into the noosphere had occurred. Although Teilhard did not specifically say, it would be logical to assume that Teilhard saw the rational and philosophical mind as being subject to more pressure, and therefore growth, in large communities (i.e. among civilized societies) than among smaller hunter/gatherer societies, and also among town dwellers than among village/country dwellers within civilized societies. A concentration of minds within a specific geographical area was necessary for radial pressure to increase.

In the same way that inorganic matter had been 'left behind' once life had manifest as cellular organisms, so other living organisms had been 'left behind' once humans had claimed the noosphere. For humans, the pressure of evolution operated solely in the noosphere. There was no longer a build-up of pressure in the human equivalent of the biosphere, which explained to him why the human form had changed little since humans became reasoning beings.

Teilhard envisioned humanity creating a network of thought across the globe, providing mutual support and exchange. As humanity increased in numbers, the network would become more and more powerful. He saw mechanization as bringing relief from mundane chores, resulting in more and more leisure time. This leisure time would be devoted to mental activities, such as research. The way forward was in the evolution of the mind and the spirit, the two non-material aspects of our being. Teilhard lived and wrote before the age of the computer and the 'World Wide Web'. One can but wonder with what glee Teilhard would have greeted these developments, which he seems so closely to have anticipated.

Teilhard followed what he believed was the course of evolution from a framework which put life (spirit) first. He claimed that humans were not the centre of Creation, as had been claimed in earlier times, but the forefront, not the leader, because no other creature would be able to follow in its path, but the unique result of an evolutionary process whose aim had always been the evolution of the mind. He saw evolution as having started at the *Alpha Point,* for which he used the analogy of either the north or the south pole. Now evolution, via humanity, was converging upon the opposite pole, which he called the *Omega Point,* reaching which was the *point* of evolution itself.

Although Teilhard's name is frequently mentioned in the literature as a prominent 20th century philosopher, his theory is rarely extrapolated in any detail. This might be expected among evolutionists of a predominantly agnostic or atheistic way of thinking, but it is somewhat surprising that his views have not been brought into more prominence by religious thinkers, particularly those who agree with Teilhard that humans are different in some profound way from any other animal.

While other writers have expanded and adapted Darwin's theory of evolution by common descent through variation and natural selection, Teilhard was the only writer of the twentieth century to propose a new theory of evolution, one concentrating on the non-physical aspects of human existence. There were some aspects of Teilhard's thinking which were in line with that of Darwin. For example, he believed that more advanced races would supplant the less advanced, even suggesting that, at some time in the not too far distant future, some acceptable form of eugenics would be found that would encourage the spread of more intelligent human beings. This thinking did not endear him to Rome, or to post war religious thinkers of any denomination, and may explain why his theories are so rarely cited in any detail. Teilhard foresaw an incredibly bright future for humankind. Within the noosphere humanity was not merely leading the evolutionary way, it was the only traveler. The outdated and outmoded would, and should, give way to the new and improved. This was the way forward; this was progress.

Chapter 31

The Mystery of the East

As already mentioned, Teilhard de Chardin was actively involved in archæological excavations which took place on the east coast of China during the first half of the 20th century. These had been very exciting times for archæologists. Many more discoveries of Neanderthal remains were made in France and even earlier remains (Heidelberg Man from Mauer) were found in Germany.

It had always been understood and accepted that the Neanderthal people predated modern humans (i.e. the Cro-Magnon people) in Europe, but not by much. The exact relationship was debated: were the Neanderthals our direct ancestors or were we both descended from a common ancestor? Were the Neanderthals a different 'race' or were they a completely different species? Whatever the answer, the Neanderthals were accepted as our immediate predecessors in Europe.

The situation became more complex when ancient fossils were found near what was then known as Peking - now known as Beijing. Darwin had surmised that humans had evolved in Africa, that being the home of the Great Apes. Had Darwin been wrong? Ape remains were now being found in parts of Asia - and so were the remains of ancient humans.

When first discovered, the finds from Trinil and Solo in Java, South East Asia and those from Peking, were called *Pithecanthropus* and *Sinanthropus pekinensis* respectively, although both have now been reclassified as *Homo erectus,* along with fossils of similar age from Africa. There was much rivalry between various teams of archæologists, each wanting to find the earliest evidence of the human race, or, even better, the 'missing link' between humans and apes. Each find in a new place was given a new species name. That this implied that they each belonged to a different species did not seem to trouble the finders. Clearly, if two of these 'species' were contemporary, only one could (possibly) be ancestral to modern humans, but nobody could know which one. With no radiometric dating, and millions of years now available into which to fit (pre)human history, archæologists were having a field day! In the end so many fossils were found that it became obvious that humans could not have been, 'speciating,' at such a rate and moves were made to bring order back into what was rapidly becoming a chaotic situation. The inclusion of both *Pithecanthropus* and *Sinanthropus* within *Homo erectus*, along with what were recognized as being contemporary finds from Africa, was a major part of this move. Those who advocated separate names for each find, on the grounds of some minor difference in osteology, became known as 'splitters', those who recommended combining finds into broader groups were 'lumpers'.

The situation was far from clear. Had there been contact between the different populations or had the groups remained separate? Did the Asian line become extinct? Had the African line given rise to both the Neanderthals (via the Levant) and the *Homo sapiens* who replaced them? It was during this time that early hominid remains were found in the Rift Valley in Kenya, and elsewhere in Africa, which proved that Darwin had been right – Africa was the birthplace of humankind.

The second half of the 20th century saw the introduction of the far more accurate radiometric dating, which brought with it some surprises, and caused a re-evaluation of the course of human evolution, particularly in relation to the fossils from Asia.

Having just mentioned the uncertainty which surrounded the accurate dating of fossil remains, this may be an appropriate place in which to give more information about dating techniques and how they were changing at this time.

Initially, dating was little more than guesswork. As artifacts and bones (both human and animal) were increasingly unearthed, the stratigraphic layer in which they were found was used to estimate their age. Allowance needed to be made for deliberate or accidental burial. Since each geologists was free to make his own estimate of the age of any particular formation (and geologists were male in those days), this method was not very accurate in an absolute sense, but was quite successful in estimating relative dates.

While tree ring and varve (ice ring) dating became increasingly sophisticated during the first half of the twentieth century, these dating methods were only applicable to certain localities and were of insufficient length to be of use in the study of human evolution. The three ages, Stone, Bronze and Iron, were of use in Europe but not elsewhere. 'Stone Age' could, and did, mean 'current' in certain parts of the world, such as Australia and Africa. In the latter country it could, and did, also mean the 'dawn of humanity'. The stone tools themselves did not distinguish between these two extremes. Some 'modern' people used ancient forms of tool, for example the Australian aborigines from Tasmania and Cape York

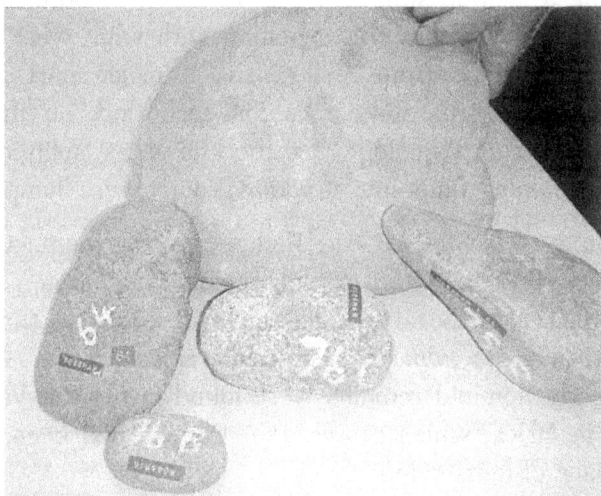

Fig 31.1: Aboriginal tools from Mossman, Cape York, Queensland.
(by permission of Bennet Walker, Kuku Yalanji people)

However, attempts were made to apply typology (changing shape, sophistication and æsthetic appeal) of stone tools in the same way that these criteria were applied to household items, such as cookware and vases, as well as weapons.

It became known that bone gradually lost nitrogen and took up fluoride and uranium, but the rate at which these changes occurred varied from place to place. This method of chemical analysis was able to be used to establish whether bones found in the same locality were, in fact, of the same age and it was the application of this method of testing which finally proved that the skull and mandible found in the quarry at Piltdown were of completely different ages, thus confirming the most famous and (temporarily) successful of all scientific hoaxes.

The second half of the 20th century saw the emergence of 'absolute' dating, its advent being of immense assistance to archæologists interested in early human evolution. A number of methods are now in use, many relying on the principle of radioactive decay. Radiocarbon dating is becoming increasingly reliable, with improved methods, such as Accelerator Mass Spectrometry (AMS) requiring extremely small amounts of organic material for testing. While this method is claimed to be applicable for dates up to 50,00 BP, its reliability decreases after 30,000 years, just the time that anatomically modern humans were replacing the Neanderthals in Europe. Tiny charcoal samples taken from cave paintings, together with charcoal samples from the floor, showed that the walls of the Chauvet Cave in France were painted approximately 30,000 years ago (Renfrew and Bahn 1996), considerably earlier than had previously been thought.

Potassium argon dating proved useful when dating volcanic material aged between 100,000 and 5 millions years old, making it particularly appropriate for the dating of human artifacts from the time of the divergence of the human and chimpanzee lines (approximately 5-7 million years ago) up to the emergence of early modern *Homo sapiens*. Unfortunately, not all skeletal remains or human artifacts were associated with such volcanic material.

Uranium dating is more restricted in time (50,000-500,000 years) but is useful in Europe where volcanic material is limited, as it is effective in areas where there is an abundance of calcium carbonate. Its time span roughly coincides with that in which humans (of whatever species) are believed to have lived in Europe. Bones found in the Pontnewydd Cave, North Wales, were dated by this method to 220,000 BP (Renfrew and Bahn). Such bones as these are particularly interesting in view of the fact that no Neanderthal bones, as such, have ever been recovered from the British Isles. (B.P. (Before Present Era) = before 1950 A.D.)

Other means, such as fission track, thermoluminescence, optical and electron spin resonance, have all contributed to the level of certainty which now exists in regard to the dating of fossils finds.

Going against the trend, some authors (for example O'Connell et al. 2002) took to referring to early African fossils as *Homo ergaster* to distinguish them from the Asian *Homo erectus*, believing that the two must represent different species, since they were contemporary. In other words, it was held that two distinct species had evolved from the Australopithecines. Since there is ongoing debate as to the exact relationship between the various Australopithecine remains, that two different Australopithecine species each gave rise to a different species of *Homo* must be considered a possibility. However, I do not consider geographical separation *alone* to be sufficient grounds to claim speciation. Quite how these early humans relocated from Africa to South East Asia is unclear but, since there are no Australopithecine fossils in South East Asia, the implication is that they found some means of crossing open water. It is possible that one species was able to do this, but not the other. However, I am unconvinced that all members of this (more advanced?) species left Africa and prefer to continue to use the term *Homo erectus* for all early humans, no matter where they lived.

Nowhere did the re-evaluation of the age of fossil remains cause a greater 'rethink' than in China.

Peking Man

Evidence as to the age and place in human history of 'Peking Man' was deliberately manipulated by its finders to obtain political advantage. The events illustrate how enthusiasm may outweigh integrity, even among scientists. These finds were discovered in the first half of the 20th century, but consideration has been held back until now because the sequel took place in the 1980s and it shows that, even at that time, whether for political or personal reasons, the desire to promote and/or maintain a particular position may dominate and control the reporting of evidence.

In 1914, Johan Gunnar Andersson, a Swedish geologist, was employed by the Chinese Government as a surveyor. He was intensely interested in the fossil record of the earliest forms of plant and animal life, but not initially interested in human fossils. Peking was surrounded to the north and west by mountainous regions which were virtually inaccessible. When he had time, Andersson liked to spend days exploring these unknown regions, knowing that he and his companions were probably the first Europeans ever to set foot there. A local man told him that there were 'dragon bones' at a site nearby and Andersson and his colleagues went to excavate the abandoned quarry at Chou K'ou Tien. They were soon rewarded with bones of pig, hedgehog, beaver, bear and many other species.

While at Chou K'ou Tien, Andersson noticed a narrow vein of quartz and it occurred to him that, had Early Man ever lived at Chou K'ou Tien, such quartz would have made useful tools (Andersson 1934: 101):

> This was the train of thought which led me … to knock on the wall of the cave deposits and say: "I have a feeling that there lie here the remains of one of our ancestors and it is only a question of your finding him. Take your time and stick to it till the cave is emptied if need be."

The Crown Prince of Sweden was a keen amateur archæologist and included Peking in his World Tour of 1926. Andersson was entrusted with the task of arranging the archæological and art presentations. He was

350

delighted to learn from one of the excavators, Zdansky, that the molar and pre-molar tooth of a creature resembling a human being had been found (Andersson 1934: 103):

> He had dug out the molar himself and identified it ... as belonging to an anthropoid ape ... *So the hominid expected by me was found. (Italics in original)*

The scientific meeting before the Prince commenced with a talk on Chinese history, which was followed by a presentation by Teilhard de Chardin, who was also working as an amateur archæologist in the area. Andersson gave the final address and, after talking about finds of fossil animals, showed a lantern slide of the 'hominid' teeth discovered by Zdansky. As a result of this unexpected revelation, Andersson was able to obtain funding from a number of organizations, including the Rockefeller Institute, for excavations to continue, to uncover further information about 'Peking Man'. Teilhard wrote to Andersson that the assumption that the teeth were human was premature (Andersson 1934: 104-105). The teeth were very large by human standards.

The next year (1927), a third tooth was found. Davidson Black, in charge of the excavations, named the hominid from which these three teeth had come *Sinanthropus pekinensis*. The following year, 1928, saw further finds – more than a score of teeth, and parts of skulls of both young and adult individuals which, unfortunately, were so embedded in the limestone that they could not be closely examined. The skulls from the Upper Cave were later determined to date from the Upper Pleistocene and not to be directly related to the finds from the Lower Caves, which were of an earlier date. Two fragments of jaw were found and, most importantly, a cranium which Andersson assumed to be approximately contemporaneous with the remains from Trinil. In 1932, another skull part was found, as well as two fragments.

Excavations were interrupted by the Japanese invasion of China, followed by World War II. Up until the time of the writing of his account, Andersson (p. 122) listed the following finds as having been made at Chou K'ou Tien: a number of teeth, several jaws, two complete and several fragmentary skulls. Like the teeth, some of the lower jaws were large, although others were of a more moderate size. Since all the cranial fragments appeared to be the remains of one type of hominid only, it was assumed that there was a pronounced degree of sexual dimorphism among these early Peking people, similar to that found in gorillas. At this time, these remains were thought to be up to a million years old and represented the first divergence of the human line from apes, so such dimorphism was not unexpected.

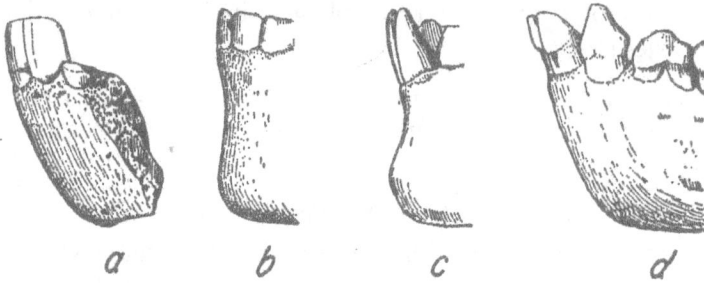

Fig.31.2: Black's drawing of a portion of lower jaw of (a) Sinanthropus; (b) stone
age child; (c) modern Chinese adult; (d) young chimpanzee.
(Andersson 1934: 109)

The lack of skeletal remains other than teeth and cranium led to the
suggestion that *Sinanthropus pekinensis* had been cannibalized. The missing
remains were presumed to have been hunted and eaten by other
Sinanthropus pekinensis. Neither Andersson, nor Weidenreich (who took over
after the sudden death of Black), suggested that the base of the skulls had
been broken in such a way as to extract brain tissue. This suggestion
appeared in later works, for example von Koeningswald (1956) and Birdsell
(1975: 305):

> All the skulls were smashed, the region round the occipital foramen was broken
> away in order to extract the brain, and there was even a femur that had been
> split lengthways to get at the marrow.

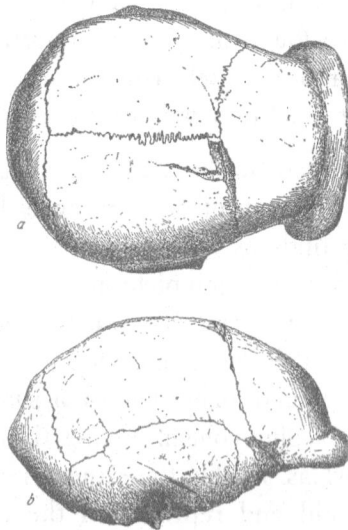

Fig.31.3: Sinanthropus pekinensis skull
(Andersson 1934: 117)

A few other skeletal remains were found: a collar bone and an *os lunatum* from the wrist of *Sinanthropus pekinensis*, resembling more closely the *os lunatum* of a modern human than that of a modern ape and four phalanges from a *Sinanthropus pekinensis* foot, one of which was believed to be the top joint of the big toe. These phalanges were different from those of modern humans, causing Black to suggest that the foot of *Sinanthropus pekinensis* deviated more from that of a modern human than did the hand.

Weidenreich (1940) attributed the remains to about forty individuals. Von Koenigswald suggested that this was an 'exaggeration', since many of these supposed 'individuals' were represented only by a few teeth, there being evidence at most for only about a dozen skulls and jaws.

While most interest was concentrated on the remains of what were then believed to be the oldest (pre)human remains found anywhere in the world, Weidenreich was struck by differences in the Upper Pleistocene remains from the Upper Cave at Chou K'ou Tien, remains which have now been dated to 100,000-130,000 B.P. The age of these fossils was at that time uncertain but they were clearly far more modern than those found in the lower excavations. Weidenreich (1938-1939) attempted to discover in these fossils evidence of what were to become the Mongoloid, Negroid and Caucasian races. Only three of the seven were sufficiently preserved for detailed analysis. The first, a male, Weidenreich considered bore resemblance to Cro-Magnon from Europe, or even to some of the earlier European Neanderthal remains. The second, a female, appeared more Melanesian in character, while the third, also a female, resembled a modern Eskimo. Weidenreich (1947) firmly believed that all human fossils, no matter in which part of the world they were found, were of the one species only. This approach was to become known as Multiregionalism.

Vallois saw both *Pithecanthropus* and *Sinanthropus pekinensis* as being so closely related that he suggested the Chou K'ou Tien fossils be renamed *Pithecanthropus pekinensis* (Boule and Vallois 1957: 142). Von Koenigswald (p. 46) noted that the continuous bony supra-orbital ridge found in *Sinanthropus pekinensis* was only seen among anthropoid apes and Coon's (1962: 437-458) analysis of the *Sinanthropus pekinensis* remains led him to the conclusion that *Sinanthropus pekinensis* was more ape-like than *Pithecanthropus*. He cited a number of features, including large teeth, bowed legs, anomalous toe bones and the skull with its continuous brow ridge, in support of his contention.

Andersson's determination to find early human remains at the Chou K'ou Tien cave led him to make a hasty (and inaccurate) evaluation of the find of the original tooth. Whether funding for further excavation would have been forthcoming had the tooth been shown as that of an ancient ape, it is not now possible to say. Having received funding to excavate for early human remains, and spurred on by his own conviction that these would be forthcoming, Andersson, his colleagues, and later teams, continued to interpret much of the evidence uncovered according to their original ideas.

In 1924, the earliest Australopithecine find in Africa had been made, that of the 'Taung child'. It had been controversial, both as to date and nomenclature. Even today, infant chimpanzees have remarkably human looking faces and what shape this infant skull would have taken as it matured, it was impossible to say. Today it is generally accepted as being that of an Australopithecine - *Australopithecus africanus*. The word *Australopithecus* means 'southern ape' and it is the term which has been accepted for remains seen to be neither fully human nor fully ape - in other words, the 'missing link'. Many have now been discovered from Africa, South-East Asia and East Asia (China) but none from West Asia or Europe. Continuing finds after the war, and the advent of radiometric dating, firmly established Africa, not Asia, as the place from which the hominine line had originated. There was an effort to stop the burgeoning classification of all new finds as new species by determining that all fossils sufficiently advanced to be considered' 'human' were to be classified as *Homo erectus*. The earliest known *H. erectus* fossil from Africa, KNM-ER 3733 from Koobi Fora, was dated to 1.7 mya. This extended the range of *H. erectus* in Africa to a possible two million years, on the grounds that it is highly unlikely that we have uncovered the remains of the very first *H. erectus* that ever lived.

It had been in October, 1891, that Eugene Dubois had discovered the skull cap that he named *Pithecanthropus* on the banks of the Solo River, near Trinil in Java. It was not until 1936 that the well preserved skull of a juvenile was discovered at Mojokerto on the banks of the Brantas River, not far from the Solo River. Two specimens of a species named *Meganthropus* were found in the late 1970s at Sangiran, further up the Solo River. Radiometric dating gave a surprising age of ~1.8 mya for the Mojokerto calvaria and ~1.6 mya for *Meganthropus*. The earliest south-east Asian fossil remains appeared to be as old, if not older than the earliest African remains, but lack of an intermediate group, such as an Asian Australopithecine, precluded any suggestion of *H. erectus* having originated in Asia. *H. erectus* must have evolved in Africa from *Australopithecus* somewhat earlier than had previously been thought if they had migrated to S.E. Asia by 1.8 mya.

In 1983 Wu and Lin published the results of a five year comprehensive investigation of the Zhoukoudian (Chou K'ou Tien) site by more than 125

Chinese scientists. They claimed that up to the year 1966, the fossil remains of more than 40 males and females had been found, along with tens of thousands of stone artefacts. Their investigations had led to the oldest fossils from the tenth layer being dated to 460,000 B.P., less than half the age previously believed. The eighth and ninth layers were dated to 420,000 ya, the seventh to 370,000-400,000 ya and the three topmost layers to 230,000 ya (Wu and Lin 1983). These dates placed the earliest occupation at Zhoukoudian nearly 1.5 million years after the earliest finds in Java and half-a-million years later than *Pithecanthropus* from Trinil. The upper layers (not to be confused with the Upper Cave whose remains were Upper Pleistocene) were dated closer to the end of the time of *H. erectus* than the beginning, as had previously been thought. Zhoukoudian became of interest now, not because it was the home of one of the earliest members of the *H. erectus* species, but one of the latest.

Following Wu and Lin's (1983) publication of the radiometric dates for Peking Man, Binford and Ho (1985) made a lengthy re-evaluation of the taphonomy of Zhoukoudian. They disputed the assumption that all the animal bones found at the site were the result of butchering by hominids, believing that much, if not most, of the skeletal remains had been brought to the cave by other carnivores, or washed into the cave after having been killed elsewhere. They questioned whether some of the 'cut' marks on some of the pieces of bone might not be gnaw marks? They suggested that 'burned' bones had become darkened by mineral staining, not burning. They questioned whether there had, in fact, been any fires at Zhoukoudian, whether the 'ash' was not organic detritus, an accumulation of guana left over the millennia by raptorial birds? They pointed out that the 'ash' was some six to seven metres thick, extremely thick to have been the remains of hearths. And if hearths, where were the remains of the meals? The only bones found in the 'ash' were those of fossil pellets ejected by the raptorial birds after their meal.

Binford and Ho believed that of all the bucketfuls of quartz pieces removed from the site, only a small number gave proof of having been worked. They did not deny that there were some stone tools; they merely asserted that most of the quartz pieces had been fractured naturally. Binford and Ho did not openly dispute the claim by Weidenreich (1949) that Zhoukoudian contained the remains of some 40-45 individuals, but they did make certain comments which clearly showed that they were not happy with the numbers given. They identified the location of the skulls, or skull fragments, which had been found, numbering them up to XIII. Most of the rest of the finds were loose teeth, with occasional fragments of mandibular, wrist or collar bones. They listed the finds from Locus B, which represented 25% of all hominid remains recovered from the site, as being (Binford and Ho p. 421):

Teeth	39
Mandibular fragments	5
Skull	1
Humerus fragment	1
Lunate fragment	1

Following the publication of this paper, Binford visited China for two months, during which time he was able to spend four days examining the Zhoukoudian material (Binford and Stone 1986). The original hominid finds, i.e. of *Sinanthropus*, had been lost at the time of the Japanese invasion but Binford had expected to be able to access most of the other fossil fauna. He was surprised that the material available was far less than had been reported. For example, fossil material corresponding to 2,000 animals of the cervid species *Megaloceros* had been reported but only 501 fragments were available and only 333 specimens of *Pseudaxis grayi* were available against 1000 claimed. Since Binford had been told that all bones had been made available to him for study, he asked to see the original records but only records from 1935-1938 were available and these he was shown but briefly.

Examination confirmed Binford's earlier suspicions that most of the bones previously reported as having been burnt were, in fact, mineral stained. Eleven bones appeared to have been burnt after they had become dry and degreased and had probably been lying on the surface when a fire passed over them. Seven specimens, all upper teeth, showed evidence of fresh burning, coming from upper deposits (Levels 3-4). Binford examined the 'ash' and confirmed to himself that it was actually organic material. He was to be allowed to take a sample for analysis in the States, but the sample was re-appropriated.

As a result of his visit, Binford concluded that the amount of material recovered from Zhoukoudian had been exaggerated, the number of stone tools had been exaggerated (some did exist), the amount of burnt bone had been exaggerated (a few such pieces did exist, although not burnt when fresh) and the 'ash' was actually guana detritus. That the caves had been inhabited by early humans was not disputed, but Zhoukoudian was not the rich source of information about the earliest hominids originally thought.

Possibly in response to Binford and Stone's article, Jia and Huang (1990) retold *"The Story of the Peking Man"*, giving their book the subtitle *"From Archæology to Mystery"*.

This book recounted the early excavations very much as they had been told by Andersson (1934) but went on to tell the history of the later excavations from the mid 1930s until the 1980s. Jia and Huang related how, during the first season of digging in 1927, 3000 cubic metres of material had been excavated, which resulted in 500 crates of animal fossils. This material also included one 'human' tooth, the tooth which Teilhard had disputed. They repeated the claim that the quartz fragments found at the site were tools (p. 7): "Each day during the operation in 1931, large basketfuls of stone artefacts were delivered twice daily to Jia Lanpo at site headquarters". They reiterated the presence of layers of ash and scorched bone, continuing to believe that the large numbers of animal remains found at the caves were the result of human predation.

Also restated was the belief that the paucity of post-cranial human remains indicated cannibalism. The remains, although fragmentary, were seen as representing "upwards of forty persons" and to "top" all other sites in the world in the abundance of stone artefacts and traces of the use of fire. They cited traces of fire at Locality 13 as "the earliest known use of fire on record" (p. 195), yet Locality 13, dated to 600,000 to 700,000 BP, contained no human remains. The earliest date given by Wu and Lin for hominid remains at Zhoukoudian was 460,000 BP, although Jia and Huang stretched this time to 500,000 B.P.

No mention was made by Jia and Huang of any of the African finds of *Homo erectus* or of *Australopithecus*. Chinese readers, dependent upon this book for their information, might believe that Zhoukoudian was almost the only known home of *Homo erectus*, except for *Pithecanthropus*, which Jia and Huang claimed to be morphologically similar to Peking Man, ignoring recent dating which showed that the fossils were separated by about half a million years.

At the time that Jia and Huang were writing, China was still under the influence of Chairman Mao, and still pursuing a very isolationist path. It is to be hoped that the spirit of international engagement and co-operation which the Chinese authorities are currently exhibiting will allow the whole "mystery" of *H. (erectus) sinanthropus* to be re-evaluated in a scientifically neutral manner.

The original *Sinanthropus* fossil remains had been 'lost' at the time of the Japanese invasion, although casts were still available. Gish (1973) was convinced that the remains were never 'lost' but deliberately destroyed to cover up the fact that the casts, which had been widely viewed, were not a faithful reproduction of the original but had been 'modified' to make them

357

appear more human than they really were. Gish came to this conclusion after reading an account of the circumstances surrounding the supposed event by a Roman Catholic priest, Rev. Patrick O'Connell, who was in China at the time of the Japanese invasion and who had read the accounts given in both the Chinese and foreign language newspapers. Gish was convinced that O'Connell was right because only the *Sinanthropus* fossil material was missing. Other related finds were in order, or so Gish concluded from an article published in 1954 in the Peking periodical *China Reconstruction*. Whether it will ever be possible accurately to determine precisely what material was recovered from Chou K'ou Tien and what went missing, must remain doubtful, but if only the *Sinanthropus* skull fragments are missing, then perhaps O'Connell and Gish may have made a valid point.

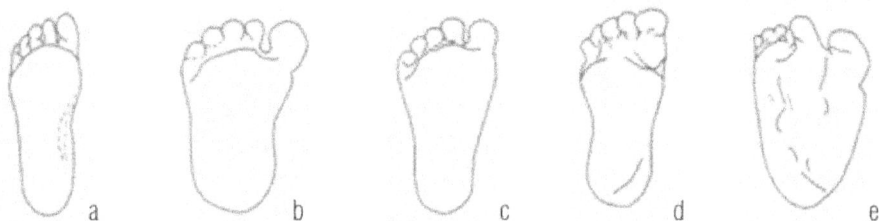

Fig.31: 4: A selection of footprints:
a) human adult; b) Neanderthal; c) Bigfoot, Russia; d) Bigfoot, Sasquatch, USA;
e) Bigfoot, Yeti, Nepal.
(Shackley 1980: 132)

Many caves in the mountainous area surrounding Peking contained organic material which was collected by peasants for use as fertilizer. It was during such a collection that a peasant, Chin Hau-Huae, found a tooth from the jaw of a giant ape, which was called *Giganthropus* (Pei 1957). A subsequent search of the area recovered the jaw bone.

Did such 'apes' continue to live until recent times? The Chinese philosopher, Hsin-Tzu, who lived around 400 B.C., stated that an ape the size of a man, covered with hair but standing erect, lived in the Yellow River Valley and other references were made to ape-men living in Tibet in the Liang Annals written between 200 B.C. and 200 A.D. (Coon 1962). A traditional Tibetan medical book, published in Peking at the end of the eighteenth century, which Coon had seen, contained illustrations of the mammals, birds, reptiles and fish used in Tibetan medicine. It included a picture of a man-ape, with the summary: "The wild man lives in the mountains, his origin [habitat] is close to that of the bear, his body resembles that of a man and he has enormous strength. His meat may be eaten to treat mental diseases and his gall cures jaundice" (Coon p. 208).

Coon assumed this ape-man to be the Yeti, or closely related thereto, and accepted its factual existence, pointing out that the relationship of the 'Abominable Snowman' to past and present Asian apes/humans had yet to be determined. Shackley (1980) agreed that it was feasible that an earlier form of human, possibly a Neanderthal, might still exist. She cited the Yeti from Sasquatch in the United States and the Almas of Mongolia, the possible existence of which had been sufficient to attract a research study grant. Footprints had been found in the Dzungaria region of Mongolia, an 'exact match' for those of Neanderthal man found in a cave in Italy. Locals described these people as being small, rather hirsute and without speech (understandable by them). These 'people' left skins at appointed places and took away in trade articles left there by the locals.

Reports of an ape-like human (or a human-like ape) occur in both Russia and America. Viewed from the North Pole, these places are not so very far apart and warmer climates in times gone by may very well have allowed travel between them. There is no 'evidence' of any such people in the Southern hemisphere. The Aboriginal people of North Australia report the existence of 'Quincan', although whether these are seen as physical or spiritual entities, I confess I am not clear.

Chapter 32

Rediscovering Our Origins

It is generally agreed that about 6 million years ago, the human line diverged from the ape line. That chimpanzees are the living apes most closely related to humans does not mean that theirs was the ape line from which human beings diverged.

There is no skeletal evidence of the earliest, or archaic, 'ape-men' for at least two million years after their presumed divergence from apes. From the 1960s onwards, there has been a steady increase in the number of fossil finds reaching back nearly 4 million years, attributed to *Australopithecus*, the descendants of this diverging form. These finds took place in Africa, either the east (notably Afar, Omo, Olduvai, Lake Rudolph, Koobi Fora) or the south (Taung, Swartkrans, Sterkfontein, Malapa), although the southern Australopithecines are of a later date (around 2.5-1.5 mya). This happened mainly because this was where excavations were taking place. There was an initial tendency to allocate each new find to a new species, but as the number of these finds increased, it became acceptable simply to describe the fossils as 'robust' or 'gracile'. However, there was a return to former practice when the find from South Africa, dated to ~1.9 mya, was given the separate species name, *Australopithecus sediba* (Berger et al. 2010). There was evidence that both the robust and gracile types co-existed in the southern part of the continent for a considerable period of time (up to 2 million years), as well as for the co-existence of the robust type with early *Homo erectus* in East Africa. Birdsell (1975: 268) did not accept either type as being directly ancestral to *H. erectus*:

> It is now evident that neither of the australopithecines were direct ancestors of more modern kinds of men ... It could be that deep in the Pliocene a form more like the gracile australopithecines did give rise to more advanced types of men.

Clearly, if evolutionary theory is correct, modern humans are descended from an earlier, animal, form. Why not either the robust or gracile Australopithecine?

Stanley (1996) suggested that early hominids could not have evolved larger brains until Australopithecines had abandoned their tree-climbing life-style. The shoulder girdle of Australopithecines indicate that they were adept climbers and it would have been necessary for their infants to cling to their mother as she ran along branches and leapt between trees. Stanley argued that the larger hominid brain came about due to continued post-natal growth which, in effect, meant that infants were born immature. Such immature infants would not have been capable of clinging sufficiently firmly to their mothers immediately after birth. Therefore, it was necessary for Australopithecines to have abandoned their arboreal life-style for enlargement of the brain to have become possible. He termed this the 'Terrestrial Imperative'.

Because of the Australopithecine's upright stance, Stanley further suggested that the infant draped itself over the mother's shoulders and wrapped its arms around her neck or clung to the long hair on the back of her head. The latter suggestion seems to assume that the mother had no hair on her body. Exactly when, or why, hominids lost their fur is not a matter which attracts much speculation in the literature compared to the amount of space devoted to the shape of the foot, the alignment of the big toe, the shape of the femur, of the pelvis, or of the foramen magnum where the head balances on the top of the spinal column. Losing the protection which fur would have afforded from the African sun is difficult to explain, as is losing the protective warmth such fur would have given on a winter's night. Only one writer, Morgan (1972), gave this matter extensive consideration.

All warm-blooded animals, be they mammal or bird, need some method of retaining heat and the three methods used are fur, fat or feathers. It is believed that the reptilian line, while continuing to exist in its own right, gave rise to two separate, warm-blooded lines: mammals and birds. This would mean that warm-bloodedness evolved separately on at least two occasions, albeit in a (geologically) short space of time. Feathers became the chosen method of thermal insulation for birds (and possibly dinosaurs) and fur/hair for mammals.

At some point, for reasons still not understood, a small group of mammals, including the whales and dolphins, returned to the sea. These mammals remained air-breathing, did not revert to gills or evolve a new method for extracting oxygen from water. However, they lost their fur, relying on subcutaneous fat for thermal regulation. Another group of mammals (including seals and sea lions) made a less complete return to an aquatic environment. They retained some fur/hair but also developed subcutaneous fat for thermal regulation. Morgan suggested that there was a third group of mammals which had briefly toyed with a return to the water (elephants, hippopotami and pigs) but which had abandoned the process, although still showing a great affinity for water and/or mud. These animals, too, had lost their fur and developed subcutaneous fat. Morgan then suggested that humans should be added to this group. Humans alone among the primates have replaced fur with fat.

This theory was not new. Morgan pointed out that it had first been proposed by Professor Sir Alister Hardy in 1960 but had received little support. To Morgan, it explained several things. It had long been a puzzle how a small, almost defenceless Australopithecine could have survived on the African Savannah. The Big Cats not appearing for several million more years could help explain the matter, but presumably there was some form of ancestral cat, or dog/hyena, which would have been a threat to so small a creature as the Australopithecine. Morgan suggested that there were no such predators on the foreshore. There might also have been some protection from the greatest of human predators, the mosquito, although Morgan did not suggest this. There would have been an ample supply of food in the shallows, fish, shellfish, crabs and so forth. Morgan even suggested that the female Australopithecine might have been the first user of a tool. Not endowed with the large canine teeth of her companion with which to open shellfish, she might have used a pebble to crack the shell – after all, 'necessity is the *mother* of invention' (p. 27) why should all inventions be attributed to men?

The loss of body hair, the retention of hair on the head, are readily explained by an extended time (those missing two million years) spent in the shallows. Morgan pointed out that the realigned pelvis might have been as beneficial for swimming as it was for walking.

Birds exclusively use feathers for thermal regulation, including penguins who have also assumed a semi-aquatic life-style. Mammals use fur/hair, with very few exceptions – those who have replaced fur with subcutaneous fat. This change has taken place in every mammal who has returned to an aquatic existence. If the hominid line at no point in time flirted with a more aquatic life-style, what is the explanation for human hairlessness and subcutaneous fat?

Morgan's book was written in a light-hearted manner, but she was very

362

critical of what she perceived as a male oriented view of human evolution. She scoffed at the suggestion of Desmond Morris that humans lost their fur because males would have become overheated during the hunt if burdened with fur (p. 16): "I don't believe it's ever been all that easy to part a woman from a fur coat". She could also have pointed out that the Big Cats have found no reason to abandon their fur and they run a lot faster than any human ever has! Morgan was a feminist and made no apology for writing from a feminist perspective. She merely asked that males be more conscious of their biased point of view.

Lips (1949: 103) took a similar attitude.

> There can be no doubt that the institution of agriculture marks one of the greatest contributions woman has made toward the welfare of mankind. It was woman who in the acquisitive economy already took care of the provision of the family with vegetal food; it was woman who, consequently, put the great new invention of sowing and planting into practice.

The Australopithecines occupy a fairly non-controversial place in the history of human evolution. Somewhat more controversial is our understanding of *Homo erectus*. We have already discussed the first *H. erectus* discovered in S.E. Asia (Trinil and Solo, Java) and China (Peking) under their original names *Pithecanthropus* and *Sinanthropus*.

At the time of the discovery of Peking Man, there had already been numerous finds of Neanderthal Man in Europe, but not of Australopithecines or *H. erectus* in Africa. It appeared that the transition from ape to hominid might have occurred in Asia, not Africa. It was assumed that the descendants of the people from Trinil, Peking and Solo, now collectively known as *H. erectus*, entered Europe and gave rise to the Neanderthals. Boule (1923: 104) expressed doubts as to whether the skull cap and the femur found by Dubois in Java, and then known as *Pithecanthropus*, actually belonged to the same individual:

> If we possessed only the skull and the teeth, we should say that we were dealing with a large Ape; if we had only the femur, we should declare we were dealing with a Man.

Boule suggested that *Pithecanthropus* might not be a direct ancestor of the present human line, but might have been a 'grand-uncle' rather than a 'grandfather', more closely related to the gibbon line than the human.

The well-known American anthropologist, Carlton Coon had written a book, *The History of Man*, on the origin of the races of Europe. For example, the peoples of Eastern Europe were easily distinguishable from those of the Mediterranean and both were as easily distinguishable from those of Nordic race, but how had this come about? He now wanted to extend his work to a study of the origin of the races of the world, but realized that, in order to complete such a work, it would be necessary for him to trace the origin of races as far back into history as was possible. He spent twenty years collecting information which lead to the publication of *The Origin of Races* (Coon 1962). Coon consulted with Mayr, Simpson and other biologists (pp. viii-x) and his theory of the evolution of human races was based on the *Evolutionary Synthesis*, as propounded by Sir Julian Huxley, which held that changes took place in peripheral populations and gradually spread through interbreeding.

At the time Coon wrote, it was generally accepted that fully modern *Homo sapiens* had come into being around 35,000 BP, approximately the time of the appearance of the Cro-Magnon people in Europe. Coon did not accept that this was sufficient time for regional differences to have developed. He could not accept that an Australian aborigine or a Pygmy from the Congo could have had a common ancestor with Europeans as recently as 35,000 years ago, particularly since they were living in a manner comparable to that of European people 100,000 years ago. He believed that our common human origin lay far deeper in time. Coon (p. 30) wrote:

> ... related populations, which in our case are subspecies, passed from species A, which is *Homo erectus*, to species B, *Homo sapiens*, at different times, and the time at which each one crossed the line depended on who got the new trait first, who lived next to whom, and the rates of gene flow between neighbouring populations ... The new critical trait responsible for speciation first appears in a few individuals ... and in the process it has become a new species of its own. It need not, however, have completely lost the gene or genes for the old trait ... After the new species has established itself ... the old trait may completely disappear. At that point a second and final threshold of speciation has been crossed ...

> It is easy to understand, then, why some populations within any polytypic species have come closer, at any given time, to the second threshold of speciation than other populations. In man some groups alive today have preserved archaic traits, diagnostic of *Homo erectus*, in a higher percentage of individuals than other populations ... There is no necessary correlation between the time at which a threshold was crossed and the rate of change that follows the crossing.

Unfortunately for Coon, the 1960s saw the emergence of Africa as a continent of independent nations, no longer one of colonies. It also saw the emergence of the Black Rights movement in America and of the Land Rights movement in both North America and Australia on behalf of their aboriginal populations. Any suggestions that some populations crossed the threshold to humanity later than others, even if the time suggested was hundreds of thousands of years ago, became unacceptable, and Coon's position came to be seen as racist.

Coon was arguing that the retention of some physical features from an earlier time was quite consistent with the acquisition of a new trait and could occur as readily among those who were the first recipients of the new trait as it could among the last. However, his suggestion that the natives of New Caledonia, with their big teeth and heavy brow ridges, had retained more ancient characteristics than had the Japanese (pp. 30-31) did not find favour in the changing post war political environment.

Writing only three years earlier, Howells (1959) reminded his readers that mammal-like reptiles crossed the line into 'mammalhood' at several different periods, and this statement caused no controversy, yet the passage being referred to was far more complex. Not only are mammals warm-blooded, but their reproductive systems are significantly different from that of reptiles, both by the presence in the female of a uterus and by the excretion of milk for the suckling of the young. It is generally accepted that the change from cold to warm bloodedness took place on two separate occasions, with the evolution of mammals and the dinosaurs/birds. Coon (1962: 92) believed Australian monotremes evolved separately from reptilian ancestors. If Coon was correct, then warm-bloodedness evolved three separate times. These major evolutionary changes are accepted as having occurred on more than one occasions at widely-spaced intervals, without any suggestion that 'priority' inferred 'superiority', yet to suggest that time was taken for an evolutionary change to spread throughout a geographically diverse human population was deemed by many to be 'racist'.

The Evolutionary Synthesis demanded change in a peripheral population and the gradual spread of that change over the entire population. Coon merely pointed out that *unimportant* physical features, such as heavy brow ridges, might have been retained within the genetic material of some populations long after the other changes, which made people uniquely 'human', had become generalised.

Nowhere does evolutionary theory suggest that all members of a population change instantaneously. Some change earlier than others – that is how evolution works. A suggestion that earlier or later change is somehow a matter for pride or shame is really quite ridiculous.

Finds, such as those at Sima de los Heusos, Atapuerca, Spain, dated to 300,000 BP, Gran Dolina, Atapuerca, dated to 78,000 BP and Dmanisi, East Georgia, dated to approximately 1.7 mya supported earlier finds, such as

Heidelberg and at Swanscombe, England, which established that *Homo erectus* had left Africa much earlier than previously thought. Acheulian type stone tools had also been found at Swanscombe, suggesting occupation as far back as 500,000 years, or more.

In the early 1960s, Brace compiled a lengthy review of the, then, past and current literature, which he entitled *The Fate of the "Classic" Neanderthals* (Brace 1964). He sought to demonstrate, not only the deep divide in opinions held by various experts, but the way in which the same evidence could be interpreted to support opposing views. Brace was of the opinion that 'the climate of opinion' at the time of the discovery of Neanderthal remains had had a profound and lasting effect upon the interpretation of that fossil material, namely that there had been a reluctance to acknowledge Neanderthals as ancestral to present day humans.

Brace argued that for modern humans successfully to have invaded Europe from another continent, they would have needed technology superior to that of the resident population, something he believed they clearly did not have. Brace concluded that it had been the "fate" of the Neanderthals to give rise to modern humans (Brace 1964: 19). His arguments were later countered by the suggestion that language was the missing superior skill (a subject which will be addressed in another chapter) and by the suggestion that the invasion of Europe had taken place after 40,000 BP, when it could be shown that new technologies were available (and which were assumed to have been fashioned by the invaders, not the resident Neanderthals).

It was generally agreed that *H. erectus* had migrated from Africa and colonized from China and Java in the East to England in the West. (No early remains or artefacts have been found in Ireland.) The well-known English archæologist, Chris Stringer, of the Natural History Museum in London agreed that early European 'archaic' *H. sapiens* resembled *H. erectus* but questioned (Stringer and Andrews 1989) whether *H. erectus sensu stricto* was known from the European fossil record. (Chris Stringer became well known, not just because of his publications, but because he was in constant demand by television producers for participation in documentaries or for comment on new finds, thus becoming known to the interested public across the world.) Although the number of early hominid remains in Europe was not large, such as had been uncovered tended to suggest that the early form had been closer to the modern type. For example, the Swanscombe skull, dated to approximately 300,000 BP, could not, as far as the occipital and left parietal bones were concerned, be distinguished from *H. sapiens* (Day 1965).

The Neanderthals were seen as a local specialization evolved from European 'archaic' *H. sapiens*, who had lived in Europe between about 800,000 and 200,000 BP (Radovcic 1985, Trinkaus and Smith 1985, Stringer 1985), occurring mostly in Europe from *circa* 120,000 to 30,000 BP, but also living in the Levant from *circa* 100,000 to 40,000 BP. This scenario made it difficult to deny the Neanderthals *sapiens* status without postulating that they had undergone a form of degeneration or 'reverse evolution'. Prior to this time, skeletal remains indicated the presence of 'pre-Neanderthals', mostly from Europe but also from northern Africa and, more controversially, from further south in Africa (Broken Hill) and east into Asia. Later, Trinkaus (Trinkaus et al. 2001) was to refer to the earlier papers cited above, saying that the stated perception that Neanderthals were significantly less effective than the modern humans which succeeded them contrasted strongly with his present view that they were technologically equivalent.

The 'Complete Replacement' hypothesis

The point of view referred to by Brace as 'Catastrophism' became known as the 'Out of Africa' or 'Complete Replacement' hypothesis because it proposed that all hominids throughout the world were replaced by a new species of hominid, early modern *H. sapiens*, who evolved in Africa ~200,000 ya, or possibly later, and spread from there to inhabit the rest of the world.

The discovery in Ethiopia in 1969 of the anatomically modern human skeletal remains known as Omo I, tentatively dated on the basis of associated mollusc shells to 130,000 BP, strengthened the case for the replacement of all *Homo* species by the more evolved new species of human which was seen to have evolved in Africa (Leakey et al. 1969), although Trinkaus and Shipman (1992) stressed that the dating of Omo I was insecure, ranging between 130,000 and 40,000 BP. Other fossils, such as those from South Africa, Tanzania and Kenya, also indicated the early presence in Africa of early modern *H. sapiens*.

Debate continued into the first decade of this century. Skeletal remains from Herto, Middle Awash, Ethiopia, dated to between 160,000 and 154,000 BP were seen to give further support to the 'Complete Replacement' hypothesis. These remains were "morphologically and chronologically intermediate between archaic African fossils and later anatomically modern late Pleistocene humans" (White et al. 2003: 742). They were classified as *Homo sapiens idaltu*, although Stringer (2003: 693) was of the opinion that the creation of a new subspecies was not justified since "the distinctive features described for *H. sapiens idaltu* might not be so unusual and could probably be found in late Pleistocene samples from regions such as Australasia".

While White et al. believed the discovery of these intermediate remains strengthened the argument for the evolution of *H. sapiens* in Africa alone, the uncovering of samples which filled a gap in human evolution, which had long been hypothesized, did not change the substance of the debate. The findings did not clarify whether the evolution in Africa towards anatomically modern humans resulted in change so profound that it produced a species separate from those of hominids evolving at the same time in Europe and Asia, or whether continuing evolution on these two latter continents resulted in different subspecies, all of which were nevertheless still capable of interbreeding with each other.

Insufficient fossil evidence precluded theorists from being clear as to whether it was only a small population of *H. erectus* within the large continent of Africa which made the transition to archaic *H. sapiens*, or whether the transition took place over the whole continent before the exodus. Uncertainty as to exact dating made it difficult to establish whether fossil remains bore evidence for rapid change over the whole continent or of gradual spread. The remains from Herto had been dated to 160,000-154,000 BP, Omo I to ~130,000 BP, Klaisies River Mouth to between 120,000 and 75,000 BP and Border Cave to 80,000 to 70,000 BP, a possible time range of 90,000 years.

Those who supported the model of global replacement of archaic by modern humans with an African origin, did not necessarily argue in favour of *complete* genetic replacement (for example Skinner 1994). Templeton (2002) argued for a minimum of two Out-of-Africa expansion events but claimed that the most recent was not a replacement event since, if it had been, the genetic signatures of older gene flows would have been completely eliminated, which was not the case. Lahr and Foley (1994) argued that, although there was no clear rubicon for modern *Homo sapiens*, their evolution had taken place in Africa. Africa, being a large land mass, had allowed the development of diverse morphologies, and these had dispersed 'Out of Africa' on several different occasions.

The publishing of a seminal paper by Cann, Stoneking and Wilson (1987) which claimed that mitochrondial DNA evidence proved that all humans alive today had a common (female) ancestor who lived in Africa some 200,000 BP, tended to harden attitudes in favour of the evolution of a new species in Africa around that time, a species which subsequently colonized the whole world.

Each cell within the body has two completely different sets of DNA. The one with which most people are familiar is the nuclear DNA, that which is found in our reproductive cells and contributes to our physical make-up. Not so well known is mitochondrial DNA (mtDNA). There is a small body within each cell, known as the mitochrondia, which is responsible for the energy requirements of each cell. Quite why each cell has these two separate nuclei is not fully understood. What is of interest here is that mtDNA is passed down only through the female.

Going backwards in time, tracing our individual hypothetical ancestral tree, the number of ancestors who may have contributed to our DNA multiplies: 2, 4, 8, 16, 32 and so on. At each generation, the proportion of DNA likely to be present reduces and, eventually, it must be assumed that none will be left from some ancestors. Against this is the fact that, until recently, humans lived in close communities and it was not unusual for a great grandparent to have contributed to the genetic make-up of a great grand-child through both the male and female line. Charles Darwin married his cousin, Emma Wedgewood. They had a common grandfather.

Since mtDNA is passed down only through the female line, do we not all have identical mtDNA? No, we don't. Siblings and close family members do, but random mutation results in differences between those of more distant ancestry. Theoretically, if we trace back far enough, our family tree will bring us to one female from whom we can all claim descent. Cann et al. nicknamed this common female ancestor 'Eve' and calculated that she had lived in Africa between 140,000 and 280,000 years ago, a time referred to as approximately 200,000 BP for the purposes of simplicity. Although Cann et al. (1987: 35) further concluded that this particular mtDNA had migrated from Africa to all other parts of the world, they stated:

> Our placement of the common ancestor of all human mtDNA diversity in Africa 140,000-280,000 years ago need not imply that the transformation to anatomically modern Homo sapiens occurred in Africa at that time. The mtDNA data tells us nothing of the contribution to this transformation by the genetic and cultural traits of males and females whose mtDNA became extinct.

Despite this caution, Cann et al. did refer in their paper to the two competing models of human evolution, 'Out of Africa' and 'Multiregional Evolution'. The similarity of the time of the existence of 'Eve' with that proposed by 'Out of Africa' theorists for the second exodus of hominids from Africa was sufficient to lead Cann et al. to propose that *Homo erectus* in Asia was replaced without much mixing with the invading *Homo sapiens* from Africa. They made no comment in relation to Europe and the Neanderthals, presumably because Neanderthals had not at the time under

369

consideration evolved into their classic form, although pre-Neanderthals were already in existence. The use of the words 'without much mixing' leaves open the possibility of some interbreeding. If even 'some' interbreeding took place, the new form would necessarily have been a subspecies, rather than a completely new species, the product of a 'speciation' event. A few months later, Cann (1987) published a paper on her own in which she referred to *Homo sapiens neaderthalensis* as a variant of archaic *Homo sapiens*, which she nevertheless considered not to have been ancestral to modern humans, who had evolved elsewhere concurrently. In this paper, Cann (p. 37) elaborated on the caution she and her co-workers had previously given:

> The mitochondrial evidence indicates that archaic *Homo sapiens* evolved into *Homo sapiens* about eighty thousand years earlier than the fossil evidence suggests. Yet it is possible that Eve herself belonged to the archaic subspecies and contributed her mitochondrial DNA to those who, perhaps many generations later, evolved into *Homo sapiens sapiens* ... In fact, Eve's descendants could have maintained their archaic form for thousands of years before taking on fully modern characteristics. There is no evidence to suggest that people today have retained any of Eve's particular physical features. We know only that we have inherited her mitochondrial DNA; she might have contributed very little to the surviving pool of human nuclear DNA ... It is important to note that unlike her biblical namesake, mitochondrial Eve was not the only woman alive during her time. She merely is the only woman of her age whose descendants have included some females in every generation. Some of her contemporaries, no doubt, also have progeny alive today who carry traces of the ancestral nuclear DNA ... That the world's races evolved from multiple lineages implies that early modern humans moved back and forth around the world several times, intermixing with one another along the way, before they settled down in sufficient isolation to develop racial features.

It would seem that while the team, of which Cann was an important member, collectively favoured the Complete Replacement hypothesis, Cann herself favoured the Multiregional model. A divergence of opinion was evident in a paper co-authored by Wilson and Cann (1992) in which it was stated (p. 22) that "modern humans arose in one place and spread elsewhere" and (p. 24) that "the farther back the genealogy goes, the larger the circle of maternal relatives becomes, until at last it embraces everyone alive ... she [Eve] did not necessarily live in a small population ... or constitute the only woman of her generation", which begs the question of what precise use is information regarding our last common ancestor? Despite his co-authorship of the above comments, Wilson came to see the mtDNA evidence as supporting the 'Out of Africa' theory so strongly that he even suggested that modern humans had language, while the Neanderthals and other early people did not, because the gene for language was carried in the mtDNA (Frayer et al. 1993: 21). It is my understanding that genetic material which influences our physical attributes is carried in nuclear DNA, not mitochondrial DNA.

In another article, Cann (1988: 136) once again stressed that "to claim all living humans can trace their mitochondrial genomes back to a single female founder is not to say that we all come from a single female ancestor ... the extent of admixture of modern human with humans who may have migrated out of Africa at an earlier stage of human evolution, even at the *Homo erectus* stage, can be judged only indirectly from the mitochondrial DNA". Depending upon the length of time between generations, 1200 years ago (800 AD) each of us had the statistical possibility of between 4 billion and 17 billion ancestors (Cann 1988: 136) but only one would have been our 'mtDNA mother'. Clearly every potential (distant) ancestor must have many places in any ancestral tree, but only from one male and from one female can an unbroken line of same gender descent be traced. Every one of us has an unbroken line of same gender descent which is traceable back through our male or female ancestral line, not just to 800 AD, but to 80,000 BP, 800,000 BP, and so on throughout evolutionary time. We each have an astronomical number of potential ancestors, but only two lines of same gender descent, all other lines being interrupted by an ancestor of the opposite gender, which nevertheless had an equal chance of contributing their nuclear DNA. Of three hundred English families who have written records of descent from William the Conquerer, only one is linked to him through direct male descent (Jones 1997: 87). Four years later, Cann, in collaboration with Wilson, reiterated her position (Wilson and Cann 1992: 24):

> ... that all humans today can be traced along maternal lines to a woman who lived about 200,000 years ago, probably in Africa ... The further back the genealogy goes, the larger the circle of maternal relatives becomes until at last it embraces everyone alive ... all human mitochondrial DNA must have had an ultimate common female ancestor ... she did not necessarily live in a small population or constitute the only woman of her generation.

This was accepted even by those promoting the Complete Replacement model, such as Stringer and McKie 1996. Common ancestors within any community are to be expected and if genealogies are traced back far enough, it is inevitable that a common ancestor will be found (Thorne and Wolpoff 1992b, Cavalli-Sforza et al. 1994, Klein and Takahata 2002). Vigilant et al. (1991: 1506) supported the "contention that all the mtDNA's found in contemporary human populations stem from a single ancestral mtDNA that was present in an African population approximately 200,000 years ago" and that "the mtDNA evidence is thus consistent with an origin of

371

anatomically modern humans in Africa within the last 200,000 years". They thus represent the school of thought that evidence for a common *mitochondrial* DNA ancestor is evidence for the origin of anatomically modern humans (common *nuclear* DNA), a position not taken by Cann et al. (1987), Cann (1987, 1988) or Wilson and Cann (1992).

Reanalysis of the mtDNA data (Templeton 1991, 1993) resulted in the finding of 10,000 trees more parsimonious by five steps than the mtDNA tree cited by Cann et al. and of 100 more parsimonious by two steps than that given by Vigilant et al., some having non-African origins. The possible range of time at which the postulated 'event' had taken place ranged from 33,000 to 675,000 years ago. Templeton and Hedges et al. (1991) interpreted the mtDNA research as supporting a model of restricted but recurrent gene flow. Klein (1994) considered the mtDNA results as flawed as the fossil record they were meant to complement and concluded that there were no statistical grounds for preferring an African to a non-African origin for humans.

At one time it had been thought that anatomically modern humans (AMH) had arrived in Europe at the time of, or shortly before, the extinction of the Neanderthals some 40,000 to 30,000 years ago. However, the discovery of skeletal remains in the Levant which were classified as AMH and dated to around 80,000 BP, and the dating of the first human occupation of Australia between 60,000 to 50,000 BP caused this time to be extended backwards.

The arrival of humans in Australia necessitated the crossing of open water by some form of watercraft, a feat not then believed to have been possible by any other than a member of our own (modern) species. Genetic research had shown that at least fifteen different groups of people populated Australia, all of African origin (Jones 1993), but there was no clear scientific evidence that all of the Aboriginal people living in Australia today were direct descendants of the first arrivals. (Jones' finding may appear to contradict the belief that the most recent arrivals, some 4,000-5,000 ya, came from India, bringing with them the dingo. It must be remembered that genetic evidence reaches far back in time and the 'Indian' arrivals may well have had African ancestry.) The assumption that archaic *H. Sapiens* was incapable of building and navigating an ocean-going watercraft led to the conclusion that the first arrivals in Australia were modern *H. sapiens* but mtDNA analysis of a 60,000 year old skeleton from Lake Mungo, New South Wales, raised some doubts.

Adcock et al. (2001) analysed the mtDNA of ten ancient skeletal remains from Australia - a Pleistocene (gracile) skeleton from Lake Mungo, western New South Wales, known as LM3 dated to 60,000 BP, three Holocene (robust) skeletons also from Lake Mungo and six individuals from Kow Swamp, Victoria, dated to 13,000 to 9,000 BP. The robust morphology of the Kow Swamp individuals was outside the range found among modern Australian aborigines. However, their mtDNA was not. In contrast, mtDNA analysis showed that the sequence from gracile LM3 (60,000 BP) and one of the robust Kow Swamp fossils (KS8) resembled that of Feldhofer, the Neanderthal type specimen, analysed by Krings et al. (1997), all belonging to a lineage which had diverged from that of modern humans prior to the time of their most recent common ancestor, at least 150,000 years ago, possibly far longer. Adcock et al. (2001: 540) felt that the data relating to KS8 was insecure and thus did not pursue a claim for a separate linage for KS8 but did conclude that mtDNA and nuclear DNA, as evidenced by anatomical features, may have different evolutionary paths.

It had previously been assumed that modern humans arrived in Australia about 30,000 ya, about the same time that they arrived in Europe as Cro-Magnon. Most early skeletal remains from Australia do date from that time but ash evidence of extensive land clearing by the use of fire, carbon dated to 45,000 B.C., indicates the earlier presence of people. It is now accepted that, when (modern) humans left Africa, they spread, not only to Europe and Mainland Asia, but across the Malay Archipeligo to Australia as well. As already mentioned, there is evidence that earlier humans *(Homo erectus)* and possibly Australopithecines likewise migrated south.

Stringer and Andrews (1988) noted that some Australian fossils looked decidedly more 'archaic' than their counterparts elsewhere. The first *H. sapiens* to arrive in Australia should have been no more archaic than *H. sapiens* people anywhere else. If they were "this would need explaining ... Perhaps Australia was a special case where local differentiation, cultural practices, or pathologies led in some cases to apparent evolutionary reversals" (p. 1267). This may have been a reference to the opinion of Thorne and Macumber (1972: 316-317) who pointed out that the Kow Swamp crania displayed archaic cranial features not seen in recent Aboriginal populations, which suggested to them that *H. erectus* might have survived in Australia until as recently as 10,000 years ago. An 'evolutionary reversal' might have explained this discrepancy. Today the robusticity of the Kow Swamp people is not viewed as an indication of their being an archaic people. Adcock et al. (2001) pointed out that just because their robust morphology has not survived does not mean that they do not have modern descendants. Their mtDNA indicates this possibility. Conversely, just because the mtDNA of the ancient Lake Mungo man has not survived does not mean that his people have no modern descendants. His gracile morphology indicates this possibility. (This is an echo of Coon's argument.)

Redating of fossils from Ngandong and Sambunmacan, Java, gave ages of 27,000 to 53,000 years, bringing forward the survival of *H. erectus* in Asia by some 250,000 years (Swisher et al. 1996). Perhaps a late survival of *H. erectus* in Australia is not so inconceivable after all. In any event, *H. erectus* was living in South-East Asia contemporaneously with *H. sapiens* in Europe.

While it was a surprise that the remains of LM3 were dated to 60,000 BP, it is generally accepted that this man from Lake Mungo was anatomically modern (Stringer and McKie 1996: 110). If this assumption is correct and the LM3 mtDNA does show that he belonged to a lineage which diverged before the time of our most recent common ancestor (150,000 or more years ago), then mtDNA of a type distinct from that of modern *H. sapiens* is not evidence of a separation of species. Just because the mtDNA of LM3 indicates descent from an ancestor who lived 150,000 or more years ago, it cannot be assumed that human occupation of Australia took place 150,000 years ago. No conclusions can be drawn as to migration patterns from mtDNA alone.

It is politically correct to assume that all human remains found in Australia are ancestral to current Aboriginal people, different mtDNA notwithstanding, and on the basis of this early dating, as well as of lithic remains found in northern Australia dated to 50,000 ya (Roberts et al. 1990), it is claimed that Australian aborigines have occupied Australia for some 60,000 years. If this different mtDNA is accepted as 'modern' on the grounds that mutations would have occurred over the intervening time, is there any reason for denying that mutations may account for the differences observed between Neanderthal and modern mtDNA? The similarity between the mtDNA of the Kow Swamp fossil, KS8, and the Neanderthal type specimen, Feldhofer, is of particular relevance in this regard.

An obvious exercise, if it were possible, would be the comparison of DNA from Neanderthals and LM3 with that of the Cro-Magnon remains. Stringer and Davies (2001) commented that it had not been possible to extract the necessary DNA from any Cro-Magnon but, even if that were to be achieved "it will be far trickier to tell whether Cro-Magnon, as opposed to Neanderthal DNA, is contaminated with our own DNA", implying that Cro-Magnon DNA would be closer to that of modern humans and therefore less able to be distinguished – which surely is to be proved, not assumed? DNA from the Abrigo do Lagar Vehlo remains, dated to ~25,000 BP, which show a mosaic of Neanderthal and early modern features (Trinkaus et al. 2001) would also be of great interest.

Chapter 33

Multiregional Evolution

The human genus (Homo) appears to have separated from Australopithecus approximately 2 million years ago. Early remains (approximately 1.7-1.8 mya) were found both in Africa and the Archipelago, although lack of later fossils indicates that this early exit from Africa was unproductive, leaving no long-term descendants.

Early archaic *Homo sapiens* remains from Europe, dated to approximately 0.7-0.8 mya, are evidence of a second exodus from Africa approximately 1 mya. Their descendants continued to occupy Europe, becoming more robust as they evolved first into *Homo heidelbergensis* and then *Homo neaderthalensis*. At what point, if any, they became a separate species was a matter of debate.

Either these people, or people from Africa, or both, travelled along the coastline to Asia, finally arriving in China, as evidenced by the fossils found at Peking (Beijing). The Himalayas proved a barrier to more northerly/westerly settlement. The time at which people first arrived in the Far East has yet to be scientifically determined. An approximate date of 300,000 ya is possible. It is to be hoped that recent advances in DNA analysis will solve this mystery.

Between 100,000 and 80,000 ya Neanderthals moved south into the Levant, as evidenced by remains from Shanidar, Iraq, and elsewhere. Here they met with early modern humans moving north out of Africa. Although some researchers have seen similarities in some African fossils with certain Neanderthal characteristics, these were of earlier date (between 100,000-200,000 ya). There is no evidence that fully developed Neanderthals ever entered Africa. Similarly, there is no evidence that early modern humans entered Europe. Only Neanderthal remains are found in Europe up until the sudden appearance of Cro-Magnon in France, 30,000-32,000 ya. Some Levant remains, particularly from Skhull and Qfzah, have been difficult to identify as either Neanderthal or early modern human, showing characteristics of both lineages, lending support to the claim that these two races interbred.

The finding of Cro-Magnon in the late 19th century, evidence of changing stone tool technology and of advanced cave art in France during the first half of the 20th century, gave rise to the theory that a more advanced 'civilization' had entered Europe, eliminating the Neanderthals. However, no equivalent fossil remains, no equivalent stone tool technology and no equivalent art, has been found anywhere else in the world. Quite from where Cro-Magnon came remains a mystery to this day. On the grounds that it would have taken time for these people to migrate across Europe from Turkey to France, a date of 50,000 years has been suggested, but there are no artifacts to support this. That they may have come from Africa, possibly during a time of lowered sea levels, does not seem to have been considered .

During the second half of the 20th century, the theory that the Neanderthals had evolved into a separate species, that they were not fully human, that they had been replaced by modern humans coming from Out of Africa became the 'dominant discourse'. There was an alternative opinion, not supported by Academia, known technically as the 'subversive discourse'. All suppressed opinions endeavor to overthrow the dominant discourse however they can, which behaviour upholders of the dominant discourse consider 'subversive'. Guerilla warfare in academia!

The opinion opposing the 'Catastrophism' of Brace, also known as the 'Out of Africa' hypothesis, was that of evolution having occurred continually throughout the populated world, with no human population having been genetically isolated to such an extent that it was unable to interbreed with other existing populations, thus becoming a new species. This theory of multiregional evolution, promoted by Weidenreich (1947), Coon (1962), Thorne and Wolpoff (1981, 1992b), Wolpoff (1989a), Wolpoff and Thorne (1991) and Wolpoff et al. (1994), was the subversive discourse.

A number of authors (such as Hublin 1985, Radocic 1985, Aiello 1993, Frayer et al. 1993, Trinkaus et al 2001) cited evidence for mosaic features in skeletal remains from Europe at the time of archaic *H. sapiens* and others (such as Arsuaga et al. 1993) argued for the presence of Neanderthal type features on remains such as the Broken Hill fossil from Africa. Wolpoff and Thorne were the main proponents of the Multiregional model, arguing that the number of

Neanderthal features which continued to be found in later European populations, even up to the present time, were evidence of the continuance of the Neanderthal heritage, even if as a minor component. Some of these were only identifiable through careful measurement; others were plainly visible, for example, the large nose found among some Middle Eastern and European people. Yet others were visible under X-ray, such as long, straight spines on the cervical vertebræ, differing from the short downward-sloping cervical processes found in most modern people (Trinkaus and Shipman 1992). One archæologist, on being X-rayed following an accident, noticed that he had such straight cervical processes, although Trinkaus and Shipman refrained from saying who that archæologist was!

In many Neanderthals, the opening of the mandibular neural canal in the lower jaw was partially covered by a broad bony ridge, but in others the ridge was absent. Fifty three per cent of known Neanderthal remains have the bridged form, as did forty four per cent of the earlier Palæolithic occupants of Europe (i.e. the pre-Neanderthals) while in Upper Palæolithic, Mesolithic and recent times, the incidence gradually dropped to below six per cent (Thorne and Wolpoff 1992, Wolpoff et al. 1994). The bridged form of neural canal is rarely found in fossils from Asia or Australia and appears to be missing from Africa at the time of the hypothetical 'Eve' speciation.

Since the bridged form of neural canal has no known evolutionary benefit, its development among the pre-Neanderthals is hard to explain, but not as hard as explaining its development twice in the same geographical area, which must be done if it is to be argued that there was no interbreeding between Neanderthals and the incoming anatomically modern humans.

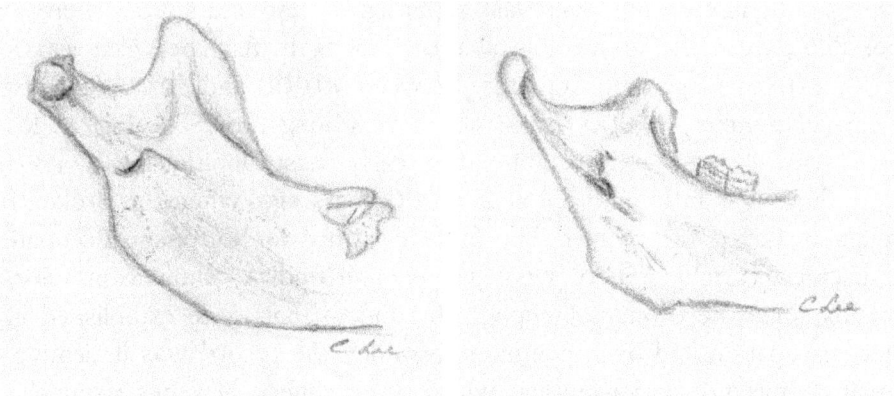

Fig. 33.1: Grooved and ridged mandibular nerve canal openings in
Neanderthals
(Caroline Lee after Thorne and Wolpoff 1992: 30)

377

The steeply sloping or depressed nasal floor present within the nasal cavity of many Neanderthals also occurred, although less frequently, in Middle Pleistocene African and Late Pleistocene non-Neanderthal populations (Skhul and Qafzeh in the Levant), as well as later Upper Palæolithic European populations, being found in approximately 10% of recent human samples (Franciscus 2003). It is not possible to establish whether the small percentage of modern humans who exhibit this particular nasal configuration do so due to a genetic inheritance from the Neanderthals or from other early populations. Reference to 'Skhul' and Qafzeh' is usually a reference to the remains known as 'Skhul 5' and 'Qafzeh 9', which are the most modern of those found in those two locations. Other remains (such as Skhul 4 and Qafzeh 6) are significantly morphologically different and 'span the spectrum' between archaic and Neanderthal forms (Frayer et al. 1993).

The literature abounds with regrets regarding the paucity of the fossil record (for example Shackley, 980, Stringer and Andrews 1988, Thorne and Wolpoff 1992b, Trinkaus and Shipman 1992, Aeillo 1993, Klein 1994, Lahr and Foley 1994, Stringer and McKie 1996, Asfaw et al. 2002, Ashton and Lewis 2002). There was at least one gap of 50,000 years in the occupation record at Chou K'ou Tien, two, if the earliest fossil teeth are accepted as human. There is very little evidence of early human occupation anywhere else in mainland Asia, it being assumed that Chou K'ou Tien was approached *via* the shore from South East Asia. After the first very early fossil find in Europe, (Dmanisi, East Georgia, dated to approximately 1.7 mya), there is nothing more for nearly 1 million years. Much the same is true of the record from S. E. Asia.

Cavalli-Sforza et al. (1994) and Stringer and McKie (1996) found it hard to accept genetic interchange over vast expanses of time and space. There was a possible gap in the archæological record of S. E. Asia between 100,000 and 50,000 years ago, and definitely between 90,000 and 60,000 B.P. Even the smaller number, 30,000 years, was a very long time. The African fossil record was likewise patchy. After the early Australopithecine/*H. erectus* record, there was little until about 200,000 years ago, when the Broken Hill cranium restarted the record. The fossil evidence for European occupation was likewise variable. There were just enough finds to show continuity of hominids somewhere but not enough to confirm that, once established in a region, occupation had been continuous. Rather, the record was indicative of a small, mobile population among which the exchange of genes would have been possible. Against this scenario was the claim of the Multiregionalists that it was lengthy periods of isolation which had resulted in regional differences.

Writers, such as Ahern et al. (2002) and Trinkaus et al. .(2001) saw evidence for the interbreeding of the Neanderthals with modern humans. The Krapina-Vinja fossils from Croatia, and the juvenile skeleton discovered in 1998 at Abrigo do Lagar Velho, Portugal, dated to ~25,000 B.P., displayed a mosaic of features which indicated intermixture between the Neanderthals and early modern humans. The late date was seen as evidence that this child was not a sterile hybrid (which would have required that at least one 'pure' Neanderthal was still in existence) but part of an ongoing mixed breeding population. Contra to this opinion, Harvati (2001) found no strong morphological similarities between the Neanderthals and later Europeans and concluded that there was no evidence that Neanderthals contributed to the evolution of modern humans. His opinion was supported by others, such as Ponce de León and Zollikofer (2001) who held that Neanderthals were a separate evolutionary lineage for at least 500,000 years, becoming a separate species.

Fossils are 'hard evidence'. They can be seen, held, weighed, measured, photographed, compared, and yet the information they yield may mean one thing to one person, and something quite different to another. Some people may change their minds in relation to some matters, but few people experience profound changes in their philosophical beliefs as a result of material evidence. Some people adhere all their lives to the belief system in which they were brought up as a child; others abandon that belief system in favour of another, often in their teenage years or twenties. Whatever belief system each one of us adopts seems to resonate very profoundly with something from deep within ourselves, something which is not easily shaken. People tend to interpret evidence in accordance with their preconceived beliefs.

Erik Trinkaus was particularly well known for his extensive work on Neanderthal remains, especially from the Levant. In 1992, in collaboration with Pat Shipman, he published *The Neanderthals*, a book which comprehensively outlined the position of the Neanderthals as it was understood in the last decade of the 20th century.

It was believed that the classic European Neanderthals had evolved from more archaic human ancestors in Europe and western Asia (the Levant) at the same time that archaic *H. Sapiens* were evolving from *H. erectus* in eastern Asia and Africa. There were similarities between early European Neanderthals and the contemporary people of Africa and eastern Asia, sometimes referred to as African or Asian Neanderthals. Having considered both the fossil and the mtDNA evidence, they concluded that modern humans had evolved from Neanderthals in Europe sometime after 100,000

B.P., when the classic Neanderthals first appeared. This was after the 'Out of Africa' event and left open the amount of interbreeding which may have taken place between the original inhabitants of Europe and the new arrivals. Trinkaus and Shipman then saw the two lines as diverging, the one line evolving into the classic Neanderthal, whose remains dated from ~80,000 B.P. to ~30,000 B.P., the other line evolving into modern humans *(Homo sapiens)*. It was their opinion that the earliest modern humans evolved from Neanderthals, or from late archaic people very like them, soon after the Neanderthals had themselves appeared about 100,000 years ago. Archæologists become used to dealing with large numbers. Whether the 20,000 year 'gap' between the proposed exit from Africa, via the Levant, into Europe and the establishment of the Classic Neanderthals be considered a large amount of time, more than sufficient for this line to become established as a separate species, or whether it be considered but a short period within extended evolutionary time, was up to the individual.

Fossil remains from Vindja, Croatia, dated from ~28,000 to 29,000 B.P. (26,000 to 27,000 B.C.) exhibit features intermediate between those of Neanderthals and modern Europeans (Ahern et al. 2002) and the skeleton of a child approximately 4-5 years old, found in 1998 at Abrigo do Lagar Vehi, Portugal, dated to ~25,00 B.P. (23,000 B.C.), also showed a mosaic of Neanderthal and early modern features (Trinkaus et al. 2001). This is more likely to be indicative of interbreeding between the two populations than the evolution of one population into the other. By this time, Stringer who had been a strong advocate of the Complete Replacement (Out of Africa) hypothesis, was acknowledging that modern humans and Neanderthals had probably been sufficiently closely related to allow hybridization, citing taurodontism among Neanderthals and the Inuit and the presence in about a quarter of Cro-Magnon people of the Neanderthal form of the mandibular foramen.

Cladogenesis

It will be remembered that when Darwin was first envisioning his theory, he conceived of a population of animals, all experiencing the same conditions. If these conditions changed, for example if the climate became colder, the population would change in unison, for example they might acquire thicker fur. This process is known as anagenesis – one population morphing into another. Darwin seems to have envisioned quite small populations. His 1844 *Essay* spoke constantly of 'islands', either physical islands or populations isolated from others by some geographical barrier. It was not until much later, some say until after he read Wallace's work, that it dawned upon Darwin that, if the population had increased such that it covered a wide area, it was possible that those at one extremity could have experienced conditions different from those at the other, that they might vary in different ways, gradually giving rise to two species in place of one. This is known as cladogenesis. Under both these scenarios, the original form ceased to exist. Eldredge and Gould, whose theory of *Punctuated Equilibrium* we will discuss later, pointed out that there was no reason why the original form should not continue, at least for some time, and might even outlast the variant form, if conditions reverted to their former state.

Stringer (1998) adopted the basic cladogenic view of evolution, suggesting that the date of the origin of one clade, in this case the Neanderthal, would date the origin of the other, *H. sapiens*.

Unlike Day (1965), who had seen the Swanscombe skull in some respects as almost indistinguishable from modern humans, Stringer (1998: 31) regarded it as that of a primitive Neanderthal "implying that the modern human clade must be of a similar antiquity". This took the separation of these two lines much further back in time than the 90,000 years which he had suggested in 1988 as the time at which the African population had split into two components: a southern population (Klaises River Mouth/Border Cave) and a northern population (Omo-Kibish/Qafzeh). It was the northern component which Stringer had suggested supplied the genetic material for future *H. sapiens*. Based on this, and other fossil material, such as that from Irhoud, Florisbad, Atapuerca Gran Dolina and Nauer, Stringer suggested that these two clades became isolated, the African isolate giving rise to *H. sapiens*. This scenario, it will be remembered from the previous chapter, was the basis of Stringer's 'Out of Africa' hypothesis. However, if the Neanderthal and modern human lines both started at the same time because they diverged from a common ancestor, then clearly they must have originated, not only at the same time, but in the same place. The suggestion must be that each clade *continued* its evolution separately. The cladistic approach implied speciation, but did that speciation occur, resulting in two separate clades, or did two populations separate and then evolve into two separate species/clades? A diagram may look simple on the page of a text book, but things are far from simple when attempts are made to work out specific details. The Multiregional approach allowed *H. erectus* to continue living in Africa after part of its population migrated to Europe, giving rise to archaic *H. sapiens, H. heidelbergensis* and *H. neanderthalensis*. *H.* erectus was seen to continue elsewhere after the emergence of archaic *H. sapiens* in Europe, as evidenced by the remains from Chou K'ou Tien.

Archaic *H. sapiens* lived in Europe before *H. heidelbergensis* and *H.*

neanderthalensis. Their craniums were more rounded and more gracile than either *H. beidelbergensis* or *H. neanderthalensis,* and more closely resembled, 'modern *H. sapiens'* than did either *H. heidelbergensis* or *H. neanderthalensis.* This can be confusing, especially, since they have been accorded *sapiens* status, unlike the two later (sub) species. It will be remembered that when Neanderthals were first discovered they were accepted as our ancestors, *Homo sapiens.* Later, they were designated a sub species, *Homo sapiens neanderthalensis.* When the idea of a 'speciation event' having taken place some 100,000 years ago took hold, they become a totally separate species within the genus *Homo, Homo neanderthalensis,* loosing their *sapiens* status. This left the unsatisfactory situation of early European remains being archaic *H. sapiens* and late European remains being modern *H. sapiens,* straddling two 'non-*sapiens'* species.

Wolpoff and Caspari (1997) disputed that the Klaises River Mouth specimens were fully modern on the grounds that the apparent chin of one of the fossils had been caused by resorption of bone following the loss of teeth. This could be taken as supporting Stringer's view that only the population of northern Africa gave rise to *H. sapiens* because it denies '*homo'* status to the southern population, which remained *Australopithicus.*

By the 1990s, it had become accepted that the human line had originated in Africa and had migrated from there on more than one occasion. It seemed clear that there had been a very early exodus, since very early remains of *H. erectus* had been found both in Europe and S. E. Asia. However, these remains were isolated and there was no evidence that these people had survived to contribute to today's human population. A later exodus, believed to have taken place about 1 million years ago, seemed to have been more successful and to have provided the basis for all future *Homo* populations. The unanswered question was whether the European line speciated, giving rise to *H. heidelbergensis* and *H. neanderthalensis,* who later became extinct. This view required that Europe, and, indeed, the rest of the world, was eventually populated by a yet later exodus from Africa, possibly as recently as 100,000 years ago. The opposing view held that there had been no speciation since that of *H. erectus,* which meant that *heidelbergensis* and *neanderthalensis* were both subspecies of *sapiens,* or, strictly speaking, that they, and we, were/are all subspecies of *H. erectus.* During a telephone conversation with Alan Thorne, he expressed the very firm view that "We are all *H. erectus".* I agree.

Stringer and McKie (1996) drew attention to work then being carried out on nuclear DNA. Chromosome 12 had been found to contain two variations among its 'junk' DNA. The first was a simple deletion present in some people. The second variation involved repetition of five bases, CTTTT. Some people have between four and fifteen copies of this 'little stammer'. People living in sub-Saharan Africa may or may not have the deletion and may have any variation of the number of CTTTT repeats. Elsewhere in the world, chromosomes with the deletion have a six-fold CTTTT repeat while non-deletion chromosomes have between five and ten repeats (p. 129):

> There is only one feasible explanation: that the small wave of settlers who set off from their African home to conquer the world was made up of a tribe or group of African *Homo sapiens* among whom only those who possessed a chromosome 12 had a sixfold CTTTT repetition. They carried this combination out to the world 100,000 years ago.

Since everyone has chromosome 12, it would appear that a printing error has occurred and the sentence should have read "who possessed a chromosome 12 deletion". It would appear that Stringer and McKie were suggesting that *H. sapiens* first evolved in sub-Saharan Africa and that part of this population migrated out of Africa, gradually populating the rest of the world. They further suggested that there would not have been merely one exodus (pp. 152-154):

> Our exodus would have occurred in trickles ... one such expansion ... spread eastward out of the Horn of Africa about 80,000 ya ... a later dispersal about 50,000 ya infiltrated North Africa, western Asia — and Europe in the form of our old friends, the Cro-Magnon.

Of interest is that Stringer and McKie were still holding to a late dispersal 'Out of Africa'. They also suggested migration eastwards (and, presumably south-east, although they do not mention Australia) before permanent settlement in northern Africa. All early human fossil remains have been recovered from East Africa, or southern parts, suggesting migration along the east coast. The 'bulk' of Africa, to the west and to the north, is surprisingly devoid of fossil remains. It would seem that they were suggesting that all of the people who left Africa, as well as some who remained in the North, who carried the same chromosome 12 pattern, had common ancestry. Even though the migrations may have taken place tens of thousands of years apart, they were all made by the population which had migrated north of the Sahara.

These findings are of great interest because they imply that the people of sub-Saharan Africa (San, Bushmen, Hunza and Hottentots) have remained genetically isolated for nearly 100,000 years. While this is a very long period of time, it is possible, since the Sahara Desert did provide an impenetrable land barrier and there is no indication that sub-Saharans ever took to the water. What is of particular interest is that these people have been proven quite capable of interbreeding with the rest of humanity. In other words, their isolation did not lead to speciation. In fact, apart from the chromosome 12 variation, the rest of their genetic make-up is remarkably similar to that of everyone else. This raises the question as to quite how long isolation has to be before a new species evolves? This work denies Stringer's hypothesis that the people from the north were the ones which later evolved into *H. sapiens*, separately from the people from the South, since the sub-Saharan people are clearly of the same lineage, being able to interbreed with any other human population.

To this day, there are a small number of tribes in Central Africa maintaining a hunter/gatherer lifestyle, making theirs the longest existing continuous culture, stretching back, not merely tens of thousands, but a hundred or more thousand years.

Chapter 34

Six Degrees of Separation

All botanists and zoologists, at least of yesteryear, longed to discover a new species, to have the honour of choosing its name. Even today, new species are still being discovered, especially in the insect world and their discovery still causes excitement. (For some reason which I do not really understand, these species are considered newly discovered, not newly evolved, although I do not see why new species should not still be evolving now, as they have done for the past hundreds of millions of years.) Now seems like an appropriate place to view the confusion which was created by indiscriminate classification of hominine fossils at the time of their discovery. As mentioned earlier, attempts are now being made to bring some order into this confusion. If it all seems too much, skip to the next chapter.

Evolutionary theorists begin with consensus: at some point in time, probably about six million years ago, the human line diverged from that of the ape. How this happened is not a matter of debate. All evolutionary theorists accept that it did. There is also consensus about what happened next: after a further period of time, about four million years, the ape-people (Australopithecines) became extinct and their place was taken by *Homo erectus*.

After *Homo erectus*, consensus breaks down. Adherents of Multi-regionalism hold that there has been no further speciation, merely different varieties (sub-species, races) of developing humans. Adherents of the Complete Replacement hypothesis hold that there have been several further speciations and that one species, our own, not only completely replaced all earlier species of human, but did so within the last 100,000 years or so.

The first writer to tackle this thorny issue was King, who made an extensive study of the few Neanderthal remains which were available in the 1860s and who, in 1864, concluded that the remains so closely resembled that of a chimpanzee that there was "doubt about the propriety of *generically* placing it with Man" (p. 96 - italics in original). In other words, he did not accept that the Neanderthals were human at all. His remained an isolated view. Indeed, for most archæologists there was no doubt, not only that the Neanderthals were human, but that they were our ancestors. It was as further discoveries were made in the 20th century, that debate erupted about quite how many species of Australopithecine there had been, and whether there had been but one, or several, species of *homo: habilis, rudolfensis, erectus, archaic homo sapiens, heidelbergensis, neaderthalensis,* early modern *sapiens,* fully modern *sapiens* and, even, *sapiens sapiens.*

There was debate also as to the degree of mental/intellectual evolution which may have been expressed as humanity passed through each stage. Those who held that Neanderthals were a separate species tended to look for intellectual differences between 'us' and 'them' and these will be discussed in another chapter. In the meantime, below is a brief record of species names which have commonly been applied to fossil finds in various places, dated to a variety of times, the use of which nomenclature implies speciation.

While there is general agreement that the split from chimpanzee to Australopithecine occurred but once, and that that event occurred in Africa, there is no agreement as to how many species of Australopithecine may have existed. Johnson and Edgar (1996) listed nine: *Ardipithecus ramidus* (included as an Australopithecine because it was placed between ape and human), *Australpithecus prægens, anamensis, bahreighazali, afarensis, africanus, crassidens, robustus* and *boisei.* By the 1970s, it was being suggested that *Homo habilis* should be added to the list of Australopithecines (Birdsell 1975) and this suggestion has received ongoing support. The discovery of *Australopithecus sediba* (Berger 2010) in the Malapa Reserve, South Africa, further increased the number of purported Australopithecine species. Some writers, such as Jurmain et al. (1997) suggested that there were at least two different genera, robust and gracile, as well as several different species living in Africa between two and four million years ago. However, only one species of one genera could have given rise to *Homo erectus* and/or *Homo ergaster* (depending upon whether or not you accept *ergaster* as a separate species), so the question of how many there were is 'academic' (i.e. irrelevant) in the context of human evolution. Furthermore, since all humans alive today belong to the same species, only *erectus* or *ergaster* could have given rise to the species which was ancestral to humans today, not both, assuming, that is, that there has been further speciation and we are not "all *homo erectus*".

As has already been pointed out, the first remains of humans earlier than the Neanderthals were found outside Europe, these being the *Homo erectus* from Asia, although some became known as *pithecanthropus* and *sinanthropus*. Some of these fossils have now been dated to approximately the same age as the earliest human fossil remains from Africa. Since the finds from Chou K'ou Tien are now known to be much later (at least 1,000,000 years) than those from South East Asia and Africa, there arises the question of whether the Chou K'ou Tien fossils were *H. erectus* descendants of the people who once inhabited Java, or whether they were the descendants of people who left Africa somewhat later and were *H. ergaster*, if this species division is considered valid. If *erectus* and *ergaster* were separate species, then human evolution, post *Australopithecus*, had been subject to double speciation. Such a double speciation would be in conformity with the concept of cladogenesis – one species ceasing to exist as two divergent species take its place. You will remember that this differed from anagenesis which saw one species changing into another, there being no increase in the number of species. A third alternative was also offered: that both the original form and one (or more) divergent forms survived. However, since, only one species of human remains alive today, this third possibility would seem to be irrelevant. The literature speaks of *H. erectus* as being ancestral to *Homo sapiens*. (It will be assumed here that *ergaster*, if it existed, became extinct without giving rise to any other species.) While the possibility of divergent forms of *H. erectus* having existed, but of one line only having survived, excited little discussion, there was ongoing debate as to whether the Neanderthals were to be included within the one line of descent, or whether there had been a branching (cladogenesis) of the original species, *H. erectus*, ceasing to exist, two other distinct species, *H. neanderthalensis* and *H. sapiens* taking their place.

The remains of a human-looking jaw from Dmanisi, East Gerogia, were dated to approximately 1.7 my (Gabuni and Vekus 1995), although this date is controversial, some believing that it may be as 'recent' as 900,000 B.P. (Fagan 1998). Whichever date is accepted, it indicates that *erectus* dwelt briefly in Eastern Europe at a time well before settlement by *H. heidelbergensis*. If the earlier date is upheld, lack of further evidence would seem to indicate that this was not a successful settlement and it is unlikely that these people were either a new species or contributed genetic material to later hominoids. Other finds have indeed extended the possibility of permanent human presence in Europe as far back as ~900,000 years and these people are generally referred to as 'archaic' *H. sapiens*. They received their name because the shape of their skull was closer to that of modern humans than was that of the Neanderthal, with its protruding occipital area. The assumption is that Neanderthals varied but that they were replaced in Europe by others who had retained the more primitive form. (The word 'primitive' means 'first'. There is no implication of any lack of sophistication. Indeed, the word 'prime' may have exactly the reverse connotation!) What is important here is that, by giving these early inhabitants of Europe separate nomenclature, there was an implication that they had undergone speciation and were in some way genetically different from *Homo erectus*.

In 1907, a mandible was found near Mauer, Germany, which was attributed to another new species, *H. heidelbergensis*. It was dated to ~400,000-500,000 B.P. For a long time it remained a solitary find, but as more fossils were found, more were attributed to this species, including finds from England: Boxgrove (~400,000 B.P.), Swanscombe (~300,000 B.P.), Ostend (Norfolk), Westbury-sur-Mendip (Somerset) and High Lodge (Norfolk), all believed to be between ~500,000 and 200,000 years old. Ward and Stringer (1997) included African remains from this era under *H. heidelbergensis*, although whether they were implying that *H. heidelbergensis* had migrated out of Europe (or the reverse) or that there had been a parallel evolution/speciation in Africa is not clear.

By contrast, Mania and Mania (1988) referred to skull fragments found in Germany, dated to 250,000-300,000 ya, as 'late *H. erectus*'. Their attitude to these early humans was quite different from that of some authors towards the Neanderthals. They interpreted lines on bone artifacts as having been deliberately engraved. Those who saw evolution from *H. erectus* to *H. sapiens* as having been continuous were more inclined to attribute a gradual, but steady accretion of capabilities on both physical and mental levels than those who espoused the view that a speciation event had occurred, resulting in a distinct divide between modern *H. sapiens* and its forebears.

Fossil remains in Europe prior to those of the Neanderthals were sparse, considering the length of time involved, although those attributed to *H. heidelbbergensis* have increased in number. In addition to the Mauer fossils, further fossils have been found at Arago, France, (300,000-250,000 ya), Eringsdorf (140,000-120,000 ya) and Fontechavade (~120,000 ya). These may be referred to as 'Pre-Neanderthal' or simply as 'Neanderthal'. The Neanderthal represents the third degree of separation from *H. erectus*.

The Neanderthals from the Levant and from Eastern Europe do not present the extreme features associated with 'classic' Neanderthals (heavy brow ridges, bun shaped occipital region, etc.). In some cases, it has been difficult for the experts to agree into which category a particular fossil or group of fossils should be placed. This confusion is not surprising since 'archaic' *H. sapiens* was so called because of their generally less robust build, smaller brow ridges and generally rounded shape of the cranium in comparison with later Neanderthals. This led some writers (for example Howell 1952) to postulate that the classic Neanderthals, whose remains are found only in Western Europe, may have become isolated after 60,000 B.P., due to deteriorating climatic conditions, and have evolved into a separate species. This would place the classic Neanderthals four degrees of separation from *erectus*. Valladas et al. (1988) dated remains from Kebara, Israel, to ~60,000 B.P. Despite the association of the fossils with a Mousterian culture, they included these fossils in the category 'Proto-Cro-Magnon', as they did remains from Qafzeh, which they dated to ~90,000-100,000 B.P. It was their contention that these were the remains of a people whose descendants later populated Europe, and their separate occupation of an area inhabited at the same time by the Neanderthals excluded a close phylogenic relationship between the two groups. It may be remembered that in recent historical times Arabs and Negroes lived in close proximity in Egypt and Ethiopia for thousands of years with little interbreeding. Lack of interbreeding does not necessarily imply separate species.

Meanwhile, back in Africa, *H. erectus* appears to have continued to exist, without speciation, for nearly two million years. The word 'appears' may be misused since the most striking feature of the African fossil record during this time is its paucity, particularly when the number of Australopithecine remains is considered. These show that the process of decomposition alone cannot be responsible for the lack of fossil survival, since Australopithecine fossils have survived for millions of years more. The two most complete skeletons are those of 'Nariokotome Boy' and the female KNM-ER 1808 from Lake Turkana (formerly known as Lake Rudolf, hence *Australopithecus rudolfensis*). Both of these were dated to ~1.6-1.7 mya, which is early in the lifetime of the species. Remains thereafter are sparse and it would be difficult to prove, by fossil evidence alone, that *H. erectus* lived in Africa continuously for nearly two million years. The cranium from Broken Hill has been provisionally dated to ~200,000 years, although it may be younger. Some have seen a similarity between it and some Neanderthal remains (for example von Keoningswald 1962, Arsuaga et al. 1993) but there is no general consensus of opinion for any speciation in Africa until that which is held by some to have occurred between 200,000 and 100,000 years ago, giving rise to early modern *H. sapiens*. Under the scenario outlined above, when these people left Africa, they would have been separated from the Neanderthals of the Levant and Eastern Europe by four degrees of speciation, and from the classic Neanderthals of Western Europe by five.

But the picture is not yet complete. It has been suggested that *H. sapiens sensu stricto* did not arrive in Europe until about 40,000 B.P. (Aiello 1993, Klein 1998). This argument is made, not only on the grounds of perceived differences between Cro-Magnon skeletal remains and those of earlier people in Europe, but also on cultural grounds, in particular the appearance of magnificent cave art and the first fired clay during the Upper Pleistocene, and the introduction of agriculture, which paved the way for the expansion of *H. Sapiens* in Holocene times. (The term 'Holocene' was introduced to describe the last 10,000 years because it was thought that by then evolution was 'whole', complete, that no new species had evolved since then. I fail to understand this logic. Perhaps it was influenced at that time by a residual feeling that humans had been 'created', or at least had evolved, not more than 10,000 years ago and that their appearance made creation 'whole' or 'complete'? This may not have been a conscious thought but it must have taken early theorists some time completely to divorce themselves from their earlier education.) Those who held these views agreed with Stringer's 'Out of Africa' model, which, as has been discussed, saw more than one migration as having taken place. What was not clear from this model was whether it was being proposed that each migration was made by a new 'people' – by a new species – or whether there were several migrations made over periods of time by descendants of the same African population.

If the implication is that *H. sapiens sensu stricto* were, indeed, a further, more highly evolved species, that would separate them from the Neanderthals by six degrees. This needs to be borne in mind when considering suggestions that the Neanderthals copied, stole or traded their Châtelperronion technology. How much interaction is likely to have taken place between two species separated to such a degree? Even if the degree of separation was not as great as six, it would still need to be at least four or five, according to current theories: *H. erectus,* archaic H. *sapiens, H. heidelbergensis*, proto-Neanderthal, classic Neanderthal and early *H. sapiens* leaving Africa and arriving in Europe sometime after 100,000 years ago, in one, two, three or more, waves. Add the suggestion that a later *H. sapiens – sensu strcto* – evolved within the last 50,000 years and this number increases just that little bit more.

There was no preliminary fine art, firing of clay or agriculture in Africa (or anywhere else). The supposition that the Cro-Magnon people had brought these skills with them when they came 'Out of Africa', rather than that they were native to Europe, appears to be supported by nothing other than an "Everything came out of Africa" paradigm, which has no evidence to support it.

Chapter 35

Punctuated Equilibrium

The theory of 'punctuated equilibrium' was a hybrid organism from its inception. It was the brain-child of two completely different personalities, the punctilious and methodical Niles Eldredge (1943-) and the outgoing and expansive Stephen J. Gould (1941-2002). It was first presented as a paper read at a symposium intended to assist graduate geologists wishing to extend their knowledge of invertebrate paleontology. (There has always been a close connection between geology and evolutionary theory because, from the very beginning, geologists used changes in fossils, mostly shells, to determine boundaries between geological periods of time.) The symposium papers were later published in book form (Schopf 1972). Gould (2002) explained that, when asked to present a paper on speciation, he realized that work being undertaken by his graduate associate, Niles Eldredge, would provide a good basis and asked that they present a joint paper. Gould wrote most of the paper, and coined the term 'punctuated equilibrium' but both were based on Eldredge's work.

Eldredge's area of speciality was trilobites, ancient sea creatures, somewhat resembling a beetle, which have been extinct for thousands of years. His theory, however, applied to all evolution, human, animal and plant, and is particularly relevant when placed alongside the 'theories' we have just considered which, offered haphazardly and without much apparent thought, implied rapid and frequent speciation within the hominid line. If evolutionary theorists were confused before Eldredge and Gould published their work, they were even more confused after! The infant theory was part gradualism, part saltation and, if there was some misunderstanding as to the position punctuated equilibrium took, that was due, in part, to the fact that Eldredge and Gould themselves appeared not to be completely in agreement. Eldredge (1979, 1985a, 1985b, 1991a, 1991b, 1995, 1999) insisted that periods of stasis were punctuated by episodes of change, which appeared swift or sudden in the geological record but which had actually taken place over thousands, if not millions, of years. Gould (1977), on the other hand, did, at times, appear to imply that saltation had occurred. The twin theories of evolution and punctuated equilibrium were the main focus of Eldredge's many books and papers, whereas Gould seemed to have very little interest in the theory once it had been launched. By the time punctuated equilibrium had 'came of age' twenty-one years later, and Eldredge and Gould published a second joint paper (1993), Eldredge's view that punctuated equilibrium was never meant as a saltational theory had prevailed.

Gould, whose writings often threw out broad theoretical suggestions with little, or no, scientific support, offered a broad generalization which he thought could be taken as a guide to the frequency with which species might evolve (2002: 768): If gestation time, for example in a human, was 1%-2% of an ordinary lifetime, "we should permit the same general range for punctuational speciation". Under this 'formula', a species which survived for 4,000,000 years may have taken 40,000-80,000 years to accomplish speciation. Such a period would definitely qualify as gradual evolution, although it might appear saltational in the geological record, which routinely dealt with periods of millions of years. From this suggestion, it can be seen that Gould was the 'ideas man', Eldredge was the man who honed the broad idea into practical shape – although, it should be noted, that this was one idea of Gould's that he made no attempt to utilize! More on some of Gould's ideas later.

The full title of Eldredge and Gould's paper was *"Punctuated Equilibrium: an alternative to phyletic gradualism"* and it was the second half of the title which caused the confusion. Darwin had insisted that all evolution took place by phyletic gradualism and any theory offered as an 'alternative' must, surely, be being offered as an alternative to Darwin's thinking? Eldredge and Gould argued that, while evolution was gradual, it was not continuous. They never argued that evolution was sudden (i.e. by saltation), but they did argue that many breaks in the fossil record were real and not the effect of any inadequacy of the fossil record. They supported the notion that evolution may have taken place in other areas, as suggested by Darwin, but held that evolution was static for long periods, followed by episodic change.

When presenting their paper as part of a book, Eldredge and Gould were not subject to the stringent word limit that they would have been had their article appeared in a journal. They took advantage of this by devoting a large part of their chapter to a consideration of the debate regarding allopatric and sympatric speciation. It was their purpose to show that rapid and episodic allopatric speciation (speciation occurring in peripherally isolated populations) would result in gaps in the fossil record, such gaps being an accurate record of historical events, not 'imperfections' as suggested by Darwin. They used their own areas of special expertise (trilobites in the case of Eldredge and snails in the case of Gould) to illustrate their ideas.

Gould had studied *Poecilozonotes bermedensis*, a land snail found only on Bermuda during the last 300,000 years of the Bermudian Pleistocene, while Eldredge had made a study of the trilobites, *Phacops rana*, found in North America during the Middle Devonian. In both cases, Eldredge and Gould argued that the fossil evidence showed that the species in question had each lived for many millions of years with very little change, that change, when it had occurred, was allopatric and occurred over a comparatively brief period of time, followed by another extended period of stasis, during which very little change occurred.

Trilobites had numerous eyes (complete eyes, not facets) and these occurred in lines slanted in a dorsal-ventral (d-v) direction. The earliest trilobites appeared to have 18 d-v lines of eyes, but the number of eyes in each line could vary, especially near the mid-line, where they were frequently reduced. Fossils of trilobites with 18 d-v had been found in the New York basin and also in the mid-west, near Ohio and the Great Lakes. It appeared that the mid-west location had dried out, due to the lowering of sea levels, and the 18 d-v trilobites living there had become extinct. However, trilobites continued to live in the area of the New York basin, but the number of their eyes diminished, possibly as the shape of the head gradually changed. The number of eyes in the most central d-v line gradually lessened until it contained only one or two eyes, or at times vanished altogether. Gradually the number of d-v lines of eyes was reduced from 18 to 17. When the Midwestern sea was re-formed millions of years later, it was repopulated by trilobites but this time they were the 17 d-v species from the New York basin. The fossil record showed an abrupt change but evolution had been gradual, occurring in a different location (Eldredge 1991b). Darwin would have been delighted with such a fine example of his thinking. Eldredge was deeply disturbed that his work was interpreted by some as showing evidence of saltation and devoted the remainder of his career to writing books and articles to rectify this error of thinking.

Because Eldredge considered allopatric speciation fully explained evolution, he became more interested in what caused extinctions, particularly mass extinctions, than in how evolutionary change had occurred.

Over the thousands of millions of years that multicellular life had existed on this Earth, fewer than a dozen biotic crises qualified as major mass extinctions (Stanley 1987). Which is more extraordinary: that these events occurred at all, or that they occurred so infrequently? It was after extinction events had occurred that diversification, even the evolution of completely different life forms, occurred, but why should forms of life which had survived, often with little change, for millions, even tens of millions of years, become extinct, let alone large numbers of them within a geologically short period of time?

The trilobites had survived two major extinction events over the 300 million years of their existence. While these events had decimated their numbers, they were able to recover, but finally fell to the third mass extinction event, which occurred 245 million years ago, at the Permian-Triassic boundary. The mass extinction claimed approximately 90% of all marine species, there being as yet no terrestrial species. After years of study, Eldredge and his colleagues concluded that all major extinctions had been brought about by cooling, either global or polar (Eldredge 1991b, Stanley 1984, 1987, Vrba 1995). It might have been expected that marine fauna would have been less affected by cooling than land fauna, but the evidence showed that shallow water marine creatures were as sensitive to climate change (cooling) as their terrestrial counterparts. In both cases, more species were lost in tropical regions than at higher latitudes. There appeared to be two reasons for this. Firstly, tropical fauna is not normally subject to large degrees of temperature change and are, therefore, far more sensitive to change than are fauna which experience large variation of temperature on a daily, let alone an annual, basis. Secondly, tropical regions tend to be populated by a large number of species, although each species is represented by less individuals. The biomass is not necessarily greater, but since extinctions are graded according to the number of species which are lost, tropical areas always rate more highly.

Eldredge believed that the first response of any form of life, be it plant or animal, to climate stress would be 'habitat tracking'. This was easier for highly mobile animals than for slow moving animals or plants, but would always be attempted. As temperatures dropped, fauna inhabiting tropical regions became trapped. There was nowhere warmer for them to go. Deep sea marine life would have been able to track, but shallow water, bottom dwelling marine fauna might be similarly trapped. Furthermore, climate change might cause sea levels to drop, reducing the amount of shoreline in areas with continental shelves, although it would increase the shoreline around volcanic islands, due to their conical shape.

Following major, or mass, extinctions there were periods of diversification, which took place as temperatures rose. Eldredge was most insistent that evolutionary diversification of surviving species was a slow occurrence, taking at least 5,000-10,000 years, possibly as long as five million years. He pointed out that environment (geology and climate) was but a small part of any niche, that most niche were provided by other living organisms. An environment denuded of so many life forms was not a blank slate waiting to be filled. It was a desert, waiting to be converted into a condition suitable to sustain new life forms.

While Eldredge was primarily interested in early extinctions, reference was also made in the literature to later events and the climate changes which accompanied them. Some of the more recent dates appeared to be significant in the history of hominid evolution. For example, a major cooling event had occurred 6.5-5 mya (Kennet 1995), the time at which it is believed that the human line diverged from that of the chimpanzee. This was followed by a period of significant warming circa 5-3 mya, during which time Vrba (1995) noted that pigs, monkeys and girrafids, dependent upon wooded environments, first appeared.

Of interest is the late arrival of monkeys in the fossil record, which appeared about the same time as the chimpanzees and the Australopithecines. Species favoured by a more open environment, such as bovids and rodents, appeared during the cooler period which followed, 2.7-2.4 mya. This time saw the disappearance of the Australopithecines in favour of *Homo habilis*, if *H. habilis* is to be considered distinct from the Australopithecines, although it should be noted that *habilis* had become extinct by ~2 mya. This was also a time of polarity reversal, although polarity change has not been shown to be a significant factor in evolution. Another period of cooling occurred about a million years later, *circa* 1.6 mya, at the time that *H. erectus* became established. Vrba concluded to her own satisfaction that climate change had been a significant factor in human evolution but was of the opinion that there were insufficient human remains for any definitive conclusion to be drawn.

Vrba's main location of interest was Africa. There was one climatic event which she did not mention, the very warm interglacial which occurred in Europe about 120,000 ya. This interglacial is relevant for two reasons. Firstly, it is the time by which *H. erectus* is believed to have been transformed into early *H. sapiens* in Africa. It is also the time at which the

Neanderthals started to evolve into their well-documented form. Remains of 'classic' Neanderthals do not appear in the fossil record until about 80,000 ya, but the pre-Neanderthal form is recognizable from about 120,000 B.P. Earlier European remains were traditionally classified as *H. Heidelbergensis*, but those post 200,000 B.P. are now classified as Neanderthal.

The inauspicious start of mammals during the Age of the Dinosaurs, when they were but little creatures scurrying in the undergrowth, is referred to in just about every book on evolution, as is the 'mammalian explosion' which took place following the dinosaurs' demise. Less frequently mentioned is another 'explosion' in mammalian life, which occurred in Africa during the Lower Pleistocene. Reference has already been made to the monkeys and the rodents. The leopard and the cheetah both appeared between 2-1.6 mya and the lion a little later, 1.5 mya (Walker 1984). Obviously, these creatures had antecedants but it was at these times that they first appeared in the fossil record in the form which we recognize today. Clearly, the hominid line was not the only one undergoing major change.

If Eldredge is correct in thinking that extinctions occurred during times of cooling and new forms appeared as the climate became warmer, then the disappearance of fauna (and flora) during the Ice Ages which occurred in Europe between 60,000-40,000 B.P. and 25,000-15,000 B.P. is to be expected. The last Ice Age, which started about 30,000 ya, did not 'officially' end until 10,000 ya, the time between 25,000-15,000 B.P. being the coldest. It is to be noted that this last Ice Age was the severest recorded and that we are still in the post-glacial 'warming' phase, temperatures today not yet having recovered to their former levels. If the pattern of extinctions followed by evolutions is being followed, then it is to be assumed that new forms will have appeared on this Earth, particularly in the Northern Hemisphere, over the last 10,000-15,000 years, that absence from the fossil record may not be due to any insufficiency in the record but to the appearance of new forms. I have commented before that I have been puzzled by an apparent lack of willingness to acknowledge that evolution is ongoing, other than in the laboratory. I am sure it is not necessary for me to point out that humans post the last Ice Age (i.e. post 10,000 - 15,000 B.P.) appear to be different from human populations pre Ice Age (i.e. pre 25,000 - 30,000 B.P.)

While Eldredge always maintained that the theory of punctuated equilibrium was one of gradualism, and therefore Darwinian, he did disagree with Darwin on another point. Darwin saw evolution as the evolution of individuals. Individuals living under similar conditions would tend to acquire similar changes in synchrony, but it was not the species *per se* which evolved. Eldredge argued that it was the species which evolved. This was the cornerstone of his argument throughout his writings. Species were born, lived and perished, as did all living things. Eldredge (1982: 42) used Mayr's example of mitosis: during cell division there is a time during which it is difficult to decide whether one is dealing with one cell or two, but once this brief phase is passed, then there are two cells where before there was only one. There was a time during speciation when it would be difficult to decide whether two populations were one species or two, as with the well-known 'ring' species already mentioned, but once speciation was complete, the two species would forever remain separate identities. As all were born, all lived, and all would eventually die.

In many ways, Eldredge's view is closer to that of Lamarck than it is to that of Darwin. Lamarck saw species existing, unchanged, for vast periods of time, major modifications occurring in response to major changes in the environment. Lamarck seemed to have a far greater grasp of the incredible changes that had occurred on the surface of the Earth over geological time than did most of his contemporaries, for example than did Cuvier. Lamarck saw the appearance, not only of new species, but of new genera, even new families, and at one point, of new Orders, as being initiated by changes in the environment, although it must be remembered that by 'environment' Lamarck was referring to other biota as well as the geological and climatic environment. Both Lamarck and Eldredge noted the appearance of new forms following changes in the environment, in the case of Eldredge particularly the climatic environment, but neither made any attempt to explain the appearance of a totally new feature or new line of evolution. No one has offered an explanation for these things. All theorists have merely noted that they happened. Darwin's theory of natural selection was a theory of the selection of features which had come into existence. It offered no explanation for how new features had evolved, other than chance.

Eldredge (1995) told of how he had started his research work looking for evidence of evolutionary change in trilobite fossils and, as his work progressed, was disturbed by the lack of evidence for change, worried that he would have nothing to report in his thesis. It suddenly dawned on him that it was the lack of change which was significant. A colleague had surveyed the lineage of over 70 species, but reported only on the three which had shown evolutionary change, an example, Eldredge believed, of the way in which pre-conceived ideas influenced, not only the research itself, but the interpretation of results. Eldredge's original work had been undertaken at the time when Kuhn's theory of paradigms, as outlined in his classic work, *The Structure of Scientific Revolutions* (1962) was gaining increased acceptance. Eldredge (Eldredge and Tattersall 1982) cited Kuhn's theory of stasis followed by sudden change as consistent with punctuated equilibrium and pointed out (1992b) that he and Gould had even quoted Kuhn in their original paper.

While Kuhn had seen paradigm shifts as the result of sudden insight by a scientist, such as Newton with the falling apple, Popper (1972) had seen scientific theory as forming - or changing - as the result of numerous pieces of accumulated evidence (facts) coming together, pointing in a certain direction. Once a theory had been formulated, more facts could confirm a theory, but not prove it, but a single opposing fact, should it come to light, would be enough to disprove (falsify) it. (It only took the sighting of one black swan to disprove the theory that "All swans are white".) Lakatos (1970) realized that overturning an established paradigm was not that simple. Rather anomalies would be noted and, if necessary, a rider would be added to the hypothesis. (All crows are black, except albinos.) Rather than abandoning an established theory in response to the first falsification, scientists would attempt to explain away the result as an aberration, possibly faulty procedure. Then they would attempt to incorporate the new result within the established theory, by means of a rider. Several riders might be added, forming 'protective belts'. This Lakatos termed 'saving the theory'. Further falsification would eventually render the theory so unstable that it became subject to a paradigm shift, as outlined by Kuhn.

In rather the same way, Eldredge saw punctuated equilibrium as an example of sudden shift, the organism having tried to maintain its position under changing circumstances, eventually failing and either becoming extinct or indulging in a major shift, the result being a new species. He did not see the theory of punctuated equilibrium itself as being a shift in paradigm, rather a rider to Darwin's original theory of natural selection relating solely to Darwin's claim of consistent gradualism.

Despite many efforts, no one had been able to come up with a definition of 'species' which was completely satisfactory under all circumstances, although many had tried. Eldredge (1985b) reviewed the attempts of major contributors, such as Dobzhansky, Mayr and Simpson, and concluded that the major difficulty was the impossibility of reproducing under laboratory conditions the circumstances which occurred in nature in relation to genetic change.

Eldredge noted that the Neo-Darwinists had rejected Goldschmidt's concept of macro-evolution *via* the 'hopeful monster' on the grounds that, even if such a viable monster were to be produced, with whom would it mate? He argued (Eldredge and Cracraft 1980: 119):

> ... the creation of a 'hopeful monster' directly implies the creation of a new species ... Indeed, the only cogent arguments against particular theories of saltation appear to be those which see grave difficulties in creating a new, sexually reproducing species through the appearance of a single individual hopeful monster.

Despite acknowledging the problem of the lack of a potential mate in this instance, neither Eldredge, nor any Neo-Darwinist, raised such objection in relation to speciation involving a change in chromosome number. Eldredge (1985: 121) went so far as to say that "minimally, only some modification of the reproductive system is required for speciation to occur". Can any modification of the reproductive system, let alone one sufficient to cause speciation, justly be described as 'minimal'?

Eldredge considered evolution at levels higher than species (or perhaps genera) to constitute macroevolution. Bock (1979: 21) agreed that "Macroevolutionary changes are usually not open to experimentation or direct observation" but nevertheless considered that "fancy breeds of goldfish and the diverse breeds of dogs surely must be regarded as major modifications". It was Eldredge's opinion that (1979: 7):

> If we assume ... that species are real entities in nature, their origins must be explained, not because species are aggregates of populations, which are themselves aggregates of individuals, the mechanics of speciation do not seem immediately reducible to the principles of genetics ... A theory is simply incomplete if it lacks either a set of statements concerning (1) changes in genes and their expression or (2) the origin of new species.

Phyletic gradualism was accounted for by microevolution resulting from gene mutation, but failed to account for the birth of species. The alternative view was that evolution was "quintessentially the origin of new taxa" which Eldredge claimed to be a superior position, since the taxic approach had the capacity to subsume the phyletic (transformational) approach.

Bock and Eldredge both agreed that the synthetic theory had not successfully reduced macroevolution to microevolution merely by concentrating attention on genetic mutation, that higher taxa must be taken into consideration. Bock took as his illustration 'night' and 'day', two distinct entities which nevertheless graded gradually into each other with no sharp demarcation, claiming that micro- and macro- evolution similarly graded smoothly into each other. Bock (p.67) concluded that "Macroevolution is simply the consequence of additive microevolutionary change", which seems to be very much the position of the Neo-Darwinists.

In 1979, Eldredge had taken a similar 'blurring' approach. First he suggested that macro-evolution referred to higher taxa, then pointed out that higher taxa were merely an invention of taxonomists. They had no existence in reality. Taxonomists may consider the lion, the leopard and the cheetah all belong to the 'Cat' family, notwithstanding that cheetahs do not have retractable claws, but the creatures themselves have no such familial recognition. They regard each other in the same way they regard other animals, such as the elephant or the hyena - prey when first born, then predator of their own young. It was argued that higher taxa did not exist, except as a convenient concept. Since macro-evolution related to higher taxa, which did not exist, macro-evolution did not really exist either. Only species (or possibly genera) truly existed. These lower taxa may be explained by micro-evolution. Therefore micro-evolution was the only explanation necessary. This is a similar argument to that used by Darwin in relation to the existence/non-existence of species/varieties when he argued for the evolution of individuals, the only true entities.

Eldredge held that the role of palæontologists was equal to, rather than subservient to, that of the geneticist in uncovering the history of evolution. Far from trying to promote punctuated equilibrium as a new theory, Eldredge was constantly at pains to point out that all evolutionary theorists, from Darwin onwards, had been well aware that some fossil lines showed little change over vast amounts of geological time. That the Neo-Darwinists had given so much attention to change had simply been because change needed explaining in a way that stasis did not. He and Gould had merely been redressing what they saw as an inbalance, created by the great strides being made in the understanding of genetics which were tending to overshadow the evidence of the fossils. That random genetic change was constantly occurring in the laboratory was being translated into constant change in nature. Mayr (2001: 270) agreed that punctuated equilibrium was "in no respect whatsoever in conflict with the conclusions of the evolutionary synthesis".

Gould's Contribution

Little has been said of Gould's contribution to the debate, because Gould contributed little. His only individual contribution, *The Return of the Hopeful Monster* (1977), served to confuse the issue rather than to clarify it. Gould found his niche publishing, over a period of years, monthly articles for the *Natural History Magazine*, which were collated and published annually in book form, under such catchy titles as *Bully for Brontosaurus*, in which he explored all manner of weird, wonderful and controversial topics, most, but by no means all, of which had some connection with evolution. Like Eldredge (1985a), Gould (1991) was critical of 'facile Just So' stories invented to explain certain adaptations yet Gould, in that very same book (pp. 297-304), was guilty of suggesting one of the strangest 'Just So' stories of them all.

Amphibians, such as frogs and toads, have a more diverse range of reproductive adaptations than most other taxa. They may lay eggs in water, and leave them unattended, guard them, wrap the eggs around their hind legs, take them into their mouths in special brood pouches until hatched as tadpoles, even brood them until they pass through the tadpole stage into that of the tiny frog. In November, 1973, two Australian scientists discovered a species of frog, *Rheobatrachus silos*, which brooded its young in the stomach of the female, giving birth to juvenile frogs through her mouth some eighteen days later. A second species *(R. vitelinus)* was later found which did the same thing.

Gould accounted for the female not eating during the brooding period by assuming that the presence of eggs/tadpoles in her stomach created a full sensation, preventing her from feeling hungry. More difficult to account for was the fact that the female's normal digestive enzymes were suppressed during this time, due to the presence of prostaglandins. The presence of prostaglandin could not be the result of 'foresight' in anticipation of this particular adaptation, because the process of natural selection depends on random, fortuitous change and does not allow 'foresight'. Gould explained (p. 304):

> One cannot seriously believe that ancestral eggs actively evolved prostaglandins because they knew that millions of years in the future a mother would swallow them and they would need some inhibitor of gastric secretion ... Prostaglandin provided a lucky break ... a historical precondition fortuitously available at the right moment ... a female *Rheobatrachus* must have swallowed its fertilized eggs (presumably taking them for food, not with the foresight of evolutionary innovation) — and the fortuitous presence of prostaglandin suppressed digestion and permitted the eggs to develop in their mother's stomach.

Even if the unlikely scenario outlined by Gould had in fact taken place, was Gould seriously suggesting that forever after all female *Rheobachtrachus* would reproduce by this new method? And if not, what was the point of suggesting the scenario at all? Gould's articles were intended for reading by the general public, were always lighthearted and generally interesting, but, even so, this piece of 'free thinking' was surely outrageous?

While Eldredge was generally circumspect in his criticisms of opinions which differed from his own, Gould, writing for a more general audience, allowed himself a freer reign. He referred to Creationists as 'Yahoos' on at least three occasions (1991: 156, 417, 427). On two of these occasions, he was referring to William Jennings Bryan, the lawyer who led the prosecution of John Scopes in 1925, which has already been discussed. You will recall that Bryan was an eminent man, not only extremely well-educated but a person who had run for the Presidency of the United States and who had held the position of Secretary of State. The word 'Yahoo' was introduced by Jonathan Swift in his famous book, *Gulliver's Travels,* published in 1735. The Yahoos lived in the land of the Houyhnhnms and were a degenerate type of creature bearing some resemblance to humans, parts of their bodies being covered with frizzled or lank hair. 'Gulliver' described the Yahoos as filthy, odious and abominable creatures, attributing their flat, broad noses to their practice of allowing their infants to lie groveling on the earth. They were the most unteachable of all brutes. The practice of denigrating the *opinions* and *ideas* of others by denigrating the person is not one to be applauded but, unfortunately, Gould was not the only writer to adopt this tactic.

Before he died, Gould did somewhat redeem himself by writing a large, serious compendium on evolution theory, *The Structure of Evolutionary Theory*, which ran to more than 1000 pages, and which showed that he was, in fact, extremely well grounded in evolutionary theory and capable of scholarly work. Let that be his legacy!

Chapter 36

Chromosomes

Before proceeding further with the consideration of fossil and artifactal evidence relating to the evolution of early humans, such as *H. erectus* and the Neanderthals, it is time to take a second look at that other line of evidence – genetics.

When I commenced my research for my thesis, my knowledge of genetics was limited to what I had learned at school – and that was a long time ago! I had truly appreciated the opportunity of extending my knowledge offered by the reading of Michael White's little handbook, *The Chromosomes*, written for the benefit of his students in 1937, which was discussed in Chapter 28. When reading material relating to evolutionary knowledge in the second half of the 20th century, I took the opportunity of reading both the 5th and 6th editions of White's book, which had by then grown considerably in size. Unlike other writers, White was interested in the chromosomes themselves, not the genetic material they carry. Whereas other writers, such as Dobzhansky, had considered evolution in the context of genetic change, inversions, translocations, duplications, deletions, which were presumed to have happened accidentally and fortuitously to have resulted in positive change, White considered the actual chromosomes themselves and how any aberration affected the zygote's possibility of survival. White's main area of interest was insects, hardly surprising considering that some have really complicated reproductive mechanisms, making the simple X/Y process adopted by mammals seem almost like kindergarten material. (Insects evolved before mammals, so was their method of reproduction always so complicated, or has it become more complicated over time?) White did discuss human chromosomes and I learned much about chromosomal abnormalities, such as XXY, XYYY, and so on. I also learned that abnormalities were by no means rare but they nearly always resulted, either in fertilization not being able to take place at all or in the destruction of the zygote after but a few divisions. However, there were a few chromosomal (as distinct from genetic) abnormalities which were able to survive to full term. The child was invariably born with severe abnormalities and usually died shortly after birth. Few could survive for a few months, or even years, but there was only one chromosomal abnormality where the infant had expectancy of reaching adulthood, Down's syndrome, or Mongolism, as it was formerly known.

What I did not learn was how chromosome numbers had changed over the course of evolution.

Cells can only grow to a certain size before their membrane becomes unsustainable. For this reason there is a limit to the number of chromosomes, and the amount of genetic material, which can be contained within any cell. Most organisms (plant or animal) have between 20 and 40 chromosomes. Some have less than 10; others have more than 60. With 46, humans are towards the higher end of the scale, along with guppies, who have the same number. This means that the millions of different species of organic life on this Earth all share quite a small possible number of chromosomes. Two species having the same number of chromosomes does not necessarily imply a close evolutionary history.

Clearly over the course of time, there must have been countless occasions when a change in chromosome number occurred but nowhere could I find any explanation of how such an occurrence could have taken place and still resulted in a viable embryo, let alone one which was 'favoured'. Experiments within laboratories under radiation had resulted in many genetic changes, the duplications, deletions, translocations and inversions mentioned above, but all of these, when viable, had produced changes which were either neutral or disadvantageous. No favourable, or new, feature had been produced such as could be perceived as leading to a new evolutionary line. Nevertheless, evolutionary geneticists clung to the theoretical possibility, not only that such an event could have happened, but that evolutionary time was so great that it could have happened time and time again, as was clearly necessary to produce the variety of living forms we see today, plus, of course, all those which have existed in the past and become extinct. This belief still persists today, despite Dobzhansky's mathematical calculations, discussed in Chapter 28, which clearly demonstrated this to be impossible.

The problem of changing chromosome numbers came to worry me more and more, as, I believe, it had White. In his first edition of *The*

Chromosomes, White had devoted his final chapter (VI) to *Chromosomes and Evolution*. In it he tried to suggest a viable mechanism but his effort was not successful. This chapter was omitted from future editions. I searched the books on genetics on the shelves of the University library but could find nothing on changing chromosomes numbers. The more I pondered, the more certain I became: chromosome numbers had changed – but not by the process we know as 'natural selection'. White (1937: 105) had come to the same conclusion:

> … in all the types of reproductory mechanisms the primary origin of new species lies in some accident of the chromosome set. The occurrence of such an accident is entirely unconnected with natural selection.

[Before reproducing, little changed, my thesis chapter on chromosomes, here are a few reminders.]

Chromosomes occur in pairs – the famous 'double helix' which separates during meiosis. Each separate strand is known as a 'haploid'. One haploid from each chromosome of one parent joins with the equivalent haploid of the other parent at conception. Each finds the partner which carries the genetically matching material. Only when it has found its 'perfect match' can the 're-zipping' process proceed and a new double helix be formed. If a haploid has been too badly damaged during separation, the whole process is aborted. Very small translocations, inversions, deletions or duplications, are occasionally tolerated. Considering, not just the millions, but, in some cases, the billions of reproductive events which are constantly occurring (how many mosquitoes breed each year?), the accuracy of the mechanism is quite amazing.

Michael White (1910–1983)

Michael White had a cosmopolitan upbringing. Born in London, his family moved to Italy during the First World War. White returned to London in 1927 to attend university, where he studied botany, turning to entomology during his third year of study. After graduation, White remained at University College, London, as assistant lecturer, gaining the post-graduate qualification of Doctor of Science (D.Sc). During this time White associated with people such as Fisher and Haldane (Atchley 1981: 6).

After the war, White spent some time (1947-1953) in America where he accepted a position as professor of zoology at the University of Texas, at which time he formed a lifelong friendship with Dobzhansky (Atchley p. 8).

In 1953, White migrated to Australia, taking up the position of senior research fellow with the CSIRO in Canberra, where he remained until 1956. He then took up the Chair of Zoology at Melbourne University, following the establishment of the Department of Genetics in 1964. White occupied this Chair until his retirement in 1975. During his career, he published several books and a large number of articles.

Chromosomes and evolution

White asserted that evolutionary theorists concentrated too much on the role of genes and DNA in the evolutionary process and too little on the role of chromosomes (White 1973b). For example, Dobzhansky (1964) in his book, *Heredity and the Nature of Man*, mentioned chromosomes only as the carriers of genetic material. Dobzhansky's interest was in proteins, amino acids, nucleic acids, DNA and RNA. He never addressed the issue of how species had acquired their differing numbers of chromosomes. Unlike most evolutionary theorists, Dobzhansky was a geneticist, making his lack of interest in chromosome change the more remarkable. The role of chromosomes themselves in evolution appeared to be of little interest, except to White.

Initially, White (1937) had thought that chromosome change was a major cause of changing phenotypes. As more was learnt about chromosome numbers, it became clear that this was not the case (White 1973a, 1973b, 1978). Some families, such as the Big Cats and the Great Apes, had the same number of chromosomes (38 and 48 respectively) although members of the family were clearly phenotypically differentiated. Conversely, some birds and mammals had different chromosome numbers but were so phenotypically similar that, prior to the establishment of their chromosome numbers, they had been considered the same species. Different species, genera, families, etc., might have the same number of chromosomes, such as humans and guppies, which both have 46. Thus, identity of chromosome number signified nothing when attempting to trace speciation. However, once a difference in chromosome numbers had been identified, then more than one species was involved, no matter how similar the species might be in appearance.

Considering the millions of species alive on the Earth today, as well as the millions of species now extinct, the variation in chromosome numbers is not large. While insects and crustaceans exhibit a wide variation in chromosome numbers, the haploid number of most vertebrates lies between $n = 6$ and $n = 20$. Thus, *Homo sapiens* with $n = 23$ is above the average number, although numbers double this are known in some vertebrates (White 1973b: 409). All primates appear to have fairly high chromosome numbers and there are more metacentric chromosomes in the gorilla and chimpanzee than in humans (White p. 449).

Although the sixth edition of *The Chromosomes* (White 1973a) and the third edition of *Animal Cytology and Evolution* (White 1973b) were large by comparison with their first editions, very little had changed as far as the understanding of the basic workings of chromosomes, in particular their role in reproduction, was concerned. The increased volume was brought about by an escalation in examples, especially in relation to insects and other invertebrates, which were White's area of expertise.

While polyploidy had clearly played a role in the evolution of some plants and non-sexually reproducing members of the animal kingdom, polyploidy was not viable in sexually reproducing plants and animals and played no part in their evolution. Chromosomes occurred in pairs and this pairing was essential for meiosis to occur. All diploid chromosome numbers in sexually reproducing species are 'even', although the haploid number may be 'odd' or 'even'. Fusion of two chromosomes had been observed under irradiation. However, fusions thus produced, even though they may survive one or two cell divisions, were invariably fatal.

All chromosomes had one centromere. During meiosis, each centromere attached to a spindle fibre, there being the same number of spindle fibres as centromeres. Thus, if a centromere was to duplicate itself by some chance, the extra centromere (and any chromosomal material which had duplicated itself along with it) would have no spindle with which to attach. If two centromeres somehow fused, there would be a spare spindle. Both these circumstances would impede meiosis. In his earlier work, White (1937) had thought that 'microchromosomes' might be centromeres discarded after fusion or that they might be 'spares' awaiting utilization when numbers were to be increased. White made no mention of the 'microchromosomes' in his later works and clearly no longer thought them of any evolutionary importance.

The terms 'metacentric' (V-shaped), 'acrocentric' (J-shaped) and 'telocentric' (rod shaped) were retained. White (1973b) was still of the opinion that telocentric chromosomes contained a small amount of genetic material on a second micro-arm, although he acknowledged that this was unproven.

Chromosome abnormalities in humans

A number of chromosomal abnormalities are known to affect the human sex chromosomes. Turner's syndrome is a condition in which there are only 45 chromosomes instead of 46, there being no Y chromosome. The person presents as female, but is sexually underdeveloped, the ovaries being merely fibrous streaks. Females with XXX (47 chromosomes) occur as frequently as 1:500 female births. These people are often mentally defective, but may be fertile, since one of their chromosomes appears to be partially 'switched off'. XXXX and XXXXX females are also known and one was known to give birth to a daughter. Unfortunately, White did not record the chromosome count for this child, but if the extra X chromosomes were 'switched off', then it may be presumed that the daughter's count was normal?

Klinefelter syndrome (XXY and XXXY) occurs in about 1:500 'male' births. These people are generally tall, long-legged, with small testes, abnormal testicular histology, atrophy of seminferous tubules and usually some degree of gynæcomastia. They may be sexually active, but sterile. A variation of Klinefelter's syndrome is the XYY male, affecting about 1:550 live male births.

Polyploidy may have played no part in mammalian evolution but that is not to say that it does not occur in humans. The most well-known polyploid condition in humans is that of Trisomy 21, otherwise known as Down's Syndrome. The condition is caused by the non-disjunction of the number 21 chromosome, either at meiosis or in one of the first cleavage mitoses after fertilization, which results in the person being born with an extra chromosome 21, making their total number 47, instead of 46, chromosomes. This is the only human trisomy which allows the affected individual to survive into adulthood, although the person is sterile, uneven numbers of chromosomes not being viable at meiosis.

Trisomy 13 may also occur on rare occasions and is known as Patau's syndrome. The individual has harelip, cleft palate, polydactyly, heart and kidney defects, extreme mental retardation and viability is very low, the individuals rarely surviving long after birth. Another rare trisomy is that of chromosome 18, which causes severe heart defects and profound mental deficiency. Those afflicted die in infancy. Other trisomies appear to be lethal in utero and are believed to be responsible for many spontaneous abortions. Extremely rarely a live birth may be recorded exhibiting complete trisomy (3n = 69) but none survive more than a few days.

None of the above named anomalies contributed in any way to human evolution, since they all result in infertility.

White believed that the human metacentric chromosome 2 was formed by the fusion of two acrocentric chromosomes from the chimpanzee. White made no suggestion here that the fusion occurred in a hypothetical common ancestor.

However, two chromosomes combining into one would have reduced the number of chimpanzee chromosomes from 48 to 47, not 46. Such an uneven number, even if the, 'person,' survived into adulthood would have rendered them sterile.

Possible scenario

None of the evolutionary theorists ever postulated a possible mechanism by which karyotypes (chromosome numbers and arrangements) could be permanently changed. Fissions and fusions were assumed to have taken place. White (1973a, 1973b) was more cautious than most in that his books are full of modifiers – 'might have', 'possibly', 'it seems that' and so on, yet he, himself, never explained precisely how fissions and fusions might have occurred. An attempt will now be made to do this by the use of hypothetical scenarios.

Meiosis, which is the production of *reproductive* cells, is more complicated than mitosis, which is the replication of somatic (body) cells. Since all chromosomes are paired, their numbers are referred to as '2n'. The 46 pairs of chromosomes in humans are referred to as 2n = 46; for chimpanzees, having 48 pairs, the number is 2n = 48. The first stage of meiosis involves the *doubling* of chromosome numbers, in chimpanzees resulting in 4n = 96. There follows a further double splitting of the potential reproductive cell, the four resulting cells each having half the original number of chromosomes. If fusion of chromosomes were to occur at the very start of the process, such that n = 23 was to be obtained, then it would be necessary for four chromosomes to fuse (not two), which is deemed unlikely. If it did occur, the further stages of meiosis would not be viable. By the second stage of meiosis, cell division has reduced the number of chromosomes in each of the two 'daughter' cells to the original number, which for the chimpanzee is 2n = 48. If fusion were to occur in one of these two daughter cells, the number would be 2n = 47, which would not be viable. By the third stage of meiosis, the number of chromatids in each of the four daughter cells has been reduced to n = 24. The fusing of two chromatids at this final stage of meiosis would result in n = 23, and this would appear to be the only stage at which such a fusion might be possible, since this is the only stage at which an uneven number is acceptable.

In the female, three out of the four 'daughter' cells produced during meiosis become 'polar bodies' and are lost. Let it be assumed that this reduction by the fusing of two chromatids took place in the fourth 'daughter' cell, which was destined to be fertilized. In the male, all four 'daughter' cells produced during meiosis become spermatozoa. However, the assumed fusion would only have occurred in one of these four cells, meaning that only one sperm out of the hundreds of thousands – millions – produced would carry the new formation. The chance of this one being the one to fertilize the female egg is extremely small.

Let it be assumed that two of the chimpanzee's telocentric (rod-shaped) chromosomes combined. For this to happen, an arm would have to dislocate itself from the centromere of both telocentrics, in order that 'sticky' ends would be available for a new attachment. If the dislodged arms merely reattached themselves to other chromosomes, a translocation would have occurred, but there would have been no reduction in chromosome number. For a reduction to occur, it is necessary to postulate that two of the long 'arms' of the rod, with their centromeres, fused, and that the two short (extremely small) 'arms' were lost.

It will be remembered that White held that telocentric (rod shaped) chromosomes did, in fact, possess a small amount of DNA material on the 'micro' arm, even though this had not yet been demonstrated. If White was correct, then the two ends of the rods would be sticky and could join, but the two centromeres would be separated in a very small degree. It is hard to envisage that these two newly joined telocentric chromosomes would not pull apart at cell division. This is what had been observed in the laboratory. This scenario could be postulated for two acrocentric (J-shaped) chromosomes, or one telocentric and one acrocentric chromosome, provided the small arm of the acrocentric chromosome was very small and the amount of the DNA material it carried was not so great that its loss would be lethal.

If White was incorrect and there was no DNA material, no second arm, however small, on a telocentric chromosome, then it must be assumed that the centromeres themselves became 'sticky' and that the two centromeres attached to each other, lying side by side. White never explained exactly why he believed that telocentric chromosomes must have had at least a very small second arm, but it may have been that he believed that this was necessary to 'seal' the end of the chromosome. If centromeres could become sticky and adhere to each other, meiosis would become a complete shambles.

On fertilization, the two telocentric chromosomes from the other parent are assumed to pair with the two arms of the newly formed metacentric

411

chromosome. The two short arms must have been lost, not joined with each other, otherwise there would have been no change in chromosome number. For a permanent reduction in chromosome numbers to occur, it is necessary to assume that the two centromeres remained attached to each other and acted as one, i.e. migrated to the one spindle, leaving one spindle with no chromosome attached. This would leave an imbalance and almost certainly be unviable, as laboratory tests had shown.

However, for the purpose of this scenario, it is necessary to assume that further cell division took place. It would have been necessary for the reproductive cell which had undergone the reduction in its number of chromosomes to have formed a gamete with a reproductive cell from a mate which had undergone an *identical* reduction, not merely a reduction involving different chromosomes. It is further necessary to assume that the loss of DNA from the abandoned short 'arm' did not cause any deformity. Some change may have occurred, possibly early in embryonic life, such that some new feature might have evolved. However, it would also be necessary to assume that this feature was not so extreme as to interfere with mate recognition, since it would have been essential for evolution that this new embryo be born healthy, be accepted by its community, mature and reproduce.

Let it be assumed that this fœtus, the first *Australopithecus*, was born healthy, with the new karyotype n = 23. Let it further be assumed that this *Australopithecus* mated with a chimpanzee (the only mate which would have been available). There is no reason to suppose that this union would not be fertile, in the same way that the union between a horse and a donkey is fertile. The first 'chimp-man-zee' would have been born. This 'chimp-man-zee' would have had 47 chromosomes and been sterile.

For an ongoing line to become established, it is necessary to assume that a second *Australopithecus* was available, in the same location, at the same time, of opposite sex, which happened to have, by chance, an *identical* chromosomal mutation. In order for this to have happened, it would have been necessary, not only for a second chimpanzee to have undergone a fusing of two telocentric/acrocentric chromosomes into one metacentric chromosome, but the fused chromosomes would have had to be the *same two* that fused to produce the first *Australopithecus*. This rare occurrence must not have happened only once – producing the first *Australopithecus* – it would have needed to occur again, in the same breeding group, to produce a viable mate, and these two would have needed to select each other as mates for there to be any chance of a new line becoming established. Such changes must have happened, not once, but millions of times to account for all sexually reproducing plants and animals.

While it is generally assumed that, in the case of the hominine line, the number of chromosomes had been reduced, in other cases evolution must be assumed to have occurred as a result of an increase in chromosome number. Centromeres are known to be capable, not merely of duplicating themselves (along with their attached DNA arms), but of over-replicating themselves in a manner that produces polyploidy. In non-sexually reproducing organisms, this may give rise to a new form, without any complications. However, in sexually reproducing organisms, polyploidy is not viable, as already explained. It sometimes happens, is always detrimental, usually fatal before birth, and never viable in the long term. The only way that chromosome numbers could be increased would seem to be for the centromere to divide itself in half. Such a division would result in two rod shaped telomeres, with no second arm, however small. White resisted this concept, since no such splitting of a centromere had ever been witnessed in the laboratory. "Although spindle attachments divide longitudinally at mitosis [and meiosis], they do not appear to be transversely divisible" (White 1937: 94). The only way for an increase in centromere numbers to occur was by polyploidy (trisomy) and, while that had been shown at times to result in a viable foetus, it had never been known to give rise to a permanently fertile line. Combinations such that a chromosome temporarily possessed two centromeres had been witnessed, but this was never viable for more than one or two cell divisions.

The method by which chromosome change is achieved, as well as what role, if any, such change plays in the process of evolution has been sadly neglected. This neglect has provided a potential opportunity for future research.

Chapter 37

The Rise and Fall of the Hapless Neanderthals

A t last it is time to consider the best known of all the former varieties of human.

Variety? Subspecies? Species? Upon one thing everybody agreed. The Neanderthals formed a separate population of humans. Where they disagreed was upon whether or not Neanderthals were our direct ancestors, whether or not the Neanderthals had interbred with another incoming population or whether they had been totally replaced by the incoming population. The first two possibilities required that the Neanderthals be recognized, not only as fully human, but as the same species as ourselves, because they could not have interbred with the incoming, fully human Cro-Magnon population if they were not the same species. Total replacement left open the question of species. None of the Neanderthal remains showed any sign of having been brutally killed. Did the incoming Cro-Magnon people out-compete them for resources? Unlikely, since no early human population appears to have been very large and there were enough resources in Europe to support a population far larger than that estimated for both populations combined. Were the Neanderthals already in decline? Possibly – but why? Ongoing fossil finds showed that the Neanderthals had lived in Europe alongside our Cro-Magnon ancestors for thousands of years as a distinct population. Whatever its cause, the demise of the Neanderthals was gradual.

During the second half of the 20th century, the 'Out of Africa' hypothesis brought to the fore the question of whether the Neanderthals were another species, diverged from descendants of the first migration out of Africa about a million years ago, or whether they remained the same species but became a distinct variety. As the 'separate species' hypothesis gained dominance, it came to be accepted that Neanderthals were not merely another species of 'human', but a species of 'subhuman'. This was the first time (since King) that the full humanity of the Neanderthals had been brought into question and occurred at the same time as Binford was questioning the relationship of Asian finds, such as Peking Man, with those of others from Europe and Africa.

While some fossil Neanderthal remains seemed to be of people who had been buried accidentally by rock falls, for example Shanidar I, II and III, the body of Shanidar IV from Iraq appeared to have been placed in a 'crypt' scooped out among the rocks in the cave and then covered with earth (Solecki 1971). Others seemed to have been deliberately buried, such as those from France, at La Ferrassie, Le Moustier and La Chapelle. The Neanderthals were the first people known to practice intentional burial (Birdsell 1975; Wolpoff and Caspari 1997). Furthermore, at least one grave at Shanidar contained pollen from eight different species of spring flowers (Solecki 1971). Digging a grave takes time and effort, even in cultivated soil with the aid of a metal spade or shovel. It is for this reason that many murder victims found abandoned in woods are but covered with leaves and branches, or buried in a shallow grave. Hard, uncultivated soil, tree roots, stones, these make graves difficult to dig, even with modern implements. It must have taken considerably more time and effort when carried out in uncultivated soil with the aid only of stone, or possibly bone, implements. While the burial in Iraq at Shanidar, was in a cave, those in France, such as at La Ferrassie, were not. Why the Neanderthals forsook the simple method of disposal of the dead by abandonment to predators in favour of this time and labour intensive method is unknown, but it was generally accepted that this was evidence of human thinking. The presence of red ochre, tools and bones in the graves, was seen by some as indicative of philosophical/religious thought.

The remains of Shanidar I were of a male who had suffered serious injuries, possibly quite early in life, including the loss of an eye and the lower half of his right arm. Despite his injuries, which would have been disfiguring and disabling, Shanidar 1 must have been accepted as a member of his community, since he lived to an old age, leading Solecki to conclude that the Neanderthals were of a compassionate and caring nature.

Previous archæological experience digging Native American burial mounds caused Solecki to take soil samples from the Shanidar burial site in Iraq as a matter of routine but he was completely unprepared for the results relayed to him by the palæobotanist with whom he consulted. Not only had pollen from eight different species of late spring flowers been retrieved, but the pollen grains were in clusters and "... appeared to be resting inside the anther, or pollen bearing part of the plant ... Mme Leroi-Gourhan deduced that no accident of nature could have deposited such remains deep in the cave" (Solecki p. 247). Solecki reasoned that the presence of flowers, along with the fact of the burial itself, was evidence of human compassion, grief and, possibly, philosophical thinking. Solecki dubbed the Shanidar Neanderthals "The First Flower People", not entirely surprising considering that he was writing in the early 1970s. At that time 'everyone was beautiful in their own way', even the Neanderthals. Never before or since have the Neanderthals been so highly regarded.

The Neanderthals not only buried their dead, but the east/west orientation of the La Ferrassie burials and the presence of red ochre, tools and bones in the graves, were seen by some as indicative of religious/philosophical thought. Others considered that the practice of burial might have been merely an hygienic measure, copied from modern *H. sapiens* in the Levant and the apparent grave goods merely backfill (McBrearty and Brooks 2000). When commenting that at least one of the Qafzeh (non-Neanderthal) burials may have been associated with grave goods, McBrearty and Brooks made no suggestion that these items were merely backfill.

In passing, I would mention that it was the practice of the Celts also to bury their dead facing west. For the Celts of Britain there was no land west of Cornwall, Ireland and the Scottish islands. It was to the land of the setting sun that souls went after death. Christians reversed the practice, burying their dead facing east, towards the rising sun. Sir Winston Churchill asked to be buried facing west in accordance with Celtic tradition.

With evaluations such as Solecki's, it was becoming increasingly difficult to differentiate between the Neanderthals and modern humans. In every detectable respect, *H. sapiens* >40,000 BP and their Neanderthal contemporaries, were indistinguishable (Stringer and Grün 1991; Klein 1998; Speth and Tchernov 1998). At one time, it was thought that the European Châtelperronian tool industry heralded the arrival in Europe of a population with superior lithic skills, but as Neanderthal remains started to be found in association with Châtelperronian tool sites, it was acknowledged that these tools, together with the beads and other ivory objects associated with them, were part of the Neanderthal culture and exclusive to them (Harrold 1983: 127; d'Errico et al 1998: S2). Reynolds (1990) concluded that much of the perceived difference between the Mousterian/Châtelperronian and Aurignacian complexes was one of classification, which gave an impression of increased complexity and innovation that was not sustainable. Reynolds suggested that the Neanderthals were associated with the process of change itself and that the Middle-Upper Palæolithic transition may have been a Neanderthal phenomenon. Others, such as Leakey and Lewin (1982) and Stringer and McKie (1996) suggested that the 'Neanderthals' had adopted the culture as a result of interaction with the new-comers, but whether that interaction took the form of copying, trading or stealing remained open. Would any of these things have happened if the Neanderthals were a completely separate species?

The earliest evidence of the presence of humans in Europe had come, not from bone, but from stone. Early hand axes, some believed to date back as far as 600,000 years, were found and the period of thousands of years over which these simple tools were produced was known as the Acheulian. No human fossil remains had been found in association with early Acheulian tools, their presence was deduced. Early fossils gradually appeared: 400,000-500,000 years old at Vretsz (Hungary), 400,000 years old at Boxgrove (England) and Pont-newydd (Wales), 300,000 years old at Swanscombe (England) and 250,000-300,000 years old at Arago (France). The Welsh remains were judged to have had Neanderthal type teeth, stretching back the time of the pre-Neanderthals. Remains between 120,000-150,000 years old have been recovered from Jersey (Channel Isles). This is not a complete list, but covers the major finds and should be sufficient to show the establishment of early humans, not only in mainland Europe, but on what are now islands (Britain and Jersey).

More advanced tools, associated with the Neanderthals rather than the pre-Neanderthals, were known as Mousterian, after Le Moustier where they were first found. These were beautifully worked flints and had been produced by persons, not only manually dexterous, but also with artistic minds, able both to plan and to execute. As the flint tools were subject to more detailed and expert analysis, differences in technique were noted The more 'advanced' of the techniques became known as the Châtelperronian and was believed to have been introduced into Europe by the earliest Cro-Magnon people whose presence in Europe was back-dated to 35,000-40,000 years on the evidence of these tools alone.

As further finds placed Châtelperronian tools at Neanderthal sites, the earliest firm evidence for the presence of Cro-Magnon in Europe became the original Cro-Magnon find from 1868. This famous fossil, which fortunately included the cranium, was not only the first 'fully modern' European found, it remained the earliest, coming to be dated 30,000-32,000 B.P. Many now consider the slightly different tool-making techniques to be indicative of nothing more than different tribes - a salutary lesson against over enthusiasm. The time during which Cro-Magnon people are believed to have inhabited Europe (32,000-17,000 B.P.) is known as the Aurignacian.

The apparently deliberate placement of bear skulls in certain caves suggested that the Neanderthals may have practiced a primitive religion or 'bear cult' (Birdsell 1975; Shackley 1980). At Arcy-sur-Cure, also in France, nearly a hundred and fifty bone and ivory tools were found, along with ornamental pieces, such as grooved/perforated tooth pendants (d'Errico 1998: S2; Farizy 1994). The only items which d'Errico et al. conceded may possibly have been traded were some ivory rings, which they saw resulting from cultural interaction similar "to that observed between neighbouring modern and submodern human groups". (The term "submodern" refers to the technology, not the people.) There were mammoths in Europe at that time and their tusks were as useful to the people of Europe as they were to the people of Africa. When considering the 'humanity' of Neanderthals in this context, it should be remembered that trade takes place between people, not between people and animals. Even if the Neanderthals differed in some essential way from the Cro-Magnon, there would have been no trade between them unless they each recognized the other as human.

Marshack (1990) drew attention to engraved bone fragments dating to 100,000 ya found at Tata, Hungary, and to even earlier (110,000 ya) pendants made from wolf bone found at Bocksteinschmiede, Germany. He noted that there was little or no evidence of artistic activity during the Middle Pleistocene period in places outside Europe. The use of ochre was well attested in Europe at sites as ancient as Terra Amata, France, which has been dated to 300,000 ya, as well as much later by Neanderthals. While the number of beads and pendants found at Neanderthal sites was less than that found in Upper Palæolithic European sites, it was more than that found in Upper Palæolithic sites anywhere else. No one suggested that Upper Palæolithic humans in Africa, Asia or Australia were devoid of speech and/or symbolic thinking merely because no evidence of art of that age had been found. Just because no 'art' has been found in the archæological record, does not mean that none existed.

Many factors govern survival of bones and artifacts, not least being farming and industrialization, which destroy buried items. In past ages, pieces of bone or stone turned over during farrowing would have been thrown away, as were items found during mining. It was not until the 19th century that the possible significance of such items came to be recognized. Finds are most likely to be made where modern humans do not live, i.e. in mountainous or rocky areas unsuitable for farming and where people have not, therefore, founded villages. As archæologists came to seek out such places for investigation, finds were made, but ancient people, some much older than the Neanderthals, may have lived on lower ground, especially by rivers, any evidence of their existence may have long been destroyed, both by the forces of nature and by later inhabitants. Even today, artistic ability is a variable human characteristic. If being artistic and/or musical came with some 'selective advantage', then I find it puzzling that these attributes have not become spread more universally among the human population. If they have no selective advantage, then I wonder why they became developed to the extent that they have (in some people). There is no correlation today between artistic ability and intelligence or verbal skills and no conclusions can, therefore, be drawn about the mental abilities of earlier people based upon artistic development alone. However, art could have been associated with increased imagination and creative ability and this would have been indicative of a continuing and expanding evolutionary advance.

As the grounds for a single African exodus about 100,000 years ago, after a speciation event possibly 150,000 to 200,000 years ago, became less secure, there was a tendency to specify Europe as the place of the 'Great Leap Forward', the place where art, pottery, musical instruments and trade became evident: "Insofar as there was any single point in time when we could be said to have become human, it was at the time of that leap" (Desmond 1991: 27). This was a return to the position which had been held prior to the postulation of the Complete Replacement Hypothesis. These events all appeared to have taken place in Europe around 35,000 to 40,000 years ago. The delay of some 50,000 years between the appearance of anatomically modern humans at Skhul and Qafzeh and the appearance of Upper Palæolithic technology raised questions as to whether early anatomically modern people were, in fact, modern at all as far as their mental capacities were concerned (Aiello 1992). What made a 'modern human' modern? Aeillo (p. 84) saw the Skhul and Qafzeh populations as being replaced by Levantine Neanderthals and, therefore, not ancestral to the modern humans who appeared in Europe: "The immediate ancestors of the European modern humans would be later modern humans that reappeared about 40,000 ya". Quite from where these people 'reappeared' was not clear.

The more the fossil record is examined, the less clear it is exactly when, and from where, fully modern humans first arrived in Europe. Tattersall (1998; 2002; Tattersall and Schwartz 2001) claimed that modern humans invaded the Neanderthal's European and Western Asian domain around 40,000 B.P., that 'moderns' arrived in Europe fully equipped with modern abilities and sensibilities, although, once again, it was not stated from where these fully equipped 'moderns' came. It seems almost as if there is reluctance to allow that fully modern humans may have emerged in Europe rather than either Africa or Asia, although quite why this should be is difficult to explain, other than by some form of 'reverse racism' – if it is possible for there to be such a thing.

Tattersall and Schwartz pointed out that there was in Africa no equivalent of the 'symbolic explosion' found in Europe, which would have been a prerequisite if Africa was to be suggested as the point of departure for these 'fully equipped' people. Tattersall (2002) listed 'sculpture, engraving, painting, body ornamentation, music, notation, subtle understanding of diverse materials, elaborate burial of the dead, painstaking decoration of utilitarian objects' as being documented at European sites more than 30,000 years old. The (comparatively) sudden appearance of these things had led earlier archæologists to assume that these skills had been imported into Europe by people who had migrated from elsewhere but more than a century of digging and excavation had failed to uncover any other areas anywhere in the world where these skills had been evidenced at an earlier time. Nevertheless, there persisted a reluctance to admit that they may be native to Europe.

The English had been very impressed by the great civilizations they discovered in India and China during their great Age of Exploration. These ancient civilizations had been established for thousands of years, years during which the English were 'wood, woad and witches'. That civilization had spread from East to West seemed quite clear and, if civilization, then why not humanity itself? Unfortunately, archæological excavations failed to find any evidence to support this idea. Other archæologists followed the fossil trail to Africa, but again were unable to find any evidence of advanced artistic expression. The cave paintings of France, dating back more than 30,000 years, had 'exploded' upon the scene. Cave paintings in other places, such as Australia, were not only far more recent, they lacked 'perspective', the perception of depth, the appearance of 'life' portrayed by these paintings. Indeed, art of historical times from places such as China, India and Egypt, although very beautiful and often very complex, was 'flat' by comparison. Painting of the calibre of these early artists did not re-appear until the Middle Ages (in Europe) when persons such as Leonardo da Vinci and Michael Angelo transformed our ideas of art forever.

These creations were presumed to be the work of the newly 'arrived' fully modern humans, not the Neanderthals, even though some of the work predates the earliest Cro-Magnon fossils. Is it possible that it was interbreeding, the mixing of Neanderthal mental capabilities with those of the 'arrivals' (whoever they were) which somehow produced a combination which allowed this 'Great Leap Forward' to take place? It is strange that, after making its appearance, this beautiful art disappeared, and its disappearance coincided with the disappearance of the Neanderthals. It was not to reappear again for thousands of years. I have often wondered whether this beautiful work was not that of an 'idiot-savant'? There are rare people who, while seeming to struggle with some everyday life skills, excel in other areas, such as mathematics, memory, art or music. There is no reason to suppose that this is a purely modern phenomenon. I have no doubt that some of the 'witch doctors' or shamans of the past exhibited some of these outstanding abilities and were revered by their peers, so why not then?

Lithic evidence of habitation in the European Arctic region as early as 40,000 BP has been found but absence of fossils leaves open the question of to whom this belonged: the Neanderthals or newly arrived Cro-Magnon people? Pavlov et al. (2001) concluded that these early occupants were *H. sapiens* since survival in such an environment would have required long-term planning and an extended social network, neither of which they believed were within Neanderthal capabilities. Mellars (2004) noted remains which had by then been found in Romania dated to 35,000 B.P., a fragmentary maxilla from Kent's Cavern in Devon, dated to ~30,000 B.P., two modern mandibles from Les Rois in western France dated to 31,000-35,000 B.P., and remains from Madec, Czech Republic, dated to 34,000-35,000 B.P., none associated with any artifacts. These finds show that modern humans were present in Europe prior to 30,000 B.P., along with the Neanderthals, but their numbers would seem to have been small. On the grounds that it was unlikely that remains of the very first modern humans to arrive in Europe had been found, Mellars extended the probable time of their arrival to between 40,000 and 50,000 ya. Were such an estimate correct, modern humans would have co-existed with the Neanderthals for between 10,000 and 20,000 years, which is a long time. There is still no firm fossil evidence for the presence of modern humans in Europe prior to 32,000 B.P. Even if there were, until fossil remains are found in the Arctic region, who made the tools remains a matter for speculation.

From the time of Boule, the Neanderthal stocky build had been associated with smallness of stature, and an assumption made that inferior stature was associated with an inferior intelligence - this despite the fact that Neanderthals had somewhat larger brains than modern humans. Coon (1962: 548) stated that the Neanderthal was "a squat, stunted fellow, about 5 ft. tall, or 155 cm.", although von Koeningswald (1962: 97) gave their height as up to 5 ft. 4 in. (155 cm.). Both were referring to the European Neanderthals. These heights were not much different from that of the average European a few centuries ago, or even of some South Mediterranean people today. Asians are also short, but a comparison with them is not as relevant in this context. There is no correlation between height and intelligence, although, for some reason, taller people are often perceived to be more competent and some writers do seem to have assumed that the Neanderthal's inferior stature was indicative of inferior intelligence. The Shanidar Neanderthals' height averaged 5 ft. 7 in. (162 cm.) (Trinkaus 1983a).

Life expectancy is another area in which Neanderthals were generally considered to have fared less well than current humans. This was pure conjecture. The inaccuracy of estimations of 'age at death' has already been discussed. All that can be said with certainty is that some Neanderthals lived to an old age.

Jonathon Leakey, the eldest son of the famous husband and wife team of archæologists, Louis and Mary Leakey, found the first fossil remains of *habilis* at Olduvai Gorge, Tanzania, in 1960. *Habilis* was designated *Homo* because evidence was found of simple tools, which were believed to have been deliberately fashioned. So simple were the tools that it was difficult at times to determine which pieces had been worked and which had flaked as the result of having been dropped. However, it was considered that, even if the witnessing of an accidental flaking had given the idea to *habilis*, the copying of this accident could only have been done by a creature with a human mind. A plethora of wonderful wild life documentaries since then has shown us that many mammals and birds make and use simple tools and this criteria is no longer considered sufficient to justify the nomenclature *homo* and *habilis* are increasingly being referred to as *Australopithecus*. The idea that a change in technology was indicative of a change in species underpinned the thinking of archæologists for many years. However, were such thinking to be valid, should not a new 'speciation event' be postulated for ~35,000 ya, which saw the advent of the cave art of Western Europe, for ~28,000 ya, which saw the first firing of clay and the production of the Venus figurines at Dolne Vestonice, again at 10,000 ya with the introduction of agriculture, ~300 ya with the Industrial Revolution and ~60 ya with the introduction of the Age of Electronics? Consistency of thinking is essential. If technological changes can happen in society today without speciation, then they could have so happened in the past.

Roth was an early ethnologist who lived among the Tasmanian Aboriginal people for a considerable period of time during the middle of the 19th century. His fascinating book, *The Aborigines of Tasmania*, published in 1899, is an account of his time with these people, which extended over many years. He was accepted into their tribe, a tribe which was still living as it had done for countless thousands of years. He reported that the men carried only two weapons: a long, thin spear of some ten to twelve feet (3 metres) and a waddy, a form of cudgel similar to a baseball bat. The spear was carried as a defense from attack by another tribe, it never being used for any other purpose than to kill another man. However, in all the time Roth was with them, he never witnessed any contact with any other tribe. He never witnessed the spear being used. It was the waddy which was their working implement. The men carried nothing but these two items, everything else, including children and any items worth the effort, being carried by the women. The women made grass bags for collecting shell fish and carrying their few belongings (see Fig.37.1).

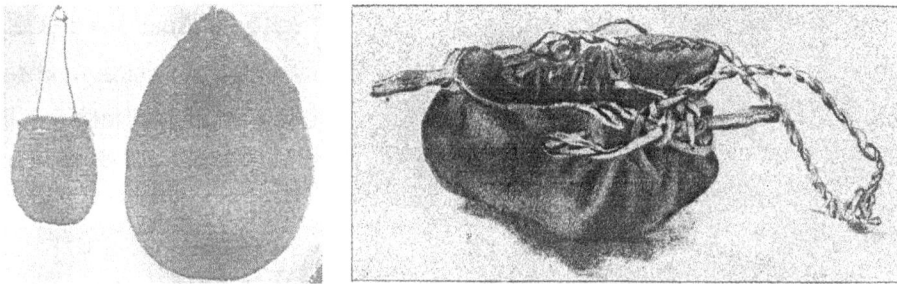

Fig. 37.1: Grass bags made by Tasmanian aboriginal women.
(Roth 1899: 142, 153)

The Tasmanians had no throwing sticks or boomerangs, hardly surprising considering how thickly wooded was Tasmania. They used flints only for cutting their flesh for ornament, keeping the wounds open by filling them with grease or ash to enhance scarification. They wore thin strips of animal fur or sinew as bands around various parts of their body, neck, ankles, arms, legs, which they sometimes coloured with red ochre, and adorned themselves with feathers and shells. I found this particularly interesting, since it reminded me of some of the natives of New Guinea, although the Tasmanians wore no clothing, so there were no grass skirts. However, it is a reminder that fossil remains can provide but part of the story. They can tell us little about cultural practices or philosophical beliefs. The Neanderthals may have been every bit as well-groomed and proud of their appearance as were the Tasmanians. Interestingly, Roth reported that these Tasmanian Aborigines appeared to have no religion or ritual, taking these two items out of consideration when defining 'human'. To the best of my knowledge neither did the Eskimo.

The Tasmanians were able to make fire and ate well. Mainland aborigines also ate well and any suggestion that people dependent upon stone tool technology ate but meagerly is pure supposition. Neither mainland nor Tasmanian people knew how to boil water but the Tasmanians made an alcoholic beverage from the sap of the "cider-tree", which was collected in a hole at the bottom near the root of the tree and allowed to remain until it fermented, being "rather intoxicating if drunk to excess" (pp. 94-95). Armed only with the evidence of their very simple tools (the spears, waddies and grass bags of the Tasmanians would have decomposed), an hypothetical future archæologist, applying

Fig. 37.2: Tasmanian shell necklace
(Roth 1899: 132)

to these people the same criteria which are applied to early hominids, might well attribute them with only meagre intellectual abilities, even less than the Neanderthals, since the Tasmanian tool kit was considerably more simple. Fortunately, we know them to have been completely human, with full linguistic capabilities and this knowledge should guide our (re)evaluation of early humans.

In 1975, Birdsell pointed out that the remains from Lake Mungo, New South Wales, then dated to about 26,000 B.P., preceded those of any skeletal remains of modern humans which had been found in Europe until that date. When more advanced dating techniques were introduced, many fossils were redated, and found to be older than previously thought in both the northern and southern hemispheres. One of the remains from Lake Mungo has now been dated to ~60,000 B.P., but mtDNA analysis has shown that his line had diverged before the time of the most recent common ancestor of living Australian aborigines and is not now considered fully modern, leaving an assumption that *H. erectus* must have reached Australia and have been capable of making some form of sea craft. More on this later. The remaining Lake Mungo remains, with modern mtDNA, are still dated to approximately the same time as the European Cro-Magnons, both being a little earlier than had previously been thought.

The Australian record continues to be difficult to reconcile with that of the arrival of fully modern humans in Europe, since both events appear to have taken place at approximately the same time, half a world away, but no fully human 'common ancestor' has been found anywhere in between. Ash evidence of extensive burning indicates that people arrived in Australia more than 40,000 ya, before there is any firm evidence of their arrival in Europe. Furthermore, it is generally assumed that these first arrivals were ancestral to modern day Australian aborigines and that they had dark skins, whereas those arriving in Europe, for similar reasons, are assumed to have been light skinned, indicating that the two populations must have already been separated for an extended period of time.

More aware, perhaps, of the Australian story than his northern hemisphere colleagues, it is not surprising that Alan Thorne, from Canberra, held so unswervingly to the theory of Multiregional evolution, since only by accepting that humanity evolved in concert over a wide area and over a long period of time can these apparent inconsistencies be reconciled.

The opinion of Howells, whose book *Mankind in the Making* was published in 1959, was influential around the middle of the 20th century. Howells pointed out that the fossil remains from the eastern side of Africa, as far south as Broken Hill, were not in any way negroid, but rather appeared to have an affinity with later European (white) people. The remains from the southern part of Africa were small and, he believed, the precursors of today's sub-Saharan Africans, the San, Hottentots and Bushmen. He claimed that there was no evidence of any negroid race before the Mesolithic (Middle Stone Age) and, even then, nothing positive. Not until the Neolithic were there found skulls which were clearly negroid and these were found in West Africa, where it was assumed that the negroid races had evolved. Frayer et al. (1993) noted the same thing and Wolpoff et al. (1994) pointed out that post-Neanderthal Europeans, living at the same time, were not 'African'. Unfortunately, no negroid remains earlier than late Mesolithic/early Neolithic have been found.

Today there is a general assumption in the literature that early humans in Africa had dark skins. Evidence to support Howells' opinion that the early people of Africa were not negroid, and may well have been light skinned, will be discussed in Chapter 42..

Going down

By the end of the 20th century, opinion of Neanderthal abilities had reached its lowest point since King, at least according to Jordan (1999: 94-95):

> Neanderthal folk may have been prone to live for the day, indeed cut out by Nature for nothing else ... their diets must always have been poor ... meat may have been largely consumed away from "home" (by hunting males?) and only a small proportion of it brought back on the bone (for women and children), otherwise faring meagerly? ... it is possible to see a picture of women and children around the cave fire preparing plant food and animal scraps with their simple tools and of men dropping by with proper cuts of meat acquired by hunting ... The males, moreover, may have often eaten much of their meat out

in the wilds and not brought so much back with them to share ... It would not really be appropriate to speak of 'home' ... for family life, with nuclear families and extended kin relationships ... would not have existed for the Neanderthalers ... Sex might have been less a prominent, certainly less a routine, aspect of Neanderthal life ... It is possible to see in meat sharing by hunting males ... [what] has been rather directly called the 'sex-for-meat' contract ... Neanderthals would indeed be much closer to our ape ancestors ... than to us in terms of behaviour.

And again (Jordan 1999: 112):

... they lived every day like the first day of their lives ... Their own bones are thick from heavy labour, as if they lacked the wit to make life easy on themselves ... they got their food ... opportunistically and stored none of it against a snowy day ... they wore no personal decoration ... a cloud hangs over their very humanity ... they were in some respects more like all their primitive forebears and indeed the ape-like ancestors of the human line than they were to us.

Jordan did acknowledge that care appeared to have been extended by Neanderthals at both La Chapelle and Shanidar, but suggested "perhaps their survival can be attributed relatively more to toleration than to active concern, for poorly individuals can scrape along even among chimpanzee groups" (p. 97).

Tattersall's opinion of Neanderthals was little better. He described burial as being merely a way Neanderthals dealt with "a rather distressing kind of clutter or ... of dealing with obscure emotions" (Tattersall 1998: 161; 2002: 123; Tattersall and Schwarz 2001: 213-214). Tattersall was prepared to admit that Neanderthals may have caught fish, "bears do, after all", but believed that they would have eaten them at point of catch, rather than bringing them back for sharing "which is typical human practice" (Tattersall 2002: 129). Tattersall (1998) considered that Neanderthals were not inferior humans because they were not humans at all. Mithen (2005) concluded that since the Neanderthals were incapable of speech, they must also have been incapable of thought. (A six-month old child is without speech but I dare anyone to tell its mother that it is without thought!) Surely thought must *precede* speech? Opinions expressed regarding the abilities of *Homo erectus* fared little better. Following a similar line of argument to that used by Mithen, Walker and Shipman (1996) concluded that Nariokotome Boy, (1.7 mya – 1.6 mya) was speechless, and therefore probably without thought. Earlier, Walker et al. (1982) determined *H. erectus* in its early stages of evolution had not yet evolved to the stage of having learned which were their correct foods, accounting for the vitamin A toxicoses evidenced in the remains of the female KNM-ER 1808, also dated to 1.6 mya (see Chapter 43).

Tattersall (2002) had a similarly poor opinion of *H. erectus*. Vultures wheeling overhead might have been an indication of the availability of material suitable for scavenging but "to go further than this and to suggest, for example, that the early hominids were reading animal spoor, probably goes much too far in the direction of viewing these creatures as junior-league versions of ourselves" (Tattersall 2002: 94-95). Surely the closer our ancestors were to their animal origins, the more developed would have been their 'animal intelligence'? Animals are very adept at reading spores and other signs when hunting their prey, of acting co-operatively when necessary. Why should it be assumed that early humans, alone of all creatures which have ever called this planet home, would have been devoid of both human and animal intelligence?

What had precipitated this change in attitude? Brace (1964) identified 'sociopolitical ideology' as being as important in the acceptance of some theories as basic biology and more than thirty years later, d'Errico et al. (1998: S22) were to lament an 'anti-Neanderthal prejudice' which they believed was hindering correct interpretation of evidence. Trinkaus and Shipman (1992: 322-324) concluded that the 1960s were a time of "outspoken moralizing ... ostensibly fighting prejudice and stereotype, [but with] a stony undertone of political correctness" which an entire generation of anthropologists soon learned not to question or transgress. "Race was not only not a fit subject to study, *it didn't even exist*" (p. 324, *italics in original*). As 'political correctness' became established, it was no longer permissible to insult another person by reference to their race, social class, sex or intellectual ability. One could not call another person 'a moron' but one could call them 'a Neanderthal'!

Wolpoff and Caspari (1997) wrote a book on the politics of social theory with particular reference to evolution, recalling how, as the generations had passed, many people had felt compelled to present their views in ways which conformed to the current paradigm, not necessarily because that was what they genuinely believed but because that was the way they wanted their views to be seen, possibly for reasons of promotion or funding. "We see things not as they are, but as we are" (p. 323), or are not, as the case may be.

The change in attitude towards the Neanderthals, and other early *Homo* people is difficult to understand, since it is not supported by physical evidence, either fossil or artifactual. On the contrary, physical evidence was narrowing the gap between the Neanderthals and modern humans. Trinkaus and Shipman (1992), Wolpoff and Caspari (1997) and d'Errico et al. (1998) identified this trend as having been motivated by a form of 'political correctness' but why 'political correctness' should be brought to bear in connection with the Neanderthals is difficult to understand.

When species classification was introduced, for both plants and animals, modern humans were given the designation *'Homo sapiens'*. Usual practice was for species to be identified according to some individualizing characteristic, whether that be their location or some other physical attribute. Two only, *Homo habilis* (handy man) and *Homo sapiens* (wise man) were named according to (presumed) mental attributes. The rules of nomenclature are such that, should the Neanderthals be accepted as a subspecies of humans, they must be known as *Homo sapiens neanderthalensis*. If they are designated a separate species, then they loose their 'wise' status and become simply *Homo neanderthalensis*, which is how they are referred to in most of the literature. If the former designation is accepted, then a means must be found of distinguishing the Neanderthal subspecies from us and the choice has been *Homo sapiens sapiens*. With the stroke of a pen, modern humans become doubly wise! That we are intelligent, I do not question; that we are wise, considering the state of the world today, I question more and more. Doubly wise? That I do not believe!

Unfortunately, 'priority' is also a strongly held principle. Once a discoverer has named a species, be it living or fossilized, it is very difficult for it to be changed. If Alan Thorne was right when he declared "We are all *Homo erectus*" (i.e. there has been no speciation since then), would that make us *Homo erectus sapiens* or them *Homo sapiens erectus*? How I wish our line could be reclassified: *Homo erectus africanus* (earliest humans), *Homo erectus asiaticus* (early Asian fossils), *Homo erectus europensis*, (pre Neanderthal Europeans), *Homo erectus australis* (Lake Mungo 3 and possibly remains from the Archipelago), *Homo erectus neanderthalensis* and *Homo erectus modernis*.

Lithic evidence of habitation in the European Arctic region as early as 40,000 ya shows that either the Neanderthals expanded much further north than previously thought or that modern humans established themselves in this northerly position very shortly after their first arrival in Europe, rather a strange choice for people who had evolved under the warmth of the African or Middle Eastern sun. Perhaps that was the only area available to them? Pavlov et al. (2000) came to this conclusion because they did not believe the Neanderthals had sufficient skills to cope with such a cold environment, a conclusion starkly in contrast with that of other theorists who claimed that Neanderthals were 'cold adapted', even 'hyperarctic' (Walker and Shipman 1996). Stocky body and short limbs were seen to be adaptations designed to conserve heat. In fact, people living in the northern regions of Europe (Scandinavia) are taller than those living in the warmer regions (Italy and Greece). The broad Neanderthal nose was seen to be necessary to warm air before it entered the lungs. Polar bears have long narrow noses to warm air. Broad noses are more likely to be found in tropical regions, especially among monkeys and apes – our line.

There is something known as the Allen's Rule which stipulates that species living in warmer climates have longer limbs than those living in cold or Arctic conditions. The short limbs of the famous Shetland pony and the long limbs of the giraffe may have given this 'Rule' popular appeal. It was estimated that Nariokotome Boy, had he lived to adulthood, would have been at least 6 ft., possibly more. This fitted with estimated temperatures of 29.2°C to 30.8°C, similar to conditions in that part of Africa today. This was pleasing to Walker because "had the result indicated an obviously incorrect temperature we would have had to go back and rethink the relationship between body build and temperature" (Walker and Shipman 1996: 161).

For 'fun', Walker then made a similar calculation for Neanderthals, concluding that they would have been 'hyperarctic', suited to a mean temperature of -1°C, which today would have placed them within the Arctic Circle. So some theorists deduced that the Neanderthals were physically cold adapted while others held that they were intellectually unable to cope with the cold. How true "One sees what one wants to see"!

Coon drew attention to the Tiwi people of the Melville and Bathurst Islands, north of Australia, who went naked, lived in flimsy shelters and who's only cutting tools were an all-purpose clam shell and a flaked stone axe. They had no spear-throwers, stone-tipped spears or boomerangs, as had mainland Australian aborigines, and their only use for their few crude flakes was gashing their foreheads at funerals. This culture would have left little evidence for a future archæologist to use to reconstruct their lifestyle, yet they had a great love of poetry, song, dance and painting, although none of their paintings would have survived the ravages of time.

Before the Neanderthals

It was not only the Neanderthals who engendered debate. The capabilities and both *H. erectus* and *Australopithecus* also divided opinion.

There are no fossil remains of Australopithecines for what is believed to have been the first two million years after their divergence from the ape line. It is assumed that they were arboreal, but by the time of the first fossil remains, they appear to have become mostly terrestrial and partially bipedal. By the time of *habilis* (2.5 mya) Australopithecines were completely bipedal. That the Australopithecines were meat eaters was shown by the number of baboons the Australopithecines appear to have killed (Birdsell 1975). The weapon of choice appears to have been zebra femur, the double ridge of these bones matching the double depressions on the baboon skulls (and some Australopithecine skulls as well). All but two blows were delivered to the left side of the victim's skull, showing that these Australopithecines were, by preference, right handed.

Writing of one of the earliest Australopithecines, *Zinanthropus*, Bunn and Kroll (1996) stated that there was no evidence that these ancient hominids actively hunted, believing that they acquired most of their meat by scavenging. However, they did find evidence for processing (defleshing) of complete carcasses of both large and small animals and concluded that the hunting of the smaller, gazelle-sized animals appeared likely because such small animals would have been eaten rapidly by larger predators, leaving little or no remains to be scavenged. While accepting that Australopithecines probably participated in co-ordinated group activity, transporting selected carcass portions back to their home site, they appeared to agree with Coon that small scale hunting of slow-moving animals, such as rats, lizards, snakes and tortoises, which could have been caught by women and children, did not really count as hunting. Coon (p. 80) commented that bones of new-born and suckling animals had been found at the *Zinjanthropus* site but commented that "... eating new-born ungulates is hardly hunting ... true hunting ... began as a way of life sometime during the Lower Pleistocene". Many predators, including the mighty lion, selectively target the newborn. Indeed, the simultaneous birthing of does within a herd is believed to have evolved specifically to accommodate this predatory habit by the producing of more infants at one time than are able to be eaten, thus ensuring the survival of some of the young. Furthermore, the Big Cats do not consider it beneath their dignity to take over some other animal's kill in preference to expending energy making a kill of their own. On the contrary, stealing another's kill is a sign of dominance. It is difficult to understand why there seemed to be such a concerted effort to portray our earliest ancestors as incompetent.

While Coon may have been somewhat scornful of the achievements of the Australopithecines, others saw them in a more positive light. Jordan (1999: 154) acknowledged that even scavenging required organization and social skills:

> Once ... embarked on a bipedal, open living, group structured, toolmaking and meat eating way of life, there was no going back on cleverness ... Even scavenging ... required organisation and social skills ... Getting meat

demanded powers of observation and interpretation, together with memory and communication of information ... It may be that the beginnings of a sort of family life were made during the habilis era ... meat sharing promotes more or less monogamous relations and family life.

Jordan's book was entitled *Neanderthals: Neanderthal Man and the Story of Human Origins.* Perhaps the sexist language in the title was deliberate because the book does seem to be more concerned with the men than with the females. The book focuses mostly on the Neanderthals but, in his penultimate chapter, Jordan did address the Australopithecines. Compare his description of probable/possible family life, with monogamous relationships, with his picture of the life of Neanderthals (see pp. 425-426) from some 50,000 ya (more than 2,000,000 years later) and you will appreciate just how complete was the dehumanizing of Neanderthals during the second half of the 20th century.

Neanderthal nemesis

So, why did the Neanderthals die out? In my thesis, I suggested the answer might lie with the Neanderthal blood type. I asked, 'Were they rhesus negative (Rh-)?'

Today, all blood groups are found throughout the world, but some are more predominant in some areas than others, leading to an assumption that in times gone by, before humans became so mobile, different populations developed different groups. The situation with rhesus blood types, positive or negative, is different. Rh- is a European phenomenon. Rh- is virtually unknown among Africans, rare among Asians, and, when found, almost certainly the result of interbreeding with a European. Even among Europeans, Rh+ is predominant. Until recently, a Rh- female married to a Rh+ male could expect to have only the one child. Today, Rh- mothers receive an injection of RhGAM to prevent the formation of antibodies which would cause their bodies to destroy any subsequent fœtus which had inherited its father's Rh+ (dominant) blood type, but this is a recent medical innovation.

If the Neanderthals were Rh-, they would have had no problems breeding among themselves, but, if they tried to interbreed with the Cro-Magnon newcomers, presumed to be Rh+, any of their females, who took a Cro-Magnon male as her mate, would probably only carry the one child. This would definitely be a 'selective disadvantage' which, over a period of several thousand years, would have had a negative effect on Neanderthal numbers.

At the time I wrote my thesis, I had not seen the subject of the evolution of different blood types addressed in the literature, although the possibility is now debated via the internet. I should mention that to date no Neanderthal remains have been shown to be Rh-.

During the second half of the 20th century, evidence for interbreeding rested with the fossils themselves and there were a few remains which seemed to carry a mix of features. There was the grooved jaw - and the spinal process mentioned earlier. DNA evidence of interbreeding had to wait for the 21st century.

Chapter 38

Lost for Words

In the previous chapter, it was mentioned that some authors had suggested that because the Neanderthals lacked language, they also lacked thought. I disagree on both counts. I disagree that Neanderthals lacked language and I disagree that the Neanderthals lacked thought.

All social animals communicate with each other, either by sound or body language. Some of these communications will be automatic, such as the yelp we give when suddenly startled, along with the slight 'jump' – body language. Even non social animals may communicate, by puffing up their body, their fur, their feathers, arching their back, when they perceive a threat. Again, the first reaction may be automatic but the animal needs to assess the situation to determine its further reaction – fight or flight? This takes a process of thought, however simple. Even deciding in which direction to fly, upon which flower to settle, may require some degree of mental processing and, as already stated many chapters ago, I feel Lamarck was on the right track when he identified the beginning of the development of a small enlargement at the superior end of the spinal cord, even in butterflies, as heralding the beginning of 'thought' – not philosophical thinking, but thought, an ability to make an independent determination in relation to some matter requiring action. I am absolutely certain that thought preceded speech. How could the person spoken to interpret the vocalization without the ability to hear, analyze and interpret?

As it became clear that there was little difference between the capabilities and life-styles of the Neanderthals and fully modern *H. sapiens* who lived in Europe after 40,000 BP, explaining the demise of the one and the survival of the other became increasingly difficult. In the absence of a viable physical or material explanation, archæologists increasingly favoured a difference in mental abilities as their explanation and the evolution of full speech (language) strengthened as the most likely candidate. That animals, from insects to birds to mammals, had communication skills, at times quite complex, had long been known. Work with chimpanzees had shown that they had capabilities beyond those utilized in the wild (Linden 1974). Laboratory experiments found that chimpanzees were capable of symbolic thought, of associating a symbol with an object to which the symbol had no relationship. Young laboratory apes, simply by observing their caretakers as they talked about what they were doing, learned to use symbols appropriately (Linden 1974, Gibson and Ingold 1993).

The fact that chimpanzees were capable of symbolic thought, although they did not appear to use this capability in the wild, lent support to Wallace's claim that mental capacities evolved *prior* to their use and were not, therefore, evolved as a result of natural selection. Wallace's arguments were rejected by his colleagues more than a century ago and are still not accepted today. This made the evolution of speech and language one of the least logically argued of all aspects of evolution. In *Introduction to Physical Anthropology*, Jurmain et al. (1997) emphasized the standard position that evolution resulted from random mutations and was not pre-planned or goal directed, yet, when discussing the many physiological and neurological changes necessary to allow the development of language, they stated (p. 276) that these changes had come about as the result of "systematic reorganizations" of the physiological structures involved. Either a change is random or it is systematic. It cannot be both.

It might have been thought that in choosing the title *"The Singing Neanderthal"* for his book, Mithen (2005) had envisioned the Neanderthals in full voice, their songs echoing across the Alps, assisted by their large lungs and huge sinuses, through which sound would have resonated. Far from it. Mithen was of the opinion that Neanderthals communicated by prosody (variation in intonation), meaning being conveyed not so much by the use of individual words as by 'melodious phrases'. This type of communication is sometimes referred to as 'IDS' (infant directed speech) because mothers have a tendency to sing, croon or in some way give rhythm or rhyme to their speech, a technique which is also frequently used when addressing animals or the elderly.

Mithen referred to this proto-language as '*Hmmmm*' - *H*olistic *m*ulti-*m*odal, *m*anipulative and *m*usical. Each '*Hmmmm*' phrase would have been an indivisible unit that had to be learned, uttered and understood as a single acoustic sequence. While acknowledging that Neanderthals had the level of respiratory control necessary for speech, Mithen was of the opinion that, since Neanderthals lived in small groups, whose lives would be much

434

the same from day to day, they would have had little need for any originality in communication (pp. 227-228): "To put it plainly, they didn't have much to say to each other that had not ... been said many times before". The same would probably have been true for many small modern hunter/gatherer tribes, but that did not prevent them from possessing complex language. Mithen was probably right in claiming that singing, being very much a communal activity, may have evolved fairly early as a human bonding experience, but it is questionable whether an ability to sing would have acted as a brake upon learning to use language, *per se*.

Even Jordan, who, as pointed out in the last chapter, had a very poor opinion of Neanderthal capabilities, acknowledged that they almost certainly had developed language, since they had all the necessary physical attributes: nerve supply to the thorax for sufficient breath control, lowered larynx and nerve supply for tongue control via the hypoglossal canal, which were fully developed by 400,000-300,000 ya (Laitman 1984). Small physical differences in the shape of the mouth cavity and windpipe might have impeded some sounds used today (but might have helped others we can no longer make?), but the number of sounds and words capable of being produced by humans is so vast that this is unlikely to have been a limiting factor in Neanderthal communication. The discovery of a Neanderthal hyoid bone at Kebarra was thought by some to settle the question in favour of the Neanderthals, but others pointed out that even pigs have similar hyoids (Stringer and Gamble 1993: 88).

It was the study of brain trauma, mainly strokes resulting from hæmorrhage within the cranium, which led to the discovery that the principal centres for speech and language were normally situated in the left hemisphere of the brain. Until these discoveries in the nineteenth century, it had not been understood that the two hemispheres of the brain operated differently. A further surprise was the discovery that the left hemisphere controlled the right side of the body and *vice versa*. Even more surprising was the finding that among left-handed people, the area for speech and language was frequently (but not always) in the right hemisphere, showing an unexpected association between these two quite separate functions, possibly because the left hemisphere controlled sequential movements (Savage-Rumbaugh and Rumbaugh 1993; Gibson 1993). The right hemisphere of the brain (in right-handed people) is associated with imagination, creative, spatial and holistic activities, including music and art. In early humans (and recent hunter/gatherer tribes) cognitive mapping of one's territory would have been an important function. Males generally being more involved in protection of territory, it was suggested that these right hemisphere spatial abilities, along with a creative imagination, may have contributed to the production by males of the first tools (Savage-Rumbaugh and Rumbaugh 1993). In general, while males perform better with visio-spatial skills including face recognition, musical composition and mathematics, females perform better with language, emotional expression and decoding, and fine motor skills.

The two hemispheres of the brain are not identical in size and shape, one hemisphere usually being larger than the other. In humans the right frontal lobe protrudes further forward than the left. Similar differences have been noticed in a number of animals, including birds, rodents and monkeys and are, therefore, phylogenetically old. Adult male song-birds may have left hemispheres three to four times larger than those of the adult females who sing less complex songs (Falk 1993). Parrots may not be song birds, but they have greater vocalization skills than any animal other than humans. Unlike most birds, they are able to use their feet independently, using one claw to stand and the other to hold food and they show a tendency to be right-handed (or, rather, right-footed). Non-human primates tend to use their right hand for manipulative tasks (Falk 1993).

Laboratory experiments have shown that when a cat is processing new information, both hemispheres of its brain are equally involved, whereas in humans one hemisphere is nearly always dominant (Sperry 1964). Laboratory rats, with their unequal hemispheres, are more adept at solving problems than the domestic cat. It would appear that specialization, allowing the storage of information and carrying out of specific functions to be allocated to one, rather than both, hemispheres, was an evolutionary solution to the problem of increasing cortical function without increasing the size of the brain.

Lateralization has been shown to be present in some form in many mammals and birds, but did not seem to be able to proceed to any marked degree in quadrupeds, since it was also associated with handedness. Only those able to use half their 'feet' as 'hands' (primates and parrots) had developed this specialization, although quite why is not clear. It might have been thought that ambidexterity such that both hands had the same degree of manipulative skills currently enjoyed by the dominant hand (or claw) would have been a selective advantage? However, it was not to be. Of equal interest was the fact that it would appear some form of vocalization and/or symbolic thinking had evolved in *all* animals which had made this change, birds as well as mammals. The association of these skills appeared to be not merely strong, but compulsive. Chimpanzees exhibit right handedness and the capacity for symbolic thought and communication, at least to the level of a two or three year old human. The New Zealand Kia parrot has shown itself extremely adept at problem solving. Is there any reason to assume that our earliest ancestors, the Australopithecines, would have been any less competent than a chimpanzee or a parrot?

The endocast from the skull of *Homo (Australopithecus) habilis* KNM-ER 1470 revealed a sulcal pattern in the left frontal lobe that looked similar to that associated with Broca's speech area in modern humans, leading Falk (1993) and Savage-Rumbaugh and Rumbaugh (1993) to conclude that *H/A. habilis* had at least the beginnings of speech. What may these beginnings have been?

It has been established that the lowering of the larynx in the throat is associated with flexion of the basicranium (Laitman 1984). Since the basicranium of *Australopithecus* showed no such flexion, it may be concluded that the larynx had not started its descent before 2 million years ago. Full flexion of the basicranium became established about 300,000 to 400,000 y.a., the same time as the hypoglossal canal became fully developed. Nevertheless, the Australopithecines were right-handed and the evidence from rodents, song-birds, parrots and chimpanzees, as well as the Broca's area of KNM-ER 1470, clearly indicate that Australopithecines had lateralization such that they would have been capable of some form of vocalization/verbalization, even if it were not the same as that of modern humans, as well as some degree of enhanced right hemisphere activity, in the form of creative/symbolic thinking.

People from the Canary Islands developed a sophisticated method of communication to convey complex messages across large distances of mountainous terrain. This was done by cupping the hands in front of the mouth and manipulating the shape of the hands and the position of the fingers, resulting in sounds not unlike those of birds, but capable of considerable variation. A system of communication which mimicked bird calls, or other animals, could have been very useful, not merely in communicating with other tribe members over great distances, but doing so without alerting members of other tribes about possible finds of food, especially if fossicking near (or on!) the neighbouring tribe's territory. It may also have deceived predators/prey.

The Australopithecines may not have had sufficient nerve supply to the rib cage and diaphragm for the breath control necessary for modern human speech, but with their larynx in the 'ape' position, they would have been capable of breathing continuously, independently of the need to swallow. Breath control may not have been such an issue for them as it is for us.

With the larynx in the standard mammal position, it may even have been easier for Australopithecines to mimic the roar of a lion than it is for us! The sound of a lion emanating from their camping area, either by day or night, would be a very effective method of keeping many predators at bay. If a bird brained parrot can mimic sounds, is there any reason why an Australopithecine could not?

Along with left lateralization would have come right hemisphere specialization. Far from reading spore being "a little league version of ourselves" as suggested by Tattersall, there is every reason to conclude that the Australopithecine spatial and tracking skills would have been greater than those of other mammals who had not developed lateralization. Their imaginative skills would also have been greater. A possible use for this imagination might be the placing of a male lion head in a conspicuous place, such as a rock outcrop, to warn off predators. Modern Arctic dwellers are able to approach their prey (caribou) by the simple method of disguising themselves by donning a caribou skin circus style, one man in the front as the head and forelegs, another behind as the body and hind legs. The caribou are not alarmed, since the shape and smell of the approaching 'animal' is comfortingly familiar. Just because this technique is used by modern people does not mean that it is necessarily a modern idea. Rubbing itself in another animal's scent to disguise its own is not an uncommon behaviour among animals and there is no reason why an imaginative Australopithecine should not have conceived this, or some similar form of deception, as an aid in hunting.

These are only suggestions. No doubt there are many other possibilities, but surely some use must have been made of hemisphere lateralization when it appeared? Why else would it have been selected and continued to evolve? Upright posture is associated with handedness, the flexion of the basicranium and the lowering of the larynx, the latter change making humans uniquely susceptible to choking. Surely this combination must have produced a highly significant beneficial effect to have compensated for such a dangerous side-effect?

The theory of *Punctuated Equilibrium* better fits human evolution than classical Darwinian gradualism, at least as far as the appearance of *H. erectus* is concerned. *H. erectus* may not have had the innervation to the rib cage and diaphragm that have modern humans, suggesting comparatively reduced language skills, but they did have partial flexion of the basicranium, indicating that the larynx had started its descent. We know that their infants were born with their larynx in 'ape' position, not just because that is the case with our infants, but because if it were not so, their infants would not have been able to suckle. Babies must be able to breathe and swallow at the same time, making choking a hazard for toddlers but not for infants.

The lowering of the larynx had two effects. The first was to make available increased air space in the throat which could be utilized, along with greater control of the tongue and vocal cords, to make a higher number of vocalizations. The second was that it prevented breathing and swallowing at the same time. Tens of thousands of people die each year by choking (Lieberman 1991). Parents are warned not to allow their young children to eat peanuts. Government legislation prohibits the sale of toys whose pieces might cause choking. According to Darwinian theory, nature selects favourable changes. Unfavourable ones are eliminated. Why was the tendency to death by choking not eliminated as an unfavourable random change? If there is no preplanning in nature, it cannot have been because some two million years in the future *Homo sapiens* was going to develop speech and language. It can only have been because there was a favourable condition associated with the change *at that time* which outweighed the disadvantage.

Tattersall (1998) claimed the simple fact that scientists had been unable to demonstrate an advantage to the lowering of the larynx, other than that of the production of speech, did not mean that no such advantage had at one time existed. Acknowledging the evolutionary problem of choking, Tattersall (2002: 166) stated that "there must be some powerful countervailing advantage in the human conformation of the vocal tract, but the ability to speak, unfortunately, is not it". Tattersall concluded (p. 167) "Maybe ... it did bestow certain advantages in the production of more archaic forms of speech ... or maybe it conferred some kind of benefit in terms of respiration". It is difficult to understand how this evolving configuration, with its concomitant danger from choking, could have initially become established because it conferred a respiratory benefit.

To Leiberman (1991) it was clear that adaptation for phonation came at the expense of respiration. He suggested (p. 56) that the adult supralaryngeal vocal tract was preadapted for a new function by changes in shape and position of the larynx and supporting structures. This explanation suggests preplanning/purpose, not allowed under natural selection.

The adult position of the modern human larynx is not achieved until puberty. The human infant is born with the larynx in the ape position. During its descent, the modern human larynx passes through all intermediary positions experienced by our ancestors. A child of between one and two years of age has difficulty pronouncing words and it may be assumed that a similar degree of descent would have placed a similar restriction on our ancestors at that degree of evolution (possibly *H. erectus*). However, against this restriction must be placed the large number of sounds able to be made by a child of that age when considered on a world-wide basis, i.e. all languages now known as well as all those passed into oblivion. By the time a child is three years old, it can articulate fluently and so we may conclude that that degree of flexion and descent is all that is required physically for the production of full language. According to Leiberman the larynx of *H. erectus* was so placed that they would still have been able to breathe and swallow at the same time. However, the human child shows a tendency to 'gag' on its food at about the same age as its first attempts to stand (about six months) in a manner never seen, for example, in a calf or a kitten, which would suggest that choking became a real possibility as soon as any lowering of the larynx took place, which was at about the time that our ancestors first experimented with bipedalism. This would seem to indicate that the ability to choke came very early, even before the ability for a high degree of vocalization, which is very difficult to explain. If flexion is a function of upright stance, choking, being a detrimental factor, should have prevented further development of the fully upright position.

Tattersall (1998: 166) approached the problem of whether or not the Neanderthals had language from the standpoint that "if they had language, they were us in a profound sense". Tattersall, who described himself as "a skeptic at heart" (Tattersall and Schwartz 2001: 13) interpreted all evidence from the point of view that the Neanderthals were not "us" but "a creature that possessed another sensibility entirely" (Tattersall 1999: 153). His conclusion that the Neanderthals did not have language was made, not on physiological evidence, but on the assumption that Neanderthals were incapable of symbolic thought (Tattersall 1998: 186): "Language both permits and requires an ability to produce symbols in the mind ... Thought as we know it depends on the mental manipulation of such symbols". Tattersall (1998: 172, 186) held that Neanderthals did not have full basicranial flexion. In the Neanderthals there may have been a reversal of laryngeal descent compared with *Homo heidelbergensis*.

Mithen had suggested that because the lives of early people were much the same day after day they would have little to talk about, yet most human daily conversation is mundane – the weather, little details about what one has done, what one had for dinner last night, whatever! Conversation probably revolved around the day's foraging and may have seen the beginning of the first 'tall story' as accounts were given of previous hunts, both narrow escapes and successful kills. That emerging right hemisphere imagination was no doubt put to good use!

Most writers agree that the Neanderthals had all the physical prerequisites for language – fully flexed basicranium, a hyoid bone, full innervation to the tongue and respiratory system as evidenced by the size of the relevant foramen, and hemispherical laterality. It is generally conceded that early reconstructions which portrayed the Neanderthal basicranium as poorly flexed were in error. The Neanderthals were not only right handed, they were right 'armed'. There is evidence that their right arms were developed more than their left in a manner associated today with tennis players (Trinkaus et al. 1994: 29). Taking all these things together, there would seem to be no logical or scientific justification for denying Neanderthals full speech, whether or not their syntax and grammar were similar to ours.

As adults, we are so used to thinking in words it is difficult to imagine thinking without them, and yet this is possible. Babies, not yet linguate, are clearly able to think, as are the deaf. Today much is done to teach speech to those born deaf, but in the past many grew up dumb as well as deaf. There is no doubt that these people had a more limited intellectual development than those not thus handicapped. The conceptual basis of education is that it stimulates the brain to make more neurone connections and that those who do not receive this stimulation do not reach their full potential. Our early ancestors would have received plenty of stimulation, although somewhat different from that which most of us experience today. At time of contact, Australian aborigines were living a 'stone age' existence, but they were not only fully linguate, they had a rich culture of story and myth (imagination). Their tracking skills amazed the white settlers, whose 'civilized' (i.e. city/town/village dwelling) ancestors had lost these abilities. The Neanderthals were alive in Europe at the time the earliest modern humans arrived in Australia, and they were living a similar existence. There is no evidence to support the suggestion that they had any less abilities, vocal or otherwise.

Part V

RAMIFICATIONS

Beyond 2000 - The First Decade

Truth - the Daughter of

Time?

Chapter 39

Palæolithic People of
Portly Proportions

Early in my studies, I chanced upon a book, *The Archæology of Disease*, by Charlotte Roberts and Keith Manchester, first published in 1993, which took a fascinating look at diseases of the past as evidenced by skeletal remains. It addressed a number of areas of pathology: congenital disease, dental, trauma, infectious, metabolic, endocrine and neoplastic, but the one that took my attention occurred in the chapter on joint disease. One of the pathologies listed was diffuse idiopathic skeletal hyperostosis, appropriately abbreviated to DISH, because it was a pathology associated with over-eating. It was found most frequently in graveyard remains associated with monasteries, supporting the legend of the overweight monk, which perhaps was not a legend at all, but fact! Although all of the other cases cited in the book referred to remains from historic times, the authors could not resist calling attention to the earliest known case of DISH, found in the remains of a Neanderthal dated to 45,000-50,000 ya.

As a retired natural therapist and psychologist, this case immediately took my attention. Eating, diet, these are issues both of the body and the mind. I realized that it should be possible to learn more about the lives of ancient people from their bones than just their species. For my Masters thesis, I investigated a number of pathologies associated with Pleistocene people and drew from them inferences about the lifestyles and attitudes of these people which supplemented the information already gathered by other means. There were not many cases. For pathology to affect the bone, it has to be long standing (chronic). Acute diseases leave no trace upon the skeleton.

Submitted in 2004, material from my Masters thesis provides the starting point for the final part of this book, which looks at new information and at changing attitudes during the first decade of the 21st century.

The Neanderthal people may have lingered beyond 30,000 ya, possibly until as recently as 26,000 B.C., but the era from 32,000-15,000 B.C. was determined to have belonged to the Cro-Magnon and was termed the Upper Palæolithic. This time was subdivided, according to tool culture, into the Châtelperronian, Aurignacian, Gravettian, Soloutrian and Magdalenian. The last subdivision, Azilian, borders the end of the Upper Palæolithic and the beginning of the Neolithic, 10,000 B.C., which also marked the beginning of the Holocene era - our own time. The last, and most severe, Ice Age lasted from 30,000 B.C. until 10,000 B.C., when it was officially deemed to be over. The coldest times were between 25,000-15,000 B.C. There is surprisingly little written about the five thousand year period between 15,000-10,000 B.C.

By this time, art in the form of drawings, paintings, beads and other ornaments had become universal across all known cultures. The art of Europe, especially the magnificent cave paintings of France, is particularly well known but, unfortunately, none of this artwork portrays any human figure, which is quite extraordinary. These paintings may be viewed on the internet and, since they do not really tell us anything about the people who painted them, other than that they were skilled artists, I will not spend time considering them any further. Rather, I turn attention to the clay figurines, often known as the Venus figurines because most of them depicted females, since, in many ways, they tell us so much more about the people of the Upper Palæolithic.

The Venus figurines were made between 25,000-15,000 ya, thousands of years after the earliest cave paintings were created. Dating from the late Aurignacian period of the Upper Palæolithic, through the Gravettian and Solutrean periods and into the Magdalenian, these figurines were a mixture of realism and stylization. Unlike their cave painting ancestors, these Upper Palæolithic people had no qualms about representing the human form. Indeed, nearly all of their representations were of the human form, only a few of animals. None were of flowers or trees. It is interesting that these figurines provide some of the oldest surviving examples of the firing of clay. It might have been expected that the earliest use of this new technology would be for the manufacture of utilitarian items, such as water pots or bowls, but our early ancestors preferred to produce items of artistic or symbolic value. It may be that the people of the Upper Palæolithic were happy with the water carriers they had, whatever they were. Treated skin? Old cranial bones? Nevertheless, that the sole use, in Europe, as far as we know, of the firing of clay should have been for the making of ornaments and objects of art and/or ritual, both human and animal forms, may tell us more about the attitudes and thinking of early humans than do the figurines themselves. The importance of representation, symbolic or naturalistic, runs very deep in the human psyche. If these figurines served any type of 'ritual' purpose, that would indicate philosophical thinking of a type different from that which may have inspired east/west Neanderthal burials since it would imply some attempt to control some aspect of life through 'displaced' activity.

Humans had been using fire for tens of thousands of years, but, nevertheless, one can imagine that those living in Ice Age Europe may have maintained their fires in a way that other people may not. Hunter-gatherers make a fire, use it, let it burn out and move on. Europeans were more settled. There is evidence of huts from far earlier times and the digging of not one, but several, graves at the one site, such as at La Ferrassie, definitely suggests permanent settlement. Keeping fires burning may have led to the accidental discovery of the process of 'firing' clay.

The most noticeable thing about these figurines is that so many of them were obese, although a few slender figurines have been found, showing that excess fat was not universal. Duhard (1991) analyzed the degree of adiposity exhibited by figurines from south-west France, dating from the Gravettian period (25,000-20,000 B.P.), choosing these for his study, not only because they were among the oldest images known of women, but because they were all carved in low relief or in the round, which allowed a better appreciation of their shape. In addition, they were not as schematic as some of the later images tended to be, notably those from the Magdalenian period. The great majority of the human depictions were of women.

Duhard found that of the twenty-three figurines he studied, eleven exhibited obesity, four showed posterior steatophygia (large deposits of fat on the buttocks), one showed hypermastia but the lowest part of the figure was missing so that it was not possible to make a full assessment, one was described as 'of extreme gynoid form and normal adiposity', while the remaining eight were slim or normal. Thus a sufficient number of the figurines were not obese to show that obesity was not universal and was not a physiological feature to which these women were born. It was a condition which developed over time in some people, a situation similar to that which exists among modern peoples today.

Figurines were made over a vast area of what is today Europe and northern Asia, from France in the west to Russia in the east, from Italy in the south to Siberia in the northeast. Dating of the figurines has proved difficult in many cases because they were chance finds, removed from their provenance and often not coming to the attention of professional archæologists until sometime later. Their manufacture spans thousands of years. Covering such an expanse of time and space, it is not to be expected that all figurines would have been made for the same reason, although there is a surprising conformity in style, including a lack of feet, possibly because the figurines were made to stand upright by being stuck into the ground, and a lack of facial features. Portly proportions were evident in many cases but were not universal, the sample study carried out by Duhard being typical of finds from other areas and other times.

Of interest here is not the reasons for the manufacture of these figurines, be they representations of individual women, tribal goddesses or anonymous representations, but the fact that so many of the figurines represented people who were overweight or obese, the obesity being mainly of the gynoid type (see Fig. 39.1). Some of the figures appear to be pregnant, although, then as now, pregnancy is not necessarily accompanied by obesity. While drooping breasts may be explained as that of multiparous and/or lactating women, multiple births alone do not cause this condition. On the contrary, many women find that their breast size reduces after breast feeding and may reduce further with each child suckled. None of the women were depicted holding a baby or young child which raises the question of how likely it is that these figurines were any form of fertility idol? What does seem apparent is that these people were well fed.

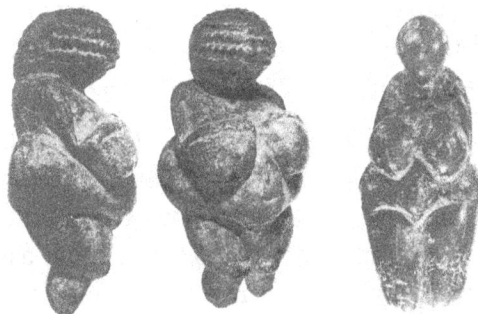

Fig 39.1: Venus figurines
(Delporte 1979: 138 & 161)

While modern Western societies promote slimness as desirable, this point of view is by no means universal. Societies, such as those of some South Pacific island communities, took pride in body weight. In most cases, it was the wives who brought honour to their husbands by the acquisition of weight, thus showing their husbands to be good providers and/or that they were elite members of their society who did not need to work. Some male rulers also boasted very heavy body weights.

A common misperception by many people in Western societies is that hunter-gatherers, who have few permanent possessions, live in a state of poverty (Sahlins 1972, Ferrara 2001). Paucity of permanent possessions is a necessity for people frequently on the move, since carrying possessions would not only be an unnecessary burden, it would also hamper a spontaneous hunt, for which these people must always be ready. An extrapolation frequently made from the assumption of poverty is that these people must also be in a permanent state of hunger or, at the very least, a state of anxiety about the acquisition of their next meal (Sahlins p.1):

> Almost universally committed to the proposition that life was hard in the Paleolithic, our text books compete to convey a sense of impending doom, leaving one to wonder not only how hunters managed to live, but whether, after all, this was living.

The negative picture is almost certainly far from the truth since hunter-gatherer peoples know when and where food will be available and the occasional day or two without a great deal of food does not concern them unduly because they know there will soon be more (Sahlins 1972; Tanaka 1976; Turnbull 1984). Roth never spoke of the Tasmanian aborigines, among whom he lived for so many years, suffering hardship from lack of food. Only when traditional grounds had been invaded by Westerners and subsistence patterns, which had been sufficient for millennia, were disturbed did these people routinely face hunger.

Since the acquisition of too much weight by many people in Western societies is a prime concern of health professionals, it might be thought that gaining weight is a simple, easy process. This is not the case.

When a person starts to eat too much, to take in more calories than their body can use immediately, the excess energy causes an increase in metabolic rate which results in some of the excess calories being burnt off (Garrow et al. 1993). Contrary to popular belief, overweight people have higher than normal metabolic rates. Younger people can eat more than they need without putting on weight until their metabolic rate has increased to its maximum. After that, weight starts to accumulate. It takes time (and effort) to become overweight or obese. Occasional over-eating on festive occasions is not sufficient. It is necessary to eat too much on a regular basis over a long period of time, at least months, usually years, to become overweight, let alone obese. There is very little evidence that inactivity in adults is the cause of obesity. However, overweight or obese people generally avoid activity because of their condition.

Although much media attention is focused on the role of fat in the diet and its role in weight increase, a distinction is not always made between the fat which occurs in foods which form a natural part of the human diet, which our bodies are able to digest, and those which have been altered by processing. Our bodies are able to metabolize meat and eggs, as would be expected in a mammal which has evolved over millions of years to consume these items. The meat of wild animals is higher in protein and lower in fat than that of domestic animals. The intake of fat by the people of the Upper Palæolithic would have been considerably lower than that of people living in Europe in Holocene times, before agriculture and husbandry became established, since they would not have consumed butter, cheese or cream obtained from domestic cattle.

It has been estimated that the palæolithic diet contained five to ten times as much fibre as the present Western diet due to the consumption of wild vegetables, including roots, which would have been high in carbohydrates (Goode 1988). However, the major contribution of carbohydrates to current diets comes from grains which were not domesticated, and did not contribute greatly to the human diet, until Holocene times. Their consumption, in association with sugar, which is a comparatively recent addition to the human diet, has vastly increased in recent times, as has the size of our collective waistline. The Upper Palæolithic diet would have consisted of meat, fish, eggs, nuts, roots, flowers, berries, honey and some fruits, which would have been wild and far less sweet than the domesticated varieties available today. Some wild grains may also have been eaten. The Holocene settlements of Eastern Europe, where cereal crops are known to have been cultivated, also left a legacy of female figurines, many of which also displayed adiposity, especially of the gynoid type (Gimbutas 1974). The possibility that wild cereals formed a substantial part of the Upper Palæolithic diet must be considered.

It is known that the consumption of alcohol will increase adiposity, especially around the middle of the body. Is there any reason to assume that people of this time did not have access to alcohol? Many bushes and trees fruit prolifically for a short time and the gathering of their fruits may well have involved the 'one for me, one for the basket' system of collection which is still prevalent today when collecting fruits, such as blackberries, from the wild. The supply of fruits would have been far greater in relation to the size of the population than today and they may well have collected more than they could eat. Left over fruit and their juices may well have fermented and been consumed, possibly on return to a home site temporarily abandoned during a seasonal move. We know that the Tasmanian aborigines made alcohol in a similar way.

450

Fermented honey (mead) was a popular beverage in Europe in mediæval times. Turnbull (1988: 243-244) wrote of the pigmy people of central Africa who, for two months of the year, forsook hunting animals to concentrate on foraging for honey, which they ate "grubs, lavæ, bees and all". Some of the honey fermented, "tasting like a bitter liquor; if eaten in any quantity, it could be highly intoxicating". If hunter-gatherers of recent times enjoyed fermented products, is there any reason to suppose that our more ancient ancestors may not have done so also?

Did these Upper Pleistocene people deliberately store fat to protect them from the cold conditions then prevailing? This idea has some immediate appeal but is impossible to substantiate. People living today in the colder parts of Europe, such as Scandinavia, are not overweight, nor are those from Siberia, northern China or northern Canada. Overweight has traditionally been more of a problem for those living in warmer climates, possibly because they are not metabolizing energy to keep warm. This makes the obese conditions demonstrated by the Venus figurines all the more surprising.

It must not be thought that, just because the climate was much cooler than it is today, Europe was a barren or desolate place. The southern parts were well vegetated and supported a wealth of animals, including the mammoth, woolly rhinoceros, horse, bison, reindeer, wolf and fox. While most of these larger animals were vegetarian, there must have been a wealth of smaller animals to support the wolves and the foxes. Omnivorous humans would have been well provided for. Indeed, the evidence would seem to suggest an over-abundance of food and too little energy expended in acquiring it. There is post-hole evidence of substantial huts having been built in Europe hundreds of thousands of years ago (see Chapter 41). A communal hut, a roaring fire, plenty of food – and drink? – life must have been good for these people. They certainly had time to sit and whittle, or make clay figurines. They may have carved wood, indeed it would be surprising if they did not, but, of course, no evidence of this remains, so this must remain speculation.

The Upper Pleistocene man may have needed to control his weight to fulfill his role as a hunter. It is possible that a plump wife brought honour to her husband and was much valued. Overweight may have been a form of protection for females against being carried off by marauding men from neighbouring tribes. Don't laugh – this was a serious matter! If weight was prized, was it a prize shared by all, or were some well-fed at the expense of others? Were these societies ones of caring and sharing or of the greedy and the needy?

Chapter 40

Dishing Up the
Fat of the Land

It is of interest to note that, as one goes back in time, it is not until about 45,000 to 50,000 B.P. that one finds the first skeletal evidence of a life-style induced disease in Pleistocene people. There are plenty of diseased skeletons dating from the Holocene epoch; indeed, as farming and the cultivation of foods, especially cereals, became more widespread, population numbers increased but the standard of health went down and the mean age-at-death became lower (Larsen 1995). While the artistic representations of humans living during the Last Glacial give evidence that some people at least 'lived well' and put on an excessive amount of weight as a result, no skeletal evidence from that time has been found of a life-style induced pathology.

This, and the next three chapters, will consider, not only life-style related pathologies evident from Pleistocene remains, but the information these pathologies provide about the probable lives of these people.

The skeletal remains to be studied now are those of a Neanderthal, known as Shanidar 1, from Shanidar Cave in the north-eastern side of the Shanidar Valley in Kurdistan, Iraq. Discovered in 1957 by Ralph Solecki, this skeleton was radiocarbon dated to 46,900 ± 3,000 years (Crubézy et al. 1992).

Shandiar 1 is a male skeleton, morphologically similar to the 'classic' Neanderthal from Europe and estimated by Trinkaus (1983) to have been not less than 30, not more than 45 at the time of his death, which was brought about by a rock fall. It is now generally agreed that aging of skeletal remains has had a tendency to be conservative. What is certain is that this was not a 'young' man. He was at least 'middle aged', probably at least 40, possibly even 50 (or more?). We cannot be sure.

During his lifetime, Shanidar 1 had suffered many traumas and degenerative conditions, including the following:

> 1. Withered right clavicle, scapular and humerus, with a partially healed osteomyelitis on the clavicle, a mid-distal shaft fracture of the humerus, resulting in a pseudo-arthrosis and/or amputation of the distal humeral epiphysis, forearm and hand;
> 2. Ante-mortem right frontal scalp wounds;
> 3. Crushing fracture to the left lateral orbit and zygoma;
> 4. Fracture of the right metatarsal 5 diaphysis;
> 5. Degenerative joint disease involving periarticular osteophytes and/or subchondral degeneration of the right knee and right distal tibiofibular, talocrural and tarsometatarsal 1 and 2 articulations;
> 6. Asymmetry of the patellae and tali;
> 7. Abnormal curvatures of left tibia and fibula, probably resulting in secondary remodelling from post-traumatic alteration of gait;
> 8. Bilateral degeneration of the temporomandibular joints and bilateral auditory exostoses.
>
> (Crubézy and Trinkaus 1992)

The damage sustained to the left orbit, which had healed, indicated that the man had probably been blinded in the left eye. Some of his injuries had been sustained in childhood. Solecki (p. 195) considered this proof that "his people were not lacking in compassion". Notwithstanding this evidence of compassion, Solecki also believed that Shanidar 1 "had been born into a savage and brutal environment". Solecki was writing before the creation of the wealth of wild-life documentaries which have graced our television screens over the last few decades. We have been privileged to witness, *via* the camera lens, the lives of many animals and have seen the care for members of their own family which many express. Caring is not a universal trait: herd animals do not seem to make friends of their companions, but those living in smaller, family groups, form strong bonds. Why should humans have been any different? Without doubt, humans are the most cruel creatures on the planet at this point in time. Has it ever been so? Possibly, but why should we be asked to assume that the Neanderthals were savage and brutal just because a small minority of people alive today behave in that way?

Trinkaus (p. 75) used words which gave a slightly different implication from those used by Solecki:

> The risk of injury appears to have been high among the Neanderthals for those individuals who lived to an old age. Every currently known, reasonably complete partial skeleton of an elderly Neanderthal shows evidence of trauma ... This elevated risk suggests that life among Neanderthals was indeed harsh and dangerous.

How many of us go through life without sustaining some trauma? I have broken my right wrist three times yet I have not led a 'harsh and dangerous' life. Many people, especially men, sustain injuries when participating in sporting events and there is no reason why Neanderthal men should not also have indulged in friendly combat, as well as in less friendly combat, of course, if the need arose. Considering the number of wars and battles being fought on the face of this planet even at this very moment, is there any reason to suppose that life among the Neanderthals was any more brutal, harsh or dangerous than that lived by many people today?

Some look back on the past with rose-coloured glasses, believing life in a village, or as a member of a tribe, to have been more simple, more contented; others look back on the past as being more cruel, more difficult. It was probably all of those things, since people are all of those things.

As if the pathologies listed above were not enough, Crubésky and Trinkaus found that the skeleton of Shanidar 1 presented a number of enthesopathic growths in a pattern consistent with a diagnosis of diffuse idiopathic skeletal hyperostosis (DISH). This skeleton provides the earliest known case of this disease to date. The skeleton known as Shanidar 4 is older and also exhibits two lesions suggestive of DISH but the remains are too fragmentary for a definite diagnosis to be made.

DISH, also known as Forestier's disease and ankylosing hyperostosis, is a disease of unknown ætiology, seen in middle-aged and elderly persons (Ustinger 1985). Although described as a skeletal disease, the pathology manifests in the ligaments and tendons, which calcify and the ossification is preserved on the anteriolateral aspect of the spinal column and in regions of ligament and tendon attachment to bone (enthesis) leading to the occurrence of enthesal ossification and bony spurs, although the ligament remains free in front of the actual vertebral bodies. The ligament becomes incorporated in the new bone which is integrated with the cortex and peripheral parts of the disc. The calcification of the ligaments gives a characteristic 'candle-wax' appearance to the spinal column, which is diagnostic. The discs between the vertebræ are not affected by this disease and the lack of fusion of the vertebræ to each other, accompanied by evidence of fusion by the bony 'overlay', is a vital diagnostic criteria. The location of the bony spurs, indicating that they did, in fact, originate as calcification of the ligaments/tendons, also aids in diagnosis.

Because the primary pathology of DISH is calcification of ligaments and tendons, not bone, its initial manifestation is stiffness, rather than pain or

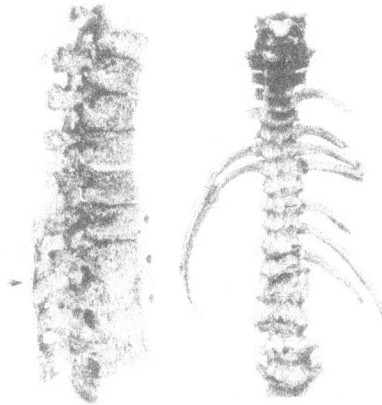

Fig 40.1: Diffuse idiopatic skeletal hyperostosis of the spine
showing diagnostic 'candle-wax' new bone formation
(Roberts and Manchester 1997: 119-120)

immobility. Stiffness occurs primarily in the area of the dorsal spine, where the pathology is most likely first to occur. It is most marked morning and evening, after a reasonably stiffness-free day. Symptoms are aggravated by inactivity and/or cold and wet weather. Pain develops as the disease progresses and the calcification begins to involve the spinal vertebræ, resulting in some degree of immobility.

As mentioned above, the ætiology (cause/development) of DISH is unknown. However, three factors have been shown to be associated with the disease: age, obesity and mature-onset (Type II) diabetes. DISH occurs at least twice as frequently in males as it does in females. It is unknown in those below 40 years of age and most commonly presents in the sixth and seventh decades of life. At one time this syndrome was known as 'senile ankylosing spondylitis/hyperostosis' but the term 'senile' was dropped when X-ray examinations, taken for another reason, showed that DISH could start developing in younger people, i.e. in their 50s, or even in their late 40s, while the patient was still asymptomatic. There is a sharp rise in incidence with increasing age.

In addition to the frequent association of DISH with obesity, there is an association with hypertension (high blood pressure), coronary artery disease and diabetes mellitus (Type II diabetes) (Denko 1994).

In view of its associations with conditions of over-weight, it is not surprising that palæopathologists have found that DISH occurred more frequently in people of higher social backgrounds - an Egyptian priest, a Saxon bishop, a mediaeval dean (Rogers et al. 1985). Today, obesity is common among people of lower socio-economic groups, fast-foods being a convenient choice of diet among many families in which both partners need to work.

The high carbohydrate diet based on various cereal crops, such as wheat, corn and rice, consumed by many people today, was not available to Shanidar 1, whose diet would have been more similar to that of the Upper Pleistocene people we considered in the last chapter – meat, eggs, nuts, roots, berries, some wild (not very sweet) fruit and honey. There is little evidence that the Neanderthals ate fish, but fish bones do not fossilize as do animal bones, so their lack cannot be taken as evidence that the Neanderthals did not catch fish. Why would they not? This high protein diet resembles the diet recommended by Dr. Atkins to lose weight, making the obesity of both this Neanderthal and Upper Pleistocene 'Venus' people all the more intriguing.

While all people who develop Type II diabetes are overweight or obese, not all people who are overweight or obese develop Type II diabetes. While the overweight condition of Shanidar 1 is confirmed by his diagnosis of DISH, it cannot be assumed that he also had Type II diabetes or heart problems. Wild fruits contain considerably less sugar than the cultivated fruits of today. The only item of his diet which may be thought to have induced an overweight condition is honey. Perhaps he became addicted? Humans today tend to like sweet foods. So do dogs – if they are given a chance – yet their natural diet is exclusively meat. I have seen documentaries showing meat-eating animals in Zoos and Reserves being given a cake on their birthday and it is surprising how willing many of them seem to be to 'tuck in'. These things are hard to explain, but, fortunately, here is not the place to try to offer explanations, merely to note the facts!

Roots, which formed a staple part of the traditional hunter-gatherer diet, are high in carbohydrate and sugars, substances required for the functioning of the brain. The sugar from cane, which features so highly in western diets today, would not have been consumed by Neanderthals, but sugar is also available from a root vegetable – sugar beet. The cultivated beet recommended by Dr. Erasmus Darwin, would not have been available either, but is mentioned again here to draw attention to the fact that roots of various vegetables do provide sugar, and carbohydrate, which is converted into a sugar by the body. Less active than his peers, perhaps Shanidar 1 munched his way through heaps of roots to while away the time?

Living as he did in what is modern day Iraq, the climate he enjoyed would have been more equitable than that experienced by his contemporary Neanderthals living in more northerly Europe. Do not think of Shanidar as Middle-Eastern desert. The Zagros mountains, in which is found Shanidar Valley, are fertile and beautiful. "Up in the higher hills are found acres of narcissi, violets, buttercups, orchids, tulips, roses and tiger lilies. There is hardly a single traveler who does not come away amazed by the wide-spread covering of flowers" Solecki (p. 78). Such a wealth of vegetation would have attracted plenty of animal life: insects, birds, snakes, lizards, rodents, wild goat, wild sheep, wild cattle, wild pig, tortoises (Solecki 1971; Morean and Kim 1998; Shea 1998). Western people tend to look down on the consumption of insects, but other cultures do not and, even today, the French eat snails – and frogs' legs! According to Binford (1985) pollen traces left on flake tools show that the Neanderthals at Combe Granal, in southwestern France, included generous amounts of aquatic plants plucked from the canyon stream in their diet. Bryant and Williams-Dean (1973) made a study of the coprolites of the people who lived in the Texas area of America from 1,300 B.C. to 800 A.D. and found evidence that flowers, leaves and even the bark of trees had been sucked, chewed, brewed or eaten. With such a wealth of flowers at Shanidar, even today, surely flowers can be added to the presumed diet of its Neanderthal inhabitants?

Whatever it was that Shanidar 1 ate, there seems to have been plenty of it!

If an animal cannot find sufficient food, it dies. It does not try to eat the wrong food. When plenty of food is available, it eats sufficient and then stops. Only humans eat inappropriate foods detrimental to their health; only humans regularly eat too much food to the detriment of their health. If the Neanderthals behaved in this way, the Neanderthals were fully human.

Before leaving the study of the remains of this very interesting individual, there is one further life-style induced pathology which needs to be considered. Trinkaus (1982) believed that the crania of both Shanidar 1 and Shanidar 5 showed evidence of having been artificially deformed. The cranium of Shanidar 5 needed considerable reconstruction but the skull of Shanidar 1 was in comparatively good condition. Besides being fragmentary, the cranial remains of Shanidar 5 showed evidence of a healed scalp wound. It was Trinkaus' (p. 198) opinion that both crania showed evidence of "frontal flattening, high parietal curvature and elevation of the parietal region associated with artificial cranial deformation".

The altering of one's appearance in some way is so widespread, both in time and place, that it may be questioned whether any human society has ever existed which has not participated in this practice. It may be by way of clothing, either utilitarian or fanciful, by body painting, ornamentation, scarification or by some other form of mutilation or deformation. It must be assumed that the Neanderthals, who lived in Europe during some of its coldest times (albeit in the warmer parts thereof) had effective, if utilitarian, clothing. We do not know if their clothing served any other purpose, whether it was decorated or decorative in any way to satisfy an æsthetic sense or to indicate superior status, either within or between groups. We do not know if they wore fancy head-dresses, although some form of head covering is highly likely, especially in the winter among the European Neanderthals, even if only to keep their ears warm. We do know that they used ochre and were by no means the first people to do so (Bednarik 2001, Marshack 1990, 1996, 1997).

There is evidence for the acquisition of objects, such as crystals, for their æsthetic value from times as early as those of *Homo erectus* (Bednarik 1992). A flaked bone point and wolf incisor from Austria, both perforated, have been dated at nearly 300,000 B.P. and a small pebble from Israel, engraved to enhance its similarity to the female shape, has been estimated to be about 200,000 years old (Chakravarty and Bednarik 1997). Pigment (ochre or hæmatite) have been found at a number of Acheulian and later sites, including Terra Amata (France), Olduvai (Tanzania), Bambata Cave (Zimbabwe) and Klaisies River Mouth (South Africa) (Bednarik 1992). The use of ochre is both widespread and ancient.

It is extremely unlikely that any people expressed their first artistic or æsthetic feelings through media as difficult to manipulate as those of bone or stone. The earliest manifestation of such expression will almost certainly have involved items such as feathers, flowers or wood, or drawings in mud or sand, which will have left no trace. While evidence for the early expression of art may seem limited, it almost certainly is meagre compared with that which came and went without trace.

Some birds appear to show an æsthetic sense when they collect items which they consider impressive, taking them back to their nests to impress the female of their choice. However, the production of art (portable, standing, mural, body decoration or any other form) is a purely human activity. More and more items are being discovered which are believed by some to be deliberate patterns (art) produced by people who definitely predated modern humans, while others believe these marks to be accidental. Art incorporating 'perspective' would seem to have made its first appearance about 30,000 ya. However, lack of ability to produce art of this calibre does not imply lack of full modern human status. There are millions of humans alive today who 'have difficulty even drawing a straight line with the help of a ruler' or whose 'musical' ability is restricted to the singing of one note! Why some people should be gifted in art, in music, or, as frequently is the case, in both, while others of us are artistically 'challenged' is difficult to understand. Why has not 'positive selection' spread these talents more universally? What is important to note here is that evidence of these abilities is evidence of the presence of a human mind; absence of evidence of these abilities is not evidence for the absence of a human mind.

If Shanidar 1 was indeed subjected to artificial cranial deformation, this would be the earliest example of this category of physical æsthetics. Other forms of bodily distortion recorded ethnographically have involved the stretching of the neck or of the lips by the use of metal rings, popular among some tribes in Africa, and deformation of the feet (of women) by binding, practiced in China. While the former does not take place until puberty, the latter must be practiced on infants or young children, before the bones are fully formed. The 'victim' had no choice. The deformation (or reformation as it may be viewed) may be undertaken purely for reason of æsthetics, or for religious or elitist reasons, to distinguish the person from another tribe or society or to assert his or her position within the society or tribe.

In his classic work, Dingwall (1931) recorded examples of artificial cranial deformation from all over the world: Europe, Asia, Africa, the Americas and the Pacific Islands, although he had found no evidence of its practice in Australia or New Zealand. However, Macgillivary (1852) spoke of Aboriginals of Cape York Peninsula (the Kowrarega and Gudang tribes) who applied pressure to the forehead and occiput of their babies, thereby flattening both areas and causing the skull to be broader and longer than it would naturally have been.

Liddell (1991: 14) recorded cranial deformation of infants which occurred among the Muralug Islanders of the Torres Strait. When the baby was but three days old, the mother would gently press the child's forehead with the palm of her hand, while it was sleeping. This practice would be continued for several hours each day until the child's forehead became flat. This was done to imitate the flat head of the local sucker fish, with which the Muralug tribe had a 'working' relationship. They attached string made of coconut fibres to the fish. When it attached itself to a turtle, they hauled in both fish and turtle.

Dingwall noted that it was not always a simple matter to distinguish between intentional deforming and accidental. A baby may be lain on its back to sleep for months, resulting in the flattening of the occipital region, but this may not have been the intention of its mother.

Among the many cases of deformation, Dingwall included the placing of the child's head in some form of cap. In many areas of Europe, these were required to keep the infant's head warm, but many were tied into place quite firmly, which had the effect of leaving permanent ridges on the skull. Some of the caps were quite substantial, being several layers thick, and they did have the effect of causing the skull to become elongated, although Dingwall was not able to determine whether the deformation was intentional, or whether the resulting head shape was considered 'normal' because everybody had it.

All over the world, it was common practice for the newborn child's skull to be massaged to eliminate any misshaping which had occurred during birth. In a few places, there were deliberate attempts to mould the child's head into a round shape. Where there was a clear attempt to deform the skull, flattening of the occiput, the forehead, or both, or the extending of the cranium outwards, or upwards, was the rule. The length to which some people were prepared to go to produce in their young the shape of head considered desirable were sometimes truly remarkable.

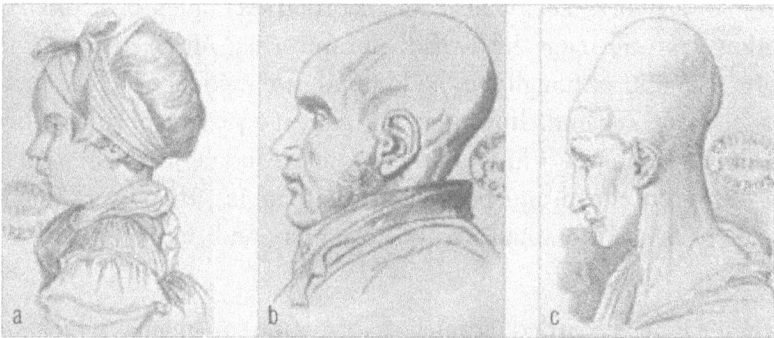

Fig.40.2, a, b and c: Results of head bandaging from France.
(Note deformed ear in c, a common concomitant of head bandaging.)
(Dingwall 1931: 46)

Dingwall repeatedly mentions the protruding eyes of babies being subjected to cranial pressure from boards or tight bandaging. There was bruising around the eye area, bleeding from the nose and reports of continuous excretion of a white mucous from the eyes, nose and ears. Some communities only bound the heads of their babies when they were asleep, removing the boards if the discomfort awakened the child.

460

In some areas, flattening was intended to broaden the forehead, in others the prime purpose was to push it backwards, so that the angle of the forehead followed that of the nose. This shape was particularly favoured in the Americas.

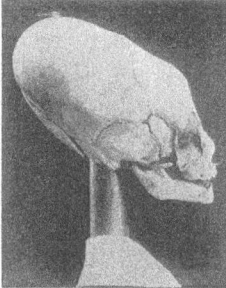
Fig.40.3: Chelebes, Indonesia
(Dingwall 1931: 170)

Fig.40.4: Chinook American
Indian
(Dingwall 1931: 170)

While cranial deformation was practiced extensively in mainland Europe during historical times, it did not appear to have been practiced in England, except possibly in London.

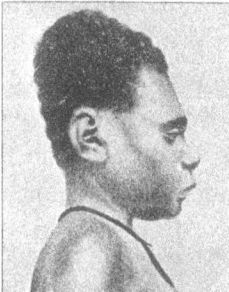
Fig.40.5: A boy from New
Britian
(Dingwall 1931: 146)

Fig.40.6: New Hebrides (now
Vanuatu)
(Dingwall 1931: 146)

The dolichocephalic head was favoured, not only in Europe, but in Egypt and the Levant. If this was the case in Neanderthal times, the Neanderthals had an advantage. If they were aware of anatomically modern humans (AMH) in the Levant at that time (and there is no reason why they should not have been), then they did not envy the more rounded shape of their contemporaries, but appear to have tried further to exaggerate their own flattened and elongated shape, which may be indicative of either elitism or separatism. In later times, it was the anatomically modern humans who appear to have tried to reproduce the shape of the Neanderthal skull, not the other way around!

An infant's skull is very pliable. Any attempt permanently to change its shape must be continued for some months, otherwise it will revert to its natural shape. Some deliberate shaping is induced by the application of a band or a board; other cultures use manual reshaping by the placing of one hand on the forehead and the other beneath the occiput. Trinkaus (p. 199) concluded that it was the manual technique which had been used at Shanidar. This would certainly be indicative of 'forward planning' and an ability, not only to plan ahead, but to persist in a plan of action over a

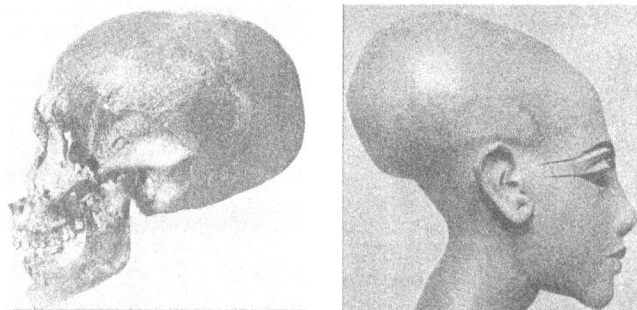

Fig.40.7: Skulls of Aknaten (a) and his daughter (b)
(Dingwall 1931: pp.102, 104)

considerable period of time. However, it is possible that the small degree of deformation evident in Shanidar 1 may have been produced by the wearing of some form of cap in infancy, especially if it was held in place by some kind of band. Against this theory is the fact that head covering would have been needed more in the colder European climate than it was in the milder one at Shanidar. However, to date, no evidence of cranial deformation has been found among European Neanderthals.

Trinkaus (1983a) noted the coincidence in time of the possible appearance of cranial deformation and of intentional burial, which was also first practiced by the Neanderthals. It cannot be known whether intentional burial was associated with religious belief or whether it was a matter of practicality. Certainly, lack of intentional burial does not signify lack of religion or philosophical thought. Some cultures cremated their dead, others left them 'exposed' to animals, and these practices continued into modern times.

Nevertheless, deliberate burial of the dead, the use of ochre and of clothing, along with evidence of the much earlier production of simple, decorated artifacts, is clear evidence that the Neanderthal people did possess an æsthetic sense and this suggests that they may have been capable of philosophic thought. The possibility that some, at least, of the

462

Neanderthal tribes practiced cranial deformation, even to a minimal degree, opens up the possibility that they may have had some form of elitist structure in their society. All of these things, together with their less than desirable eating habits, make it difficult to argue that these people were not fully human. The question which still remained unanswered was: were they the same species as us, or another species of the same genus, homo?

Chapter 41

Where There's Fire,
There's Smoke

Attention now turns to France, to the remains of La Ferrassie 1, who, living between 50,000 to 75,000 B.P., pre-dated the people of Shanidar by 10,000 to 15,000 years. While his exact age at death cannot be determined, it is accepted that this man was old when he died. He is believed to be the oldest 'classic' Neanderthal whose remains have been uncovered, both in respect to his age and in the time at which he lived.

His remains were discovered in September, 1910, by Peyrony and Capitan in a rock shelter in the Dordogne region of France. Subsequently, a further five skeletons were uncovered between August 1912 and June 1921, World War I having halted excavations for several years. The burials appear to have been ritualistic, the people having been placed in a crouched position, the cave debris of calcareous rubble and light brown deposits of flint waste being accompanied by food debris, burnt bone and ash (Oakley et al. 1971). Today the area is occupied by a small hamlet, La Ferrassie, after which the skeletons were named.

La Ferrassie 1 is one of the most complete Neanderthal skeletons preserved. The large amount of skeletal material recovered, together with the unusual and characteristic pattern of its pathologies, enabled Fennell and Trinkaus (1997) to make a diagnosis of hypertropic pulmonary osteoarthropathy (HPO). During their differential diagnosis, they eliminated periostisis, tuberculosis, mycoses, treponemal disease (pinta, yaws and syphilis), metabolic and nutritional factors, tumours of the bone, congenital heart disease and primary hypertropic osteoarthropathy as possible causes.

Although the primary lesion of HPO is pulmonary (in the lung), the resultant cell damage causes abnormal blood platelets to be released into the blood stream. These abnormal blood platelets release platelet derived growth factor. This results in a strictly symmetrical periosteal bone deposition on the shafts of the long bones of the extremities (Ortner and Putscher 1981). The periosteal new bone formation is a distinct and unmistakable lesion. The bones most frequently affected are the long bones of the distal extremities (radius, ulna, tibia and fibula). There is no involvement of the medullary cavity, the axial skeleton or the skull. Clubbing of the digits occurs in advanced cases but this affects soft tissue only, leaving the bones of the terminal phalanges unchanged. This is believed to be the result of the associated abnormal blood platelet clumps restricting circulation in the digital extremities, causing thickening of the tissues.

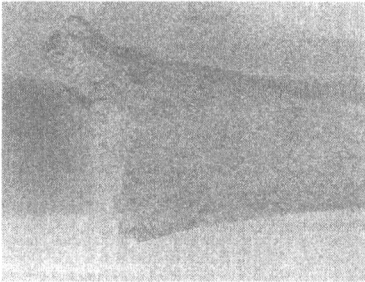

Fig.41.1: La Ferrassie 1 distal left femoral diaphysis (Fennell and Trinkaus 1997: 987)

Other features commonly associated with HPO are bronchogenic carcinoma and tumours of the pleura (Aderaye 1996), as well as lung abscesses and cirrhosis of the liver (Fennell and Trinkaus). In cases where periosteal changes are secondary to the pulmonary neoplasm, their presence may be noticed months, or even years, before the pulmonary tumour is diagnosed. However, not all cases of HPO are tumour-related, some resulting from chronic pulmonary infection. It cannot, at this stage, be known which preliminary pathology precipitated the onset of the hypertrophic pulmonary osteoarthropathy suffered by La Ferrassie 1, but this is irrelevant for this study, because all have some form of air pollution as their primary cause and it is this primary cause which is of interest.

The pathology is slow to develop. Ongoing pollution over a considerable period of time is necessary and, as he became increasingly sick before he died, La Ferrassie 1 would have needed the care of his companions for him to have survived long enough for the pathology to develop to the extent that it did. Initial manifestation of the disease would have involved painful swelling of the joints, knees, ankles, wrists and sometimes fingers. Pain would have been aggravated by movement.

Until the beginning of the 20th century, primary neoplasms of the lung were rare (Osler 1892). By the end of the 20th century, they had become one of the most common forms of cancer, attributed almost solely to the smoking of cigarettes, either directly or indirectly through passive smoking. The connection between the inhalation of smoke and the development of cancer of the lung is so strong that even the rich and powerful cigarette companies, with almost unlimited funds to retain the finest lawyers, found it extremely difficult to refute claims by litigants. Although air pollution is believed to be the primary cause of lesions of the lung, not all people exposed to such pollution develop cancer. It would seem that some people have a genetic susceptibility (Jones 2002). Other pollutants, such as asbestos and silica, have also contributed to lung pathologies, but these cannot have contributed to La Ferrassie 1's problem.

Osler drew attention to the high incidence of pulmonary tumours occurring in workers in the Schneeberg cobalt mines. Low levels of naturally occurring background radiation are present everywhere on the surface of the Earth, but potentially hazardous levels are only emitted when the substrate is granite. People living in granite houses in parts of the British Isles, especially Cornwall and Scotland, may be exposed to excessive radiation through exposure to the gas radon which leaks from the granite. The amount may even be more than that to which workers in nuclear power stations are exposed (Jones). La Ferrassie is not such an area. On the contrary, being an area high in iron, La Ferrassie has low levels of radiation. Natural radiation can, therefore, be eliminated as a probable contributing factor in the development of this pathology in this instance.

The tissues of the lung are extremely resilient. They need to be able to resist wind-blown dust, which in some areas can be considerable. Many animals, whose lung tissues are similar to ours, live in close proximity to the ground, constantly sniffing, and presumably inhaling small particles of dust. Even when a pathology has started to manifest, if the pollutant is removed (i.e. if the person gives up smoking) the lungs have an amazing ability to recover.

The pollutant most likely to have caused HPO in La Ferrassie 1 is smoke from constant fires. Since open fires in winter have been a feature of European homes throughout recorded history without causing lung cancers, it would seem that the fires would have needed to have been burning constantly throughout the year and to have been burning in an enclosed environment to have produced sufficient smoke to have caused the observed pathology. La Ferrassie lived during one of the coldest phases of the Late Pleistocene. Some form of shelter in which a fire could burn at all times would seem to be the only circumstance which would have supported the necessary conditions. As early as 130,000 B.P. early European people were supplementing their cave shelters by the addition of skin drapes, as has been demonstrated at La Grotte de Lazaret.

A nurse friend, who had spent some time in Kenya, told me that people there built quite substantial huts in which they kept fires burning at night.

There was no ventilation, except for a low door for entry. The people suffered constantly from bronchial and eye problems. When she asked them why they did not allow ventilation at the apex of the roof, as did the Native Americans, she was told that the main purpose of the fires was not warmth, but smoke - to keep out the mosquitoes! Mosquitoes have been a constant bane in the life of human beings since time immemorial. Pritchard (1940: 66) recorded that the Nuer people of the Sudan used smoke as a measure to deal with the problem of an over-abundance of insect life:

> Mosquitoes swarm in the rains, their ravages being terrible from July to September, when, as soon as the sun sets, men and beasts have to take refuge in huts and byres. The doors of the huts are tightly closed, the air-holes blocked, and fires lit ... in the centre burn large dung fires which fill them with smoke so dense that one cannot even see the cattle ... if the smoke clears ... they ... pile on more fuel.

Mosquitoes are as troublesome in the Arctic regions as they are in the equatorial. Binford (1978: 256) told of the suffering of the Inuit of northern Alaska during the short Arctic summer:

> It is hard to describe the mosquitoes. They begin to appear in mid-June and by the warm days of July a walk, the mundane act of defecation ... all intruded on by swarms of mosquitoes which attack the smallest patch of exposed skin ... My tent was moved during the night by the mosquitoes trying to fly out through the roof!

The clouds of mosquitoes in the Arctic are so great, and they suck so much blood from the caribou, that these large animals are reduced to skin and bone. Some even collapse and die from loss of blood. There is no reason to suppose that insects would have been any less of a problem in the past. A combination of winter cold and summer insects may have contributed to the problem of smoke inhalation from which La Ferrassie 1 seems to have suffered. It may be difficult for us today fully to comprehend the insect problems of the past. Daubing the body with mud and clay probably served a useful protective purpose long before it served a decorative one. Fires no doubt helped to keep large animals at bay but the role of fires in repelling insects should not be overlooked. In the former case, it was the open flame which was the deterrent; in the latter, it was the smoke.

Europe today is so heavily populated, and so little soil remains undisturbed, that the chance of finding evidence of constructions from Middle Palæolithic times is much reduced. However, evidence of post-holes dating back some 380,000 years have been found at Terra Amata, Nice, in south-eastern France (de Lumley 1969a, de Lumley and Boone 1976). De Lumley (1969b) also found evidence for the construction of some form of shelter in the cave at Lazaret dating to 130,000 B.P. It is noted that the reconstruction illustrated showed a vent in the roof to allow for the escape of smoke but, bearing in mind the construction of huts in Africa in modern times, mentioned earlier, the inclusion of this refinement may be open to question.

While there is no doubt that the Neanderthals did make use of caves, it is unlikely that these were their soles means of shelter. Skin and (brush)wood shelters are not difficult to construct. The Neanderthals exhibited considerable dexterity in the manufacture of their stone tools. Is there any reason to suppose that they would not have been capable of constructing some form of shelter, as had other European people thousands of years before them? Caves have limited value. They are dark and dank, except near the entrance. Over time they would have become cluttered with debris. Animals either move, or maintain their nests in spotless condition. It is possible that the Neanderthals kept rats and cockroaches at bay by assiduous housekeeping. It is more likely that they moved!

Shanidar 1 showed us that some, at least, of the Neanderthals were well fed. Combined with the evidence of the Venus figurines, a picture is emerging of a more comfortable life having been lived by our ancestors than has traditionally been portrayed. Our ancestors may not have lived a harsh and dangerous existence, constantly worried about where their next meal was coming from. Now, La Ferrassie 1 adds another component. Maybe not luxury, 5-star accommodation, but a home, warm and secure. Why should not ancient humans have been able to build themselves homes? Birds can. Beavers can. Bees can. Why, so often as we go back in time, are our ancestors portrayed as being, not only destitute of human intelligence, but destitute of animal intelligence as well?

That both these Neanderthal men lived long enough for their disease to progress to the stage that it did, is evidence that they received long-term help and care from their companions. They were not only human, they were humane.

Chapter 42

The Light End of

the Spectrum

The hominid skull known as E 686 from Broken Hill, Zambia, thought to be an adult male, was uncovered in 1921 during mining operations. It was found in a limestone cave which contained many fossils but no absolute date was established. It was believed to be between 130,000 and 250,000 years old, which placed it within the timespan proposed for the second exodus 'Out of Africa' by proponents of that model of human evolution. Following Cann and her colleagues this period will be referred to as 'approximately 200,000 years' for reasons of convenience.

If one assumes the catastrophism/Out of Africa model to be correct, there is no way of knowing whether this individual was one of those who had already made the transformation into a new species and whose progeny would one day spread all over the world, or whether he was one of the unlucky ones about to be replaced. It is not possible, therefore, to make an absolute attribution of (sub)species identification. It may be *Homo sapiens rhodesiensis*, a form of archaic *Homo sapiens*, *Homo heidelbergensis* or *Homo rhodesiensis* (Rightmire 1990, Montgommery et al. 1994). The attribution by some of the nomenclature '*Homo heidelbergensis*' reflects the thought that pre-Neanderthals evolved in Africa as well as in Europe, but that the people of Africa evolved somewhat differently from the early Europeans, resulting in the development of archaic *Homo sapiens* on the former continent and of the Neanderthals on the latter. It will have been noticed that there is disagreement as to whether or not these remains were '*sapiens*'. Holding, as I do, to the Multiregional hypothesis, I do not believe the human line has speciated since the emergence of *Homo erectus*. I am more in agreement with the first option offered above, which allows for these remains to be '*sapiens*' .

In the early 1990s, this skull was re-examined by Montgomery et al.

(1994). The evidence for dental caries, unique in its extent for a fossil hominid of this antiquity, may offer insight into some aspects of archaic *Homo sapiens* life-style, although this aspect will not be considered any further here.

Montgomery et al. disagreed with the findings of earlier examiners of the cranium, who had suggested that it showed evidence for pathologies such as acute mastoiditis, chronic middle ear suppuration, mastoid abscesses, metastatic abscess to the left temporal region and cholesteatoma (Yearsley 1928; Wells 1964; Price and Molleson 1974). They proposed a pathology which had not been suggested previously.

They agreed with the earlier writers that the lesion (see Fig. 42.1) had not resulted from trauma. The middle meningeal artery would have been located at the site of the lesion and injury would have resulted in extensive bleeding and death before a healing reaction could have taken place. The lesion had rounded edges suggesting an ante-mortem (healing) reaction of the skeletal tissue. They considered fifteen possible conditions in their differential diagnosis, coming to the conclusion that the lesion was most likely the result of an intradiploic dermoid or an eosinophilic granuloma.

Fig 42.1: Broken Hill cranium showing site of lesion
(Rightmire 1990: 211)

As a Natural Therapist, the treatment of skin cancers was not within my province, for legal reasons. My knowledge of these pathologies came from theoretical studies during my training and from knowledge gleaned from patients and other sources. However, it was my understanding that these pathologies resulted from over-exposure to ultra-violet (UV) rays and that they rarely affected people with dark skin pigmentation. The term 'over-exposure' carries with it an implication of being more than intended by Nature/Natural Selection. In other words, it resulted from life-style choices – spending too much time in the sun!

I consulted a skin specialist from the Department of Medicine at the University of Melbourne, who was of the opinion that the possibility of secondary cancer had not received enough consideration as a potential cause. Cancers of the bowel, lung, stomach, thyroid gland, prostate or kidney are the ones most likely to metastasize to bone (Roberts and Manchester 1997). Since cancers of the soft tissues are usually fatal without

470

leaving any definitive indications in surviving skeletal remains, paleopathologists are unable to determine the frequency with which ancient people were affected by malignant tumours or, indeed, whether they were affected at all. Malignant neoplasms of bone, either primary or secondary, are rarely found in ancient skeletal remains. The earliest example of osteosarcoma recognized and cited by Manchester and Roberts was that found in a humerus bone of Iron Age date from Munsingen, Switzerland. Myelomatosis (multiple myeloma) affects mostly people over 40 years of age and this may be a reason why it is not found in archaic populations, the earliest case in paleopathology being from the fourth century B.C. from Kentucky. A primary cancer of the nasopharynx was found in an Egyptian skull of III-V Dynasty; the earliest perforating lesions characteristic of metastatic cancer occurred in the skull of a female from Mokrin, Yugoslavia, dated to the Bronze Age. All these cases were cited by Roberts and Manchester (1997). The apparent rarity of cancers, either primary or secondary, from Early Holocene as well as late Pleistocene times, let alone 150,000 earlier, may well have been justification enough for the possibility of secondary metastasis to have been passed over in favour of the more likely causes considered by Montgomery et al.

From the information available to him, the specialist agreed with Montgomery et al. that this lesion had most probably resulted from basal-cell carcinoma or eosinophilic granuloma. Eosinophilic granuloma most commonly presents as a solitary round or oval lesion of the cranial vault, with a beveled edge, and there may be destructive involvement of the mandible (Ortner and Putscher 1981). Unfortunately, the mandible of E 686 was not recovered so it was not possible to ascertain whether there was any associated mandibular pathology.

Basal cell carcinomas originate in the skin and are slow growing, but as they progress, especially if left unattended (as they presumably would have remained in Pleistocene times) they can erode sub-cutaneous tissue. However, they do not spread to other areas of the body and can, therefore, grow for many years without causing death. They are common in elderly people with light skins, especially males affected by baldness, as they result from damage to the skin by UV light. The specialist wrote, "They are very, very rare in people with black skin ... The assumption would have to be that it might have occurred in a person who did not have black skin, but whose skin was white".

Chimpanzees have light skin beneath their black hair and it is generally assumed that our earliest ancestors, who first branched off from the ape linage to form that of the Australopithecines, likewise had light skins and

471

were covered in dark hair. However, it is also generally agreed that the loss of body hair, with the acquisition of sweat glands, took place when the hominoid line left the shelter of the forest and took up residence in the savannah. "The evolution of a naked, darkly pigmented integument occurred early in the evolution of the genus *Homo*" (Jablonski and Chaplin 2000: 57).

The discovery of a cranium from central Africa, showing a pathology associated with 'white' skinned people suffering from an over-exposure to UV rays over many years, raises some interesting possibilities, not only as to their community life-style, but as to their physical appearance, i.e. skin colour. Were all the ancient people living in Africa, including *Australopithecus* and *Homo erectus* dark skinned after all? Dark colours absorb more heat than do light and it may be questioned why a light-skinned species, coming from the shelter of the forest into the heat of the sun, should have evolved a skin colour designed to retain heat. Why did not our ancestors, like the lion and the tiger, keep their covering of hair to protect them from the elements?

It is melanin which now protects us from the UV rays of the sun and which causes the skin to develop a darker, tanned, appearance. I have often wondered why nature could not have evolved an alternative protective chemical, which was light in colour, able both to protect and reflect? If nature could develop a white sclera for the eyes, which withstands the effect of light, why could it not evolve a lighter skin, resistant to UV? Nevertheless, it does seem apparent that the evolution of dark skin was nature's response to the loss of hair under the African sun, although the reason for the loss of hair remains a mystery.

As has already been pointed out, skin cancers do not develop quickly. They occur in older people, who are likely to be passed the age of reproduction, and would not, therefore, be a factor in evolutionary survival. These cancers occur more frequently in men than they do in women and men can continue to reproduce much later in life than their partners; nevertheless, it is doubtful that they would have occurred often enough, early enough, to have been a factor in 'selection'.

Forensic pathologists/archæologists have no difficulty distinguishing between Caucasian/Mongoloid/Negroid human skeletal remains. Not only does robusticity differ, but the shape of the cranium and of the jaw (prognathism) is also distinctly different. Experts can identify remains of mixed heritage. Thoma (1973) pointed out that none of the early fossil remains from Africa carried any negroid characteristics. He even went so far as to suggest that negroid characteristics, which first appeared in remains less than 100,000 years old, i.e. after the speciation event postulated by the 'Out-of-Africa' theorists, had developed as the result of the intermingling of African peoples with Neanderthals. While he makes a good point - we should not assume that migrations always took place in one direction only - nevertheless I am unconvinced by his argument since, if it were true, should not negroid characteristics have developed in Europe? The point to be noted here is that E 686 was not negroid. Even its latest possible dating precedes the evolution of the negroid typology.

Not being a geneticist, I have no idea whereabouts on our genome lie the genes relating to skin colour, nor do I have any idea of whether or not they are irrevocably associated in any way with any other physical characteristic, such as robusticity. What is important here is to realize that ancient skeletal remains, of themselves, do not provide us with any information about skin colour. We do not know the skin colour of Neanderthals. Neanderthals were robust and had a degree of prognathism, but their crania were a completely different shape from the negroid. All pictures which I have seen portray the Neanderthals with white skins. I assume this is because the Neanderthals lived in Europe and Europeans are white? But were not the Neanderthals supposed to have evolved from a diaspora out of Africa, via the Levant? Were they white then? The pathology of E686 provides evidence that the skin colour of people living in Africa prior to ~100,000 years ago was white.

Or does it?

There is another possible explanation for the pathology of E 686. That person may have been albino.

There are today in the northern regions of Tanzania and Kenya, a group of people who are Negroid in every way except that they lack melanin. They are albino (Lookingbill et al. 1995; McBride and Leppard 2002). Is it possible that there were albino people living in Africa all those years ago? Why not? Lack of melanin is not an evolutionary disadvantage since it does not affect either life span or reproductive ability and there is no reason of which I am aware why this genetic abnormality should not have an ancient history.

The first study to assess the possible evolutionary effects of UV rays on skin colouration was undertaken by Jablonski and Chaplin (2000). They noted that the adverse effects of UV radiation on the skin itself did not manifest until well into, or even after, reproductive age and agreed that external (skin) pathologies alone were not enough to explain the occurrence of dark skin in regions with greatest exposure to UV radiation. They

473

suggested that photolysis (breakdown caused by sunlight) of folate might be a more cogent reason. They pointed out the important role of folic acid as an essential nutrient for DNA synthesis. Lack of folic acid brings about macrocystic megaloblastic anæmia, since folate is required for the maturation of bone marrow and the development of red blood cells They stated that males tended to have darker coloured skins than females and that males needed increased protection against photolysis since folic acid was an important factor in spermatogenesis. If this is the case, then perhaps our ancestors developed dark skins very early, even though they did not have a negroid appearance? Perhaps the light skin colour of Europeans was a later development? Against this is the fact that dark skinned people, Australian as well as African, have pale soles and palms. Is it not possible that these are remnants of an earlier light coloured skin, since 'white' skinned people do not have any dark skin remnants?

While Jablonski and Chaplin concluded that there was a correlation between skin colour and latitude, it should be pointed out that their data did not include people from the most northern latitudes, such as the Eskimo/Inuit and those from northern Scandinavia and northern Russia (Siberia), who generally have dark hair and a dusky, almost Mediterranean, skin colour. The Tasmanian aborigines were very dark yet lived in latitudes similar to those of Portugal, Spain, Italy, Greece, Turkey, Macedonia, Bulgaria, Iran, Uzbekestan, North China, North Korea and the northern Islands of Japan, none of which have been home to darkly pigmented people in known historical times. Tasmania is believed first to have been occupied about 30,000 years ago. The island has been separated from the mainland of Australia since the end of the last Ice Age, about 12,000 years ago. If skin colour was dependent upon climate (UV), should not the Tasmanians have started, at least, to develop light coloured skin?

Skin does not fossilize and when it is preserved as the result of mummification, desiccation or immersion in a peat bog, it becomes discoloured. The pathology suffered by archaic *H. sapiens* E 686 200,000 years ago may be the only clue we will ever have as to the skin colour of early humans and it would seem that it was white. If Thoma was right and negroid types remains do not appear until after 100,000 B.P., then, far from being the 'connecting link' between early humans and 'modern' humans, as was once thought by some, it would seem that the negroid people could be the most recently evolved members of our species.

I will close this chapter by referring back to the comments of Howell, Frayer et al. and Wolpoff et al. (p. 425) on the late appearance of negroid cranial features.

Chapter 43

How Human(e) were
H. Erectus?

The final pathological remains to be considered here are the only female ones, a female who was the youngest in age to die but the oldest in time to live. She was not Neanderthal; she was *H. erectus*, believed to have lived some 1.6 million years ago. Her designation is KNM-ER 1808.

In 1982, Walker, Zimmerman and Leakey published a paper in *Nature* which gave a differential diagnosis for the pathology shown by this female *H. erectus* skeleton, which had been recovered from Koobi Fora, East Lake Turkana, Kenya.

As a species, *H. erectus* is believed to have come into existence ~1.8-1.7 mya, meaning that these remains were of a very early member of the species.

The skeletal remains were comparatively complete, including, as they did, parts of the all-important cranium. They showed a strange pathology. After a period of lengthy examination, Walker et al. came to the conclusion that the only known condition which could account for all the preserved pathology was hypervitaminosis A - the consumption of too much vitamin A. During their lengthy deliberations, they eliminated as possible explanations hypervitaminosis D, hyperparathyroidism, fluorosis, hypovitaminosis C (scurvy), infantile cortical hyperostosis, generalized cortical hyperostosis, osteomyelitis variolosa, syphilitic osteomyelitis and hypertrophic osteoarthropathy.

In order to affect the bone, and thus to be preserved in a fossilized condition, over-dosing with vitamin A must be continued for a very long time, not months, but years. The lesions were confined to the outer cortex of the bone shafts, over which new bone had grown locally. This sharply demarcated, coarse-woven bone contained enlarged, sub-spherical and randomly placed lacunæ. There was no evidence of abnormal remodeling of the underlying bone. It was the dense mineralization of the bone which accounted for its preservation but it precluded satisfactory X-ray examination.

I rejected Walker et al.'s suggestion that this over-consumption of vitamin A had occurred during a period when the dietary habits of *H. erectus* were changing to include a greater proportion of raw meat. My understanding of evolution is that any change in diet must occur very slowly *in conjunction* with the production by the body of whatever changed/new enzymes, etc., are necessary for the digestion of the revised diet. I accept that a sudden shift might take place if there was some sudden change in DNA, due to a 'fortuitous error'. However, if such a change in digestion had occurred, then *H. erectus* would have been able satisfactorily to digest whatever meat they were eating. I just cannot envisage generations of *H. erectus* eating an indigestible substance, raw meat, in quantity, while they waited for their digestive systems to catch up with their newly acquired taste. I believe that if the body cannot digest something, that item does not taste good. All of our digestive systems are slightly different and all of us have slightly different food preferences.

I may have disagreed with Walker et al. in that regard, but in another I was in complete agreement. They saw KNM-ER 1808's pathology as a sign of concern and compassion beyond anything established by any other hominid species to that date, since her companions would have needed to care for her and support her over a long period of time as her symptoms gradually developed and intensified. They considered this evidence of progression towards true humanity, complementary to that provided by the anatomical evidence of the bones themselves.

Vitamin A is one of the fat soluble vitamins and it is found in association with lipids which are absorbed into the body with dietary fats. Not being water soluble, vitamin A is not excreted in urine but is retained in the liver, which stores 95% of the body's vitamin A reserves, sufficient to meet requirements for one or more years (Burton 1976). Carnivores which feed on the livers of their prey build up stores of vitamin A in their own livers. Vitamin A is essential for maintenance of visual purple in the retina of the eye and therefore plays an important role in visual function. It also aids in the building of strong bones and teeth, growth in general, and the formation of healthy blood. It helps maintain cell membrane development and stability, increases immunity to disease and is important for reproduction. Vitamin A is essential for healthy skin and hair, protects the mucous membranes of the nose, sinuses, lungs, eyelids, mouth, throat, stomach, intestines, vagina and uterus and promotes the secretion of gastric juices necessary for the digestion of protein.

Vitamin A, as retinyl ester, is available directly from animal sources, such as meat, and related animal products, especially liver, eggs and fish liver oil. Humans can also synthesize vitamin A in the wall of the intestines or in the liver from precursor carotenoids of vegetable origin, of which more than 600 are now known. Thus, vitamin A occurs in two forms: preformed vitamin A from animal sources and provitamin, or precursor vitamin A from vegetable sources. One hundred grams of modern herbivore liver contains 44,000-50,000 IU (International Units) of vitamin A, whereas 100 grams of carnivore liver contains $1.3-1.8 \times 10^6$ IU (Walker et al.), the result of the storage of vitamin A ingested when consuming the livers of prey. Despite the prevalence of meat in current Western human diets, normal human liver levels of vitamin A are within the herbivore range.

Acute vitamin A toxicity is very rare. According to Keene (1985), native Alaskans traditionally consumed as much as 12,000 IU/day and the Inuit people of Greenland as much as 50,000 IU/day, without any apparent toxic effects. It has been found that amounts of 100,000 IU daily may be needed to produce acute toxicity, although 50,000 IU/day may produce toxic effects if continued over a period of time. When ingestion of food, usually the liver of polar carnivores, does induce toxic effects, the symptoms have their onset within a few hours and include drowsiness, irritability, vertigo, severe headache, vomiting and diarrhoea (Cleland and Southcott 1969a). However, it should be pointed out that these are the assumed effects of too much vitamin A from consuming carnivore liver since all of the recorded cases occurred in polar regions which meant that none received medical treatment; none of the victims were tested to determine their vitamin levels, or, indeed, any pathology at all and, of course, no remains of the suspect meal were tested either.

Vitamins were first discovered just before the First World War. While some work was carried out during the 1920s, it was during the 1930s that the importance of vitamins in maintaining health came to be realized and the literature emphasized the need for an adequate daily intake of these substances. It was the aim of the medical profession, not only to uncover the role of each vitamin, but the *minimum* amount necessary to maintain health. This information was important, not so much for the Western nations, but for those eating sub-standard diets in other parts of the world, principally Africa, where many European nations still had colonies. After the Second World War, emphasis continued to be placed in this direction. The invention of food processors and juice extractors saw many Western people, already eating an adequate diet, daily consuming large amounts of extracted juice. Oranges and carrots were particularly favoured, both being cheap, easy to obtain and producing an attractively coloured and flavoured drink. Those advertising these machines, and promoting this practice, and those, such as natural therapists and some alternatively minded medical practitioners, were quite correct in claiming that one glass contained far more nutritious material than any one meal ever could, because the human body would not 'stomach' eating a dozen or more of each of these items at the one time. Unfortunately, no one seemed to consider whether the body needed, or wanted, to ingest such large amounts. Articles began to appear in medical journals recording symptoms of overdosing, especially of the fat soluble vitamin A. By the end of the 1990s, the symptoms were well known, doctors were aware of the problem, and reports in the literature, which would merely have been repetitive, ceased.

It would be very unusual for any adult to consume large amounts of vitamin A by mastication. Acute toxicity did occur in adults who drank large quantities of carrot juice, especially during the 1970s when juicers first became a common household item. When this happened, the skin acquired an orange tinge and the problem was usually recognized. Symptoms quickly disappeared when overdosing was stopped.

Chronic vitamin A toxicity develops slowly. It results in hair loss, hepatosplenomegaly, hyperostosis, cessation of menstruation, blurred vision, bone fragility, increased cranial pressure (headaches) and elevated blood alkaline phosphatase levels. The chronic condition, affecting, as it does, the skeletal system, is painful and debilitating. There is bone tenderness/pain, especially in the distal extremities, which may be accompanied by weakness, making walking difficult, fragility, œdema both of the brain and of the legs, and ulceration (Bergen and Roels 1965; Di Benedetto 1967; Burton 1976; Feldman and Schlezinger 1970; Ziegler 1992; Kirschmann 1998, Ganong 1999).

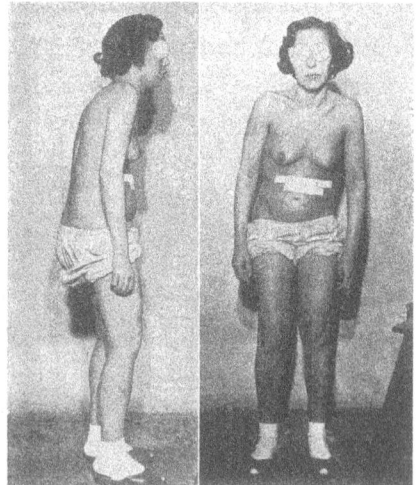

Fig. 43.1 : Patient with vitamin A toxicity
(Gerber et al. 1954: 734)

Severity of symptom is variable. While some patients complained only of joint tenderness or suffered no bone related symptoms at all, one ten-year-old boy had been unable to stand, straighten his

478

legs, or walk following an overdosing with vitamin A supplements (Frame et al. 1974) and Gerber et al. (1954) reported a case of a 21-year-old female who suffered multiple joint pains, was unable to walk straight, having a shuffling gait and generally limited joint movement. In juveniles, short stature, flexion contractures and discrepancies in the length of the lower extremities were noted (Pease 1962).

It should be pointed out that, while the taking of vitamin supplements by the general public was becoming increasingly popular, the number of tablets which needed to be swallowed each day to induce toxicity was large. Many of the cases reported in the literature arose from the over-enthusiastic prescription/injection of vitamin A by medical practitioners, who, not being aware that vitamins could have any detrimental effect since reports were only slowly appearing in their literature, prescribed large doses of vitamin A for their patients, most of whom were suffering from intractable skin problems, vitamin A having been identified as being very beneficial for the skin. Between the wars, there had been some reporting of joint problems in laboratory rats, mainly relating to the back legs, but reports of adverse effects in humans did not appear in the literature until after WWII (Pease 1962).

Cavallo (1990) cited observations of baboons scavenging tree-cached leopard kills and the taking and eating of leopard cubs to suggest that *H. erectus* may have reversed the usual predator-prey relationship. However, it should be noted that the baboons were only eating the cubs, whose livers would not have been toxic. While the idea that humans may have poached kill stored by the leopard on tree branches does not seem unreasonable, it must be remembered that leopards prey on herbivores and the livers of their kill would not have been toxic. It seems unlikely that *H. erectus* would have selectively hunted any of the Big Cats when there was a wealth of herbivore prey grazing the African plains, to say nothing of smaller prey, such as snakes, lizards and tortoises.

Walker et al. (p.248-250) attributed the chronic hypervitaminosis A suffered by KNM-ER 1808 to "a high intake of animal liver, most probably that of carnivores, during a period when the dietary habits of *Homo erectus* were changing". They suggested that, over the 200,000 years which had elapsed since the *H. erectus* species had diverged from the Australopithecine, there had been a major shift in diet towards meat eating and that "it may have taken some time to learn which parts of which carcasses were poisonous". They agreed that the condition was unlikely to have been caused by the ingestion of herbivore liver but was due "quite possibly to a diet containing carnivore liver". This woman's diet would not only have needed to 'contain' carnivore liver, it would have needed to be predominantly carnivore liver, if liver was the only food responsible for the observed pathology.

Although not relevant to my thesis, I became interested in the suggestion that Mertz had died during the 1911-1914 Australian Antarctic Expedition due to vitamin A poisoning caused by being forced to eat Huskie liver after the sledge carrying their food supplies fell into a crevasse. Cleland and Southcott (1969a, 1969b) based their hypothesis on two cases of presumed vitamin A poisoning, although neither case was confirmed, both having occurred in the Antarctic, beyond the reach of pathological investigation. In the first case, the liver had been stored in a kerosene tin used for swabbing decks and the second involved a case of exfoliative dermatitis which occurred six days after the eating of seal liver. No acute symptoms had been reported. I disagreed with this assumption and attached an Appendix to my thesis, which I reproduce here as Appendix VI. A shortened version of this Appendix was published in 2004 by the *Medical Journal of Australia.*

There is a possibility that this female was obliged to eat carnivore liver for ritual reasons. A friend told me that he remembered reading about a female shaman/witch doctor who had been required, as part of her religious duties, to eat the liver of all animals killed by her tribe for food. The liver resembles the placenta and her tribe believed her eating of the liver would increase the fertility of the other females. Unfortunately, he could not remember where he had read this and our searches were unable to track it down. Nevertheless, this anecdotal 'evidence' does open the door to the concept of KNM-ER 1808 eating large amounts of liver for ritualistic purposes. However, this would only have happened if she, and her people, were capable of religious/philosophical thought – an ability believed to be uniquely human. The symptoms are so unpleasant that it is hard to understand why she would be required to continue eating something which was making her so ill, unless, of course, the connection was not made. One member of the tribe eating different food, that one member of the tribe becoming sick, the connection would seem to have been obvious, but humans today routinely eat unsuitable food which makes them sick. Even knowing the connection, many continue to do so. Such behavior is uniquely human. Hypervitamainosisn A causes cessation of menstruation – a symptom of pregnancy. This may have contributed to the 'myth' – if it existed.

While carnivore liver is somewhat tougher, raw herbivore liver is tender and easy to eat compared with other uncooked meat, there being no evidence at this stage that early H. erectus cooked their food. The possibility must be considered that KNM-ER 1808 suffered from some type of infection in her mouth such as stomatitis, ulcers or pyrrhœa. Since eating vitamin A is necessary for healthy mucous membranes, eating vitamin A rich foods would, at least initially, have had a beneficial effect. The mandible was partially preserved and Walker et al. made no mention of any abnormality of the teeth or jaw. It would seem unlikely, therefore, that there was a pathological condition of the mouth or teeth present such that a very restricted diet of liver alone could be tolerated for an extended period of time.

During times of famine, one may eat foods outside the usual diet. Under such circumstances, *H. erectus* could be expected to eat any animal carcass they found. However, during times of starvation one does not develop an *excess* of any nutrient. In addition, the condition of the bones of KNM-ER 1808 were such that she must have consumed her unsuitable diet for a long time, years rather than months. Neither of these factors indicate that starvation was the reason for her pathology.

Carnivore liver is not the only food substance rich in vitamin A. Fish liver is also very rich in vitamin A. As a child I received daily doses of 'cod liver oil and malt'. Today fish oil capsules are available on every super-market shelf. However, fish livers are not very big – at least, not those of fresh water fish, which is all that would have been available for these people. The eating of fish livers is unlikely to have caused her pathology.

Eggs are another rich source of vitamin A and may have been eaten by *H. erectus* in large quantities. Speaking of his many months living with the Inuit people of northern Canada, Bruemmer (1993: 49, 73) told of the large number of eggs consumed at certain times of the year:

> At the mission in Hopedale there are old, velum-bound diaries kept in meticulous Gothic script. They spoke of the Inuit collecting eider [duck] eggs in early summer, as many as 2500 per day.

> Dozens of eider ducks nest on coastal rocks. Each nest averaged four large olive eggs. We rushed from island to island collecting eggs all day. Each egg in volume equals nearly two hen's eggs. Many Inuit ate 6-10 eggs at each meal.

Since all water birds nest on the ground, as do many of the larger species of bird, it is reasonable to assume that many birds nested in the area of the lake and river systems of Lake Turkana. There is no reason to suppose that *H. erectus* should not have robbed nests then as voraciously as the Inuit do today.

Some fruits and vegetables are also rich sources of vitamin A, especially those of an orange or yellow colour. Flowers may be included in this category. Bryant et al. (1975) analyzed human coprolites from a rock shelter in south-west Texas dated to between 500 B.C. and 800 A.D. and found that these people had consumed the flowers of a number of plants. In some instances, the proportion of pollen was as high as 80%-90%. Besides flowers, they identified various animal remains, including fragments of grasshoppers and other insects and the bones of small mammals, such as field mice, reptiles (probably lizards) and fish the size of minnows. However, there is no record of any people suffering chronic vitamin A toxicity as the result of eating these substances, even in large quantities.

I came across two cases of humans being adversely affected by the consumption of vitamin A as a 'normal' part of their diet. The first (Lonie 1950) followed the eating of shark liver. The second occurred in twin girls who, at age 7 months, developed symptoms after being fed chicken liver twice daily for four months, their mother believing chicken liver to be nutritious and easy to digest (Mahoney et al. 1980). This second case was the only case of chronic symptomology following ingestion of high vitamin A rich food as part of a 'normal' diet. All other chronic cases followed supplementation, by vitamin pills or injection, often medically prescribed (Sulzberger and Lazer 1951; Shaw and Niccoli 1953; Gerber et al 1954; Hillman 1956; Stimson 1961; Pease 1962; Bergen and Roels 1965; Di Benedetto 1967; Jowsey and Riggs 1968; Feldman and Schlezinger 1970; Muenter et al, 1971; Katz and Tzagournis 1972; Furman 1973; Frame et al. 1974; Ruby and Mital 1974; Russell et al. 1974; Smith and Goodman 1976; Strange et al. 1978).

The reason for KNM-ER 1808's pathology remains a mystery. However, she clearly consumed food which was unhealthy over a prolonged period of time and this is a uniquely human activity. The evidence is that *H. erectus* were human, even if their brains were not as large as ours. Quite what it is that makes humans different from animals is still a matter for debate. Some deny that there is any difference. The Bible teaches that it is our ability to distinguish between good and evil which separates us. Others, such as Lamarck, suggested it was the human ability for artistic/philosophic thought which made the difference. Perhaps we should add to the list the ability to make stupid decisions?

Whatever it was, *H erectus* had it!

Chapter 44

Irreducible Complexity

In Chapter 29, brief mention was made of the work of Csányi who, like Teilhard before him, had extended his theory of evolution to include the inanimate as well as the animate. He had made no attempt to explain the origin of either life or consciousness. (Plants are seen to be alive but not conscious; these two states must be considered as having evolved separately.) Nor did he offer any explanation as to why, after approximately 1.5 billion years of prokaryotic life (single cell replication by mitosis), the more complicated forms of eukaryotic (multicellular) reproduction by meiosis emerged.

Concern with the role that natural selection may, or may not, have played in the evolution of early cellular life, and its place in the evolution of complex biomolecular structures, engaged the attention of the biochemist, Michael Behe, whose work is the subject of this chapter. Behe's controversial ideas first attracted attention following the publication of his book, *Darwin's Black Box*, published in 1996, but consideration has been held over until this final part of the book because Behe's work, and, indeed, Behe himself, played a pivotal role in legal action which took place in the United States in 2005, which will be considered in the next chapter. This chapter sets the scene for the next.

The drawing of *philosophical* conclusions from *scientific* observations is not new. Indeed, it must be one of the oldest activities of the human mind. Fact: This Earth exists. Conclusion: Someone made it. The most notable thing about Paley's *Natural Philosophy*, published in 1802 and discussed in Chapter 8, was the fact that he found it necessary to write it at all. Although the first stirrings of Secular Humanism, which was to become so powerful in the 20th century, were already evident, the movement at that time was not so much away from the concept of 'God' itself, as away from

Christianity and the Christian idea of God. Paley drew upon observations of the natural world to draw a philosophical conclusion – that God does exist.

Although Darwin described himself as a 'theist', and always insisted that there was nothing in his doctrine which precluded the existence of God, the very fact that he found it necessary to make these assertions bears witness to the fact that some people very quickly drew the philosophical conclusion from Darwin's work that belief in God was unnecessary. Evolution had taken place 'naturally' in accordance with 'Natural Law', although I have yet to see explained from where this 'Natural Law' came. Michael Behe is a biochemist and a Roman Catholic. He drew upon his work in the laboratory, not only to support his claim that expressions of life on this planet are the work of an Intelligent Designer, but that Darwin's theory of evolution by Natural Selection is untenable.

Darwin had suggested that all forms of life had gradually evolved, bit by bit, another piece being added to an already workable form to create a more evolved form which was an improvement in some way over that which went before. Behe argued that the subcellular forms which he studied under the microscope were so complex that to remove even one part would render the whole unworkable. He called this 'irreducible complexity'.

Before discussing Behe's work further, it is necessary to note the meanings attributed by him to the terms 'evolution' and 'irreducible complexity'.

Evolution

> Evolution means a process whereby life arose from non-living matter and subsequently developed entirely by natural means. That is the sense which Darwin gave to the word, and the meaning that it holds in the scientific community. And that is the sense in which I use the word evolution throughout this book (Behe 1996: xi).

Behe made it clear that, as a biochemist, his interest lay in how the 'prebiotic soup' became the 'biotic soup', the subject which had so interested Csányi. The phrase 'subsequently developed entirely by natural means' should be noted.

Still discussing evolution in general terms, Behe stated:

> Many people think that questioning Darwinism must be equivalent to espousing creationism. As commonly understood, creationism involves belief in an earth formed only about ten thousand years ago, an interpretation of the Bible that is still very popular. For the record, I have no reason to doubt that the universe is the billions of years old that the physicists say it is. Further, I find the idea of common descent (that all organisms share a common ancestor) fairly convincing and have no particular reason to doubt it ... Although Darwin's

mechanism — natural selection working on variation — might explain many things, however I do not believe it explains molecular life (p. 5).

Irreducible complexity

By *irreducible complexity* I mean a single system composed of several well-matched, interacting parts that contribute to the basic function, whereas the removal of any one of the parts causes the system to effectively cease functioning (p. 39).

It is clear from the above that Behe's position had far more in common with the Neo-Darwinists than it did with the Creationists. He accepted the great age of the Universe, common descent and evolution by natural means, all denied by the Creationists. His only point of agreement with the Creationist position was his belief in the existence of some form of Superior Intelligence, which initiated both creation and evolution. The latter happened 'entirely by natural means' because it occurred in accordance with 'natural law', which was God's law. The only disagreement between Behe and the Neo-Darwinists was that he held to a belief in a Superior Intelligence, while the Neo-Darwinists believed that life, consciousness and intelligence had all risen spontaneously from inert matter.

Behe pointed out that, for evolution to have taken place solely in accordance with Darwin's hypothesis, every step in the process must have been viable. Each changed (improved) condition must have been complete and workable at the time the change occurred, with no account being taken of possible future benefits. Behe gave three micro-biological examples to illustrate irreducible complexity: the bacterial flagellum, the clotting of blood and the immune system. Each was discussed in some detail. It was his example of the bacterial flagellum which received the most attention from his critics and which will be concentrated upon here.

In Darwin's time, the cell was considered to be the smallest possible somatic component. It is now known that, at a minimum, a cell must contain DNA, mRNA, tRNA, rRNA, amino acylating enzymes, polymerases, sources of energy and electrons, lipoprotein membranes and ion channels within the cell wall (Shermer 2006). "A typical cell contains thousands and thousands of different kinds of proteins ... A protein chain typically has anywhere from about fifty to about one thousand amino acid links" (Behe p. 52). "The flagellum found in many bacterial cells contains more than fifty parts. Cilia and flagella found in eukaryotic cells ... are even more complex, probably containing as many as 250 distinct proteins" (Miller 1999: 133). Even the DNA of a bacterium, such as E-coli has around 4,000,000 nucleotides (Dennett 1995: 156).

The mathematical probability of even as 'simple' a thing as a bacterium having come into existence by chance was so high that it was negligible (Shapiro 1986; Barham 2004; Meyer 2004). This had led scientists to consider the possibility that the first living (self-replicating) entity must have been simpler. Hereditary particles (DNA, RNA) are formed of nucleic acids. Viruses are relatively short lengths of nucleic acid wrapped in protein. It was suggested that small pieces of nucleic acid had at one time formed 'naked genes', i.e. existed without a protein coat. However, even a virus has a complicated form (Shapiro p. 156):

> One of the large viruses, one called T2 ... has a hexagonal head, a complex neck shaft, and six spindly jointed legs, all made of proteins. More than fifty different proteins are used to build the structure. Within the head is tucked a length of DNA ... T2 is elaborate, as viruses go, and contains more than 100,000 nucleotides on each chain of DNA.

When the T2 DNA is injected into a host cell, in this case a bacterium, it uses the bacterium's ribosomes and enzymes to take over the cell. An enzyme is produced that destroys the DNA of the host bacterium, allowing the T2 DNA to take over the cell for its own use. Other viruses are smaller. Q , for example, has only 6,500 nucleotides, but the principle is the same.

It is not clear how viruses replicated before more complex host cells evolved. Viroids are even smaller than viruses, having but a few nucleotides, and are 'naked', having no protein coat. Viroids do not code for protein and are, therefore, not 'genes', but they can replicate within certain plants, causing disease. Once again, they need the assistance of another, more complex, organism, to replicate. Nevertheless, the assumption is that viroids, or something similar, were once able to self-replicate, which would entitle them to be considered as having some form of 'life'.

Viruses, viroids, DNA, RNA are all crystalline in form, apart from the protein coat when present (Dennett 1995). Some silica crystals, found in clay, are able to replicate and it has been suggested that these may have played an early role in evolution (Shapiro).

Behe, as a biochemist working in the 1990s, was well aware of all of the above facts. It was his contention that, if natural selection was a universal law applicable to all evolution since the appearance of the first cell, or first form of self-regulating life, it must be as applicable to the smallest forms of life as it was to the largest. The complexity of life seemed, if anything, to increase, rather than decrease, the further microscopic studies advanced. Behe was unable to accept the proposition that all this complexity was the result of 'chance', even aided by 'selection'.

A different view was taken by Dennett (p. 203) who claimed:

> Once we are allowed simply to postulate organized complexity, if only the organized complexity of the DNA protein replicating engine, it is relatively easy to involve it as a generator of yet more organized complexity.

> Love it or hate it, phenomena, like this [RNA] exhibit the heart of the power of the Darwinian idea. An impersonal, unreflective robotic, mindless little scrap of molecular machinery is the ultimate basis of all the agency and hence meaning and hence consciousness, in the universe.

It is unclear who has the authority to 'simply allow' such a proposition but all Darwinian theorists take advantage of this latitude.

Miller (pp. 48-49) claimed that "genes generally encode proteins ... proteins can do just about everything required to produce an organism ... mutations can duplicate, delete, invert and rewrite any part of the genetic system in any organism, they can produce any change that evolution has documented". Unlike Behe, Miller was not concerned by how the proteins and nucleic acids themselves came to be formed and organized into 'living' entities. He was content to await further scientific evidence.

Behe claimed that over the previous few years, thousands of papers had been written about the cilia and the flagellum. Only two had suggested a possible model for the evolution of the cilium, but these suggestions had been made in a general way and lacked detail, which, in Behe's opinion, made them totally inadequate as a scientific explanation for the evolution of the cilia. Most of Behe's opponents (see, for example, Dembski and Ruse 2004; Forrest, 2004; Forrest and Gross 2004) accused Behe of misleading his readers on this point, claiming that hundreds of papers had been written outlining possible pathways of biochemical evolution. Forrest and Gross cited a paper by Musser and Chan (1998) which described the 6-part cytochrome proton pump found in eukaryotic cells and showed that parts of the complex consisting of as few as two of the six proteins could form a complex and full biochemical function. This did not address Behe's challenge to show how the pump could function in any way with one part removed, nor, for that matter, how the parts came into being in the first place.

It was Behe's contention that the removal of any one component would make the cilia/flagellum inoperable. No one accepted Behe's challenge to suggest how the removal of any one part of either the cilium or the flagellum could result in an operative, and 'selectively useful', component of an organism. It was suggested that groups of parts may have evolved separately and come together to produce a new part (Miller 2004).

In order to inject toxins into their victim cell, bacteria use something resembling an hypodermic syringe. One such implement is known as the type II secretory system (TTSS). Some of the proteins in the TTSS are homologous to proteins in the basal portion of the flagellum, leading some scientists to suggest that the flagellum itself should be regarded as a type II secretory system. Miller (pp. 87, 88) explained his position:

> It is to be expected that the opportunism of evolutionary processes would mix and match proteins in order to produce new and novel functions. According to the doctrine of irreducible complexity, however, this should not be possible. If the flagellum is indeed irreducibly complex, then removing just one part, let alone ten or fifteen, should render what remains 'by definition non-functional'. Yet the TTSS is indeed fully functional, even though it is missing most of the parts of the flagellum.

> ... that only the complete flagellum can be favoured by natural selection, not any of its component parts. However, if the flagellum contains within it a smaller functional set of components such as the TTSS, then the flagellum itself cannot be irreducibly complex — by definition ... Now that a simpler functional system (the TTSS) has been discovered among the protein components of the flagellum, the claim of irreducible complexity has collapsed and with it any 'evidence' that the flagellum was designed.

Miller's 'explanation' failed to explain how the TTSS became a flagellum. Did the conversion take place gradually, as Darwin always insisted, piece by piece? In which case, in what order were all the pieces assembled? Remember, to satisfy the requirements of natural selection, each additional component must have resulted in the creation of either a new entity, which was not only functional but which performed a new and useful function, or be an improved version of the existing part, otherwise it would not have been 'selected'. Alternatively, two functioning components may have fused, one being the TTSS and the other being ...? If Miller could have suggested another fully functioning assemblage, present in a bacterium, made of the extra pieces, which might have fused with the TTSS, then his case would have been stronger, although Behe would then, no doubt, have asked him to explain how the TTSS and the other part formed in the first place!

Irreducible Complexity (IC) never claimed that every system was created *de nouveau* without any of its component parts being found elsewhere in the body. Indeed, in his definition of evolution, cited earlier, Behe made it clear that he did not dispute the concept of common descent, only how it had happened. The fact that TTSS had been discovered among the protein components of the flagellum was not the issue. In fairness, it should be noted that Miller did address the issue of proteins and agree that their evolution supported Behe's position.

Richard Dawkins, whose writings will be discussed shortly, was dismissive of Behe's work, claiming (2006: p. 128) that Behe had made no attempt to illustrate irreducible complexity and that (p. 131) "Without a word of justification, explanation or amplification, Behe simply *proclaims* the bacterial flagellar motor to be irreducibly complex" *(italics in original).*

To be accepted as a scientific theory, IC must be presented to the scientific community in the form of a refutable hypothesis. This is the scientific method (Popper 1972). While Behe did provide a definition of irreducible complexity, he did not formulate an hypothesis. Darwinism, or Neo-Darwinism, offers the hypothesis that "All evolution has taken place as a result of numerous, successive, slight modifications, selected by nature for their usefulness, which have occurred following small, random genetic mutations". The hypothesis is subject to falsification, of which truth Darwin (1859: 146) was only too aware:

> If it could be demonstrated that any complex organ existed which could not possibly have been formed from numerous, successive, slight modifications, my theory would absolutely break down.

If Behe were to offer IC, not as a scientific theory in its own right, but as a falsification of Neo-Darwinism, it is possible that he would then be on firmer (scientific) ground. He would only need to prove one example of confirmed irreducible complexity to refute Darwin's theory. The situation for the Neo-Darwinists would be more difficult. Even were they to prove to Behe and his supporters to their complete satisfaction that, for example, the flagellum is not irreducibly complex, they would merely have supported their hypothesis, not proven it. Behe could then put forward another claim of irreducible complexity in relation to another complex organism and the Neo-Darwinists would have to defend that, and so on. That is the scientific method and that is the challenge posed by Irreducible Complexity.

Forrest (2005: 79) quoted comments made by Ussery during a personal conversation: "Mike Behe is absolutely right to point out that Darwinian gradualism can't fully explain the origin of complex systems ... and that it is indeed a 'contribution' - but this does not mean that his conclusion - 'Intelligent Design' - is scientific ... Behe is right that gradualism can't explain everything - but that doesn't mean IC should be admitted as an equally viable alternative". Ussery was correct in his statement that failure of gradualism does not prove Intelligent Design, which will be discussed below. However, Ussery was on less secure ground when he claimed that the failure of gradualism did not mean that irreducible complexity should not be admitted, if not as an equally viable alternative, then at least as a matter of scientific debate.

Miller (pp. 87-88) outlined the case somewhat differently:

> ... the entire point of the design argument, as exemplified by the flagellum, is that only the entire biochemical machine, with all its parts, is functional. For the Intelligent Design argument to stand, this must be the case, since it provides the basis for their claim that only the complete flagellum can be favoured by natural selection, not any of its component parts.

Miller's statement was accurate, until the last six words. Behe never suggested that a component part, found in the flagellum, might not be serviceable elsewhere under some other circumstance or condition. Many of the proteins found in the flagellum were found elsewhere and this fact was never an issue.

Miller (p. 95) once again changed the definition of irreducible complexity when he argued:

> If we are able to find contained within the flagellum an example of a machine with fewer protein parts that serves a purpose *distinct from mobility,* the claim of irreducible complexity is refuted. *(Italics added)*

Miller (p. 88) conceded that "until we have produced a step by step account of the evolutionary derivation of the flagellum, one may indeed invoke the argument from ignorance for this and every other complex biochemical machine'. The 'argument from ignorance', as I understand it, is simply that 'we may not know *how* something happened, we just *know* that it did happen in the way we believe it did. We have absolute confidence that, at some point in the future, scientific evidence will emerge to support our position. In the meantime, we are entitled to assert that our position is the correct one'. Miller (p. 95) stated:

> ... the claim of "irreducible complexity" is scientifically meaningless, constructed as it is upon the flimsiest of foundations ... the assertion that because science has not yet found selectable functions for the components of a certain structure, it never will.

Dawkins (2006: 139-141) appeared to agree with both Dennett and Miller, giving the following explanation for the origin of life:

> The origin of life was (or could have been) a unique event which had to happen only once ... We can deal with the unique origin of life by postulating a very large number of planetary opportunities. Once that initial stroke of luck has been granted ... natural selection takes over: and natural selection is emphatically not a matter of luck.

> Nevertheless, it may be that the origin of life is not the only major gap in the evolutionary story bridged by sheer luck ... The origin of the eukaryotic cell ... was an even more momentous, difficult and statistically improbable step than the origin of life. The origin of consciousness might be another major gap whose bridging was of the same order of improbability.

> Natural selection ... needs some luck to get started ... Maybe a few later gaps in the evolutionary story also need major infusions of luck ... But whatever else may be said, design does not work as an explanation for life. *(Italics in original)*

The 'scientific method' does not allow future expectations to be used as if they were current proof, although it does allow them to be used as a working hypothesis, until disproven by falsification. In a further chapter, Dawkins' claim that statistics justifies a scientific belief in 'luck' will be further examined.

The question is: Does the admission made by Ussery, Miller, Dennett and Dawkins confirm Behe's claim that the Neo-Darwinists have not proven that all evolution occurred by a process of "numerous, successive, slight modifications" as claimed by Darwin on page 146 of *On the Origin of Species*? It must be remembered that Behe has always allowed gradual change, by means of natural selection, a *role* in evolution. He merely denies that natural selection is the *sole* means of all evolutionary change. What a difference a letter makes!

Intelligent Design

Scientists, no less than any other people, are entitled to hold a philosophical position. Furthermore, they are entitled to draw upon their knowledge to support their philosophical opinion, using 'an inference to the best explanation' (Meyer 2004: 371). However, a philosophical opinion is not scientific knowledge. If a philosophical opinion is proven to be correct, it ceases to be 'philosophy' and becomes 'science'

Many Neo-Darwinists have extrapolated from their scientific knowledge the philosophical opinion that Neo-Darwinism has shown God to be 'unnecessary', 'irrelevant' or 'non-existent'. Prominent among these are Dawkins, Ruse, Dennett and Forrest and Gross. The proposition that 'God does not exist' is not falsifiable. Therefore, it is not a scientific hypothesis; it remains a philosophical opinion.

Similarly, those who find evidence of irreducible complexity in biota on Earth are entitled to make, 'an inference to the best explanation' which, for them, is that life as we know it is the result of Intelligent Design by some Higher Intelligence. As explained by Demski and Ruse (2004: 3):

> It is not necessarily the case that a commitment to Intelligent Design implies a commitment to a personal God or indeed to any God that would be acceptable to the world's major religions.

Some form of theistic belief is held by more than 80% of the population of both America and Europe (Dawkins 2009). The proportion may be higher in, for example, Asia and South America, where many people are either Hindu or Roman Catholic. Unfortunately, no statistics are available for areas other than America and Europe. However, if the term 'Intelligent Design' is extended to cover the philosophy of all people who believe in some form of Creator, then it would clearly encompass a very large percentage of the world's population. Included under this umbrella term would be 'Creationists' who believe that the world was created by God in six days (He rested on the seventh) and that, while some living forms may have disappeared from the planet, those that remain are essentially the same as they were when created some 6,000 years ago.

It is important to distinguish between Behe's position that evolution has happened, that it has happened under the guidance of a Superior (Intelligent) Being, from that of the Creationists who do not accept that evolution has happened at all.

It might be expected that those of secular persuasion would be drawn to Neo-Darwinism and those of religious persuasion to Intelligent Design, but that is not always the case. Experimental biologist, Kenneth Miller is a practicing Roman Catholic. He is also one of the strongest critics of Intelligent Design (Demski and Ruse 2004: 4). His personal belief in God makes his stand against Irreducible Complexity/Intelligent Design all the more noteworthy. Miller (1999: 268) explained his position thus:

> To people of faith, what evolution says is that nature is complete. God fashioned a material world in which truly free, truly independent beings, could evolve. He got it right the very first time … various objections to evolution take a narrow view of the capabilities of life — but they take an even narrower view of the capabilities of the Creator. They hobble His genius by demanding that the material of His creation ought not to be capable of generating complexity. They demean the breadth of His vision by ridiculing the notion that the materials of His world could have evolved into beings with intelligence and awareness. And they compel Him to descend from Heaven onto the factory floor by conscripting his labor into the design of each detail of each organism that graces the surface of our living planet.
>
> … If the Creator uses physics and chemistry to run the universe of life, why wouldn't He have used physics and chemistry to *produce* it too? *(italics in original)*

From the above, it would appear that Miller's concept of God is that of a Being existing somewhere far above the Earth, with the working of which He is very little concerned. This impression is confirmed (Miller 2004: 94-95):

> Their [IC/ID] view requires that behind each and every novelty of life, we find the direct and active involvement of an outside Designer whose work violates

the very laws of nature that He had fashioned.

... the struggles of the Intelligent Design movement are best understood as clamorous and disappointing double failures ... rejected by science because they do not fit the facts and having failed religion because they think too little of God.

Referring to Creationists "like Johnsons, Behe and Demski", Miller continued (1999: 215, 217):

They ridicule the notion of a passive, deist God whose activity comes to an end once the essential constants of the universe are fixed ...

Creationists ... ridicule the deistic notion of a designer God who's been snoozing ever since his great work was finished.

It would seem that Miller did not distinguish between 'Creationists' who held to the account of creation as given in the Bible and theists who accepted the reality of evolution by 'natural means' – as well as by Divine design. As Miller pointed out, 'natural means' (physics and chemistry) operate in accordance with God's Law. Everything which happens in accordance with 'God's Law' happens in accordance with 'Natural Law', and vice versa, for they are one and the same thing.

Eastern philosophy

Unlike Dawkins, who contained most of his arguments against the existence of God to criticism of the Christian religion, or the three Abrahamic religions, Miller (1999: 196) did briefly address the subject of Eastern philosophy.

Many Eastern religions take the view that reality is subjective and that man can never truly separate himself from the nature he wishes to understand. Whatever the contemplative value of these ideas, the ancient Eastern intellectual was thereby relieved of any feeling that the workings of nature might reflect the glories of the Lord ... Hindu philosophers were left to contemplate the ever-changing dance of life and time, while Western scholars, inspired by the one true God of Moses and Mohammad, developed algebra, calculated the movements of the stars, and explained the cycle of the seasons.

The ancient Hindu scriptures taught that the material world gave the appearance of solidity, but it was an illusion: "All is *maya*". 'Matter' was 'solidified energy'. Energy was the only reality. This was difficult for Westerners to comprehend before the dawning of the atomic era, when all became clear. Eastern philosophy had become Western science. The ceaseless, ever-changing, never-changing Dance of Life was the ceaseless movement of energy within the atom, itself made up of particles of energy – energy which the Hindus understood as being the material manifestation of the Mind of God.

The ancient Hindus were also great astronomers - and astrologers!

The Hindus not only understood the cycle of the seasons, they understood the cycle of the Ages. Here is not the place to describe the various *Yugas, Dwapara Yuga, Kali Yuga, Treta Yuga, Satya Yuga* and so on. Anyone interested can gain a basic understanding by reading Swami Sri Yukteswar's book, *The Holy Science*, first published in 1949 and written for the general Western reader.

The Eastern philosopher has always had a very much more extended view of time than the Western. It is not only human beings (or animals) who live, die and re-incarnate. Whole universes do as well, ceasing to exist (disappearing) when Brahma breathed in and being re-created when Brahma breathed out. And as for God "descending from Heaven" to create this world, how could He when He was already within it, within the very essence of each atomic particle? Did not Jesus say "The Kingdom of God is within"?

The existence/non-existence of God will never be able to be proven scientifically to the satisfaction of the other side. Both sides will continue to interpret the evidence in accordance with their own philosophical beliefs.

Two friends disagreed. One upheld Intelligent Design, the other did not. They joined forces to co-edit a book, *Debating Design*, (Dembski and Ruse 2004) in which some chapters supported ID, the other's rejected it. For those seeking a balanced presentation of the arguments, for and against, this is a 'must read' book. Behe was a contributor.

In *Reply to My Critics,* Behe (2001) discussed a number of empirical/philosophical matters: is Intelligent Design falsifiable? Is Intelligent Design equivalent to invoking a miracle? and much, much more. For myself, I am content to keep the two areas, science and philosophy, separate, as they were for hundreds of years. To me, science relates to the created universe. I am unclear as to whether the laws of mathematics, physics and chemistry operated *before* the 'Big Bang' or whether they came into operation that instant. There would seem to be little point in laws relating to the material Universe existing prior to the Universe itself existing, but, on the other hand, I am unclear as to how the 'Big Bang' took place in accordance with these laws if they did not exist. There are many things about which I am unclear!

However, there is one thing upon which I do think I am very clear: the English speaking world uses the word 'God' to signify some energy/consciousness/intelligence which *inexplicably* existed prior to the 'Big

Bang', prior to anything else. 'Inexplicably' is the operative word. It is impossible for 'science', which discipline relates solely to the material (created) world, to 'explain' that which existed prior to creation – if anything did.

Chapter 45

Reversing 'Scopes'

The conclusion of the trial of John Scopes in 1925 (see Chapter 26) had been perceived as an occasion in which everybody had won. Bryan had achieved the result he wanted: the law which had been passed had been upheld. The doctrine that humans had descended from a lower form of animal would not be permitted to be taught in public schools within the State of Tennessee. Bryan may have been exhausted, but one may think he died a happy man. The American Civil Liberties Union had achieved the result it wanted: a conviction that it intended to appeal and which appeal it was confident would lead to the removal of the law from the statute books. Judge Raulston was happy because he had achieved the publicity he was seeking before standing for re-election and George Rappelyea had seen thousands of people flock to Dayton and had, no doubt, reaped a sizable reward for his financial outlay.

The overturning of the verdict on a technicality had had a profound effect. There was no appeal and the statute remained unchanged. Evolution remained a proscribed subject within Tennessee public schools and many other parts of America.

A comprehensive analysis was made by Grabiner and Miller (1974) of the effect which the Scopes trial had had on education in America. They contended that an impression had been given in the general literature that the Dayton trial had resulted in a *de facto* victory for the evolutionists, which they concluded was not the case. They undertook an extensive analysis of the text books used in public schools in America both before and after the trial and found that the teaching of evolution at high schools had declined after the trial.

Before 1925, school text books generally contained a short introduction to

Darwin's theory, such as that contained in the text book used at Dayton High. After the trial, the word 'evolution' disappeared from the indexes and glossaries, replaced in some cases by the word 'development' or 'change'. Grabiner and Miller stressed that the books they reviewed were not specially 'expurgated' text books for use in southern schools, but were those in use throughout America, including New York State. Not only had the evolutionists' lobby lost at Dayton "they did not even know they had lost" (p. 836).

Grabiner and Miller drew attention to an unexpected long-term consequence of the Dayton trial. After the Second World War, many German scientists earned immunity from prosecution for War Crimes by agreeing to work for either the Russians or the Americans. By the 1960s, most of these scientists were coming to the end of their careers. Russia had beaten America in the race to put the first satellite into space. The Americans feared they were slipping behind in the training of their future scientists and undertook a review of the state school science curriculum, which included a review of biology. It was clear that if new biology texts, which included evolution, were to be approved for use in state schools, the legislation enacted in several states during the 1920s would have to be repealed.

In 1968, following a court case in Arkansas, the U.S. Supreme Court ruled that the prohibition of the teaching of evolution was unconstitutional because it arose from a conflict with a particular religious doctrine (Scott 2004). Creationists responded that their children should not be required to be taught something (evolution) which conflicted with their core beliefs. Their children were not only required to attend such classes but to answer examination questions professing something in which they did not believe, in order to acquire a pass mark in the subject (Numbers 1982). If the teaching of evolution could no longer be prevented, then, they reasoned, equal time should be given to the presentation of an alternative view.

By the early 1980s, twenty-seven states had introduced 'equal time' legislation, most of which was rejected, Arkansas and Louisiana being the exceptions (Shapiro 1986; Scott 2004). Their legislation was overturned by the Supreme Court, the Arkansas case being heard first and, having been decided, the Louisiana case was but a formality.

The Arkansas legislation had called for equal time to be given to Evolution Science and Creation Science and these two sciences had needed to be defined This had been done in six points. Those for Creation Science included sudden creation of the Universe, relatively recent appearance of the Earth (within the last 10,000 years) and the insufficiency of natural selection to account for the development of all living things from a single cell. Evolution Science was defined by the same points in reverse, the Universe had achieved its current state over an extended period of time, Earth had come into existence (considerably) more than 10,000 years ago and natural selection was sufficient to account for all evolutionary change.

The Justices made their decision based upon their interpretation of the First Amendment of the American Constitution. The Religion Clause stated that "Congress shall make no laws respecting the establishment of religion, nor inhibiting the free exercise thereof"; the Establishment Clause prohibited the State from promoting religion and the Free Exercise Clause prohibited the State from inhibiting or restricting religion. In an earlier case, the Justices had ruled that "to withstand the strictures of the Establishment Clause there must be a secular legislative purpose" for an act to be constitutional, which ruling seemed to overshadow the Free Exercise clause which required "a primary effect that neither advances nor inhibits religion". Since Creation Science postulated a Creator, such legislation was deemed to have no secular purpose and therefore to be unconstitutional. The decision also found that the Act failed by 'not advancing or inhibiting religion', which it was seen to advance. The secular approach to natural selection was not seen to fail the 'inhibiting' criteria.

During the middle of the 20th century, Humanists claimed that Humanism was their religion (Lamont 1949/1965, LaHaye 1980). This claim was dropped when their *Manifesto* was revised in 1973. The United States Supreme Court had, in 1961, recognized that a belief in God was not a necessary component of religion: "Among religions in this country which do not teach what would generally be considered a belief in the existence of God are Buddhism, Taoism, Ethical culture, Secular Humanism and others" (LaHaye p. 128). Withdrawing the claim that Humanism was the Humanists' religion from their manifesto did not change the legal definition of Secular Humanism as a religion.

A law similar to that which had been enacted in Arkansas in regard to 'equal time' had been passed in Louisiana in 1982. This was also appealed on the grounds that it was impossible to teach 'Creation Science' unless a religious view was also taught. The Court of Appeals agreed and the case was referred to the Supreme Court in 1987, which ruled that "The preeminent purpose of the Louisiana Legislature was clearly to advance the religious view point that a supernatural being created humankind" (Scott p. 109). Following this ruling, 'equal time' was no longer a legal option in State schools in America.

Positions on evolution fell roughly into three groups. The first group, the Creationists, upheld the biblical account of Creation, as told in *Genesis*, that the world was made by God only a few thousand years ago within the period of six days. These people generally accepted micro-evolution (variation) but denied macro-evolution (that species could change into different species, genera, families, etc.). The second group believed that the Universe had been created by a Superior Being, accepted that the world and the Universe were of great age and that evolution had occurred, either gradually or intermittently, under Divine guidance. This was the position of 'Intelligent Design'. The third group were atheists or Secular Humanists who denied the existence of any Superior Being and who held that the existence of the Universe, and any life in it, had resulted from natural forces operating over long periods of time, without any plan or purpose.

Traditionally the second and third groups, who upheld evolution, had allied themselves against the first, who did not. Following the restrictions placed on the teaching of alternative views during the 1980s, the first two groups, both professing belief in a Creator, combined forces to oppose the third group, which denied the existence of any creative force or Superior Being (Menuge 2004). Attempts were made to introduce legislation so worded that it would avoid reference to religion. For example, in 1996, a Bill was proposed in Ohio which required that (Scott p. 129):

> Whenever a theory of the origin of humans, other living things, or the universe that might commonly be referred to as 'evolution' is included in the instructional program provided by any school district or educational service center, both evidence and arguments supporting or consistent with the theory and evidence and arguments problematic for, inconsistent with, or not supporting the theory shall be included.

The Bill was defeated.

In 2004, the Dover Area School District agreed to representations by some parents that the required biology textbook be supplemented by another (Creationist) textbook. In November of that year, the School Board issued a statement to be read to all ninth-grade biology classes at Dover High (Shermer p.102):

> The Pennsylvania Academic Standards require students to learn about Darwin's theory of evolution and eventually to take a standardized test of which evolution is a part.

> Because Darwin's Theory is a theory, it is still being tested as new evidence is discovered. The Theory is not a fact. Gaps in the theory exist for which there is no evidence. A theory is defined as a well-tested explanation that unifies a broad range of observations.

> Intelligent design is an explanation of the origin of life that differs from Darwin's view. The reference book, Of Pandas and People, is available for students to see if they would like to explore this view in an effort to gain an understanding of what intelligent design actually involves.
>
> As is true with any theory, students are encouraged to keep an open mind. The school leaves the discussion of the origins of life to individual students and their families. As a standards-driven district, class instruction focuses upon preparing students to achieve proficiency in standards-based assessments.

No attempt was made to introduce Creationism into the classroom. A book was made available which could be accessed by any student interested enough to read it. Nevertheless, this policy was immediately challenged through the Courts.

On 14th December, 2004, eleven parents filed suit against the District, with the backing of the American Civil Liberties Union (ACLU) and Americans United for Separation of Church and State (AUSCS). Dover High School was defended by the Thomas More Law Centre (TMLC) which, since its founding in 1999, had challenged the ACLU on a number of issues ranging from assisted suicide, pornography, gay marriage to nativity scenes and Ten Commandment displays (Shermer pp. 100-102). The case was heard between 26th September and 4th November, 2005, the decision being handed down on 20th December, 2005.

At the time of the Dover trial, 'Creationism' received a new legal definition. It was deemed no longer to refer only to the belief that the Earth was merely 6,000 years old and had been created in six days. 'Creationism' now referred to any concept of a supernatural force having created the Universe, thus now including the position of 'Intelligent Design'. A new definition of 'Special Creationism' was introduced to refer to the belief that God created living things in their present form (Scott p. 51).

The purpose of the prosecution in the Dover case was to show that Intelligent Design was just another name for Creationism, which had already been banned from curricula of State Schools. The prosecution showed that the text book, Of Pandas and People had first appeared in 1983 under the title Creation Biology, had been renamed Biology and Creation in 1986 and retitled yet again in 1987 when it appeared as Biology and Origins. The word 'Creation' had been replaced by 'Intelligent Design' (Shermer pp. 102-103). In his ruling, Judge Jones stated that Intelligent Design could not be uncoupled from Creationism and thus from an implication of religious doctrine. Judge Jones further found that there was nothing anti-ethical about the theory of evolution itself, but upheld that the teaching of any theistic point of view within the State School system was contrary to the Constitution. Judge

Jones was scathing in his appraisal of supporters of Intelligent Design theory, deeming the changed words to be a surreptitious attempt to evade the law (Shermer p. 105):

> The citizens of the Dover area were poorly served by the members of the Board who voted for the ID policy. It is ironic that several of these individuals, who so staunchly and proudly touted their religious convictions in public, would time and again lie to cover their tracks and disguise the real purpose behind the ID policy.

Following the Scopes trial, alterations to school texts, such as the replacement of the word 'evolution' by 'development', had been accepted as compliance with the law, not an evasion of it. Such tolerance was not repeated here.

During the trial a witness had given evidence that a significant majority of Americans thought Creation Science should also be taught in schools. The Judge ruled that whether the proponents of an Act constituted the majority or a minority was irrelevant: "No group, no matter how large or small, may use the organs of government ... to foist their religious beliefs on others" (Scott p. 198). Nevertheless, it would seem that the interpretation of the law had been influenced by the growth of secularism within American society. It is not unusual for the dominant position in society to be held by a numerical minority, as, for example, the European after the settlement in Africa. Indeed, numerical inferiority may lead to more stringent enforcement of the dominant position.

Judge Jones agreed with Judge Overton - any form of 'Creationist' teaching "advanced religion" and was, therefore, contrary to The First Amendment. Judge Jones' decision was based on a point of law in accordance with the American Constitution, as it had been interpreted by Judge Overton in Arkansas twenty years earlier. Judge Raulston's decision 80 years earlier had likewise been based solely upon the law. This is as it should be, indeed must be. It is not the place of the Court to adjudicate between different philosophical, religious or scientific opinions. Its role is solely to interpret and enforce the law. The problem lies with 'interpretation'. It is not unusual for judges to disagree. Indeed, the whole process of appealing to a higher Court is based on this expectation. It is also not unusual, when judges sit as a 'Bench', for the one or two dissenting judges to publish their 'dissenting opinion'. Where there is only one judge, there is no dissenting opinion.

Dawkins (2009: 429) cited figures from a Gallup poll conducted in America in 2008 which showed that 36% of those polled held that humans had evolved under the guidance of God, 14% believed that God had no part in the process and 44% believed humans were much the same now as they were when created within the last 10,000 years. Dawkins further stated that Gallup Polls conducted in 1982, 1993, 1997, 1999, 2001, 2004, 2006 and 2007 had shown similar results. While Dawkins had been prepared for a high percentage of what he referred to as 'History Deniers' in America, he had been less prepared for findings from a poll conducted in 2005 in thirty-two European countries. Asked whether they believed that human beings had developed from earlier species of animals, 20% or more respondents in twenty-three of the countries surveyed responded in the negative (pp. 452-453). The highest number (51%) was from the Islamic country of Turkey. These results showed that evolution was far from being universally accepted. It was Dawkins' opinion that with the spread of Islamic influence into Western countries, as well as an increase in Christian fundamentalism, it may be anticipated that negative responses would increase rather than decrease.

You will recall that during the Scopes trial in 1925, Clarence Darrow had been supported by his 'second chair', Mr. Dudley Malone. I drew attention to Malone's impassioned speech, in which Malone argued that both points of view should be taught: "give the next generation all the facts ... all the theories ... let the children have their minds kept open ... Make the distinction between theology and science. Let them have both. Let them both be taught. Let them both live ... The truth is no coward. The truth does not need the law ... The truth is imperishable, eternal, and immortal, and needs no human agency to support it".

In a postscript to his account of the trial, Scopes (pp. 276-277) summarized the feelings which he still had forty years after the trial. He wrote of the repulsiveness of any law restricting the constitutional freedom of teachers, of how such limitations would make robot factories out of schools. He claimed that tolerance was essential, that (pp. 276-277):

> ... we, as individuals and as a society, must respect the other man's point of view, no matter how far out he seems and no matter how vigorously we disagree with him ... there is more intolerance in higher education than in all the mountains of Tennessee. There is a tendency for educated people to insist that others less schooled should think as they themselves think ... the Tennessee hillbilly and the Harvard professor have the same rights to their viewpoints as I, whether theirs coincide with mine or not.

Scopes wrote his autobiography just as the new legislation was being introduced. Was this a coincidence? The statements with which Scopes finished his book give cause to wonder, had Scopes been called to the stand in later trials, for which side would he have spoken?

The Court upheld the prohibition of the teaching in any State funded institution, including Universities, of any teaching which implied the existence of any Divine Creator or Superior Intelligence – of 'God' – based on its interpretation of the Establishment Clause of the Constitution.

This in a country founded 'Under God' and whose motto is 'In God we trust'!

Chapter 46

Dominant Genes

A lready several mentions have been made of the name 'Dawkins' and it is now time to consider Dawkins' work. Of all the theorists mentioned in this book, Richard Dawkins (1944-) is the person most likely to be known to the reader through his numerous appearances in television documentaries on the subject of evolution or wild life. Dawkins wrote primarily for the general public, the number of his books exceeding the number of his papers, because he chose to write in an informal manner, without adhering to the strictly scientific terminology and standard of proof which are required for the publication of articles in scientific journals. He first caught the public attention in 1976 following the publication of his book, *The Selfish Gene*. Although he never hid his Humanist position, it was not until the 21st century that Dawkins wrote actively supporting the atheist position, using natural selection to support his philosophical views rather than his philosophical views to support natural selection.

I have decided to divide consideration of his work into two chapters. The first, this chapter, concentrates on his publications from last century and is, therefore, chronologically out of place in this Part. However, I felt it would be easier to keep the two chapters together and the chapter relating to Dawkins' philosophical writings would have been out-of-place in Part IV, so take yourself back in thought a few decades as we begin the study of this prolific writer's ideas.

It will quickly become clear that I do not subscribe to Dawkins' thinking.

There were many people who thought that Darwin had taken too narrow an approach to natural selection by placing its principal action at the level of the individual. Even when tackling the difficult problem of the social insects, Darwin's solution had been to treat colonies as if they were individuals. For Darwin, it was the swiftest, the strongest, the most attractive *individual* which survived and passed on its inheritance to future generations. Others saw selective competition occurring between groups of living things, plant or animal, such that certain species, genera, families, were successful. Weismann had taken a different approach, seeing the unit of reproduction as being that which ultimately survived and multiplied.

Weismann's view was, rather surprisingly, eclipsed, rather than enhanced, by the rediscovery of Mendel's work and the establishment of the gene as the unit of inheritance. Mendel had chosen the sweet pea as the subject of his experiments because of the clearly separate inheritance of certain features, such as colour of flower and colour and form of seed, and because the plants, being self-fertilizing, leant themselves to artificial control by the experimenter. However, even as it was realized that inheritance was particulate, human skin colour alone was enough to show that individual particles of inheritance could, and often did, act in a manner which resulted in some form of blending.

The mathematical approach of Fisher, Haldane and Wright caused as many problems as it solved. Equations which showed that a 1% (or even less) selective advantage could spread through a population within a certain number of generations were balanced by others showing that proportions of alleles remained constant, making it difficult for genes, especially those which were recessive, to increase in any population, unless accompanied by strong selective pressure. Not all characteristics were subject to 'reproductive' (i.e. life/death) selection and it became increasingly popular to advocate that more trivial characteristics 'piggy-backed' other characteristics by being associated with them on certain chromosomes or as part of certain gene complexes. By the middle of the 20th century, the Modern Evolutionary Synthesis was well established but the survival/increase of some characteristics with no direct survival value was becoming increasingly difficult to justify.

The survival of the gene once again attracted attention.

Sociobiology

Hamilton (1964) was the first to redirect attention towards the gene itself as a unit of selection, but it was Wilson (1975) and Dawkins (1976) whose work brought this concept into mainstream thought. Wilson (pp. 3-4) named group selection *Sociobiology*, arguing that kinship encouraged altruistic behaviour between "two organisms of common descent and if the altruistic act by one organism increases the joint contribution of these genes to the next generation, the propensity to altruism will spread through the gene pool". It was not only among social insects that there was co-operative behaviour. Many animals hunted in packs, or herded/shoaled together for safety. Many could not survive alone. The survival of the group was of more importance than the survival of the individual. Earlier, Williams (1966/1992: 187-188) had been impressed by the fact that among many herding mammals, an orphan trying to suckle from another female was rejected, although in obvious distress. It seemed to him that, despite appearances, these mammals were more concerned with the survival of their own genes than they were with the survival of the species. Thus the survival of the gene pool, indeed of the gene itself, came to be seen as the aim of natural selection.

Wilson (1975:3) commenced his book on *Sociobiology* with the following statement:

> In a Darwinist sense the organism does not live for itself ... Its primary function is not even to reproduce other organisms; it reproduces genes and it serves as their temporary carrier ... Natural selection is the process whereby certain genes gain representation in the following generations superior to that of other genes ... the individual organism is only their vehicle, just an elaborate device to preserve and spread them ... the organism is only DNA's way of making more DNA.

Although the rest of Wilson's book was devoted to the evolution of populations, this first paragraph contained a synopsis of the 'Selfish Gene' theory Richard Dawkins (1976) published the following year. During the late 1960s, Dawkins had taken some undergraduate lectures while Professor Tinbergen was on sabbatical leave and used this opportunity to introduce his ideas (Brown 1999: 26-27):

> I wanted to explain what was wrong with group selection ... I wanted to go back to the fundamentals of natural selection. It was at that point that I made up all the rhetoric of genes leaping down generations and casting bodies aside as they go. I used it again in my lectures in Berkeley, California ... when I returned to Oxford, I thought it would be a good idea to put it into a book.

It would seem that Dawkins considered his lectures as sufficient evidence of independence of thought not to need to make any acknowledgement of Wilson's comments, cited above, but he did make three references to Wilson's work (pp. 101-104 and 116) in disagreement with Wilson's definition of kin selection as a special case of group selection.

It was in the discipline of ethnology (the study of animal behaviour from an evolutionary point of view) that Dawkins submitted his doctoral thesis, *Selective pecking in the domestic chick*, which investigated mechanisms that might account for the way chicks selectively pecked at various stimuli (McGrath 2005). Although Dawkins' many writings have concentrated on the role of the gene, specifically DNA, Dawkins never qualified as a geneticist. Writing outside one's area of academic expertise is not uncommon among archæological/evolutionary theorists, but is rare in the other sciences.

The Selfish Gene

Dawkins' concept of the selfish gene caught the attention of the public and the profession alike. He started by advising his reader to approach his book as though it were science fiction. He took the precaution of reminding his reader (p. 95) that "talking about genes as if they had conscious aims was a linguistic ploy" and that "we could translate our sloppy language back into respectable terms if we wanted to". Dawkins never did want to and used the same license in all his books. Previous authors, Dawkins claimed (p. 2) had "got it totally and utterly wrong" when they "made the erroneous assumption that the important thing in evolution was the good of the species (or group) rather than that of the individual (or the gene)". He explained that the predominant quality to be expected in a successful gene was ruthless selfishness and that selfishness in the gene would usually give rise to selfishness in individual behaviour, although this selfishness was sometimes best served by a limited degree of altruism.

The original replicator was some form of DNA or RNA, which only needed to arise once since its novel characteristic was its ability to replicate itself, although through mutation these replicators came to vary. They increased their own success by constructing containers, "*survival machines* for themselves to live in ... They are in you and me, body and mind, and their preservation is the ultimate rationale for our existence ... they go by the name of genes, and we are their survival machines" (Dawkins p. 21). The need to stimulate possible scenarios culminated in the evolution of consciousness, which made the survival machines "executive decision-takers, emancipating them to some extent from their 'masters' – the genes" (Dawkins p. 63). Genes, insisted Dawkins, exert ultimate power over behaviour, they being the 'policy makers', our brains merely acting as 'executives'. This was materialistic humanism in its most extreme form.

Sometimes a gene promoted its own survival by promoting the welfare of another 'vehicle' containing the same gene, for example (Dawkins p. 95):

> The albino gene should be quite happy if some of the bodies which it inhabits die, provided that in doing so they help other bodies containing the same gene to survive. If the albino gene could make one of its bodies save the lives of ten

albino bodies, then even the death of the altruist is simply compensated by the
increased numbers of albino genes in the pool.

Quite how an albino gene in one vehicle was aware that it was
sacrificing itself for the welfare of albino genes in other (more numerous)
vehicles was not clear. Albino genes are recessive and Dawkins (p. 107) had
this to say about recessive genes:

Incest taboos testify to the great kinship-consciousness of man ... the genetical
advantage of an incest taboo ... is presumably concerned with the injurious
effects of recessive genes which appear with inbreeding.

Nevertheless, Dawkins seemed to be proposing that natural selection had,
at least in some cases, been advanced by the selection of injurious recessive
genes over non-injurious dominant genes by some form of altruistic
behaviour on the part of some recessive genes which 'happily' sacrificed
themselves for the good of the other injurious recessive genes, of whose
existence it is difficult to understand that they could possibly have been
aware.

Dawkins received strong support from the geneticist, Steve Jones (1999).
The female carpenter bee from Israel rears its young in pairs, usually a
mother and daughter. The mother eats the eggs laid by her daughter, but
the daughter, whose sole job it is to raise her mother's young, 'gains' by
helping her mother rear more young because "copies of her own genes,
contained as they are within her young sister, profit" (Jones p. 170). Jones
explained that natural selection acted through DNA, rather than the flesh of
those that bear it, kinship leading an individual to reduce its own chances if
such behaviour improved the prospects of other members of its family.
Such altruistic behaviour appears to be quite common among insects, but
not so common among, for example, birds, which not infrequently eject a
smaller nestling from the nest to increase their own chance of survival.

Referring to species, such as the horse and the donkey, the goose and the
duck, which can be encouraged to interbreed in captivity even though they
would not do so in the wild, Jones (p. 185) stated that "we must hence look
at sterility not as an indelible characteristic, but as one capable of being
removed by domestication", a truly Darwinian perspective.

In *The Blind Watchmaker* Dawkins (1986) concentrated on persuading his
reader that apparent design came about by gradual change: "No matter how
improbable it is that an X [a certain feature] could have arisen from a Y in a
single step, it is always possible to conceive of a series of infinitesimally
graded intermediates between them". An inability to conceive such an
infinitely small progression was a failure of the imagination, not of the

theory. As an example, Dawkins (pp. 85-86) followed Paley and Darwin in considering the eye. Having suggested how a light sensitive spot developed a little cup to aid in assessing the direction from which light came, Dawkins supposed the cup became deeper and:

> When you have a cup for an eye, almost any vaguely convex, vaguely transparent or even translucent material over its opening will constitute an improvement, because of its slight lens-like properties. Once such a crude proto-lens is there, there is a continuously graded series of improvements, thickening it and making it more transparent and less distorting, the trend culminating in what we would all recognize as a true lens.

It is difficult to understand how any matter, however vaguely transparent or translucent, could constitute an improvement unless it was totally transparent, crystal clear. That this extraneous matter should have gradually become 'less distorting' implies that it was 'distorting' in the first place. Quite why such a blurring, light absorbing, distorting piece of acquired matter should have been 'selected' is difficult to understand. Sight is more than the eye. It also involves nerve connections from the rods and cones of the retina, via the optic chiasma to the visual cortex at the rear of the brain. It is an extremely complex function.

Dawkins (1986: 233) admitted that most mutations observed in the laboratory were injurious but expressed surprise that he had "met people who think that this is an argument *against* Darwinism" (italics in original).

Terminology has been an ongoing difficulty in discussing evolutionary theory, none more so than that referring to 'grades' of evolution: micro, macro and mega. Like the originators of the Evolutionary Synthesis, Dawkins maintained Darwin's original position that all evolution took place gradually, but admitted that there may be "special occasions when macroevolutions are incorporated into evolution" (Dawkins 2003: 86). As an example of macroevolution, Dawkins had chosen the different number of vertebræ in various species of snake, which vary between 200 and 350. When a species of snake gained (or lost) one or more vertebral segments, that was macromutation (Dawkins 1996: 93). Dawkins (2003: 86-87) took this issue up again:

> I find it plausible, for instance, that the invention of segmentation occurred in a single macromutation leap, once during the history of our own vertebrate ancestors and again once in the ancestry of anthropods and annelids.

Here Dawkins was following closely in the footsteps of the Master, since Darwin also assumed that any change which he found 'plausible', or with which he 'had no difficulty' was necessarily, not only viable, but had actually occurred and could be taken as proven. Dawkins (2004: 502) admitted that it was difficult to imagine how the first creature having two segments instead of one could have survived, let alone found a mate and reproduced, but "it evidently happened" and that was all the 'explanation' needed. Previously Dawkins had written (2003: 86-87):

> The first segmented animal is born: a freak, a monster none of whose detailed bodily features equip it to survive its new segmented architecture. It should die. But by chance ... the segmented monster finds itself in a virgin part of the world where living is easy and competition is light ... it survives ... its descendants survive ..

He does start the next paragraph by saying:

> This is the kind of speculation in which we should indulge only as a last resort. The argument stands that only gradualistic, inch-by-inch walking through the genetic landscape is compatible with the sort of cumulative evolution that can build up complex and detailed adaptation.

Like the Master before him, Dawkins attempted to cite support for both sides of the argument, not for the first time.

In *The Blind Watchmaker* (p. 119) Dawkins had brought up the subject of chromosomes:

> Chimpanzees have 24 pairs of chromosomes and we have 23. We share a common ancestor with chimpanzees, so at some point in either our ancestry or chimps' there must have been a change in chromosome number. Either we lost a chromosome (two merged) or chimps gained one (one split). There must have been at least one individual who had a different number of chromosomes from his [sic] parents.

Dawkins wrote of *pairs* of chromosomes, making the reduction from 24 pairs to 23 *pairs* seem not unreasonable. In fact, chimpanzees have 48 chromosomes and humans have 46 and, as pointed out in my chapter on chromosomes, two chromosomes combining would reduce the number from 48 to 47, not 46, which would not be viable. It would be necessary for two pairs to unite to reduce the number to 46. Surely such a change would qualify as 'macro', even 'mega', mutation, not the gradualism of Natural Selection? And with whom would this 'monster' mate? On page 231 of the same book, when discussing punctuated equilibrium, Dawkins wrote:

> There are very good reasons for rejecting all such saltationist theories of evolution. One rather boring reason is that if a new species really did arise in a single mutational step, members of the new species might have a hard time finding mates.

For Goldschmidt, macroevolution was a complete change, such as from a two or three valved heart to a three or four valved one, or the appearance of the first hair or feather, the appearance of the first nerve tissue, able to transmit sensation occurring in one part of the body to another part, and so on. I consider a change in chromosome number macroevolution, but it would seem that Dawkins does not. Lack of a clear understanding and agreement of terminology is an ongoing problem.

A greater understanding of embryology, particularly in relation to Hox genes which regulate how DNA, which is the same in every somatic cell, manifests differently according to the position of the cell in the body, has allowed the duplication of body parts or segments, such as an extra leg, antenna or vertebra, along with associated blood vessels, nerves, etc., to be considered a minor mutation, or microevolution. Some people are born with an extra cervical, thoracic or lumbar vertebra, others are born with extra digits. Is this a mutation or an anomaly? Would it be an anomaly if it occurred only in one, or a few individuals (even if hereditary as in polydactyly) and a mutation if it spread throughout the whole population? Depending upon the circumstances, extra vertebræ could be considered an example of micro, macro or mega evolution. Dawkins defined macromutation as "change in a single generation" but nevertheless considered all change gradual, even that which was punctuational (Dawkins 1996: 93) because of the co-operative nature of genetic activity during development of the embryo. Presumably, change in chromosome numbers would be an example of macromutation under Dawkin's definition, since it is difficult to envisage such a change taking place over more than one generation. Until the profession is able to agree upon such definitions, much discussion will be at cross purposes.

Dawkins' ability to manipulate words and arguments to suit his ends was illustrated when he claimed (1982: 174) that "If you sleep in the sun with your hand over your chest, a white image of your hand will be imprinted on your otherwise tanned body. This image is an acquired characteristic". It is doubtful if any other evolutionary theorist would consider this image an acquired characteristic, any more than they would a scratch or a pimple, a further example of the urgent need for agreement upon the use of terminology. Dawkins (1996: 4, 208) continued to maintain that people who thought they saw evidence of purposeful design in nature were "wrong" and that "All questions about life have the same answer ... natural selection". He (1992: 174) defended himself against accusations of dogmatism by claiming that his views were not dogmatic, but based on reason.

Dawkins (2003: 22) argued that humans are apes: "There is no natural category that includes chimpanzees, gorillas and orangs but excludes humans". He invited his readers to imagine that humans and chimpanzees may once have formed a ring species (2004: 295) but differing numbers of

511

chromosomes would not have allowed this. However, the three species of zebra have different chromosome numbers, all less than 50, while all other members of the equine family have chromosome numbers greater than 60. Clearly, closely related genera/species are able to have different chromosome numbers, although exactly how this came about is not clear. The issue here is that Dawkins, as well as many other evolutionary theorists, does not appear to consider the chromosome question at all, when it clearly needs explaining.

Dawkins evinced an ambivalent attitude towards 'evidence' when he stated (1986: 313):

> Even if the evidence did not favour it [evolution by cumulative natural selection], it would still be the best theory available!.

Chapter 47

Is God a Delusion?

Dawkins had never disguised his atheism but over the years his hatred of all things Christian, indeed of all things theistic, became more pronounced and in 2006 he finally wrote a book, not about evolution itself, but about how Darwinian evolution had shown the concept of 'God' to be superfluous. This book, *The God Delusion*, was the first he had written for the express purpose of converting his reader to atheism, which he considered to be a 'consciousness raising' exercise (pp. 1-2). Not only was atheism a brave, splendid and realistic aspiration, atheism nearly always indicated a healthy independence of mind and, indeed, a healthy mind (p. 3).

Concentrating on the Abrahamic religions, because these were the ones with which his reader would most likely be familiar, Dawkins (p. 36) made his position clear:

> I decry supernaturalism ... I am not attacking any particular version of God or gods, I am attacking God, all gods, anything and everything supernatural, whenever and wherever they have been or will be invented.

Attacking certain beliefs, such as the doctrine of the Trinity, the Virgin Birth, the Resurrection, praying towards Mecca, etc., are not disproving the existence of God, *per se*. A belief in God does not carry with it the necessary supposition that the Supreme Being has such attributes as being infinite, perfect, immutable, suprapersonal, unqualifiedly omnipotent or omniscient (Tennant 1930/1968). Dawkins disputed certain such divine attributes generally accepted within the Abrahamic religions, such as omnipotence and omniscience, goodness, forgiveness of sins, willingness to listen to prayer - an innumerable quantity of them all at the same time, but answering only some - etc.

Dawkins pointed out that, if God were omniscient, then he would know all things that were going to happen, which would mean that he could not change them, which would mean that he was not omnipotent. This was an expansion of the old debate regarding predestination *versus* free will which had troubled Christian philosophers for centuries. Many Eastern philosophies accept that God himself may be evolving, may not (yet) be perfect and are, therefore, not troubled by this dilemma. Some Western philosophers, such as Schelling (1775-1854) agreed with this line of thinking (Gutmann 1936: 85):

> God is life, not mere being. All life has a destiny and is subject to suffering and development … Being is only aware of itself in becoming … in actualization there is necessarily a becoming … a distant future when God will be all in all … when he will be completely realized.

Dawkins considered that natural selection explained the whole of life and was the only parsimonious, plausible and elegant solution, the only workable alternative to chance that had ever been suggested. Previously, people, such as Paley, had argued that either the incredible diversity of biota which populated this Earth had appeared 'by chance' or it had been purposefully created. 'Natural Selection' allowed 'purposeful creation' to be discarded without recourse to 'chance' alone because nature, without thought or purpose, 'selected' that which was most serviceable.

Dawkins was derisively dismissive of personal religious/spiritual experience (pp. 88, 92):

> You say you have experienced God directly? Well, some people have experienced pink elephants … Peter Sutcliffe, the Yorkshire Ripper, distinctly heard the voice of Jesus telling him to kill women … George W. Bush says that God told him to invade Iraq … Individuals in asylums think they are Napoleon or Charlie Chaplin, or that the entire world is conspiring against them or that they can broadcast their thoughts into other peoples' heads.

> That is really all that needs to be said about personal 'experience' of gods or other religious phenomena. If you've had such an experience, you may well find yourself firmly believing that it was real. But don't expect the rest of us to take your word for it.

It was for this reason that Dawkins chose to title his book *The God Delusion*. He maintained that all people who believed in God were deluded. As a former psychologist, it is my professional opinion that not all religious/spiritual experiences are negative and not all people who hear voices or have visions are certifiable.

The Anthropic Principle

The *Anthropic Cosmological Principle* (Barrow and Tipler 1986) stated that we cannot but view the Universe through human eyes (understanding) because that is all we have. Enhanced as our senses may be by instruments, our comprehension of the Universe cannot encompass anything outside our understanding. On page six, they asked:

> There is only one Universe … where do we find the other possible universes against which to compare our own in order to decide how fortunate it is that all those remarkable coincidences that are necessary for our evolution actually exist?

The remarkable coincidences to which they were referring were the 'universal constants' necessary for life, such as the relative strengths of the nuclear and electromagnetic forces which allow the existence of the carbon atom, and the ratio of the number of photons to protons which must lie within a very narrow range to allow carbon-based life to arise (p. 5).

Dawkins (pp. 137-138) called upon the Anthropic Principle to justify the concept that, however statistically unlikely it was that a certain factor, or series of factors, would occur such that life, as we know it, could evolve, the fact that it has evolved showed that life was 'statistically' probable, however long the odds.

> It is estimated that there are between 1 billion and 30 billion planets in our galaxy and about 100 billion galaxies in the universe … A billion is a conservative estimate of the number of available planets in the universe. Now suppose the origin of life, the spontaneous arising of something equivalent to DNA, really as a quite staggeringly improbable event. Suppose it was so improbable as to occur on only one in a billion planets … life will still have arisen on a billion planets … a chemical model need only predict that life will arise on one planet in a billion to give us a good and entirely satisfying explanation for the presence of us here.

The existence of life on a planet may be a one in a billion chance but, since life does exist, that is evidence that a one in a billion chance is sufficient to explain the existence of life. To me, this argument appears tautological.

Other people continued to follow Paley, arguing that the design apparent in nature could not have come about by chance. Tennant (1930/1968) held that Nature 'forcibly suggested' that it was the outcome of intelligent design but that such an interpretation did not require that every detail was pre-ordained, merely the processes involved. Intelligent Design theorists, such as Behe, cited examples of complex design which they claimed were 'irreducible', to illustrate this point. Dawkins (p. 125) called this creating a 'God of the gaps' – conjuring up a God to explain any gap in current knowledge. Following on from the argument cited above, Dawkins (p. 139) claimed:

> This statistical argument completely demolishes any suggestion that we should postulate design to fill the gap ... Even so big a gap as this [origin of life] is easily filled by statistically informed science, while the very same statistical science rules out a divine creator.

Dawkins (p. 155) argued that to suggest a 'great unknown' was responsible for something existing rather than nothing was a total abdication of the responsibility for finding an explanation: 'a dreadful exhibition of self-indulgent, thought-denying sky-hookery'. As far as the argument from design was concerned, Darwin had 'blown it out of the water' (p. 79). While an admission of ignorance was an essential spur for scientific exploration, ignorance was welcomed by religious people such as those existing 'among today's less educated classes' (p. 77). He equated this with belief in the Tooth Fairy or Father Christmas, acceptable in the young and ignorant but deplorable in the educated adult.

Dawkins (pp. 139-141) gave the following explanation for the origin of life; already cited but which bears repeating:

> The origin of life was (or could have been) a unique event which had to happen only once ... We can deal with the unique origin of life by postulating a very large number of planetary opportunities. Once that initial stroke of luck has been granted ... natural selection takes over ... Nevertheless, it may be that the origin of life is not the only major gap in the evolutionary story bridged by sheer luck ... The origin of the eukaryotic cell ... was an even more momentous, difficult and statistically improbable step than the origin of life. The origin of consciousness might be another major gap whose bridging was of the same order of improbability.

> ... Natural selection ... needs some luck to get started ... Maybe a few later gaps in the evolutionary story also need major infusions of luck ... But whatever else may be said, design does not work as an explanation for life.

Dawkins replaced a 'God of the gaps' by a 'Science of the gaps'. He denied that he was a fundamentalist (pp. 282-283):

> I may well appear passionate when I defend evolution against fundamentalist creationists, but this is not because of a rival fundamentalism of my own. It is because the evidence for evolution is overwhelmingly strong and I am passionately distressed that my opponents can't see it ... The truths of evolution ... are so engrossingly fascinating and beautiful; how truly tragic to die having missed out on all that! Of course it makes me passionate. How could it not?

> Fundamentalists know they are right because they have read the truth in a holy book ... The book is true, and if the evidence seems to contradict it, it is the evidence that must be thrown out, not the book. By contrast, I, as a scientist, believe ... not because of reading a holy book, but because I studied the evidence.

Dawkins approached his atheism, his faith in the power of natural selection, as if it were a religion and deliberately used religious terminology to illustrate his point whenever he thought appropriate. The sub-title to *The Ancestors' Tale* was *A Pilgrimage to the Dawn of Life*. In *Climbing Mount Improbable* one chapter was titled *The Message from the Mountain* and another *The Forty-fold Path to Enlightenment.*

Of course, Dawkins was not the only person to despise Creationists or to use statistics to argue the position of natural selection. The geneticist, Steve Jones, also called Creationists 'bigots' (1993: 118) and the Creationist movement part of a 'triumphal new ignorance' (1999: 2). Jones claimed that the Creationists did, in fact, believe in evolution, even if they would not admit it to themselves, because they accepted the evolution of the AIDS virus. Jones undermined his own argument by then stating that these people believed the AIDS virus to be 'a useful illustration of God's wrath'. Clearly if Creationists attributed the AIDS virus to the action of God, they did not attribute it to the action of natural selection. Jones (1999: 8) drew attention to the fact that the AIDS virus mutated (evolved) in response to treatment with the drug Ritonavir:

> Every resistant virus, from London to San Francisco, has an identical mix of four mutations … What is more, and although they occur at random, evolution utilizes them in the same order each time … Each point in the viral RNA has an error rate of around one in ten thousand. The chance of the four changes happening at once is that figure, multiplied by itself four times … one in ten million billion, a total greater than the number of particles made in the entire course of an illness. It could never be reached by the accidents of mutation in a single individual, let alone within the hundreds who have evolved resistance. Evolution triumphs because it turns to natural selection, the plodding accumulation of error. It gives the virus the ability to generate the same improbable results each time it is challenged. AIDS is Darwinism unadorned.

I have no doubt that Creationists would use the selfsame figures cited by Jones to argue that there was plan and purpose behind these constantly similar mutations. According to Darwinism, nature can only select a result of random chance. How likely is it that random chance would give "the virus the ability to generate the same improbable results each time it is challenged"? The statistics of themselves prove nothing. Interpretation is everything.

Dawkins evinced a tendency to see things in black or white. He wrote as if people were either Creationists or Darwinists, as if they either believed in God or in evolution. Although it is true that one cannot be both a theist and an atheist (although one can be an agnostic), it is not true that one cannot believe in both God and evolution. Millions do. Nevertheless,

Dawkins tended to speak and write in dichotomous terms: educated/ignorant, good/evil, rational/delusional. Behe (p. 250) cited an interview printed in the *New York Times* (9.4.1989) in which Dawkins reportedly said: "Anyone who denies evolution is either ignorant, stupid or insane (or wicked) but I'd rather not consider that". I firmly believe in evolution but I have known some highly intelligent, well educated people, who do not. I do not understand their position, but I respect it. All of these people have belonged to one or another Christian sect and all were very good people – far more 'good' than many others I have known!

In 2003, Dawkins published *A Devil's Chaplin*. As with *The God Delusion*, this book was aimed at discrediting religion. He invited his reader to imagine a world without religion. There might have been "no suicide bombers, no 9/11, no 7/7, no Crusades, no witch-hunts, no Gunpowder Plot, no Indian partition, no Israel/Palestinian war, no Serb/Croatia/Muslim massacres, no persecution of Jews" and the list goes on (2003: 189; 2006: 1-2). Dawkins pointed out that Hitler was not an atheist, having been brought up a Catholic and keeping strong ties with Catholic Italy throughout his rule. Having disposed of Hitler, Dawkins was strangely silent about Stalin, who undoubtedly was an atheist, as was Chairman Mao. After the fall of communism, papers were released during the 1990s which showed that Stalin's regime may have been responsible for the death of up to 100 million people, approximately ten times more than Hitler. No one knows how many people died during China's Cultural Revolution, or since, but it is likely that the death toll was closer to that under Stalin than that under Hitler. It seems likely that two major atheists were responsible for more imprisonment, torture and death in fifty years during the 20th century than the three Abrahamic religions had been during the more than two thousand years of their entire existence. A certain human personality type seems to find itself a 'cause', be that 'cause' religious, political or territorial, and uses that cause to advance themselves to a position of power, caring nothing for the suffering of others. They are known as sociopaths. Fortunately, not all sociopaths have illusions of world domination!

For Dawkins, the question was not whether evil individual human beings were religious or not, but "whether atheism systematically *influences* [his italics] people to do bad things. There is not the smallest evidence that it does" (2006, p. 273). Dawkins justified his campaign to have all vestiges of religion excluded from American State Schools on the grounds that religion exerted an evil influence on the young. Surely the history of the 20th century – not forgetting Pol Pot and the Khmer Rouge in Cambodia – would suggest that there is just as much evidence of evil among atheists?

It is neither religion nor politics which are responsible for evil deeds. People gravitate towards philosophical (religious/political) doctrines which are closest to their inner beliefs. If none are available, most people hold their opinions close within their hearts but a few are passionate enough to found new organizations, either religious or political. Modern dictators, from Cromwell to Napoleon, Hitler, Mussolini, Stalin, Chairman Mao, Idi Amin, Jomo Kenyatta, Pinochet, Peron, Hussein, Ayatollah Khomeini and many more, may have claimed a religious or political agenda to justify their actions, but all were driven by the force of their own personality and inner (self) belief.

If religion were to be cited as the underlying cause of undesirable (evil) behavior, it must also be cited as the underlying cause of desirable (good) behavior. Hundreds of thousands of people give credit to their religion for their charitable works. It is here argued that these good works flow from the inner nature of these people, who are drawn to religion to satisfy an inner need. 'Religion' can be good or evil; so can 'science'. Some scientists devote their lives to the relief of human suffering; others make weapons of mass destruction.

Like Dawkins, Alister McGrath undertook both under- and post-graduate studies at Oxford University but, by contrast, McGrath's studies led to his abandonment of atheism and to his becoming an ordained priest within the Church of England. McGrath was invited to pen a response to Dawkins' first book, *The Selfish Gene,* but, being a recent graduate at the time and newly converted, McGrath did not feel he was competent to undertake the task. Eventually, in 2005, McGrath published *Darwin's God,* chiefly in response to Dawkins' *The Blind Watchmaker* and *Climbing Mount Improbable.* Had he waited a little longer, he could have responded to *The God Delusion* at the same time. McGrath's book is a far more comprehensive disputation of Dawkins' atheistic views than can be undertaken here, but the main thrust of his argument was that Dawkins tended to define certain concepts, such as 'faith' or 'God' in a manner which suited him, then proceed to demolish these concepts even though the definitions Dawkins used bore little resemblance to the definitions used by other philosophers. He objected to what he saw as Dawkins' portrayal of 'religious folk' as "dishonest, liars, fools and knaves, incapable of responding honestly to the real world and preferring to invent a false, pernicious, and delusionary world in which to entice the unwary, the young and the naïve" (McGrath p. 9).

Having rejected the Christian religion in which he was raised, Dawkins rejected God. He make statements such as "All questions about life have the same answer: natural selection" (1996: 298) and "Ever since Darwin we have known why we exist .." (2003: 191), allowing no place whatsoever for any 'Superior' intelligence or 'Creative' force. However, the Christian concept of God is not the only one. Understandably, there is some similarity of thought between the three Abrahamic religions. For example, all three conceive of 'God' as being 'jealous'. There is only one God who does not tolerate the acknowledgment of any other 'deity'. He receives into Heaven only those who worship Him alone and follow His edicts. The Jews have interpreted this doctrine as an indication that they, as the Chosen People, should keep themselves separate from other people, keeping their own customs and frowning upon intermarriage with gentiles. The Christians adopted a different interpretation. They saw a need to convert as many people as possible to Christianity, preferably by persuasion, but at times by force, since only Christians would inherit eternal life. The Muslims, over the centuries, have mostly kept themselves to themselves but, when the occasion arose, could, and did, insist that people either convert to Islam or be killed. They, too, held that only those accepting the teachings of a particular individual, in this case the Prophet, would enter Heaven. These three religions dominated Western thought. Since Dawkins was writing for the Western reader, he directed his criticisms against the three Abrahamic religions, mostly Christianity.

Eastern religion, Hinduism and its derivatives, including Buddhism, was completely different. I have already spoken of the Hindu understanding that the solidity of matter was but an illusion, that the only reality was energy. While acknowledging the existence of 'God beyond Creation' - Atman - within Creation, the Consciousness of God, the Mind of God (Brahma), manifested as energy, from which sprang all Creation, all Universes, past, present and future. Thus the illusion, or delusion, is matter, not God! Rather in the same way as water can manifest, at a higher rate of vibration, as steam and, at a lower rate of vibration, as ice, so God can manifest as Spirit, soul (emotion) and body (matter). These vibrations manifest in many forms, principally sound and light. Both exist in forms undetectable to the human eye or ear - although scientists are now in possession of wonderful technology which is expanding our understanding. In the same way that the 'one light', which we call 'white light', can be split into the seven colours of the rainbow (and other 'colours', such as infra-red and ultra-violet that we cannot see), so the 'one light' of God split into other 'gods', 'spirits', each vibration having a particular part to play in the coming into existence of our world, with its living matter. There was constant dividing and branching of these energy rays until they eventually manifested as individual souls. Before this final division, there were group souls. The

Native Americans also recognized, for example, the spirit of the bear or the spirit of the eagle. The Great Bear Spirit watched over bears. Lesser Bear Spirits watched over groups of bears, or individuals. It was the same with humans. I sympathize with Dawkins' difficulty in imagining the one God trying to listen to, and fulfill, the conflicting prayers of millions of people all at once. Happily, there is another solution. Lesser 'deities' helping out! This was the belief of the early Church, and still is of the Roman Catholic Church of today. Rather than pray to God the Father, Catholics usually pray to Mary, or Jesus, or one of the saints. Each individual has a guardian angel – or two – who constantly watches over them. In this idea, Catholicism and Hinduism are of one accord.

I have already spoken of the Hindu belief in re-incarnation, of individuals and of universes. For the Hindu, change is constant; everything is changing, evolving, moving forward, even God, who 'learns' and 'grows' through his experiences as his creation. Everything we feel, God feels. Indeed, we could feel nothing if God did not feel it.

In the same way that Christians bring to mind different aspects of the Divine by referring to the Good Shepherd, the Sacred Heart, the Light of the World, and so on, so Hindus bring to mind different aspects of God by different names: Ishwara, the manifestation of God as light, Agni, the manifestation of God as fire, and so on. Very sensibly, recognizing that death is a part of life, they even had a god (or goddess) of destruction – Kali. The Hindu pantheon had 108 such names and it mattered not to which manifestation one chose to address one's prayers and offerings, since all are One.

Wallace, who you may remember came to accept Spiritualism, also believed that different energies, working together, were responsible for the world as we know it. In this respect, Wallace's view was diametrically opposed to that of Darwin. The (Western) Darwinian view was that consciousness and life arose from matter. The (Eastern) Wallacian view was that matter and consciousness arose from life. 'Life' (energy) is eternal; matter is not. One can but wonder how different may have been the doctrine of evolution by natural selection, and the results of its incorporation into the philosophical and political thinking of the 20th century (which will be discussed in the next chapter) had Wallace asserted his opinions more forcefully.

Chapter 48

A Never-Ending Story

Notwithstanding the many criticisms which have been levelled at Darwin's theory of evolution by natural selection, it continues to enjoy a high standing among the scientific community. Indeed, its standing may be as high now as it has ever been, if the decision taken at Dover is indicative of opinion in other parts of the Western world. Dennett (1995: 21) wrote: "If I were to give an award for the single best idea anyone has ever had, I'd give it to Darwin, ahead of Newton and Einstein and everyone else". Zimmer (2001: p.336) was more cautious, merely claiming that "the theory of evolution stands as one of the greatest scientific accomplishments of the past 200 years". However, when one considers the magnitude of scientific accomplishments over the past 200 years, this is still some claim! Add the assertion made by people, such as Dawkins, that natural selection *alone* accounts for *all* evolution, and the general public may be forgiven for thinking that the story has been told, the plot has been uncovered, there is nothing left to add except a few minor details – rather like the final scene in some television detective drama when points are clarified for the viewer as the two detectives walk serenely away.

How justified is this impression?

The number of atheists in the community is rising, according to statistics cited by Scott (2004: 56) which showed that in 1996, nine per cent of Americans identified themselves as atheists, while by 2001 this figure had risen to fourteen per cent. This still leaves well over 80% of Americans professing a belief in some sort of God, or Superior Intelligence, although not by any means all Christian. The strong fundamentalist movement in America may have resulted in a greater number of 'believers' there than in other countries.

Falls in Church attendance do not necessarily indicate falls in belief in any type of Divine Being. Since the 1960s, there has been an increased awareness of Eastern philosophies, associated with the interest in yoga and the 'Hippie' movement which took place at that time. The smoke of incense became merged with the smoke of marijuana, meditation replaced prayer in the search for Oneness with the Universe. Beliefs were changing but they were not necessarily fading. Some of the apparent increase in atheism may even have stemmed from an increasing willingness by survey participants to identify themselves, whereas before persons of no particular ideology may have identified themselves as 'Christian' as a formality.

With so many people, among them scientists such as Thaxton et al. (1984), Behe (1996), Miller (1998) and Wolf (1999), holding beliefs in a Divine Being, how firmly established is the atheistic (scientific) approach to natural selection?

Prebiotic evolution

Evolutionary theorists may 'draw the line' on their area of interest at the commencement of life, but can all scientists? If the laws of physics and chemistry are the same throughout the universe, and have been the same since the beginning of time, when did they start and why did they come into existence? Did they start before the Universe appeared, the Universe coming into being in accordance with these pre-existing laws? Why should laws pre-exist for something non-existent - and unplanned? Or did the laws come into existence gradually, evolving, as the Universe itself was unfolding? If so, which was the first law from which all others devolved?

Thaxton et al. (1984: 2-3) pointed out that at some point life appeared *de novo* out of *lifeless*, inert matter and that a pre-biological stage of evolution, such as chemical evolution, must be included in any general theory of evolution. Did chemistry come into existence at the same time as physics, or does physics ante-date chemistry - i.e. did energy exist before matter? How much do energy patterns influence biological evolution? Haldane (1932: 4-6) pointed out that each one of us passes from a state of "unconsciousness" as a gamete to that of "consciousness and reason", claiming that, because this passage is an everyday occurrence, it is 'normal', not 'miraculous'. Merely stating that some process is 'normal' does not in any way explain how it came to occur in the first place.

According to quantum physics, not only does energy pass through space between planetary bodies, it also exists as fluctuating positive/negative energy fields within each atom, in the space between sub-atomic particles

(Wolf 1999: 96). Energy fields, such as radiation and magnetism, are emitted by 'inert' (non-living) minerals, and there may be other energy fields emitted by crystals which do not find their place in routine texts. Crystals are certainly capable of 'receiving' and 'transmitting' energy, as witnessed by the earlier 'crystal' radio sets and, of course, the 'silicon chip'.

As has been pointed out, RNA, DNA and viroids are all crystalline in structure. Quite why these particular crystal structures should have the ability to self-replicate, albeit only in association with an organic cell, is not yet understood. In a statement which foreshadowed the thinking of Dawkins, Simpson (1950: 14) wrote:

> Our concern here is with the record of evolution, and there is no known record bearing closely on the origin of life … there is no theoretical difficulty … in the chance organization of a complex carbon-containing molecule capable of influencing or directing the synthesis of other units like itself … this first form of life was a "prototype" which, after the chance basic chemical combination into an organization was capable of reproduction and of mutation …

As the second half of the twentieth century progressed, simple postulation of such 'chance' events became less acceptable. Attempts were made to produce living organisms from various forms of pre-biotic 'soup', but none were successful. The 'protocells' produced did not have internal structure, such as DNA or enzymes (Thaxton et al.). There has to date been no successful attempt to produce a genuine form of life in the laboratory. Dennett (1995: 155, 203) may have claimed:

> Once we are allowed simply to postulate organized complexity, if only the organized complexity of the DNA protein replicating engine, it is relatively easy to involve it as a generator of yet more organized complexity …

> Love it or hate it, phenomena, like this [RNA] exhibit the heart of the power of the Darwinian idea. An impersonal unreflective robotic, mindless little scrap of molecular machinery is the ultimate basis of all the agency and hence meaning and hence consciousness, in the universe.

… but this is yet to be proven. Like Dawkins, Dennett appeared to be relying on some measure of luck by asking to be allowed 'simply to postulate' such an important step.

During the 1990s and the first decade of this century, modern technology allowed sequencing of Neanderthal mtDNA to be attempted. This was not without difficulties, due to the amount of handling the specimens have received. Krings et al. (1997) first isolated mtDNA from the original Neanderthal type specimen, now referred to as Feldhofer, and calculated that Neanderthals had evolved separately from modern humans for between 550,000 and 690,000 years. However, they did caution that, while their results indicated that Neanderthals had not contributed mtDNA to modern humans, it could not therefore be assumed that they did not contribute nuclear DNA. It was considered expedient for all modern mtDNA recovered during this work to be considered 'contamination'; only residual mtDNA, not found in modern humans, could definitively be considered endemic to the Neanderthals (Krings et al. 1997, Krings et al. 2000, White et al. 2003; Green et al. 2006). As a non-geneticist, I was puzzled by these results. It seemed to me that if all traces of modern human mtDNA were excluded from consideration on the grounds that they were most likely 'contamination', how could the results show anything other than that the Neanderthal mtDNA was not modern?

A second mtDNA sequence from a Neanderthal child from the Mezmaiskaya Cave in the Northern Caucasus, dated to ~42,000 B.P., was found to be similar to that of the type specimen (Krings et al. 2000). Krings et al. found that this mtDNA was no more similar to that of modern humans from Europe than it was to mtDNA from any other part of the world, strengthening their belief that the Neanderthal population had been an isolated one. However, they did repeat their caution in regard to extrapolating from mtDNA to nuclear DNA, reiterating that Neanderthals may have contributed to the contemporary human gene pool in a way not shown by mtDNA.

Ovchinnikov et al. (2000) undertook a further analysis of the Neanderthal mtDNA from Mezmaiskaya Cave, which they reported had been reliably dated to ~29,000 B.P. (27,000 B.C.), making it one of the latest known Neanderthals. They estimated that divergence from modern humans had occurred between 365,000 and 853,000 B.P. and further estimated that the most recent common ancestor of the eastern (their sample) and western (Feldhofer specimen) Neanderthal lived between 151,000-352,000 years ago. This is a long time for genetic isolation over what is quite a small geographical area and would indicate that the Neanderthals were very insular. Hawks and Wolpoff (2001) believed the Mezmaiskaya Cave burial to have been intrusive, since the remains dated to 29,000 ya but were found in a Mousterian layer with animal bones dated to >45,000 ya. They argued that the child had few Neanderthal features and any similarity of mtDNA with that of the Neanderthals argued more strongly for the occurrence of interbreeding than it did for the Neanderthal population having been an isolated one, as claimed by Krings et al (2000) and Ochninikov et al.

Further analysis of sequences from eleven Neanderthals found contamination with modern mtDNA in amounts varying between 99% and 1%, the 1% being that of one of the bones dated to 38,000 BP from the

Vindija Cave, Croatia (Green et al. 2006). This one bone was used for further analysis. The estimated time of divergence of the Neanderthals from archaic humans was 461,000 to 825,000 years. It was further estimated that the ancestral population had between 0 and 12,000 persons. Considering ancestral alleles (those found in humans and chimpanzees), they identified ~30% common to Neanderthals and modern humans and concluded this high level of derived alleles in the Neanderthal was incompatible with the simple population split model. They further concluded that their results could suggest gene flow between modern humans and Neanderthals. The conclusion that some interbreeding had taken place between the Neanderthals and the newly arrived Cro-Magnon people *(Homo sapiens)* was confirmed by Green et all. (2010).

Analysis has shown that anatomical features and the mtDNA of particular individuals may have different evolutionary paths, and some nuclear gene lineages have genealogical and/or geographical patterns that are different from those of mtDNA. This difference limits the use of ancient DNA in tracing human evolutionary history. Thus the work of Krings et al. (1997) which showed that mtDNA extracted from the Neanderthal type specimen was outside modern human mtDNA variation, and that of White et al. (2003) which supported an early divergence of the Neanderthal line from that of modern humans, do not mean that the Neanderthals did not contribute other genetic material to the modern human gene pool.

Let it be remembered that, if one group of ancient people overran another and took the females of the conquered people to wife, it would have been the mtDNA of the conquered people which would have been passed on to future generations of such unions.

Research into nuclear genetic variability between humans today had shown small, but measurable, differences between geographical groups, sufficient to trace probable migrations of peoples around the globe (Jones 1993, Sykes 2001). Some populations are now known to have travelled thousands of miles to settle in completely new homelands; others appear hardly to have moved, or even interbred with neighbouring tribes. The pattern which has emerged is one of long-term stability, punctuated by intermittent migration. This pattern was expected by proponents of the theory of Multiregional Evolution, since it allowed a sufficient degree of genetic isolation for recognizable regional characteristics to have evolved, while retaining enough movement of people/genetic material to maintain unity of species.

Templeton (2002) argued that certain 'significant genetic signatures' would have been eliminated had complete replacement taken place and the continued presence of these signatures was a strong indication that any 'Out of Africa' expansion event which may have taken place had been accompanied by interbreeding. He interpreted the genetic evidence as indication of at least three major expansions of humans out of Africa, including the original exodus ~1.7 mya, although there may have been more.

Studies of the male Y chromosome estimated the time of the latest common ancestor to be ~188,000 years (51,000 to 411,000 years) with a long term population of 10,000 (Hammer 1995; Jones 2002). Hammer concluded that such a population size was too small for Europe, Africa and Asia to have been continuously inhabited by so few individuals, thus undermining the Multiregional position. All three areas have long gaps in their fossil record. The exact locality of human beings throughout the approximate two million years of their existence is not known, whether they migrated en masse or splintered into different populations. That the Y chromosome should give results so closely approximating those of the female mtDNA is not surprising. We all have one female and one male parent, and that has been true since bi-sexual reproduction was adopted by some primitive life form. The amount of time which has elapsed since we all had a common female ancestor must be the same as that for our last common male ancestor.

Not only does internal environment influence the expression of the gene and its DNA, it is now becoming clear that the external environment can also have an effect. It has been found that plants and animals which live in an impoverished environment will not only be stunted in their own growth, they will give rise to undersized offspring, be they plant or animal, even if the offspring are raised in a nutritionally normal environment (Jones 1999; 2002). The effect may continue for several generations, raising the possibility that Kammerer's work may soon be revisited.

Even more extraordinary than some of Kammerer's work was the case of fruit fly eggs, treated with ether, producing mutants with two pairs of wings (Hollick 2006). After several generations of treatment, untreated eggs from this stock produced a percentage of flies with two sets of wings, showing that this characteristic had been inherited. Furthermore, eggs from flies whose ancestors had never been treated started to hatch with a proportion carrying the same mutation! This led Hollick (p. 202) to suggest that genetic mutations may be influenced by something which he called a 'morphic' field.

Morphic fields

Physics recognizes several distinct energy fields, magnetism and gravity being the two most commonly known. The field of gravity has exerted more influence on living matter than any other environmental factor. Every living thing, from the blade of grass pushing upwards through the soil, to the snail crawling across the surface of the Earth, to the eagle soaring in the sky, must adapt to the requirements of gravity before it can exist as a living entity. Even microscopic beings are subject to gravity, although largely cushioned from its immediate effects by the fluid in which they live. A vacuum is now referred to as a 'vacuum field' and some physicists believe this field to be so intense that one cubic centimetre of space contains more energy than all the matter in the known universe (Hollick p. 123). This is not the place to consider matter/anti-matter, subatomic particles or quantum theory, but all of these are essential to any theory of evolution, human or otherwise, since without energy fields there can be no matter, and without matter there can be no evolution. "Matter is like flecks of foam on the seething surface of the ocean of space" (Hollick p. 124).

The concept of energy fields influencing evolution may be traced back to Smuts (1926) but the 'Holistic Selection' he proposed was purposeful, in some ways reminiscent of the purposeful selective ability in Nature (of a Being of great penetrative insight) that Darwin had proposed in his *Foundations* essay. Sheldrake (1988) proposed that when a chemical substance crystallized for the very first time, it formed around itself a morphic field. Each time crystals of this chemical were formed, the field became stronger and crystals of the same form were more likely to develop. It was known that when a new chemical compound was formed in the laboratory, it was usually difficult to crystallize but subsequent crystallization become easier - all over the world! (Sheldrake p. 131, Hollick p. 133). Sheldrake did not accept that experience of the scientist alone was responsible. He postulated the existence of the morphic field with which that particular chemical resonated, wherever it formed.

On a larger scale, certain birds (tits) throughout Britain rapidly learned to pierce foil milk bottle tops once birds in one area had learned the skill, and isolated groups of monkeys started to throw grain into water to separate it from sand once this practice had emerged in one group (Hollick p. 216). Sheldrake (p. 197) suggested that there were morphic fields for behaviour and memory and that we may resonate with the thoughts and behaviours of people similar to ourselves who had lived in the past. This is somewhat similar to Jung's (1963) Universal Consciousness and Dawkins'

(1976) memes. Dawkins postulated the existences of 'bits' of memory which operated like 'bits' of matter (for example DNA), which could be passed on (spread) throughout a community, gradually changing thought patterns of groups of people.

Crystals, as well as other 'inert' substances, are known to resonate with certain energy fields. A change in the environment, however caused, would change the morphic field of the environment in which any population of any species lived. A simple shift in geological strata, such as following an earthquake or flood, would affect the total energy field for a certain area. Climate change would affect flora and fauna. Indeed, there would be a 'snowball' effect in changing morphic fields as certain species either changed or became extinct, for example during Ice Ages. If these changes were able to influence the crystalline DNA, would that offer an explanation of how DNA mutations in the same direction could occur at the same time in a given population?

Since the time of Fisher and his fellow evolutionary theorists, whose work resulted in the establishment of the Modern Evolutionary Synthesis, it has been known that DNA mutations are genetically recessive, i.e. are not expressed morphologically, and are not, therefore, available to be acted upon by natural selection. As explained earlier, during the 1920's Fisher established mathematically that, all other things being equal, the gene pool would remain stable: DD, Dr, Dr, rr. The only way in which the new recessive gene could become established and take on a dominant role was for it to be 'selected', to be more reproductively successful. In order for this to happen in sexually reproducing organisms, the recessive gene had to pair with another recessive gene being carried, at the same time and in the same place, by an organism of the opposite sex, with which it chanced to mate. However, its offspring would most likely mate with a DD organism, since in the early stages the new r gene would be extremely rare in that particular population. While the recessive gene would survive in its Dr state, its recessive characteristics would not be expressed morphologically. In order to become established, it would seem, the recessive gene needed already to be established! This difficulty is similar to that of the establishment of changing numbers of chromosomes.

If all organisms in a given population were subjected to the same change in morphic fields, and if morphic fields affected crystalline DNA, would it be possible for sufficient members of a population to mutate in the same direction for there to be suitable mates available to enable the new form to become established? Possibly. Whether this scenario would be valid in the case of chromosome mutation is doubtful.

Darwin extrapolated from domesticated plants and animals to wild flora and fauna to support his hypothesis. His work was continued in the twentieth century with laboratory animals, in particular the fruit fly (*drosophila*). Now laboratory work has descended to an even smaller level, that of micro-organisms. In support of evolution by natural selection, Miller (1999: 52) cited two experimental results, the first being the 'evolution' of a new strain of bacterium resistant to a particular antibiotic. In the laboratory the bacteria's genes were randomly mutated and those which showed resistance to the antibiotic were selected. A selected gene was then 'chopped' into small pieces, allowed to combine randomly into new sequences which were then re-inserted into new cells. This randomized swapping was seen to be close to the gene shuffling which takes place during sexual reproduction, even though the bacteria upon which the experiment was conducted do not reproduce sexually. After three repetitions of this procedure, a mutant protein 32,000 times more effective against the antibiotic than the original was produced. Considering the amount of interference by the experimenter, how justified was it to call this a replication of natural selection?

The second experiment cited by Miller (p. 52) utilized a similar technique to find a "powerful and efficient DNA enzyme that would cut RNA", which result was achieved "in just a couple of days". This was particularly striking since, according to Miller, no DNA enzymes are known in nature. The first genetic material is believed to have appeared 3,000,000,000 years ago (Jones 1993: 87). If, in all that time, no DNA enzyme had been produced by natural means, how similar to natural processes was the one that produced this result in a couple of days? Dennett (p. 156) stated that short strings of RNA could replicate themselves, without assistance from enzymes, but that longer strings could not replicate without a "retinue of [enzyme] helpers". One would need to be a geneticist to understand the exact nature of the assistance these enzyme helpers provided.

Studying the 'fit' between a hormone 'key' and its receptor 'lock', scientists broke off part of a receptor, so that it no longer fitted the lock (Miller 1999: 144):

> Now the researchers let evolution take over ... they randomly mutated the coding regions for five amino acids in the growth hormone, generating roughly 10 million different mutant combinations. These mutant growth hormones were then selected to find ones that could bind to the previously mutated receptor ... Their random mutation strategy generated a new version of the hormone that fit[ted] the mutated receptor nearly one hundred times higher than the nonmutant version.

Interesting as these results are, surely they could be considered examples of some form of 'guided' evolution as much as evolution by random processes?

Jones (1993; 1999) argued along the same lines as Miller. As his example, Jones (1999: 2) used the AIDS virus:

> Real bigotry had to wait for modern times. The creationist movement is part of a triumphal New Ignorance … In fact, the majority of those determined to tell lies to children believe in Darwin's theory … Creationists find it easy to accept the science of AIDS. Its arrival so close to the millennium and the Last Judgement is a useful illustration of God's wrath. Homosexuals, they claim, have declared war on nature … fundamentalists admit the evolution of a virus as nature's revenge.

This is not correct. Fundamentalists consider AIDS God's revenge, not Nature's. What they tell their children is consistent with their core beliefs.

As pointed out earlier, Jones (1999: 8-9) used the evolution of the AIDS virus, and its mutations, in response to the drug Ritonavir, as his example:

> Every resistant virus … has an identical mix of four mutations. What is more, and although they occur at random, evolution utilizes them in the same order each time … Each point in the viral RNA has an error rate of around one in ten thousand. The chance of the four changes happening at once is that figure, multiplied by itself four times — one in ten million billion, a total greater than the number of particles made in the entire course of an illness. It could never be reached by accidents of mutation in a single individual, let alone within the hundreds who have evolved resistance. Evolution triumphs because it turns to natural selection, the plodding accumulation of error. It gives the virus the ability to generate the same improbable result each time it is challenged.

It is not clear why the natural processes of 'plodding accumulation of error' should be capable of such a swift, sure and effective reaction. No doubt Fundamentalists would consider these changes in resistance had happened as the result of some form of Divine guidance. That they happened is not in dispute, but protagonists on both sides would be able to draw conclusions from the same facts to support their position.

The bombardier beetle

Continually, all sides agree as to fact, but disagree as to interpretation. Bombardier beetles squirt a mixture of two chemicals from their rear when threatened. Creationists claim that the storing of these two chemicals in separate chambers, and their subsequent explosive combination on expulsion, could not have come about by chance. Why would these two chemicals, inert on their own, be brought into juxtaposition if not for the 'purpose' of creating an explosion of noxious, caustic and boiling hot

(Dawkins 2004: 490) liquid with which to repel its foes? Dawkins related how he proved to an audience of children the error of this argument by combining the two chemicals together in a test tube. Nothing happened! For the reaction to occur, a catalyst was needed (Dawkins p. 390): "In nature, the beetle provides the catalyst, and would have had no difficulty in gradually and safely increasing the dose over evolutionary time". Quite why the need for a third reagent should make the process less purposeful was not explained.

Entropy

When life departs, the physical form breaks down, decays. This is true of all life, be it animal or vegetable. Opponents of atheistic evolution point out that bodies gradually becoming more complex is contrary to the second law of thermo-dynamics, that of entropy. Even the mineral kingdom is subject to breakdown - erosion - as pointed out by Lyell and other early Earth scientists. Some force, usually heat and/or pressure, is required for the production of a new substance, which is then subject to entropy. This entropy is the basis of radio-active decay.

Supporters of evolution by natural means have pointed out that the second law of thermo-dynamics, entropy, applies to closed systems. Since the Earth/Universe is not a closed system, entropy does not apply (Thaxton et al. 1984: 183; Shapiro 1986: 231; Shermer 2006: 81).

Chimera

If there is a heaven, no doubt Mendel smiled down as the twentieth century unfolded, justifying and elaborating his basic theory. If Mendel was smiling, no doubt Sir Robert Peel was laughing, as the science of genetics was put to forensic use. DNA evidence was seen by the Courts as being incontrovertible. As the twenty-first century dawned, a small chink appeared in the DNA armour.

That each body contained more than one type of DNA had long been known - mitochondrial DNA was different from nuclear DNA - but in December 2002, something strange occurred in America (Wolinsky 2007, Lam 2010). Routine DNA tests were carried out as part of the processing of a claim by single mother, LF, for state benefits, since she was now separated from the children's father, whom she had never married. The state required a paternity test and in the spirit of 'equal rights - equal responsibilities', the mother's DNA was also sampled. The results showed that the father was indeed the father, but that the mother was not the mother! The Department became convinced that the two were working some type of scam, to receive benefits for the father's children even though he was separated from the real mother and now in a relationship with LF, in order that she could claim benefits, despite the protestations of LF's gynaecologist, who had witnessed the births of both her children.

It so happened that LF was pregnant with a third child. The court ordered that someone be present to witness the birth and ordered that blood samples should be taken from the baby immediately after its birth. The results were the same. LF was not the mother! The court then opined that LF was a surrogate mother.

Meanwhile – in Boston, Massachusetts, KK was in need of a kidney transplant and her family's DNA was tested in the hope of finding a suitable donor. She, too, was shown not to be the mother of her own children! This time, the hospital sought answers. They tested a sample of her thyroid tissue, stored from a previous operation, and found that she had a second set of DNA in her body! She was a 'chimera'. This phenomenon is believed to occur when fraternal twins, developing in utero, fuse to form one person. This case being reported in a medical journal, came to the attention of people handling LF's case and DNA tests of her extended family showed that her brother had the 'missing' DNA.

Mitochondrial analysis was undertaken to confirm the identity of remains believed to be those of the murdered Russian royal family by comparing their mtDNA with that of known living relatives. It was found that the Tsar had two distinct types of DNA (Jones 1999: 120). Jones did not specify whether the two distinct types were nuclear or mitochondrial but it is to be assumed that it was the latter because he went on to comment "... the United States Army records the genes of its soldiers and sometimes has cause to compare them with those of their relatives. Again change is rapid, with a mitochondrial mutation in every forty parent-child comparisons". It would seem that not only is nuclear DNA not an infallible source of ancestor/descendant information, but even mitochondrial information may be more subject to change than previously thought.

It will never be known how common is the chimera phenomenon, since the only way of knowing would be to test every organ in the bodies of large numbers of people to make sure all had matching DNA – or not. If false negatives can occur between mother/child, they must also be able to occur between father/child. How many false negatives have proved that the 'accused' man was not the father of a child when, in fact, he was?

Be that as it may, if DNA is not a completely reliable indicator between one generation of parent/child, how reliable an indicator is it over thousands, or tens of thousands, of years?

The existence of chimera within the animal kingdom is being increasingly recognized. Kroher (2002) mentioned the rare male tortoiseshell cat, fertile because it carries both male and female chromosomes as a result of the fusion of a male and a female sibling in utero. This example, like that of the two mothers cited above, are 'accidents' of nature. Other creatures, such as the marine hydroid, *Hydractina echin*ata and the star fish, *Luidia sarsi,* emerge from lava, which do not die as most lava do, but continue to live independent lives (Kroher 2000). The lava and the independent adult form have separate genetic identities.

That two different DNA molecules should co-exist within one entity, be that entity one which retains its primary form throughout life or one which metamorphoses, is no more extraordinary than that all the different organs and tissues of the body should contain the same DNA yet develop into such different forms. The early understanding of genes as individual 'beads' on a chromosomal 'string', each responsible for one piece of development, was soon abandoned when it came to be realized how dependent genetic material was on its immediate environment for its ongoing development and the final form of the tissues being developed. Some genes were seen to be 'control' genes. Calling some genes 'Hox genes' may be a convenient way of indicating the role these genes play within the developing embryo, but the name itself does not have any explanatory power as to how these genes came to acquire their important role. The body's immune system is able to recognize foreign DNA/tissues, although quite why it was so necessary for the body to be able to repel foreign DNA may not immediately be clear, since the introduction of foreign DNA into a body is something which has occurred only recently due to medical advances. Blood cells contain no nucleus and are unable to replicate themselves and blood is the one substance which is able to be transferred from one individual to another without side effects, provided that the two individuals are of matching blood type. Blood is the substance most likely to be accidentally introduced into another body in the wild and it might have been expected that an adverse immune reaction to the transfer of this substance would have been the first to evolve, if nature deemed an adverse immune reaction necessary at all.

Chimpanzees have only A and O blood groups; gorillas have only B (Jones 1993: 128). If humans evolved from chimpanzees, how, when, where and why did they acquire the B group? Different ethnic human populations tend to have different preponderances of certain blood groups and these differences must have evolved quite recently, if the Complete Replacement hypothesis is correct. Rhesus negative blood is most common in Europe, although it does also occur elsewhere (Jones p. 255). I have already raised the possibility of Neanderthals having had rhesus negative blood, of whether this was the reason why it is found most frequently among European peoples and whether this might have explained, if interbreeding did occur, why so few Neanderthal babies survived? The continuance of this blood group, despite its potentially lethal consequences, does not comply with the assumption of many early evolutionary theorists that negative variations would quickly be eliminated.

Hobbits

While some microbiologists in the northern hemisphere were seeking the answer to evolutionary puzzles beneath the microscope, archæologists in the southern hemisphere were adding another piece to the jigsaw puzzle of human evolution out in the field.

The October 2004 issue of *Nature* contained two articles, one co-authored by seven people who had undertaken an excavation in 2003 on the Indonesian island of Flores (Brown et al. 2004). It contained the announcement of the discovery of a female hominine skeleton which was believed to have belonged to a species never before seen and which was named *Homo floresiensis*. This female, nicknamed "The Hobbit", was very small, being only about one metre tall and having a cranial capacity of 380 cm². A second article, co-authored by fourteen people, gave an account of the context of the find and its implications (Moorwood et al. 2004). The team had been led to dig in this area following the finding of archæological evidence of Early Pleistocene occupation some 840,000 years ago (Morwood et al. 1998, Morwood et al. 1999). The bones of the skeleton found in 2003 were closely associated, but not in alignment, showing evidence of slipping. The arm bones had not been found but it was hoped that they would be uncovered when the dig was resumed the next season. These remains had been dated to 18,000 ya, and those of a premolar from another individual, found in an older deposit, to 38,000 ya. An abundance of stone tools was also found.

Having satisfied themselves that these remains were not those of an exceptionally small *H. sapiens*, pathological or not, Brown et al. and Morwood et al. suggested that these were the remains of a descendant of *H. erectus,* the species having become dwarfed due to prolonged isolation on an island, as had happened with the dwarf stegadon also found on Flores (Lieberman 2005: 957-958).

Later, further articles relating to Flores appeared in Nature,: Morwood et al. (2005), Dalton (2005, 2006) Kemp (2005) and Lieberman (2005). The principle article (Morwood et al. 2005) described further work undertaken during the dry season in 2004 (before the publication of the first articles in October 2004) which had uncovered the partial remains of a further eight individuals, as well as the arm bones of the original female skeleton. The arm bones were long, more akin to an Australopithecine than *H. erectus*. Detailed analysis had shown that these remains carried a mosaic of Australopithecine and *erectus* features, but that they were predominantly Australopithecine. This conclusion was supported by further work carried out on the *H. floresiensis* wrist bones (Tocheri et al. 2007) which found that the wrist bones of *H. floresiensis* were descended from a hominine ancestor that migrated out of Africa before the evolution of the wrist morphology shared by *H. erectus*, the Neanderthals and *H. sapiens*. That no similar remains have ever been found anywhere else may be accounted for by the depth at which the Flores remains were found, six or more metres, compared with the African fossils, which were found quite close to the surface, or protruding from rock escarpments.

The Indonesian island of Flores lies east of the Wallace line. These people must have arrived by sea. The sophistication of the stone tools, the use of fire and the use of watercraft, all fall within expectations for humans living between 12,000 and 38,000 ya (the dates of the Flores remains). However, lithic evidence of occupation of Flores as early as 840,000 ya stretched the time at which watercraft were first used considerably further back than had been assumed to account for the arrival of people in Australia/New Guinea, which was estimated at not more than 60,000 years. Now the possibility was being raised that an Australopithecine type people had crossed the sea into Flores. The implication was that they had survived somewhere, probably in Africa, for about one million years longer than had previously been thought and that they had migrated out of Africa, probably at least 850,000 years ago, but possibly longer, and survived in their Australopithecine form until a mere 18,000 years ago, or even later. These Australopithecine type people had retained their small cranial capacity for nearly two million years after they had been thought to be extinct, and they, with this tiny brain capacity, had produced tools and controlled the use of fire, in a way commensurate with other stone tool cultures surviving into the nineteenth/twentieth centuries. The Southern Hemisphere continued to disturb the smooth pattern of human evolution which had seemed to be apparent in the Northern Hemisphere.

Recently, fossil remains first discovered in 1989 in Maludong (Red Deer) Cave in Yunnan Province, China, have been re-examined and dated to

~11,500 B.P. These remains have very primitive features, although their craniums were of 'moderate' size, and had been assumed to have been much older. Some researchers are claiming these remains to be of a primitive species, which like *H. floriensis*, survived until comparatively recent times; others are unconvinced, believing them to be but a regional variation. Time will tell.

In Ethiopia, remains have been found of an Australopithecine, dated to ~3.5-3.3 million ya, considerably earlier than Lucy – the famous *H. habilis* who has now been reassigned to *A. afarensis* – interestingly at a site a mere 35 km from Hadar, where Lucy had been found. Inevitably, it has been given a different name: *Australopithecus deyiremeda*. In 2015 stone tools, securely dated to 3.3 mya, were found in Kenya, near a skull, which was named *A. kenyanthropus*.

While Australopithecines are still increasing in number, evidence is mounting that archaic humans were all capable of interbreeding, at least those from Europe were, which would mean, surely, that at some point they will have to be reclassified as the same species?

DNA analysis of remains from Denisova Cave in Siberia are proving interesting. There appears to be evidence, not only that there was interbreeding between the Denisovians and the Neanderthals, but that Australian Aborigines share about 4% of their DNA with the Denisovians, an amount similar to that shared between Europeans and Neanderthals. Investigations are ongoing and it would not be appropriate to make any further comment at this point.

It would seem that there may well have been an interbreeding link between Denisovians, *H. heidelbergensis* and the Neanderthals, and between the Neanderthals and the Cro-Magnon people, who have always been accepted as our direct ancestors. Evidence is mounting favouring the Multi-regional model of human evolution. Thorne and Wolpoff were right all along!

Flores postscript

Coon (1962) in a footnote on page 112, mentioned that in 1955 two Dutch anthropologists had found six or more fossil skeletons of small people in a cave in the island of Flores. Were these also Hobbits?

Chapter 49

Onwards and Upwards

This has been a long journey, always interesting but often disturbing. Much of that which had been written, both books and journal articles, was very informative and it is hoped that an impression has not been given that all of the literature on evolution, human or otherwise, is in need of criticism. My purpose was to draw attention to what appeared to me to be problems with the established position. As I explained at the start, I am a firm believer in evolution. I have no doubt that change has happened, and is continuing to happen. However, the more I read, the more it seemed to me that certain basic premise had been accepted without sufficient supporting evidence and that, once established, had simply been repeated, over and over again.

I do not have a scientific background. I have no tertiary qualifications in physics, chemistry, genetics or biochemistry. Before coming to the study of archæology, my qualifications were in psychology and the natural therapies. My knowledge of science remains very much at the 'secondary' level, which makes me one of countless millions. I have struggled to understand the logic of what I have read. I have sought help at times, for example from a chemist and a geneticist among others, but received responses such as "That's interesting. I never thought of that before" or "I don't know", which were not very helpful. I never received a satisfactory response to my major question, how does natural selection explain chromosome change? It is my earnest hope that this book may stimulate further discussion and research.

Great strides have been made. Dating techniques have improved beyond the wildest hopes of archæologists a century ago. Then genetics was a new science. Now the human genome has been mapped. Scientists now know that certain 'diseases' are not externally caused but are the result of genetic mutation and are even able, in some cases, to make genetic modifications to mitigate the problem. Other scientists are able genetically to modify certain items of our food.

However, although archæologists have made use of many scientific advances in their endeavours to uncover the story of human evolution, the advances themselves have been made in disciplines other than archæology.

I had been both surprised and delighted when I discovered that the course in which I had enrolled, archæology, was not restricted to Egyptian pyramids and arrowheads, but included the humans who made them. In America, 'archæology' is restricted to the study of artifacts; anthropology is the study of humans, ancient and modern. Under the British system, to which Australia subscribes, 'anthropology' covers the behavior of modern humans within different societies. It became established when the study of tribal behaviours grew in popularity during the 19th and 20th centuries and efforts were made to record and understand 'non-Western' ways of living before they disappeared forever. Inevitably, comparisons were drawn between modern hunter-gatherer people and the possible lives of pre-historic people and the two disciplines, archæology and anthropology, experienced a degree of overlap. Botanists and zoologists might hypothesize about the evolution of plants and animals, but the story of human evolution remained the province of the archæologist.

This evolution of the various disciplines has resulted in certain people becoming extremely knowledgeable about human fossils who were not experts in other ancillary fields. This problem, of course, is not unique to archæology as scientists become more and more specialized. Knowledge is being accumulated and dispersed at such a rate that no one human being can possibly know it all. Inter-disciplinary co-operation is essential. How, then, has archæology itself progressed as a discipline over the past two hundred years in its attempts to uncover the story of human evolution?

Looking back

It is now more than two hundred years since Lamarck published his epic work. He was the first person to write comprehensively on the evolution of fauna and to postulate clearly the evolutionary connection between the great apes and humans. Lamarck acknowledged that a characteristic could not be regarded as being subject to inheritance until it had become generalized throughout a population, recognizing the group as the evolutionary unit, not the individual. How this was achieved he did not know. It was difficult to know which was more disturbing: the fact that Lamarck's work was so persistently misrepresented or that the occasional acknowledgments that "Lamarck did not say what people generally think he said" was made without any further explanation. I found no book, no section of a book, which I felt adequately explained Lamarck's theory – so I wrote one myself (Carrington-Smith 2015).

As disturbing was the general disregard of Darwin's position on the inheritance of acquired characteristics, which I hope I have shown underpinned his entire theory. He called it *pangenesis* and outlined his theory in detail in *The Descent of Man*. Darwin knew nothing of genetics. He could not, therefore, propose that a fortuitous genetic mutation was passed on to subsequent generations. He firmly believed that our bodies could, and did, change in accordance with our reactions to the environment in which we found ourselves – stronger muscles, shorter arms (!) – and that these changes, induced in the individual during its lifetime, were inherited by its offspring. Every effort seemed to have been made to divorce Darwin's name from this 'error' and to implicate, instead, Lamarck. Will this situation change now that it has been found, at least at the bacterial level, that change occurring in one generation, induced by external circumstances, can be, and often is, passed on (*epigenesis*)?

Not only was the name and character of a highly respected scientist (Kammerer) besmirched in order that a political variation of Darwin's theory (Social Darwinism) could be further forwarded in Germany, it seems likely that that scientist was, in fact, murdered in order to prevent him pursuing his work in Russia. That this unfortunate instance apparently took place during the 1920s, when Hitler and his Nazi movement were establishing their authority, makes the incident plausible, but murder is not the only means of stifling unwanted opinion. The genteel and civilized process of 'peer review' can ensure that no journal articles are printed which argue in favour of a subordinate position if that article is sent for review to people known to be of dominant persuasion. Alan Thorne, from the Australian National University in Canberra, told me that he found it almost impossible to find any journal willing to publish material supporting his (and Wolpoff's) theory of Multiregional Evolution, leaving the 'Out of Africa' theorists to claim that all the (published) evidence supported their position.

Enthusiasm outweighed caution in Peking. The announcing of the finding of an ancient 'human' tooth and, therefore, of an ancient human species, at Chou K'ou Tien, was premature. For many years, further excavations were carried out with the express intention of finding evidence to support the 'humanness' of this find, biasing all conclusions. The picture which emerged from this area was also clouded by missing fossils, inaccurate records and the publishing of a book which did not take into account the latest scientific findings. The whole picture of early humans in the Far East, the use of fire, etc., was based on very shaky evidence. It seems more likely that, rather than being the home of the earliest modern humans, this area was one of the last refuges of *H. erectus*.

The establishment of a particular version of Darwinism, Social Darwinism, via Hækel and the Monist League and Hitler's Nazi movement, led to two World Wars. Another version of Darwinism, that espoused by Marx and Lenin, that external (social) environments brought about change, led to the Russian Revolution and the Chinese Cultural Revolution. All of these events resulted in massive loss of life and unbelievable misery. It is, of course, possible, if not probable, that had Darwinism not been available, the leaders of these conflicts would have found themselves some other ideology to invoke to promote their cause. Nevertheless, as history stands, it would seem evident that those promulgating any theory of evolution which may have profound social effects bear a great burden of responsibility.

Whether or not Darwin foresaw the use to which his theory would be put is not the issue. What is at issue is the fact that Darwin's theory of natural selection has been used to justify political/philosophical agendas which have impinged upon the life, liberty and happiness of hundreds of millions of people without, it is here claimed, the necessary care having been taken to ensure that the theory was sufficiently sound to justify such extrapolation. Darwin's two basic premises – the inheritance of acquired characteristics and *Natura non facit saltum* remain unproven. The assumption that new species, having a different number of chromosomes from the parent species, can evolve by *natural means*, is unsubstantiated.

Darwin further claimed that variation under domestication was but an example of the variation produced by nature over long periods of time, but no new species has ever come into being under domestication. Many varieties have been produced, especially among dogs, but no new species. Dogs are still dogs. If what humans can accomplish by domestication is taken as the benchmark for what Nature can accomplish by itself, then it must be concluded that nature, of and by itself, can produce no new species. Genetic manipulation by a scientist in a laboratory is 'external'. If it is an example of anything, it is an example of a 'superior intelligence' manipulating 'design', with much 'forethought' and 'planning' for a specific 'purpose'.

Darwin also claimed that there was no such thing as a 'species', only 'well-marked varieties'. We now know that distinct species do exist. Hybridization by similar, but distinct, species under domestication is rare and produces sterile offspring (mules). I am unaware of it being known to happen in the wild, which is the province of *natural* selection.

None of Darwin's basic tenets which he offered in support of his theory that evolution was the result of random change and natural selection have been proven. If anything has been proven, it is that Darwin's ideas were wrong.

On one thing I do believe Darwin was right - change (evolution) has happened - but this, of course, was not a novel proposition. Many people, above all Lamarck, had written extensively upon this subject before Darwin. Another point on which I agree: Darwin was unsure whether all life had evolved from but one simple organism, or from several. I have no idea either.

Unaware of genetics, Darwin was spared the necessity of trying to explain, not only why single celled organisms, which had happily been reproducing by the simple method of cell division for countless millions of years, took it upon themselves to become multi-cellular and to reproduce sexually, by meiosis. However, Neo-Darwinists cannot hide behind the veil of ignorance. It may not be necessary for them to explain *why* this happened, but they do need to explain *how* it happened, if they wish natural selection to be accepted as the mechanism.

Present position

Today, increasing emphasis is being placed upon the study of the formation of the Universe, the Earth, its early atmosphere, how simple (!) atomic particles became more complex substances, how molecules were formed, how crystals came into existence and, most importantly for evolution, how some of these crystals became able to self-replicate. How this was achieved in the absence of biological cells, upon which the most simple viroids depend, is still unknown and unexplained.

Scientists sometimes refer to the Earth as the 'Goldilocks planet' - not too hot, not too cold. There are six 'constants', all of which have to occur together, within very narrow limits, for life, such as that on Earth, to occur. The Anthropic principle states that, although the chances of all these variations coming together at one place in space must be measured as 'billions to one', the existence of billions of stars and planets means that the chance of their coming together in one place, Earth, is not non-existent.

Williams (1966/1992) claimed that there was no evidence that God had any expertise in engineering, arriving at his conclusion, among other things, on the basis of the poor design of the eye with its blind spot and of the meandering male *vas deferens*. He considered the whole biological process to be "abysmally stupid" and condemned a system in which the ultimate purpose of life seemed to be to be better than one's neighbours. But is this the purpose of life? Other people think it is not, Teilhard de Chardin, Behe and Hollick, for example. Under the microscope, it is seen that each snowflake is different from any other, but its six arms are all identical with each other. "It is still a mystery how the billions of molecules in each arm know what the others are doing" (Hollick 2006: 75).

Jones (1993) claimed that natural selection had never produced a turbine blade, or even a wheel, let alone a work of art. Behe would no doubt claim that the flagellum was just about as good as a turbine blade or a wheel – and a lot more complicated. As for art – what about the tail of the peacock or the wing of a butterfly? If humans can produce art but natural selection cannot, where does that leave humans in relation to it? Outside it? Superior to it?

Those who believe passionately in a philosophical position find it difficult to understand the thinking of those who hold an opposing point of view and it is understandably difficult for them to witness anyone, let alone children, being indoctrinated into what appears to them to be an indefensible religion/philosophy. This has always been the case and in the past has led to much repression, often imprisonment, torture and death. Many people settled in America with the express purpose of escaping religious persecution. The Inquisitors remained within the Catholic Church until the 20th century but, after the horrors of the First World War, there was a concerted effort throughout the Western world to enshrine personal freedoms, firstly under the League of Nations and later by the Declaration of Human Rights under the United Nations. It is to be regretted that Darwin's theory of evolution by natural selection appears to be being used to infringe these rights.

The saga started by the Scopes trial is by no means at an end. Now that the matter has been brought before the American courts, it will not be possible for the situation there to change without further legislation. I argue that, while it might not be right for persons of one religious philosophy (the Creationists) to try to force their beliefs on other people (by having Creationism given 'equal time' in state funded educational establishments), surely it is equally wrong for people of another philosophy (Atheism/Humanism) to force *their* beliefs upon children via the school syllabus? To mandate that Darwin's theory of evolution by natural selection (without God) be taught when, I hold, it has not been proven, and may even be untenable, seems to me to be a miscarriage of justice. Should not school

543

biological text books simply state the facts, as far as they are known, without drawing any philosophical conclusions? I am with Scopes - or, rather, with Dudley Malone: "give the next generation all the facts ... let the children have their minds kept open". More than ten years have elapsed since the Dover trial and there is no sign that the Creationist/Intelligent Design lobby is preparing to contest the Dover ruling by appealing to the second part of the Religion Clause of the First Amendment to the American Constitution, that stating that nothing should be done to inhibit the teaching of religion. An appeal to the International Court, citing the International Convention on Human Rights and/or the Declaration of the Rights of the Child, may be a possibility.

The year 2009 produced the expected tributes to mark both the 200th anniversary of Darwin's birth and the 150th anniversary of the publication of *The Origin*, both in the print and electronic media. These tended, possibly inadvertently, to promote the impression that we, in the West, are indebted to Darwin for the concept of evolution itself, not for that of evolution by natural selection. For example, the words 'natural selection' did not appear in an 'Opinion' article by Elshakry published in *Nature* and the misrepresentation was confounded by the statement that "advocates of the intelligent design theory battle to have evolution removed from the classroom". I wrote before of how, for the purpose of the trial at Dover, 'Intelligent Design' had been subsumed under 'Creationism'. Here the situation has been reversed. 'Creationism' has been subsumed under 'Intelligent Design'. They are two separate positions.

Looking forward

Where do we go from here? In a new direction, I trust. It is my sincere hope that attention will be directed towards the role of chromosomes in evolution, human or otherwise. We are told that the great apes evolved between 25,000,000 and 30,000,000 years ago. How many *billions* of gorillas, orangs and chimpanzees have lived in that time? All, we are led to believe, had 48 chromosomes. Something happened. About 6,000,000 years ago the Australopithecine line, which led to the human line, split from the great apes. How many *billions* of these have there been - and all with 46 chromosomes. Clearly a change in chromosome numbers is a very rare event, but, it would seem, it happens. I am not disputing this. I am just desperate to understand *how*. Will manipulation of genetic material in the laboratory ever throw any light upon how this may have happened naturally? I hope so!

It is also my sincere and earnest hope that more understanding will be extended between those holding opposing views. We do need to listen to each other and mandating the suppression of one point of view is not consistent with our much vaunted 'democratic' system which supposedly upholds freedom of religion, speech and expression. When I was at school, we learned the art of debating – two speakers on each side and then questions from the floor. At times, the teachers made us speak for the side which we did not support. They deemed it good for us to research and present arguments against the position we actually held. That was a tall ask – but it was a good exercise! Would Dawkins arguing *for* Intelligent Design and Behe arguing *for* Neo-Darwinism be a 'dream' debate – or a nightmare? I will leave that for your imagination.

Having within their province an area of expertise with such profound and far reaching consequences for human thought and behavior, it behoves archæologists to examine with care, not only the physical evidence, but also the biases brought to bear upon it.

Epilogue

And so I came to the end of my journey and found that I had come full circle. I was back in the position from which I had started!

That which I believed at the beginning, I believed at the end, if anything, even more strongly than before. I believed:

1) that evolution is fact;

2) that natural selection is fantasy;

3) that the workings of the human mind are beyond comprehension!

How about you?

What do you think?

Appendices

Extracts from The World of Life: A Manifestation Of Creative Power, Directive Mind and Ultimate Purpose

Alfred R. Wallace (1910)

Adaptations to Drought

The plant figured on the next page, like many others of the campos, has its roots swollen and woody, forming a store of water and food to enable it to withstand the effects of drought and of the campo-fires. The old stems show where they have been burnt off, and the figures of many other plants with woody roots or tubers, figured by Mr. Warming, show similar effects of burning.

Still more remarkable is the tree figured on p. 69 (Fig. 7), which is adapted to the same conditions in a quite different way, as are many other quite unrelated species.[1] The group of plants shown is really an underground tree, and not merely dwarf shrubs as they at first appear to be. What look like surface-roots are the upper branches of a tree, the trunk of which and often a large part of the limbs and branches are buried in the earth. The stems shown are the root-like branches,

[1] The following species have a similar mode of growth: *Anacardium humile, Hortia Brasiliensis* (Rutaceæ), *Cochlospermum insigne* (Cistaceæ), *Simaba Warmingiana* (Simarubaceæ), *Erythroxylon campestre* (Erythroxylaceæ), *Plumiera Warmingii* (Apocynaceæ), *Palicourea rigida* (Cinchonaceæ), etc.

which are 4-5 inches diameter, while the growing shoots

FIG. 6.— *CASSELIA CHAMÆDRIFOLIA*, nat. size (Verbenaceæ).

are from 2 to 3 feet high. The whole plant (or tree) is from 30 to 40 feet diameter. As the branches approach the centre they descend into the earth and form a central trunk. A French botanist, M. Emm. Liais, says of this species : "If we dig we find how all these small shrubs, apparently distinct, are joined together underground and form the extremities of the branches of a large subterranean tree which at length unite to form a single trunk. M. Renault of Barbacena told me that he had dug about 20 feet deep to obtain one of these trunks." The large subterranean trees

FIG. 7.—*ANDIRA LAURIFOLIA* (Papilionaceæ).

with a trunk hidden in the soil form one of the most singular features of the flora of these campos of Central Brazil.

The above facts are from Mr. Warming's book, supplemented by some details in a letter. They are certainly very remarkable ; and it is difficult to understand how this mode of growth has been acquired, or how the seeds get so deep into the ground as to form a subterranean trunk. But perhaps the cracks in the dry season explain this.

Appendix II

Alfred Russel Wallace

'On the Law which has Regulated The Introduction of New Species'. Volume 16 (2nd Series), Annals and Magazine of Natural History,

September 1855 16(93): 184–196

Geographical Distribution Dependent on Geologic Changes.

EVERY naturalist who has directed his attention to the subject of the geographical distribution of animals and plants, must have been interested in the singular facts which it presents. Many of these facts are quite different from what would have been anticipated, and have hitherto been considered as highly curious, but quite inexplicable. None of the explanations attempted from the time of Linnaeus are now considered at all satisfactory; none of them have given a cause sufficient to account for the facts known at the time, or comprehensive enough to include all the new facts which have since been, and are daily being added. Of late years, however, a great light has been thrown upon the subject by geological investigations, which have shown that the present state of the earth and of the organisms now inhabiting it, is but the last stage of a long and uninterrupted series of changes which it has undergone, and consequently, that to endeavour to explain and account for its present condition without any reference to those changes (as has frequently been done) must lead to very imperfect and erroneous conclusions. The facts proved by geology are briefly these:- That during an immense, but unknown period, the surface of the earth has undergone successive changes; land has sunk beneath the ocean, while fresh land has risen up from it; mountain chains have been elevated; islands have been formed into continents, and continents submerged till they have become islands; and these changes have taken place, not once merely, but perhaps hundreds, perhaps thousands of times:- That all these operations have been more or less continuous, but unequal in their progress, and during the whole series the organic life of the earth has undergone a corresponding alteration. This alteration also has been gradual, but complete; after a certain interval not a single species existing which had

554

lived at the commencement of the period. This complete renewal of the forms of life also appears to have occurred several times:- That from the last of the geological epochs to the present or historical epoch, the change of organic life has been gradual: the first appearance of animals now existing can in many cases be traced, their numbers gradually increasing in the more recent formations, while other species continually die out and disappear, so that the present condition of the organic world is clearly derived by a natural process of gradual extinction and creation of species from that of the latest geological periods. We may therefore safely infer a like gradation and natural sequence from one geological epoch to another.

Now, taking this as a fair statement of the results of geological inquiry, we see that the present geographical distribution of life upon the earth must be the result of all the previous changes, both of the surface of the earth itself and of its inhabitants. Many causes, no doubt, have operated of which we must ever remain in ignorance, and we may, therefore, expect to find many details very difficult of explanation, and in attempting to give one, must allow ourselves to call into our service geological changes which it is highly probable may have occurred, though we have no direct evidence of their individual operation.

The great increase of our knowledge within the last twenty years, both of the present and past history of the organic world, has accumulated a body of facts which should afford a sufficient foundation for a comprehensive law embracing and explaining them all, and giving a direction to new researches. It is about ten years since the idea of such a law suggested itself to the writer of this essay, and he has since taken every opportunity of testing it by all the newly-ascertained facts with which he has become acquainted, or has been able to observe himself. These have all served to convince him of the correctness of his hypothesis. Fully to enter into such a subject would occupy much space, and it is only in consequence of some views having been lately promulgated, he believes, in a wrong direction, that he now ventures to present his ideas to the public, with only such obvious illustrations of the arguments and results as occur to him in a place far removed from all means of reference and exact information.

A Law deduced from well-known Geographical and Geological Facts.

The following propositions in Organic Geography and Geology give the main facts on which the hypothesis is founded.

Geography

1. Large groups, such as classes and orders, are generally spread over the

whole earth, while smaller ones, such as families and genera, are frequently confined to one portion, often to a very limited district.

2. In widely distributed families the genera are often limited in range; in widely distributed genera, well marked groups of species are peculiar to each geographical district.

3. When a group is confined to one district, and is rich in species, it is almost invariably the case that the most closely allied species are found in the same locality or in closely adjoining localities, and that therefore the natural sequence of the species by affinity is also geographical.

4. In countries of a similar climate, but separated by a wide sea or lofty mountains, the families, genera and species of the one are often represented by closely allied families, genera and species peculiar to the other.

Geology

5. The distribution of the organic world in time is very similar to its present distribution in space.

6. Most of the larger and some small groups extend through several geological periods.

7. In each period, however, there are peculiar groups, found nowhere else, and extending through one or several formations.

8. Species of one genus, or genera of one family occurring in the same geological time, are more closely allied than those separated in time.

9. As generally in geography no species or genus occurs in two very distant localities without being also found in intermediate places, so in geology the life of a species or genus has not been interrupted. In other words, no group or species has come into existence twice.

10. The following law may be deduced from these facts:- Every species has come into existence coincident both in space and time with a pre-existing closely allied species.

This law agrees with, explains and illustrates all the facts connected with the following branches of the subject:- 1st. The system of natural affinities. 2nd. The distribution of animals and plants in space. 3rd. The same in time, including all the phaæomena of representative groups, and those which Professor Forbes supposed to manifest polarity. 4th. The phænomena of rudimentary organs. We will briefly endeavour to show its bearing upon each of these.

The Form of a true system of Classification determined by this Law.

If the law above enunciated be true, it follows that the natural series of affinities will also represent the order in which the several species came into existence, each one having had for its immediate antitype a closely allied species existing at the time of its origin. It is evidently possible that two or three distinct species may have had a common antitype, and that each of these may again have become the antitypes from which other closely allied species were created. The effect of this would be, that so long as each species has had but one new species formed on its model, the line of affinities will be simple, and may be represented by placing the several species in direct succession in a straight line. But if two or more species have been independently formed on the plan of a common antitype, then the series of affinities will be compound, and can only be represented by a forked or many branched line. Now, all attempts at a Natural classification and arrangement of organic beings show, that both these plans have obtained in creation. Sometimes the series of affinities can be well represented for a space by a direct progression from species to species or from group to group, but it is generally found impossible so to continue. There constantly occur two or more modifications of an organ or modifications of two distinct organs, leading us on to two distinct series of species, which at length differ so much from each other as to form distinct genera or families. These are the parallel series or representative groups of naturalists, and they often occur in different countries, or are found fossil in different formations. They are said to have an analogy to each other when they are so far removed from their common antitype as to differ in many important points of structure, while they still preserve a family resemblance. We thus see how difficult it is to determine in every case whether a given relation is an analogy or an affinity, for it is evident that as we go back along the parallel or divergent series, towards the common antitype, the analogy which existed between the two groups becomes an affinity. We are also made aware of the difficulty of arriving at a true classification, even in a small and perfect group;- in the actual state of nature it is almost impossible, the species being so numerous and the modifications of form and structure so varied, arising probably from the immense number of species which have served as antitype for the existing species, and thus produced a complicated branching of the lines of affinity, as intricate as the twigs of a gnarled oak or the vascular system of the human body. Again, if we consider that we have only fragments of this vast system, the stem and main branches being represented by extinct species of which we have no knowledge, while a vast mass of limbs and boughs and minute twigs and scattered leaves is what we have to place in order, and determine the true position each originally occupied with regard to the others, the whole difficulty of the true Natural System of classification becomes apparent to us.

We shall thus find ourselves obliged to reject all those systems of classification which arrange species or groups in circles, as well as those which fix a definite number for the divisions of each group. The latter class have been very generally rejected by naturalists, as contrary to nature, notwithstanding the ability with which they have been advocated; but the circular system of affinities seems to have obtained a deeper hold, many eminent naturalists having to some extent adopted it. We have, however, never been able to find a case in which the circle has been closed by a direct and close affinity. In most cases a palpable analogy has been substituted, in others the affinity is very obscure or altogether doubtful. The complicated branching of the lines of affinities in extensive groups must also afford great facilities for giving a show of probability to any such purely artificial arrangements. Their death-blow was given by the admirable paper of the lamented Mr. Strickland, published in the "Annals of Natural History," in which he so cleverly showed the true synthetical method of discovering the Natural System.

Geographical Distribution of Organisms.

If we now consider the geographical distribution of animals and plants upon the earth, we shall find all the facts beautifully in accordance with, and readily explained by, the present hypothesis. A country having species, genera, and whole families peculiar to it, will be the necessary result of its having been isolated for a long period, sufficient for many series of species to have been created on the type of pre-existing ones, which, as well as many of the earlier-formed species, have become extinct, and thus made the groups appear isolated. If in any case the antitype had an extensive range, two or more groups of species might have been formed, each varying from it in a different manner, and thus producing several representative or analogous groups. The Sylviadæ of Europe and the Sylvicolidae of North America, the Heliconidæ of South America and the Euploeas of the East, the group of Trogons inhabiting Asia, and that peculiar to South America, are examples that may be accounted for in this manner.

Such phænomena as are exhibited by the Galapagos Islands, which contain little groups of plants and animals peculiar to themselves, but most nearly allied to those of South America, have not hitherto received any, even a conjectural explanation. The Galapagos are a volcanic group of high antiquity, and have probably never been more closely connected with the continent than they are at present. They must have been first peopled, like other newly-formed islands, by the action of winds and currents, and at a period sufficiently remote to have had the original species die out, and the modified prototypes only remain. In the same way we can account for the separate islands having each their peculiar species, either on the supposition that the same original emigration peopled the whole of the islands with the same species from which differently modified prototypes were created, or that the islands were successively peopled from each other, but that new species have been created in each on the plan of the pre-existing ones. St. Helena is a similar case of a very ancient island having obtained an entirely peculiar, though limited, flora. On the other hand, no example is known of an island which can be proved geologically to be of very recent origin (late in the Tertiary, for instance), and yet possess generic or family groups, or even many species peculiar to itself.

When a range of mountains has attained a great elevation, and has so remained during a long geological period, the species of the two sides at and near their bases will be often very different, representative species of some genera occurring, and even whole genera being peculiar to one side, as is remarkably seen in the case of the Andes and Rocky Mountains. A similar phænomena occurs when an island has been separated from a continent at a very early period. The shallow sea between the Peninsula of Malacca, Java, Sumatra and Borneo was probably a continent or large island at an early epoch, and may have become submerged as the volcanic ranges of Java and Sumatra were elevated. The organic results we see in the very considerable number of species of animals common to some or all of these countries, while at the same time a number of closely allied representative species exist peculiar to each, showing that a considerable period has elapsed since their separation. The facts of geographical distribution and of geology may thus mutually explain each other in doubtful cases, should the principles here advocated be clearly established.

In all those cases in which an island has been separated from a continent, or raised by volcanic or coralline action from the sea, or in which a mountain-chain has been elevated in a recent geological epoch, the phænomena of peculiar groups or even of single representative species will not exist. Our own island is an example of this, its separation from the continent being geologically very recent, and we have consequently scarcely a species which is peculiar to it; while the Alpine range, one of the most recent mountain elevations, separates faunas and floras which scarcely differ more than may be due to climate and latitude alone.

The series of facts alluded to in Proposition (3), of closely allied species in rich groups being found geographically near each other, is most striking and important. Mr. Lovell Reeve has well exemplified it in his able and interesting paper on the Distribution of the Bulimi. It is also seen in the Hummingbirds and Toucans, little groups of two or three closely allied species being often found in the same or closely adjoining districts, as we have had the good fortune of personally verifying. Fishes give evidence of a similar kind: each great river has its peculiar genera, and in more extensive genera its groups of closely allied species. But it is the same throughout Nature; every class and order of animals will contribute similar facts. Hitherto no attempt has been made to explain these singular phænomena, or to show how they have arisen. Why are the genera of Palms and of Orchids in almost every case confined to one hemisphere? Why are the closely allied species of brownbacked Trogons all found in the East, and the green-backed in the West? Why are the Macaws and the Cockatoos similarly restricted? Insects furnish a countless number of analogous examples;- the Goliathi of Africa, the Ornithopterae of the Indian Islands, the Heliconidæ of South America, the Danaidæ of the East, and in all, the most closely allied species found in geographical proximity. The question forces itself upon every thinking mind,- why are these things so? They could not be as they are had no law regulated their creation and dispersion. The law here enunciated not merely explains, but necessitates the facts we see to exist, while the vast and long-continued geological changes of the earth readily account for the exceptions and apparent discrepancies that here and there occur. The writer's object in putting forward his views in the present imperfect manner is to submit them to the test of other minds, and to be made aware of all the facts supposed to be inconsistent with them. As his hypothesis is one which claims acceptance solely as explaining and connecting facts which exist in nature, he expects facts alone to be brought to disprove it, not a priori arguments against its probability.

Geological Distribution of the Forms of Life.

The phænomena of geological distribution are exactly analogous to those of geography. Closely allied species are found associated in the same beds, and the change from species to species appears to have been as gradual in time as in space. Geology, however, furnishes us with positive proof of the extinction and production of species, though it does not inform us how either has taken place. The extinction of species, however, offers but little difficulty, and the modus operandi has been well illustrated by Sir C. Lyell in his admirable "Principles." Geological changes, however gradual, must occasionally have modified external conditions to such an extent as to have rendered the existence of certain species impossible. The extinction would in most cases be effected by a gradual dying-out, but in some instances there might have been a sudden destruction of a species of limited range. To discover how the extinct species have from time to time been replaced by new ones down to the very latest geological period, is the most difficult, and at the same time the most interesting problem in the natural history of the earth. The present inquiry, which seeks to eliminate from known facts a law which has determined, to a certain degree, what species could and did appear at a given epoch, may, it is hoped, be considered as one step in the right direction towards a complete solution of it.

High Organization of very ancient Animals consistent with this Law.

Much discussion has of late years taken place on the question, whether the succession of life upon the globe has been from a lower to a higher degree of organization. The admitted facts seem to show that there has been a general, but not a detailed progression. Mollusca and Radiata existed before Vertebrata, and the progression from Fishes to Reptiles and Mammalia, and also from the lower mammals to the higher, is indisputable. On the other hand, it is said that the Mollusca and Radiata of the very earliest periods were more highly organized than the great mass of those now existing, and that the very first fishes that have been discovered are by no means the lowest organised of the class. Now it is believed the present hypothesis will harmonize with all these facts, and in a great measure serve to explain them; for though it may appear to some readers essentially a theory of progression, it is in reality only one of gradual change. It is, however, by no means difficult to show that a real progression in the scale of organization is perfectly consistent with all the appearances, and even with apparent retrogression, should such occur.

Returning to the analogy of a branching tree, as the best mode of representing the natural arrangement of species and their successive creation, let us suppose that at an early geological epoch any group (say a class of the Mollusca) has attained to a great richness of species and a high organization. Now let this great branch of allied species, by geological mutations, be completely or partially destroyed. Subsequently a new branch springs from the same trunk, that is to say, new species are successively created, having for their antitypes the same lower organized species which had served as the antitypes for the former group, but which have survived the modified conditions which destroyed it. This new group being subject to these altered conditions, has modifications of structure and organization given to it, and becomes the representative group of the former one in another geological formation. It may, however, happen, that though later in time, the new series of species may never attain to so high a degree of organization as those preceding it, but in its turn become extinct, and give place to yet another modification from the same root, which may be of higher or lower organization, more or less numerous in species, and more or less varied in form and structure than either of those which preceded it. Again, each of these groups may not have become totally extinct, but may have left a few species, the modified prototypes of which have existed in each succeeding period, a faint memorial of their former grandeur and luxuriance. Thus every case of apparent retrogression may be in reality a progress, though an interrupted one: when some monarch of the forest loses a limb, it may be replaced by a feeble and sickly substitute. The foregoing remarks appear to apply to the case of the Mollusca, which, at a very early period, had reached a high organization and a great development of forms and species in the testaceous Cephalopoda. In each succeeding age modified species and genera replaced the former ones which had become extinct, and as we approach the present aera, but few and small representatives of the group remain, while the Gasteropods and Bivalves have acquired an immense preponderance. In the long series of changes the earth has undergone, the process of peopling it with organic beings has been continually going on, and whenever any of the higher groups have become nearly or quite extinct, the lower forms which have better resisted the modified physical conditions have served as the antitypes on which to found the new races. In this manner alone, it is believed, can the representative groups at successive periods, and the rising and fallings in the scale of organization, be in every case explained.

Objections to Forbes' Theory of Polarity

The hypothesis of polarity, recently put forward by Professor Edward Forbes to account for the abundance of generic forms at a very early period and at present, while in the intermediate epochs there is a gradual diminution and impoverishment, till the minimum occurred at the confines of the Palaeozoic and Secondary epochs, appears to us quite unnecessary, as the facts may be readily accounted for on the principles already laid down. Between the Palaeozoic and Neozoic periods of Professor Forbes, there is scarcely a species in common, and the greater part of the genera and families also disappear to be replaced by new ones. It is almost universally admitted that such a change in the organic world must have occupied a vast period of time. Of this interval we have no record; probably because the whole area of the early formations now exposed to our researches was elevated at the end of the Palæozoic period, and remained so through the interval required for the organic changes which resulted in the fauna and flora of the Secondary period. The records of this interval are buried beneath the ocean which covers three-fourths of the globe. Now it appears highly probable that a long period of quiescence or stability in the physical conditions of a district would be most favourable to the existence of organic life in the greatest abundance, both as regards individuals and also as to variety of species and generic group, just as we now find that the places best adapted to the rapid growth and increase of individuals also contain the greatest profusion of species and the greatest variety of forms,- the tropics in comparison with the temperate and arctic regions. On the other hand, it seems no less probable that a change in the physical conditions of a district, even small in amount if rapid, or even gradual if to a great amount, would be highly unfavourable to the existence of individuals, might cause the extinction of many species, and would probably be equally unfavourable to the creation of new ones. In this too we may find an analogy with the present state of our earth, for it has been shown to be the violent extremes and rapid changes of physical conditions, rather than the actual mean state in the temperate and frigid zones, which renders them less prolific than the tropical regions, as exemplified by the great distance beyond the tropics to which tropical forms penetrate when the climate is equable, and also by the richness in species and forms of tropical mountain regions which principally differ from the temperate zone in the uniformity of their climate. However this may be, it seems a fair assumption that during a period of geological repose the new species which we know to have been created would have appeared, that the creations would then exceed in number the extinctions, and therefore the number of species would increase. In a period of geological activity, on the other hand, it seems probable that the extinctions might exceed the creations, and the number of species consequently diminish. That such effects did take place in connexion with the causes to which we have imputed them, is shown in the case of the Coal formation, the faults and contortions of which show a period of great activity and violent convulsions, and it is in the formation immediately succeeding this that the poverty of forms of life is most apparent. We have then only to suppose a long period of somewhat similar action during the vast unknown interval at the termination of the Palæozoic period, and then a decreasing violence or rapidity through the Secondary period, to allow for the gradual repopulation of the earth with varied forms, and the whole of the facts are explained. We thus have a clue to the increase of the forms of life during certain periods, and their decrease during others, without recourse to any causes but these we know to have existed, and to effects fairly deducible from them. The precise manner in which the geological changes of the early formations were effected is so extremely obscure, that when we can explain important facts by a retardation at one time and an acceleration at another of a process which we know from its nature and from observation to have been unequal,- a cause so simple may surely be preferred to one so obscure and hypothetical as polarity.

I would also venture to suggest some reasons against the very nature of the theory of Professor Forbes. Our knowledge of the organic world during any geological epoch is necessarily very imperfect. Looking at the vast numbers of species and groups that have been discovered by geologists, this may be doubted; but we should compare their numbers not merely with those that now exist upon the earth, but with a far larger amount. We have no reason for believing that the number of species on the earth at any former period was much less than at present; at all events the aquatic portion, with which geologists have most acquaintance, was probably often as great or greater. Now we know that there have been many complete changes of species; new sets of organisms have many times been introduced in place of old ones which have become extinct, so that the total amount which have existed on the earth from the earliest geological period must have borne about the same proportion to those now living, as the whole human race who have lived and died upon the earth, to the population at the present time. Again, at each epoch, the whole earth was no doubt, as now, more or less the theatre of life, and as the successive generations of each species died, their exuviæ and preservable parts would be deposited over every portion of the then existing seas and oceans, which we have reason for supposing to have been more, rather than less, extensive than at present. In order then to understand our possible knowledge of the early world and its inhabitants, we must compare, not the area of the whole field of our geological researches with the earth's surface, but the area of the examined portion of each formation separately with the whole earth. For example, during the Silurian period all the earth was Silurian, and animals were living and dying, and depositing their remains more or less over the whole area of the globe, and they were probably (the species at least) nearly as varied in different latitudes and longitudes as at present. What proportion do the Silurian districts bear to the whole surface of the globe, land and sea (for far more extensive Silurian districts probably exist beneath the ocean than above it), and what portion of the known Silurian districts has been actually examined for fossils? Would the area of rock actually laid open to the eye be the thousandth or the ten-thousandth part of the earth's surface? Ask the same question with regard to the Oolite or the Chalk, or even to particular beds of these when they differ considerably in their fossils, and you may then get some notion of how small a portion of the whole we know.

But yet more important is the probability, nay almost the certainty, that whole formations containing the records of vast geological periods are entirely buried beneath the ocean, and for ever beyond our reach. Most of the gaps in the geological series may thus be filled up, and vast numbers of unknown and unimaginable animals, which might help to elucidate the affinities of the numerous isolated groups which are a perpetual puzzle to the zoologist, may there be buried, until future revolutions may raise them in their turn above the waters, to afford materials for the study of whatever race of intelligent beings may then have succeeded us. These considerations must lead us to the conclusion, that our knowledge of the whole series of the former inhabitants of the earth is necessarily most imperfect and fragmentary,- as much so as our knowledge of the present organic world would be, were we forced to make our collections and observations only in spots equally limited in area and in number with those actually laid open for the collection of fossils. Now, the hypothesis of Professor Forbes is essentially one that assumes to a great extent the completeness of our knowledge of the whole series of organic beings which have existed on the earth. This appears to be a fatal objection to it, independently of all other considerations. It may be said that the same objections exist against every theory on such a subject, but this is not necessarily the case. The hypothesis put forward in this paper depends in no degree upon the completeness of our knowledge of the former condition of the organic world, but takes what facts we have as fragments of a vast whole, and deduces from them something of the nature and proportions of that whole which we can never know in detail. It is founded upon isolated groups of facts, recognizes their isolation, and endeavours to deduce from them the nature of the intervening portions.

Rudimentary Organs

Another important series of facts, quite in accordance with, and even necessary deductions from, the law now developed, are those of rudimentary organs. That these really do exist, and in most cases have no special function in the animal economy, is admitted by the first authorities in comparative anatomy. The minute limbs hidden beneath the skin in many of the snake-like lizards, the anal hooks of the boa constrictor, the complete series of jointed finger-bones in the paddle of the Manatus and whale, are a few of the most familiar instances. In botany a similar class of facts has long been recognised. Abortive stamens, rudimentary floral envelopes and undeveloped carpels, are of the most frequent occurrence. To every thoughtful naturalist the question must arise, What are these for? What have they to do with the great laws of creation? Do they not teach us something of the system of Nature? If each species has been created independently, and without any necessary relations with pre-existing species, what do these rudiments, these apparent imperfections mean? There must be a cause for them; they must be the necessary results of some great natural law. Now, if, as it has been endeavoured to be shown, the great law which has regulated the peopling of the earth with animal and vegetable life is, that every change shall be gradual; that no new creature shall be formed widely differing from anything before existing; that in this, as in everything else in Nature, there shall be gradation and harmony,- then these rudimentary organs are necessary, and are an essential part of the system of Nature. Ere the higher Vertebrata were formed, for instance, many steps were required, and many organs had to undergo modifications from the rudimental condition in which only they had as yet existed. We still see remaining an antitypal sketch of a wing adapted for flight in the scaly flapper of the penguin, and limbs first concealed beneath the skin, and then weakly protruding from it, were the necessary gradations before others should be formed fully adapted for locomotion. Many more of these modifications should we behold, and more complete series of them, had we a view of all the forms which have ceased to live. The great gaps that exist between fishes, reptiles, birds, and mammals would then, no doubt, be softened down by intermediate groups, and the whole organic world would be seen to be an unbroken and harmonious system.

Conclusion

It has now been shown, though most briefly and imperfectly, how the law that "Every species has come into existence coincident both in time and space with a pre-existing closely allied species," connects together and renders intelligible a vast number of independent and hitherto unexplained facts. The natural system of arrangement of organic beings, their geographical distribution, their geological sequence, the phænomena of representative and substituted groups in all their modifications, and the most singular peculiarities of anatomical structure, are all explained and illustrated by it, in perfect accordance with the vast mass of facts which the researches of modern naturalists have brought together, and, it is believed, not materially opposed to any of them. It also claims a superiority over previous hypotheses, on the ground that it not merely explains, but necessitates what exists. Granted the law, and many of the most important facts in Nature could not have been otherwise, but are almost as necessary deductions from it, as are the elliptic orbits of the planets from the law of gravitation.

Sarawak, Borneo, Feb 1855

Wallace, A.R. (1858) On the tendency of varieties to depart indefinitely from the original type: instability of varieties supposed to prove the permanent distinction of species. Journal of the   Proceedings of the Linnean Society of London 3: 53-62)

On the Tendency of Varieties to depart indefinitely from the Original Type

One of the strangest arguments which has been adduced to prove the original and permanent distinctness of species is, that varieties produced in a state of domesticity are more or less unstable, and often have a tendency, if left to themselves, to return to the normal form of the parent species; and this instability is considered to be a distinctive peculiarity of all varieties, even of those occurring among wild animals, and to constitute a provision for preserving unchanged the originally created distinct species.

In the absence of scarcity of facts and observations as to varieties occurring among wild animals, this argument has had great weight with naturalists, and has led to a very general and somewhat prejudiced belief in the stability of species. Equally general, however, is the belief in what are called 'permanent or true varieties, - races of animals which continually propagate their line, but which differ so slightly (although constantly) from some other race, that the one is considered to be a variety of the other. Which is the variety and which is the original species there is generally no means of determining, except in those rare cases in which in which one race has been known to produce an offspring unlike itself and resembling the other. This, however, would seem quite incompatible with the 'permanent invariability of species,' but the difficulty is overcome by assuming that such varieties have strict limits, and can never again vary further from the original type, although they may return to it which, from the analogy of the domesticated animals is considered to be highly probable, if not certainly proved.

It will be observed that this argument rests entirely on the assumption that varieties occurring in the state of nature are in all respects analogous to or even identical with those of domestic animals, and are governed by the same laws as regards their permanence of further variation. But this is the object of the present paper to show that this assumption is altogether false, that there is a general principle in nature which will cause many varieties to survive the parent species, and to give rise to successive variations departing further and further from the original type, and which also produces in domesticated animals, the tendency of varieties to return to the parent form.

The life of wild animals is a struggle for existence. The full exertion of all their faculties and all their energies is required to preserve their own existence and to provide for that of their infant offspring. The possibility of procuring food during the least favourable seasons, and escaping the attacks of their most dangerous enemies, and are primary conditions which determine the existence both of individuals and off entire species. These conditions will also determine the population of a species; and by a careful consideration of all the circumstances we may be enabled to comprehend, and in some degree to explain, what at first sight appears so inexplicable – the excessive abundance of some species, while others allied to them are very rare.

The general proportion that must obtain between certain groups of animals is readily seen. Large animals cannot be so abundant as small ones; the carnivore must be less numerous than the herbivore; eagles and lions can never be so plentiful as pigeons and antelopes; the wild asses of the Tartarian deserts cannot equal in numbers the horses of the more luxurious prairies and pampas of America. The greater or less fecundity of an animal is often considered to be one of the chief causes of its abundance or scarcity; but a consideration of these facts will show us that it really has little or nothing to do with the matter. Even the least prolific of animals would increase rapidly if unchecked, whereas it is evident that the animals population of the globe must be stationary, or perhaps, through the influence of man, decreasing. Fluctuations there may be, but permanent increase, except in restricted localities, is almost impossible. For example, our own observations must convince us that birds do not go on increasing every year in a geometrical ratio, as they would do, were there not some powerful check to their natural increase. Very few birds produce less than two young ones each year, while many have six, eight, or ten; four will certainly be below the average; and if we suppose that each pair produce young only four times in their life, that will also be below the average, supposing them not to die either by violence or want of food. Yet at this rate how tremendous would be the increase in a few years from a single pair A simple calculation will show that in fifteen years each pair of birds would have increased to nearly ten millions! whereas we have no reason to believe that the number of birds of any country increases at all in fifteen or in one hundred and fifty years. With such powers of increase the population must have reached its limits, and have become stationary, in a very few years after the origin of each species. It is evident, therefore, that each year an immense number of birds must perish – as many in fact as are born; and as in the lowest calculation the progeny are each year twice as numerous as their parents, it follows that, whatever be the average number of individuals existing in any given country, twice that number must perish annually, – a striking result, but one which seems at least highly probable, and is perhaps under rather than over the truth. It would therefore appear that, as far as the continuance of the species and the keeping up of average number of individuals are concerned, large broods are superfluous. On the average all above one become food for hawks and kites, wild cats and weasels, or perish of cold and hunger as winter comes on. This is strikingly proved by the case of particular species; for we find that their abundance in individuals bears no relation whatever to their fertility in producing offspring. Perhaps the most remarkable instance of an immense bird population is that of the passenger pigeon of the United States which lays only one, or at most two eggs, and is said to rear generally but one young one. Why is this bird so extraordinarily abundant, while others producing two or three times as many young are much less plentiful? The explanation is not difficult. The food most congenial to this species and on which it thrives best, is abundantly distributed over a very extensive region, offering such differences of soil and climate, that in one part or other of the area the supply never fails. The bird is capable of a very rapid and long continued flight, so that it can pass without fatigue over the whole of the district it inhabits, and as soon as the supply of wholesome food begins to fail in one place be able to discover a fresh feeding-ground. This example strikingly shows us that the procuring of a constant supply of wholesome food is almost the sole condition requisite for ensuring the rapid increase of a given species, since neither the limited fecundity, nor the unrestricted attacks of birds of prey and of man are here sufficient to check it. In no other birds are these peculiar circumstances so strikingly combined. Either their food is more liable to failure, or they have not sufficient power of wing to search for it over an extensive area, or during some season of the year it becomes very scarce, and less wholesome substitutes have to be found; and thus, though more fertile in offspring, they can never increase beyond the supply of food in the least favourable seasons. Many birds can only exist by migrating, when their food becomes scarce, to regions possessing a milder, or at least a different climate, though, as these migrating birds are seldom excessively abundant, it is evident that the countries that they visit are still deficient in a constant and abundant supply of wholesome food. Those whose organization does not permit them to migrate when their food becomes periodically scarce can never attain a large population. This is probably the reason why woodpeckers are scarce with us, while in the tropics they are among the most abundant of solitary birds. Thus the house sparrow is more abundant than the redbreast, because its food is more constant and plentiful, - seeds of grasses being preserved during the winter, and our farm-yards and stubble yards furnishing an almost inexhaustible supply. Why, as a general rule, are aquatic, and especially sea birds, very numerous in individuals? Not because they are more prolific that the others, generally the contrary; but because their food never fails, the sea-shore and river banks daily swarming with a fresh supply of small Mollusca and crustacean. Exactly the same law applies to mammals. Wild cats are prolific and have few enemies; why then are they never as abundant as rabbits? The on.ly intelligible answer is that their supply of food is more precarious. It appears evident, therefore that so long as a country remains physically unchanged, the numbers of its animal population cannot materially increase. If one species does so, some others requiring the same kind of food must diminish in proportion. The numbers that die annually must be immense; and as the individual existence of each animal depends upon itself, those that must die must be the weakest - the very young, the aged, and the diseased, - while those that prolong their existence can only be the most perfect in health and vigour - those who are best able to obtain food regularly, and avoid their numerous enemies. It is, as we commenced by remarking, 'a struggle for existence,' in which the weakest and least perfectly organized must always succumb.

Now it is clear that what takes place among the individuals of a species must also occur among the several allied species of a group, - viz that those which are best adapted to obtain a regular supply of food, and to defend themselves against the attacks of their enemies and the vicissitudes of the season must necessarily obtain and preserve a superiority in population ; while those species which from some defect of power or organization are the least capable of counteracting the vicissitudes of food supply, &., must diminish in numbers, and, in extreme cases, becomes altogether extinct. Between these extremes the species will present various degrees of capacity for ensuring the means of preserving life; and it is thus we account for the abundance or rarity of species. Our ignorance will generally prevent us from accurately tracing the effect to their causes; but could we become perfectly acquainted with the organization and habits of the various species of animals, and could we measure the capacity of each for performing the different acts necessary to its safety and existence under all the varying circumstances by which it is surrounded, we might be able to calculate the proportionate abundance of individuals which is the necessary result.

If now we have succeeded in establishing these two points - 1st, that the animal population of a country is generally stationary, being kept down by a periodical deficiency of food, and other checks and, 2nd, that the comparative abundance or scarcity of the individuals of the several species is entirely due to their organization and resulting habits, which, rendering it more difficult to procure a regular supply of food and to provide for their personal safely in some cases than in others, can only be balanced by a difference in the population which have to exist in a given area- we shall be in a condition to proceed to the consideration of varieties, to which the preceding remarks have a direct and very important application.

Most or perhaps all the variations from the typical form of a species must have some definable effect, however slight, on the habits or capacities of the individuals. Even a change of colour might, by rendering them more or less distinguishable, affect their safety; a greater or less development of hair might modify their habits. More important changes, such as an increase in the power or dimensions of the limbs or any of the external organs, would more or less affect, either favourably or adversely, the powers of prolonging existence. An antelope with shorter or weaker legs must necessarily suffer more from the attacks of the feline carnivora, the passenger pigeon with less powerful wings would sooner or later be affected in its powers of procuring a regular supply of food; and in both cases the result must necessarily be a diminution of the population of the modified species. If, on the other hand, any species should produce a variety having slightly increased powers of preserving existence, that variety must inevitably in time acquire a superiority in numbers. These results must follow as surely as old age intemperance, or scarcity of food produces an increased mortality. In both cases there may be many individual exceptions; but on average the rule will invariably be found to hold good. All varieties therefore fall into two classes - those which under the same conditions would never reach the population of the parent species, and those which would in time obtain and keep a numerical superiority. Now let some alteration of physical conditions occur in the district - a long period of drought, a destruction of vegetation by locusts, the irruption of some new carnivorous animals seeking 'pastures new' - any change in fact tending to render existence more difficult to the species in question and taking its utmost powers to avoid complete extermination, it is evident that, of all the individuals composing the species, those forming the least numerous and most feebly organized variety would suffer first, and, were the pressure severe, must soon become extinct. The same causes continuing in action, the parent species would next suffer, would gradually diminish in numbers, and with a recurrence of similar unfavourable conditions might also become extinct. The superior variety would then alone remain, and on a return to favourable circumstances would rapidly increase in numbers and occupy the place of the extinct specie3s and variety.

The variety would now have replaced the species of which it would be a more perfectly developed and more highly organized form. It would be in all respects better adapted to secure its safety, and to prolong its individual existence and that of the race. Such a variety would not return to the original form; for that form is an inferior one, and could never compete with it for existence. Granted , therefore, a 'tendency' to reproduce the original type of species, still the variety must ever remain preponderant in numbers, and under adverse physical conditions again alone survive. But this new, improved, and populous race might itself, in course of time, give rise to new varieties, exhibiting several diverging modifications of for, any of which, tending to increase the faculties for preserving existence, must, by the same general law, in their turn become predominant. Here, then, we have progression and continued divergence deduced from the general laws which regulate the existence of animals in a state of nature, and from the undisputed fact that varieties do frequently occur. It is not, however, contended that this result would be invariable; a change in physical conditions in the district might at times materially modify it, rendering the race which had been the most capable of supporting existence under the former conditions now the least so, and even causing the extinction of the newer and, for a time, superior race, while the old or parent species and its first inferior varieties continued to flourish. Variations in unimportant parts might also occur, having no perceptible effect of the life-preserving powers; and the varieties so furnished might run a course parallel with the parent species, either giving rise to further variations or returning to the former type. All we argue for is, that certain varieties have a tendency to maintain their existence longer than the original species, and this tendency must make itself felt, for although the doctrine of chances or averages can never be trusted to on a limited scale, yet, if applied to high numbers, the results come nearer to what theory demands, and, as we approach to an infinity of examples, becomes strictly accurate. Now the scale on which nature works is so vast, the numbers of individuals and periods of time with which she deals approach so near to infinity, that any cause, however slight, and however liable to be veiled and counteracted by accidental circumstances, must in the end produce its full legitimate results.

Let us now turn to domesticated animals, and enquire how varieties produced among them are affected by the principles here enunciated. The essential difference in the condition of wild and domestic animals is this – that among the former, their well-being and very existence depends upon the full exercise and healthy condition of all their senses and physical powers, whereas, among the latter these are only partially exercised, and in some cases are absolutely unused. A wild animal has to search, and often to labour, for every mouthful of food – to exercise sight, hearing, and smell in seeking it, and in avoiding dangers, in procuring shelter from the inclemency of the seasons, and in providing for the subsistence and safety of its offspring. There is no muscle of its body that is not called into daily and hourly activity; there is no sense or faculty that is not that is not strengthened by continual exercise. The domestic animals, on the other hand has food provided for it, is sheltered and often confined, to guard against the vicissitudes of the seasons, is carefully secured from the attacks of its natural enemies, and seldom even rears its young without human assistance. Half of its senses and faculties are quite useless; and the other half are but occasionally called into feeble exercise while even its muscular system is only irregularly called into action.

Now when a variety of such an animal occurs, having increased power or capacity in any organ or sense, such increase is totally useless, is never called into action, and may even exist without the animal ever becoming aware of it. In the wild animal, on the contrary, all its faculties and powers being brought into full action for the necessities of existence, any increase becomes immediately available, is strengthened by exercise, and must even slightly modify the food, the habits, and the whole economy of the race. It creates as it were a new animal, one of superior powers, and will necessarily increase in numbers and outlive those inferior to it.

Again, in the domesticated animal all variations have an equal chance of continuance, and those which would decidedly render a wild animal unable to compete with its fellows and continue its existence are not disadvantaged whatever in a state of domesticity. Out quickly fattening pigs, short legged sheep, pouter pigeons, and poodle dogs could never have come into existence in a state of nature, because the very first step towards such inferior forms would have led to rapid extinction of the race; still less could they now exist in competition with their wild allies. The great speed but slight endurance of the race horse, the unwieldy strength of the ploughman's team, would both be useless in a state of nature. If turned wild on the pampas, such animals would probably soon become extinct, or under favourable circumstances might each lose those extreme qualities which would never be called into action and in a few generations would revert to a common type, which must be that in which the various powers and faculties are so proportioned to each other as to be best adapted to procure food and secure safety, - that in which by the full exercise of every part of his organization the animal can alone continue to live. Domestic varieties, when turned wild, must return to something near the type of the original wild stock, or become altogether extinct.

We see, then, that no inferences as to varieties in a state of nature can be deduced from the observations of those occurring among domestic animals. The two are so much opposed to each other in every circumstance of their existence, that what applies to the one is almost sure not to apply to the other. Domestic animals are abnormal, irregular, artificial; they are subject to varieties which never occur and never can occur in a state of nature; their very existence depends altogether on human care; so far are many of them removed from that just proportion of faculties, that true balance of organization by means of which alone an animal left to its own resources can preserve its existence and continue its race.

The hypothesis of Lamarck - that progressive changes in species have been produced by the attempts of animals to increase the development of their own organs, and thus modify their structure and habits- has been repeatedly and easily refuted by all writers on the subject of varieties and species, and it seems to have been considered that when this was done the whole question had been finally settled but the view here developed renders such an hypothesis quite unnecessary, by showing that similar results must be produced by the action of principles constantly at work in nature. The powerful retractile talons of the falcon - and the cat tribe - have not been produced or increased by the volition of those animals but among different varieties which occurred in the earlier and less highly organized forms of these groups, those always survived longer which had the greatest facilities for seizing their prey. Neither did the giraffe acquire its long neck by desiring to reach the foliage of the more lofty shrubs and constantly stretching its neck for the purpose, but because any varieties which occurred among its anti-types with a longer neck than usual at once secured a fresh range of pasture over the same ground as their shorter-necked companions, and on the first scarcity of food were thereby enabled to outlive them. Even the peculiar colours of many animals, especially insects, so closely resembling the soil or the leaves or the trunks on which they habitually reside, are explained by the same principle; for though in the course of ages varieties of many tints may have occurred, yet those races having colours best adapted to concealment from their enemies would inevitably survive the longest. We have also here an acting cause to account for that balance so often observed in nature, - a deficiency in one set of organs always being compensated by an increased development in some others - powerful wings accompanying weak feet, or great velocity making up for the absence of defensive weapons; for it has been shown that all varieties in which an unbalanced deficiency occurred could not long continue their existence. The action of this principle is exactly like that of the centrifugal governor of the steam engine, which checks and corrects any irregularities almost before they become evident, and in like manner no unbalanced deficiency in the animal kingdom can ever reach any conspicuous magnitude, because it would make itself felt at the very first step, by rendering existence difficult and extinction almost sure soon to follow. An origin as is here advocated will also agree with the peculiar character of the modifications of form and structure which obtain in organized beings - the many lines of divergence from a central type, the increasing efficiency and power of a particular organ through a succession of allied species, and the remarkable persistence of unimportant parts such as colour, texture of plumage and hair, form of horns or crests, through a series of species differing considerably in more essential characters. It also furnishes us with a reason for that 'more specialized structure' which Professor Owen states to be the characteristic of recent compared with extinct forms, and which would evidently be the result of the progressive modification of any organ applied to a special purpose in the animal economy.

We believe that we have now shown that there is a tendency if nature to the continued progression of certain classes of varieties further and further from the original type - a progression to which there appears no reason to assign any definite limits - and that the same principle which produces this result in a state of nature will also explain why domestic varieties have a tendency to revert to the original type. This progression, by minute steps, in various directions, but always checked and balanced by the necessary conditions, subject to which alone existence can be preserved by organized beings, their existence and succession in past ages, and all the extraordinary modifications of form, instinct, and habits which they exhibit.

Ternate, February 1858

Darwin's letter to Asa Gray 1857

Darwin, F., (1887) *The Life and Letters of Charles Darwin* (Vol. 1) (pp. 120-125)

C. Darwin to Asa Gray.

Down, Sept.* 5th [1857].

MY DEAR GRAY,—I forget the exact words which I used in my former letter, but I dare say I said that I thought you would utterly despise me when I told you what views I had arrived at, which I did because I thought I was bound as an honest man to do so. I should have been a strange mortal, seeing how much I owe to your quite extraordinary kindness, if in saying this I had meant to attribute the least bad feeling to you. Permit me to tell you that, before I had ever corresponded with you, Hooker had shown me several of your letters (not of a private nature), and these gave me the warmest feeling of respect to you; and I should indeed be

* The date is given as October in the 'Linnean Journal.' The extracts were printed from a duplicate undated copy in my father's possession, on which he had written, "This was sent to Asa Gray 8 or 9 months ago, I think October 1857."

576

ungrateful if your letters to me, and all I have heard of you, had not strongly enhanced this feeling. But I did not feel in the least sure that when you knew whither I was tending, you might not think me so wild and foolish in my views (God knows, arrived at slowly enough, and I hope conscientiously), that you would think me worth no more notice or assistance. To give one example : the last time I saw my dear old friend Falconer, he attacked me most vigorously, but quite kindly, and told me, "You will do more harm than any ten Naturalists will do good. I can see that you have already *corrupted* and half-spoiled Hooker !!" Now when I see such strong feeling in my oldest friends, you need not wonder that I always expect my views to be received with contempt. But enough and too much of this.

I thank you most truly for the kind spirit of your last letter. I agree to every word in it, and think I go as far as almost any one in seeing the grave difficulties against my doctrine. With respect to the extent to which I go, all the arguments in favour of my notions fall *rapidly* away, the greater the scope of forms considered. But in animals, embryology leads me to an enormous and frightful range. The facts which kept me longest scientifically orthodox are those of adaptation—the pollen-masses in asclepias — the mistletoe, with its pollen carried by insects, and seed by birds—the woodpecker, with its feet and tail, beak and tongue, to climb the tree and secure insects. To talk of climate or Lamarckian habit producing such adaptations to other organic beings is futile. This difficulty I believe I have surmounted. As you seem interested in the subject, and as it is an *immense* advantage to me to write to you and to hear, ever so briefly, what you think, I will enclose (copied, so as to save you trouble in reading) the briefest abstract of my notions on the means by which Nature makes her species. Why I think that species have really changed, depends on general facts in the affinities, embryology, rudimentary organs, geological history, and geographical distribution of organic beings. In regard to my Abstract, you must take immensely on trust, each paragraph occupying one or two chapters in my book. You will, perhaps, think it paltry in me, when I ask you not to mention my doctrine; the reason is, if any one, like the author of the 'Vestiges,' were to hear of them, he might easily work them in, and then I should have to quote from a work perhaps despised by naturalists, and this would greatly injure any chance of my views being received by those alone whose opinions I value. [Here follows a discussion on "large genera varying," which has no direct connection with the remainder of the letter.]

I. It is wonderful what the principle of Selection by Man, that is the picking out of individuals with any desired quality, and breeding from them, and again picking out, can do. Even breeders have been astonished at their own results. They can act on differences inappreciable to an uneducated eye. Selection has been *methodically* followed in Europe for only the last half century. But it has occasionally, and even in some degree methodically, been followed in the most ancient times. There must have been also a kind of unconscious selection from the most ancient times, namely, in the preservation of the individual animals (without any thought of their offspring) most useful to each race of man in his particular circumstances. The "roguing," as nursery-men call the destroying of varieties, which depart from their type, is a kind of selection. I am convinced that intentional and occasional selection has been the main agent in making our domestic races. But, however this may be, its great power of modification has been indisputably shown in late times. Selection acts only by the accumulation of very slight or greater variations, caused by external conditions, or by the mere fact that in generation the child is not absolutely similar to its parent. Man, by this power of accumulating variations, adapts living beings to his wants—he *may be said* to make the wool of one sheep good for carpets, and another for cloth, &c.

II. Now, suppose there was a being, who did not judge by mere external appearance, but could study the whole internal organisation—who never was capricious—who should go on selecting for one end during millions of generations, who will say what he might not effect!   In nature we have some *slight* variations, occasionally in all parts: and I think it can be shown that a change in the conditions of existence is the main cause of the child not exactly resembling its parents; and in nature, geology shows us what changes have taken place, and are taking place.   We have almost unlimited time: no one but a practical geologist can fully appreciate this: think of the Glacial period, during the whole of which the same species of shells at least have existed; there must have been during this period, millions on millions of generations.

III. I think it can be shown that there is such an unerring power at work, or *Natural Selection* (the title of my book), which selects exclusively for the good of each organic being. The elder De Candolle, W. Herbert, and Lyell, have written strongly on the struggle for life; but even they have not written strongly enough.   Reflect that every being (even the elephant) breeds at such a rate that, in a few years, or at most a few centuries or thousands of years, the surface of the earth would not hold the progeny of any one species.   I have found it hard constantly to bear in mind that the increase of every single species is checked during some part of its life, or during some shortly recurrent generation.   Only a few of those annually born can live to propagate their kind.   What a trifling difference must often determine which shall survive and which perish!

IV. Now take the case of a country undergoing some change; this will tend to cause some of its inhabitants to vary slightly; not but what I believe most beings vary at all times enough for selection to act on. Some of its inhabitants will be exterminated, and the remainder will be exposed to the mutual action of a different set of inhabitants, which I believe to be more important to the life of each being than mere climate. Considering the infinitely various ways beings have to obtain food by struggling with other beings, to escape danger at various times of life, to have their eggs or seeds disseminated, &c. &c., I cannot doubt that during millions of generations individuals of a species will be born with some slight variation profitable to some part of its economy; such will have a better chance of surviving, propagating this variation, which again will be slowly increased by the accumulative action of natural selection ; and the variety thus formed will either coexist with, or more commonly will exterminate its parent form. An organic being like the woodpecker, or the mistletoe, may thus come to be adapted to a score of contingencies ; natural selection, accumulating those slight variations in all parts of its structure which are in any way useful to it, during any part of its life.

V. Multiform difficulties will occur to every one on this theory. Most can, I think, be satisfactorily answered,— "Natura non facit saltum" answer some of the most obvious. The slowness of the change, and only a very few undergoing change at any one time answers others. The extreme imperfections of our geological records answer others.

VI. One other principle, which may be called the principle of divergence, plays, I believe, an important part in the origin of species. The same spot will support more life if occupied by very diverse forms : we see this in the many generic forms in a square yard of turf (I have counted twenty species belonging to eighteen genera), or in the plants and insects, on any little uniform islet, belonging to almost as many genera and families as to species. We can understand this with the higher animals, whose habits we best understand. We know that it has been experimentally shown that a plot of land will yield a greater weight, if cropped with several species of grasses, than with two or three species. Now every single organic being, by propagating rapidly, may be said to be striving its utmost to increase in numbers. So it will be with the offspring of any species after it has broken into varieties, or sub-species, or true species. And it follows, I think, from the foregoing facts, that the varying offspring of each species will try (only few will succeed) to seize on as many and as diverse places in the economy of nature as possible. Each new variety or species when formed will generally take the place of, and so exterminate its less well-fitted parent. This, I believe, to be the origin of the classification or arrangement of all organic beings at all times. These always *seem* to branch and sub-branch like a tree from a common trunk; the flourishing twigs destroying the less vigorous—the dead and lost branches rudely representing extinct genera and families.

This sketch is *most* imperfect; but in so short a space I cannot make it better. Your imagination must fill up many wide blanks. Without some reflection, it will appear all rubbish; perhaps it will appear so after reflection.

C. D.

Paper read by Sigmund Freud before the Society for Psychiatry and Neurology, Vienna, 21st April, 1896

(Masson 1984: 251-282)

GENTLEMEN. When we set out to form an opinion about the causation of a pathological state such as hysteria, we begin by adopting the method of anamnesis investigation; we question the patient or those about him in order to find out to what harmful influences they themselves attribute his having fallen ill and having developed these neurotic symptoms. What we discover in this way is, of course, falsified by all the factors which commonly hide the knowledge of his own state from a patient – by his lack of scientific understanding of ætiological influence, by the fallacy of post hoc, propter hoc, by his reluctance to think about or mention certain noxæ and traumas. This in making an anamnestic investigation of this sort, we keep to the principal of not adopting the patient's belief without a thorough critical examination, of not allowing them to lay down our scientific opinion for us on the ætiology of neurosis. Although we do, on the one hand, acknowledge the truth of certain constantly repeated assertions, such as that the hysterical state is a long-persisting after-effect of an emotion experienced in the past, we have, on the other hand, introduced into the ætiology of hysteria a factor which the patient himself never brings forward and whose validity he only reluctantly admits – namely, the heredity disposition derived from his progenitors. As you know, in the view of the influential school of Charcot, heredity alone deserves to be recognized as the true cause of hysteria, while all other noxæ of the most various nature and intensity only play the part of incidental causes, of "agents provocateurs".

You will readily admit that it would be a good thing to have a second method of arriving at the ætiology of hysteria, one in which we should feel less dependent on the assertions of the patient themselves. A dermatologist, for instance, is able to recognize a sore as luetic from the character of its margins, from the crust on it and of its shape, without being misled by the protestations of his patient who denies any source of infection for it; and a forensic physician can arrive at the cause of an injury, even if he has to do without any information from the injured person. In hysteria, too, there exists a similar possibility of penetrating from the symptoms to a knowledge of their causes. But in order to explain the relationship between the method which we have to employ for this purpose and the older method of anamnesis enquiry, I should like to bring before you an analogy taken from an advance that has in fact been made in another field of work.

Imagine that an explorer arrives in a little-known region where his interest is aroused by an expanse of ruins, with remains of walls, fragments of columns, and tablets with half-effaced and unreadable inscriptions. He may content himself with inspecting what lies exposed to view, with questioning the inhabitants – perhaps semi-barbaric people – who live in the vicinity, about what tradition tells them of the history and meaning of these archæological remains, and with noting down what they tell him – and he may then proceed on his journey. But he may act differently. He may have brought picks, shovels and spades with him, and he may set the inhabitants to work with these implements. Together with them he may start upon the ruins, clear away the rubbish and, beginning from the visible remains, uncover what is buried. If his work is crowned with success, the discoveries are self-explanatory; the ruined walls are part of the ramparts of a palace or a treasure house; the fragments of columns can be filled out into a temple; the numerous inscriptions, which, by good luck, may be bilingual, reveal an alphabet and a language, and, when they have been de3cipheredand translated, yield undreamed of information about the events of the remote past, to commemorate which the monuments were built. Saxa loquuntur!

If we try, in an approximately similar way, to induce the symptoms of a hysteria to make themselves heard as witnesses to the history of the illness, we must take our start from Joseph Breuer's momentous discovery: the symptoms of hysteria (apart from stigmata) are determined by certain experiences of the patient's which have operated in a traumatic fashion and which are being reproduced in his physical life in the form of mnemic symbols. What we have to do is to apply Breuer's method – or one which is essentially the same – so as to lead the patient's attention back from the symptoms to the scene in which and through which the symptoms arose; and, having this located the scene, we remove the symptoms by bringing about, during the reproduction of the traumatic scene, a subsequent correction of the psychical course of events which took place at that time.

It is no part of my intention to-day to discuss the difficult technique of this therapeutic procedure or the psychological discoveries which have been obtained by its means. I have been obliged to start from this pointy only because the analysis conducted on Breuer's lines seem at the same time to open up the path to the cause of hysteria. If we subject a fairly large number of symptoms in a great number of subjects to such an analysis, we shall, of course, arrive at a knowledge of a correspondingly large number of traumatically operative scenes. It was in these experiences that the efficient causes of hysteria came into action. Hence we may hope to discover from the study of these traumatic scenes what the influences are which produce hysterical symptoms and in what way they do so.

The expectation proves true; and it cannot fail to, since Breuer's theses, when put to the test in a considerable number of cases, have turned out to be correct. But the path from the symptoms of hysteria to its ætiology is more laborious and leads through other connections than one would be imagined.

For let us be clear on this point. Tracing a hysterical symptom back to a traumatic scene assists our understanding only if the scene satisfies two conditions: if it possess the relevant suitability to serve as a determinant and if it recognizably possesses the necessary traumatic force. Instead of a verbal explanation, here is an example. Let us suppose that the symptom under consideration is hysterical vomiting. In that case we shall feel we have been able to understand its causation (except for a certain residue) if the analysis traces the symptoms back to an experience which justifiably produced a high amount of disgust - for instance, the sight of a decomposing dead body. But if, instead of this, the analysis shows us that the vomiting arose from a great fright, e.g. from a railway accident, we shall feel dissatisfied, and will have to ask ourselves how it is that the fright has lead to the particular symptom of vomiting. The derivation lacks suitability as a determinant. We shall have another instance of an insufficient explanation if the vomiting is supposed to have arisen from, let us say, eating a fruit which had partly gone bad. Here, it is true, the vomiting is determined by disgust, but we cannot understand ho, in this instance, the disgust could have become so powerful as to be perpetuated in a hysterical symptom; the experience lacks traumatic force.

Let us now consider how far the traumatic scenes of hysteria which are uncovered by analysis fulfil, in a fairly large number of symptoms, the two requirements which I have named. Here we meet with our first great disappointment. It is true, indeed, that the traumatic scene in which the symptoms originated does in fact occasionally possess both qualities -

suitability as a determinant and traumatic force – which we require for an understanding of the symptoms. But far more frequently, incomparably more frequently, we find one of the three other possibilities realized, which are so unfavourable to an understanding. Either the scene to which we are led by analysis and in which the symptom first appeared seems to be unsuited for determining the symptom, in that its content bears no relation to the nature of the symptom, or the allegedly traumatic experience, though it does have a relation to the symptoms, prove to be an impression which is normally innocuous and incapable as a rule of producing any effect; or, lastly, the "traumatic scene" leaves us in the lurch, in both respects, appearing at once innocuous and unrelated to the character of the hysterical symptoms.

(Here I may remark in passing that Breuer's view of the origin of hysterical symptoms is not shaken by the discovery of traumatic scenes which correspond to experiences that are insignificant in themselves. For Breuer assumed – following Charcot – that even an innocuous experience can be heightened into a trauma and can develop determining force if it happens to the subject when he is in a special psychical condition – in what is described as a hypnotic state. I find however, that there are often no grounds whatever for presupposing the presence of such hypnotic states. What remains decisive is that the theory of hypnoid states contributes nothing to the solution of the other difficulties, namely that the traumatic scenes so often lack suitability as determinants.)

Moreover, Gentlemen, this first disappointment we meet with in following Breuer's method is immediately followed by another, and one that must be especially painful to us as physicians. When our procedure leads, as in the case described above, to findings which are insufficient as a n explanation both in respect to their suitability as determinants and to their traumatic effectiveness, we also fail to secure any therapeutic gain; the patient retains his symptoms unaltered, in spite of the initial result yielded by analysis. You can understand how great the temptation is at this point to proceed no further with what is in any case a laborious piece of work.

But perhaps all we need is a new idea in order to help us out of our dilemma and lead to valuable results. The idea is this. As we know from Breuer, hysterical symptoms can be resolved if, starting from them, we are able to find the path back to the memory of a traumatic experience. If the memory which we have uncovered does not answer our expectations, it may be that we ought to pursue the same path a little further; perhaps behind the first traumatic scene there may be concealed the memory of a second which satisfies our requirements better ad whose reproduction has

585

a genuine therapeutic effect; so that the scene that was first discovered only has the significance of a connecting link in the chain of associations. And perhaps the situation may repeat itself; inoperative scenes may be interpolated more than once, as necessary transitions in the process of reproduction, until we finally make our way from the hysterical symptoms to the scene which is really operative traumatically and which is satisfactory in every respect, both therapeutically and analytically. Well, Gentlemen, this suggestion is correct. If the first discovered scene is unsatisfactory, we tell our patient that the experience explains nothing, but that behind it there must be hidden a more significant, earlier, experience, and we direct his attention by the same technique to the association thread which connects the two memories – the one that has been discovered and that one that has still to be discovered. A continuation of the analysis then leads in every instance to the reproduction of new scenes of the character we expect. For example, let us take once again the case of hysterical vomiting which I selected before, and which lacked suitability as a determinant. Further analysis showed that the accident has aroused in the patient the memory of another, earlier, accident, which, it is true, he had not himself experienced but which had been the occasion of his having a ghastly and revolting sight of a dead body. It is as though the combined operation of the two scenes made the fulfilment of our postulation possible, the one experience supplying, through fright, the traumatic force and the other, from its content, the determining effect. The other case, in which the vomiting was traced back to eating an apple which had partly gone bad, was amplified by the analysis somewhat in the following way. The bad apple reminded the patient of an earlier experience while he was picking up windfalls in an orchard he had accidentally come upon a dead animal in a revolting state.

I shall not return any further to these examples, for I have to confess that they are not derived from any case in my experience but are inventions of mine. Most probably, too, they are bad inventions. I even regard such solutions of hysterical symptoms as impossible. But I was obliged to make up fictitious examples for several reasons, one of which I can state at once. The real examples are all incomparably more complicated: to relate a single one of them in detail would occupy the whole period of this lecture. The chain of association always has more than two links; and the traumatic scenes do not form a simple row, like a string of pearls, but ramify and are interconnected like genealogical trees, so that in any new experience two or more earlier ones come into operation as memories. In short, giving an account of the resolution of a single symptom would in fact amount to the task of relating an entire case history.

But we must not fail to lay special emphasis on one conclusion to which analytic work along these chains of memory has unexpectedly led. We have learned that no hysterical symptoms can arise from a real experience alone, but that in every case the memory of earlier experiences awakened in association to it plays a part in causing the symptoms. If –as I believe – this proposition holds good without exception, it furthermore shows us the basis on which a psychological theory of hysteria must be built.

You might suppose that the rare instances in which analysis is able to trace the symptom back direct to a traumatic scene that is thoroughly suitable as a determinant and possesses traumatic force, and is able, by the tracking it back at the same time to remove it (in the way described in Breuer's case history of Anna O) – you might suppose that such instances must, after all, constitute powerful objections to the general validity of the proposition I have just put forward. It certainly looks so. But I must assure you that I have the best grounds for assuming that even in such instances there exists a chain of operative memories which stretches far back behind the first traumatic scene, even though the reproduction of the latter alone may have been the result of removing the symptom.

It seems to me really astonishing that hysterical symptoms can only arise with the co-operation of memories, especially when we reflect that, according to the unanimous accounts of the patients themselves, these memories did not come into their consciousness at the moment when the symptom first made its appearance. Here is much food for thought; but these problems must not distract us at this point from our discussion of the ætiology of hysteria. We must rather ask ourselves: where shall we go to if we follow the chains of associated memories which the analyst has uncovered? How far do they extend? Do they come anywhere to a natural end? Do they perhaps lead to experiences which are in some ways alike, either in their content or the time of life at which they occur, so that we may discern in their universality similar factors the ætiology of hysteria of which we are in search?

The knowledge that I have so far gained already enables me to answer these questions. If we take a case which presents several symptoms, we arrive by means of the analysis, starting from each symptom, as a series of experiences the memories of which we linked together in association. To begin with, the chains of memories lead backwards, separately from one another, but, as I have said, they ramify. From a single scene two or more memories are reached at the same time, and from these side-chains proceed

587

whole individual links may once more be associatively connected with links belonging to the main chain. Indeed, a comparison with the genealogical tree of a family whose members have also intermarried, is not at all a bad one. Other complications in the linkage of the chains arise from the circumstances that a single scene may be called up several times in the same chain, so that it has multiple relationships to a later scene and exhibits both a direct connection with it and a connection established through intermediate links. In short, the concatenation is far from being a simple one; and the fact that the scenes are uncovered in a reversed chronological order (a fact which justifies our comparison of the work with the excavation of a stratified ruined site) certainly contributes nothing to a more rapid understanding of what has taken place.

If the analysis is carried further, new complications arise. The associative chains belonging to the different symptoms begin to enter into relation with one another; the genealogical tress become intertwined. Thus a particular symptom in, for instance, the chain of memories relating to the symptoms of vomiting, calls up not only the earlier links in its own chain, but also a memory from another chain, relating to another symptom, such as a headache. This experience belongs to both series, and in this way it constitutes a nodal point. Several such nodal points are to be found in every analysis. Their correlation in the clinical picture may perhaps be that from a certain time onwards both symptoms have appeared together, symbolically, without in fact having any internal dependence on each other. Going still further back, we come upon nodal points of a different kind. Here the separate associative chains converge. We find experiences from which two more symptoms have proceeded, one chain has attached itself to one detail of the scene, the second chain to another detail.

But the most important finding that is arrived at if an analysis is thus consistently pursued is this. Whatever case and whatever symptoms we take as our point of departure, in the end we infallibly come to the field of sexual experience. So here for the first time we seem to have discovered an ætiological precondition for hysterical symptoms.

From previous experience I can foresee that it is precisely against this assertion or against its universal validity that your contradiction, Gentlemen, will be directed. Perhaps it would be better to say, your inclination to contradict; for none of you, no doubt, have as yet any investigations at your disposal which, based upon the same procedure might have yielded a different result. As regards the controversial matter itself, I will only remark that the singling out of the sexual factor in the ætiology of hysteria springs at least from no preconceived opinion on my part. The two investigators as whose pupil I began my studies of hysteria, Charcot and Breuer, were far from having any such presupposition; in fact, they had a personal disinclination to it which I originally shared. Only the most laborious and detailed investigations have converted me, and that slowly enough, to the view I hold to-day. If you submit my assertion that the ætiology of hysteria lies in the sexual life to the strictest examination, you will find that it is supported by the fact that in some eighteen cases of hysteria I have been able to discover this connection in every single symptom, and, where the circumstances allowed, to confirm it by therapeutic success. No doubt you may raise the objection that the nineteenth or the twentieth analysis will perhaps show that hysterical symptoms are derived from other sources as well, and thus reduce the universal validity of the sexual ætiology to one of eighty per cent. By all means let us wait and see; but, since these eighteen cases are at the same time all the cases on which I have been able to carry out the work of analysis and since they were not picked out by anyone for my convenience, you will find it understandable that I do not share such an expectation but am prepared to let my belief run ahead of the evidential force of the observations I have so far made. Besides, I am influenced by another motive as well, which for the moment is of merely subjective value. In the sole attempt to explain the physiological and psychical mechanism of hysteria which I have been able to make in order to correlate by observations, I have come to regard the participation of sexual motive forces as an indispensable premise.

Eventually, then, after the chains of memories have converged, we come to the field of sexuality and to a small number of experiences which occur for the most part at the same period of life - namely, at puberty. It is in these experiences, it seems, that we are to look for the ætiology of hysteria, and through them that we are to learn to understand the origin of hysterical symptoms. But here we meet with a fresh disappointment and a very serious one. It is true that these experiences, which have been discovered with so much trouble and extracted out of all the mnemic material, and which seemed to be the ultimate traumatic experiences, have in common two characteristics of being sexual and of occurring at puberty; but in every other respect they are very different from each other in from each other in kind and in importance. In some cases, no doubt, we are concerned with experiences which must be regarded as severe trauma - an attempted rape which reveals to the immature girl at a blow all the brutality of sexual desire, or the involuntary witnessing of sexual acts between parents, which at one and the same time uncovers unsuspected ugliness and wounds childish and moral sensibilities alike, and so on. But in other cases the experiences are astonishingly trivial. In one of my women patients it turned out that her neurosis was based on the experience of a boy of her acquaintance stroking her hand tenderly and, at another time, pressing his knee against her dress as they sat side by side at table, while his expression let her see that he was doing something forbidden. For another young lady, simply hearing a riddle which suggested an obscene answer had been enough to provoke the first anxiety attack and with it to start the illness. Such findings are clearly not favourable to the understanding of the causation of hysterical symptoms. If serious and trifling events alike, and if not only experiences affecting the subject's own body but visual impressions too and information received through ears are to be recognized as the ultimate traumas of hysteria then we may be tempted to hazard the explanation that hysterics are peculiarly constituted creatures – probably on account of some hereditary disposition or degenerative atrophy – in whom a shrinking form of sexuality, which normally plays some part at puberty, is raised to a pathological pitch and is permanently retained; that there are, as it were, people who are psychically inadequate to meeting the demands of sexuality. But even without blatant objections such as that, we should scarcely be tempted to be satisfied with this solution. We are only too distinctly conscious of an intellectual sense of something half-understood, unclear and insufficient.

Luckily for our explanation, some of these sexual experiences at puberty exhibit a further inadequacy, which is calculated to stimulate us into continuing our analytic work. For it sometimes happens that they, too, lack suitability as determinants – although this is much more rarely so than with the traumatic scenes belonging to later life. Thus, for instance, let us take the two women patients whom I have just spoken of as cases in which the experiences at puberty were actually innocent ones. As a result of those experiences the patients had become subject to peculiar sensations in the genitals which had established themselves as the main symptoms of the neurosis. I was unable to find indications that they had been determined either by the scenes at puberty or by later scenes; but they were certainly not normal organic sensations nor signs of sexual excitement. It seemed an obvious thing, then to say to ourselves that we must look for determinants of these symptoms in yet other experiences, in experiences which went still further back – and that we must, for that second time, follow he saving notion which had earlier led us from the first traumatic scenes to the chains of memories behind them. In doing so, to be sure, we arrive at the period of the earliest childhood, a period before the abandonment of a sexual ætiology. But have we not a right to assume that even the age of childhood is not is not wanting in slight sexual excitations, that later sexual development may perhaps be decisively influenced by childhood experiences? Injuries sustained by an organ which is as yet immature, or by a function which is in the process of developing, often causes more severe and lasting effects than they could do in maturer years. Perhaps the abnormal reaction to sexual impressions which surprises us in hysterical subjects at the age of puberty is quite generally based on sexual experiences of this sort in childhood, in which case those experiences must be of a similar nature of one another, and must be of an important kind. If this is so, the prospect is opened up that what has hitherto led to be laid at the door of a still unexplained hereditary predisposition may be accounted for as having been acquired at an early age. And since infantile experiences with a sexual content could after all only exert a psychical effect through their memory traces, would not this view be a welcome amplification of the finding of psycho-analysis which tells us that hysterical symptoms can only arise with the co-operation of memories?

II

You will no doubt have guessed, Gentlemen, that I should not have carried this last line of thought so far if I had not wanted to prepare you for the idea that it is this line alone which, after so many delays, will lead us to our goal. For now we are really at the end of our wearisome and laborious analytic work, and here we find the fulfilment of all the claims and expectations upon which we have so far insisted. If we have the perseverance to press on with analysis into early childhood, as far back as a human memory is capable of reaching, we invariably bring the patient to reproduce experiences which, on account both of their peculiar features and of their relations to the symptoms of his later illness, must be regarded as the ætiology of his neurosis for which we have been looking. These infantile experiences are once more sexual in contest, but they are of a far more uniform kind than the scenes at puberty that hd been discovered earlier. It is now no longer a question of sexual topics having been aroused by some sense impressions or other, but of sexual experiences affecting the subject's own body - of sexual intercourse (in the wider sense). You will admit that the importance of such scenes needs no further proof; to this may now be added that, in every instance, you will be able to discover in the details of the scenes the determining factors which you may have found lacking in the other scenes - the scenes which occurred later and were reproduced earlier.

I therefore put forward the thesis that at the bottom of every case of hysteria there are one or more occurrences of premature sexual experience, occurrences which belong to the earliest years of childhood but which can be reproduced through the work of psycho-analysis in spite of the intervening decades. I believe that this is an important finding, the discovery of a caput nili in neuropathology; but I hardly know what to take as a starting point for a continuation of my discussion of this subject. Shall I put before you the actual material that I have obtained from my analyses? Or shall I rather try to meet the mass of objections and doubts which, I am surely correct in supposing have now taken possession of your attention. I shall choose the latter course; perhaps we shall then be able to go over the facts more calmly.

a) No one who is altogether opposed to a psychological view of hysteria, who, unwilling to give up the hope that some day it will be possible to trace back in symptoms to "finer anatomical changes" and who has rejected the view that the material foundation of hysterical changes are bound to be of the same kind as those of our normal mental processes – no one who adopts this attitude will, of course, put any faith in the result of our analyses; however, the difference in principal between his premises and ours absolves us from the obligation of convincing him on individual points.

But other people, too, although they may be less adverse to psychological theories of hysteria, will be tempted, when considering our analytic findings, to ask what degree of certainty the application of psycho-analysis offers. Is it not very possible either that the physician forces such scenes upon his docile patients, alleging that they are memories, or that the patients tell the physician things which they have deliberately invented or have imagined and that he accepts those things as true? Well, my answer to this is that the general doubt about the reliability of the psycho-analytic method can be appraised and removed only when a complete presentation of its technique and results is available. Doubts about the genuineness of the infantile sexual scene can, however, be deprived of their force here and now by more than one argument. In the first place, the behavior of patients while they are reproducing these infantile experiences is in every respect compatible with the assumption that the scenes are anything else than a reality which is being felt with distress and reproduced with the greatest reluctance. Before they come for analysis the patients know nothing about these scenes. They are indignant as a rule if we warn them that such scenes are going to emerge. Only the strongest compulsion of the treatment can induce them to embark upon a reproduction of them. While they are recalling these infantile experiences in consciousness, they suffer under the most violent sensations, of which they are ashamed and which they try to conceal; and, even after they have gone through them once more in such a convincing manner, they still attempt to withhold belief from them, by emphasizing the fact that, unlike what happens in the case of other forgotten material, they hae no feeling of remembering the scenes.

This latter piece of behavior seems to provide conclusive proof. Why should patients assure me so emphatically of their unbelief, if what they want to discredit is some something which - from whatever motive - they themselves have invented?

It is less easy to refute the idea that the doctor forces reminiscences of this sort on the patient, that he influences him by suggestion to imagine and reproduce them. Nevertheless it appears to me equally untenable. I have never yet succeeded in forcing on a patient a scene I was expecting to find, in such a way that he seemed to be living through it with all the appropriate feelings. Perhaps others may be more successful in this.

There are, however, a whole number of other things that vouch for the reality of infantile sexual scenes. In the first place there is the uniformity which they exhibit in certain details, which is a necessary consequence if the preconditions of these experiences are always of the same kind, but which would otherwise lead us to believe that there were secret understandings between the various patients. In the second place, patients sometimes describe as harmless events whose significance they obviously do not understand, since they would be bound otherwise to be horrified by them. Or again, they mention details without laying any stress on them, which only someone of experience in life can understand and appreciate as subtle traits of reality.

Events of this sort strengthen our impression that the patients must really have experienced what they reproduce under the compulsion of analysis as scenes from their childhood. But another and stronger proof of this is furnished by the relationship of the infantile scenes to the content of the whole of the rest of the case history. It is exactly like putting together a child's picture-puzzle; after many attempts, we become absolutely certain in the end which piece belongs in the empty gap; for only that one piece fills out the picture and at the same time allows its irregular edges to be fitted into the edges of the other pieces in such a manner as to leave no free space and to entail no overlapping. In the same way, the contents of the infantile scenes turn out to be indispensable supplements to the associative and logical framework of the neurosis, whose insertion makes its course of development for the first time evident, or even, as we might often say, self-evident.

Without wishing to lay special stress on the point, I will add that in a number of cases therapeutic evidence of the genuineness of the infantile scenes can also be brought forward. There are cases in which a complete or partial cure can be obtained without our having to go as deep as the infantile experiences. And there are others in which no success at all is obtained until the analysis has come to its natural end with the uncovering of the earliest traumas. In the former cases we are not, I believe, secure against relapses, and my experience is that a complete psycho-analysis implies a radical cure of the hysteria. We must not, however, be led into forestalling the lessons of observation.

There would be one other proof, and a really unassailable one, of the genuineness of childhood experiences - namely, if confirmed by someone else, whether under treatment or not. These two people will have had to have taken part in the same experience in their childhood - perhaps to have stood in some sexual relationship to each other, Such relations between children are, as you will hear in a moment, by no means rare. Moreover, it quite often happens that both of those concerned subsequently fall ill of neuroses; yet I regard it as a fortunate accident that, out of eighteen cases, I have been able to obtain an objective confirmation of this sort in two. In one instance, it was the brother, who had remained well – who of his own accord confirmed - not, it is true, his earliest sexual experiences with his sister (who was the patient) - but at least scenes of that kind from later childhood, and the fact that there had been sexual relations dating further back. In the other instance, it happened that two women whom I was treating had as children had sexual relations with the same man, in the course of which certain scenes had taken place á trios. A particular symptom, which was derived from these childhood events, had developed in both women, as evidence of what they had experienced in common.

(b) Sexual experiences in childhood consisting in stimulation of the genitals, coitus-like acts, and so on, must, therefore, be recognized, in the last analysis, as being the traumas which lead to a hysterical reaction to events at puberty and to the development of hysterical symptoms. This statement is certain to be met from different directions by two mutually contradictory objectives. Some people will say that sexual abuses of this kind, whether practiced upon children or between them happen too seldom for it to be possible to regard them as the determinant of such a common neurosis as hysteria. Others will perhaps argue that, on the contrary, such experiences are very frequent - much too frequent for us to be able to attribute an ætiological significance to the fact of their occurrence. They will further maintain that it is easy, by making a few enquiries, to find people to remember scenes of sexual seduction and sexual abuse in their childhood years, and yet who have never been hysterical. Finally, we shall be told, as a weighty argument, that in the lower strata of the population hysteria is certainly no more common than in the highest ones, whereas everything goes to show that the injunction for the sexual safeguarding of childhood is far more frequently transgressed in the case of the children of the proletariat.

Let us begin our defence with the easier part of the task. It seems to be certain that our children are far more often exposed to sexual assaults than the few precautions taken by parents in this connection would lead us to expect. When I first made enquiries about what was known on the subject, I learnt from colleagues that there are several publications by pædiatricians which stigmatize the frequency of sexual practices by nurses and nursery maids, carried out even on infants in arms; and in the last few weeks, I have come across a discussion of "Coitus in Childhood" by Dr. Stekel (1895) in Vienna. I have not had time to collect other published evidence, but even if it were only scanty, it is to be expected that increased attention to the subject will very soon confirm the great frequency of sexual experiences and sexual activity in childhood.

Now let us turn to the other objection which is based precisely on an acknowledgment of the frequency of infantile sexual experiences and on the observed fact that many people who remember scenes of that kind have not become hysterics. Our first reply is that the excessive frequency of an ætiological factor cannot possibly be used as an objection to its ætiological significance. Is not the tubercle bacillus ubiquitous and is it not inhaled by far more people than are found to fall ill of tuberculosis? And is its ætiological significance impaired by the fact that other factors must obviously be at work too before the tuberculum, which is is specific effect, can be evoked? In order to establish the bacillus as the specific ætiology it is enough to show that tuberculosis cannot possibly occur with its playing a part. The same doubtless applies to our problem. It does not matter if many people experience infantile sexual scenes without becoming hysterics, provided only that all the people who become hysterics have experienced scenes of that kind. The area of occurrence of an ætiiological factor may be freely allowed to be wider than that of its effect, but it must not be narrower. Not everyone who touches or comes near a smallpox patient develops smallpox; nevertheless infection from a smallpox patient is almost the only known ætiology of the disease.

It is true that if infantile sexual activity were an almost universal occurrence, the demonstration of its presence in every case would carry no weight. But, to begin with, to assert such a thing would certainly be a gross exaggeration; and secondly, the ætiological pretensions of the infantile scenes rest not only on the regularity of their appearances in the anamnesis of hysterics but, above all, on the evidence of there being associative and logical ties between those scenes and the hysterical symptoms – evidence which, if you were give the complete history of a case, would be as clear as daylight to you.

What can the other factors be which the "specific ætiology" of hysteria still needs in order actually to produce the neurosis? That, Gentlemen, is a theme in itself, which I do not propose to enter upon. To-day I need only indicate the point of contact at which the two parts of the topic – the specific and the auxiliary ætiology – fit into one another. No doubt a considerable quantity of factors will have to be taken into account. There will be the subject's inherited and personal constitution, the inherent importance of the infantile sexual experiences, and, above all, their number; a brief relationship with a strange boy, who afterwards becomes indifferent, will leave a less powerful effect on a girl than intimate sexual relations of several years' standing with her own brother. In the ætiology of the neuroses quantitative preconditions are as important as qualitative ones; there are threshold values which have to be crossed before the illness can become manifest. Moreover, I do not myself regard the ætiological series as complete; nor does it solve the riddle of why hysteria is not more common among the lower classes. (You will remember, by the way, what a surprisingly large incidence of hysteria was reported by Charcot among working-class men.) I may also remind you that a few years ago I myself pointed out a factor, hitherto little considered, to which I attribute the leading role in provoking hysteria after puberty. I then put forward the view that the outbreak of hysteria may almost invariably be traced to a psychical conflict arising through an incompatible idea setting in action a defence on the part of the ego and calling up a demand for repression. What the circumstances are in which a defensive endeavor of this kind has the pathological effect of actually thrusting the memory which is distressing to the ego into the unconsciousness and of creating a hysterical symptom in its place I was not able to say at that time. But to-day I can repair the omission. The defence achieves its purpose of thrusting the incompatible idea out of consciousness if there are infantile sexual scenes present in the (hitherto normal) subject in the form of unconscious memories, and if the idea that is to be repressed can be brought into logical or associative connection with an infantile experience of that kind.

Since the ego's efforts at defence depend upon the subject's total moral and intellectual development, the fact that hysteria is so much rarer in the lower classes than its specific ætiology would warrant is no longer entirely incomprehensible.

Let us return once again, Gentlemen, to the last group of objections, the answering of which has led us such a long way. We have heard and have acknowledged that there are numerous people who have a very clear recollection of infantile sexual experiences and who nevertheless do not suffer from hysteria. This objection has no weight, but if provides an occasion for making a valuable comment. According to our understanding of the neurosis, people of this kind ought not to be hysterical at all, or at any rate, not hysterical as a result of the scenes which they consciously remember. With our patients, those memories are never conscious, but we cure them of their hysteria by transforming their unconscious memories of their infantile scenes into conscious ones. There was nothing which we could have done or needed to do about the fact that they have had such experiences. From this you will perceive that the matter is not one of the existence of the sexual experiences as unconscious memories; only so long as, and in so far as, they are unconscious are they able to create and maintain hysterical symptoms. But what decides whether those experiences produce conscious or unconscious memories – whether that is conditioned by the content of the experiences, which we shall prudently avoid. Let me merely remind you that, as its first conclusion, analysis has arrived at the proposition that hysterical symptoms are derivatives of memories which are operating unconsciously.

c) Our view then is that infantile sexual experiences are the fundamental precondition for hysteria, are, as it were, the disposition for it and it is they which create the hysterical symptoms, but that they do not do so immediately, but remain without effect to begin with and only exercise a pathogenic action later, when they have been aroused after puberty in the form of unconscious memories. If we maintain this view, we shall have come to terms with the numerous observations which show that a hysterical illness may already make its appearance in childhood and before puberty. This difficulty, however, is cleared up as soon as we examine more closely the data gathered from analyses concerning the chronology of the infantile experiences. We then learn that in our severe cases, the formation of hysterical symptoms begins – not in exceptional instances, but, rather, as a regular thing – at the age of eight, and that the sexual experiences which show no immediate effect invariably date further back, into the third or fourth, or even the second year of life. Since in no single instance does the chain of effective experiences break off at the age of eight, I must assume that this time of life, the period in which the second dentition takes place, forms a boundary line for hysteria, after which the illness cannot be caused. From then on, a person who has not had sexual experiences earlier can no longer becomes disposed to hysteria; and a person who has had sexual experiences earlier, is already on the other side of the boundary line (that is, before the age of eight) may be interpreted as a phenomenon of precocious maturity. The existence of this boundary line is very probably connected with developmental processes in the sexual system. Precocity of somatic sexual development may often be observed, and it is even possible that it can be promoted by too early sexual stimulation.

In this way we obtain an indication that a certain infantile state of the physical functions, as well as of the sexual system, is required in order that a sexual experience occurring during this period shall later on, in the form of a memory, produce a pathogenic effect. I do not venture as yet, however, to make any more precise statement on the nature of the psychical infantilism or on its chronological limits.

d)   Another objection might arisen from exception being taken to the supposition that the memory of infantile sexual experiences produces such an enormous pathogenic effect, while the actual experience itself has left none. And it is true that we are not accustomed to the notion of powers emanating from a mnemic which we absent from the real impression. You will moreover notice the consistency with which the proposition that the symptoms can only proceed from memories is carried through in hysteria. None of the later scenes, in which the symptoms arise, are the effective ones; and the experiences which are effective have at first no result. But here we are faced with a problem which we may very justifiably have kept separate from our theme. It is true that we feel impelled to make a synthesis, when we survey the number of striking conditions that we have come to know; the fact that in order to form a hysterical symptom a defensive effort against a distressing idea must be present, that this idea must exhibit a logical or associative connection with an unconscious memory through a few or many intermediate links, which themselves, too, remain unconscious at the moment, that the unconscious memory must have a sexual content, that its content must be an experience which occurred during a certain infantile period of life. It is true that we cannot help asking ourselves how it comes about that this memory of an experience that was innocuous at the time it happened, should posthumously produce the abnormal effect of leading a physical process like defence to a pathological result, while it itself remains unconscious.

But we shall have to tell ourselves that this is a purely psychological problem, whose solution may perhaps necessitate certain hypotheses about normal psychical processes and about the part played in them by consciousness, but that this problem may be allowed to remain unsolved for the time being, without detracting from the value of the insight we have so far gained into the ætiology of hysterical phenomena.

III

Gentlemen, the problem, the approaches to which I have just formulated, concerns the mechanism of the formation of hysterical symptoms. We find ourselves obliged, however, to describe the causation of those symptoms without taking that mechanism into account, and this involves an inevitable loss of completeness and clarity in our discussion. Let us go back to the part played by the infantile sexual scenes. I am afraid that I may have misled you into over-estimating their power to form symptoms. Let me, therefore, once more stress the fact that every case of hysteria exhibits symptoms which are determined, not by infantile but by later, often by recent, experiences. Other symptoms, it is true, go back to the very earliest experiences and belong, so to speak, to the most ancient nobility. Among these latter are above all to be found the numerous and diverse sensations and parasthesias of the genital organs and other parts of the body, these sensations and parasthesias being phenomena which simply correspond to the sensory content of the infantile scenes, reproduced in a hallucinatory fashion, often painfully intensified.

Another set of exceedingly common hysterical phenomena – painful need to urinate, the sensation accompanying defecation, intestinal disturbances, choking and vomiting, indigestion and disgust at food – were also shown in my analysis (and with surprising regularity) to be derivatives of the same childhood experiences and were explained without difficulty by certain invariable peculiarities of those experiences. For the idea that these infantile sexual scenes is very repellent to the feelings of a sexual normal individual; they include all the abuses known to debauched and impotent persons, among whom the buccal cavity and the rectum are misused for sexual purposes. For physicians, astonishment at this soon gives way to a complete understanding. People who have no hesitation in satisfying their sexual desires upon children cannot be expected to jib at finer shades in the methods of obtaining their satisfaction, and the sexual impotence which is inherent in children inevitably forces them into the same substitutive actions as those to which adults descend if they become impotent. All the singular conditions under which the ill-matched pair conduct their love-relations – on the one hand adult, who cannot escape his share of the mutual dependence necessarily entailed by sexual relationship, and who is yet armed with complete authority and the right to punish, and can exchange the one role for the other to the uninhibited satisfaction of his moods, and on the other hand the child, who in his helplessness is at the mercy of his arbitrary will, who is prematurely aroused to every kind of sensibility and exposed to every sort of disappointment, and whose performance of the sexual activities assigned to him is often interrupted by his imperfect control of his natural needs – all these grotesque and yet tragic incongruities reveal themselves as stamped upon the later development of the individual and his neuroses, in countless permanent effects which deserve to be traced in the greatest detail. Where the relation is between two children, the character of the sexual scenes is none the less of the same repulsive sort, since every such relationship between children postulates a previous seduction one of them by an adult. The physical consequences of these child-relations are quite extraordinarily far-reaching; the two individuals remain linked by an invisible bond throughout the whole of their lives.

Sometimes it is the accidental circumstances of these infantile sexual scenes which in later years acquire a determining power over the symptoms of the neurosis. Thus, in one of my cases the circumstances that the child was required to stimulate the genitals of a grown-up woman with his foot was enough to fixate his neurotic attention for years on his legs and their function, and finally to produce a hysterical paraplegia. In another case, a woman patient suffering from anxiety attacks which tended to come on at certain hours of the day could not be calmed unless a particular one of her many sisters stayed by her side all the time. Why this was so would have remained a riddle if analysis had not shown that the man who had committed the assaults on her used to enquire at every visit whether this sister, who he was afraid might interrupt them, was at home.

It may happen that the determining power of the infantile scenes is so much concealed that, in a superficial analysis, it is bound to be overlooked. In such instances we imagine that we have found the explanation of some particular symptom in the content of one of the later scenes – until, in the course of our work, we come upon the same content in one of the infantile scenes, so that in the end we are obliged to recognize that after all, the later scene only owes its power of determining symptoms to its agreement with the earlier one. I do not wish because of this to represent the later scenes as being unimportant; if it was my task to put before you the rules that govern the formation of hysterical symptoms, I should have to include as one of them that the idea which is selected for the production of a symptom is one which has been called up by a combination of several factors and which has been aroused from various directions simultaneously. I have elsewhere tried to express this in the formula: hysterical symptoms are over determined.

One thing more, Gentlemen. It is true that earlier I put the relation between recent and infantile ætiology aside as a separate theme. Nevertheless, I cannot leave the subject without overstepping this resolution at least with one remark. You will agree with me that there is one fact above all which leads us astray in the psychological understanding of hysterical phenomena, and which seems to warn us against measuring psychical acts in hysteria and in normal people with the same yardstick. The fact is the discrepancy between psychically exiting stimuli and psychical reactions which we come upon in hysterical patients. We try to account for it by assuming the presence in them of a general abnormal sensitivity to stimuli, and we often endeavor to explain on a physiological basis, as if in such patients certain organs of the brain which serve to transmit stimuli were in a peculiar chemical state (like the spinal centres of a frog, perhaps, which have been injected with strychnine) or as if these cerebral organs had withdrawn from the influence of higher inhibiting centres (as in animals being experimented on under vivisection). Occasionally one or other of these concepts pay be part of the phenomena; I do not dispute this. But the main part of the phenomenon – of the abnormal exaggerated, hysterical reaction to psychical stimuli – admits of another explanation, an explanation which is supported by countless examples from the analyses of patients; it is bound to appear exaggerated to us because we only know a small part of the motives from which it arises.

In reality, this reaction is proportionate to the exciting stimulus; thus it is normal and psychologically understandable. We see this at once when the analysis has added to the manifest motives, of which the patient is conscious, those other motives which have been operative without his knowing about them, so that he could not tell us of them.

I could spend hours demonstrating the validity of this important assertion for the whole range of psychical activity in hysteria, but I must confine myself here to a few examples. You will remember the mental "sensitiveness" which is so frequent among hysterical patients and leads them to react to the least sign of being depreciated as though they had received a deadly insult. What would you think, now, if you were to observe this high degree of readiness to feel hurt on the lightest occasion. If you came across it between two normal people, a husband and wife perhaps? You would certainly infer that the conjugal scene you had witnessed was not solely the result of the latest trifling occasion, but that inflammable material had been piling up for a long time and that the whole heap of it had been set alight by the final provocation.

I would ask you to carry this line of thought over on to hysterical patients. It is not the latest slight – which, in itself, is minimal – that produces the fit of crying, the outburst of despair or the attempt at suicide, in disregard of the axiom that an effect must be proportionate to its cause; the small slight of the present moment has aroused and set working the memories of very many, more intense, earlier slights, behind all of which thee lies in addition the memory of a serious slight in childhood which has never been overcome. Or again, let us take the instance of a young girl who blames herself most frightfully for having allowed a boy to stroke her hand in secret, and who from that time on has been overtaken by a neuroses. You can, of course, answer that puzzle by pronouncing her an abnormal, eccentrically disposed and over-sensitive person; but you will think differently when analysis shows you that the touching of her hand reminded her of another, similar touching, which had happened very early in her childhood and which formed part of a less innocent whole, so that her self-reproaches were actually reproaches about that old occasion. Finally, the problem of the hysterogenic points is of the same kind. If you touch a particular spot, you do something you do not intend; you awaken a memory which may set off a convulsive attack, and since you know nothing of this psychical intermediate link, you refer the attack directly to the operation of your touch. The patients are in the same state of ignorance and therefore fall into similar errors. They constantly establish "false connections" between the most recent cause, which they are conscious of, and the effect, which depends on so many intermediate links. If, however, the physician has been able to bring together the conscious and unconscious motives for the purpose of explaining a hysterical reaction, he is almost always obliged to recognize that the seemingly exaggerated reaction is appropriate and is abnormal only in its form.

You may, however, rightly object to this justification of the hysterical reaction to psychical stimuli and say that nevertheless the reaction is not a normal one. For why do healthy people behave differently? Why do not all their excitations of long ago come into operation once more when a new, present day, excitation takes place? One has an impression, indeed, that with hysterical patients it is as if all their old experiences – to which they have already reacted so often and, moreover, so violently, have retained their effective power; as if such people were incapable of disposing of their psychical stimuli. Quite true.

Gentlemen, something of the sort must really be assumed. You must not forget that in hysterical people when there is a present-day precipitating cause, the old experiences come into operation in the form of unconscious memories. It looks as though the difficulty of disposing of a present impression, the impossibility of transforming it into a powerless memory, is attached precisely to the character of the psychical unconscious. You see that the remainder of the problem lies once more in the field of psychology – and what is more, a psychology of a kind for which philosophers have done little to prepare the way for us.

Top tis psychology, which has yet to be created to meet our needs – to this future psychology of neurosis – I must also refer you when, in conclusion, I tell you something which at first make you afraid that it may disturb our dawning comprehension of the ætiology of hysteria. For I must affirm that the ætiological role of infantile sexual experience is not confined to hysteria but holds good equally for the remarkable neurosis of obsessions, and perhaps also, indeed, for the various forms of chronic paranoia and other functional psychoses.

I express myself on this with less definiteness, because I have as yet analysed far fewer cases of obsession neurosis than of hysteria; and as regards paranoia, I have at my disposal only a single full analysis and a few fragmentary ones. But what I discovered in these cases seemed to be reliable and filled me with confident expectations for other cases. You will perhaps remember that already, at an earlier date, I recommended that hysteria and obsessions should be grouped together under the name of "neurosis of defence", even before I had come to know of their common infantile ætiology. I must now add that – although this need not be expected to happen in general – every one of my cases of obsessions revealed a substratum of hysterical symptoms, mostly sensations and pains, which went back precisely to the earliest childhood experiences. What, then, determines whether the infantile sexual scenes which have remained unconscious will later on, when the other pathogenic factors are super added, give rise to hysterical or to obsessional neurosis or even to paranoia? This increase in our knowledge seems, as you see, to prejudice the ætiological value of these scenes, since it removes the specificity of the ætiological relation.

I am not yet in a position, Gentlemen, to give a reliable answer to this question. The number of cases I have analysed is not large enough nor have the determining factors in them been sufficiently various. So far, I have observed that obsessions can be regularly shown by analysis to be disguised and transformed self-reproaches about acts of sexual aggression in childhood and are more often met with in men than in women, and that men develop obsessions more often than hysteria. From this I might conclude that the character of infantile scenes - whether they were experienced with pleasure or only passively - has a determining influence on the choice of the later neurosis; but I do not want to underestimate the significance of the age at which these childhood actions occur, and other factors as well. Only a discussion of further analyses can throw light on these points. But when it becomes clear which are the decisive factors in the choice between the possible forms of the neuro-psychoses of defence, the question of what the mechanism is in virtue of which that particular form takes shape will once again be a purely psychological problem.

I have now come to the end of what I have to say to-day. Perhaps as I am to meet with contradiction and disbelief, I should like to say one thing more in support of my position. Whatever you may think about the conclusions I have come to, I must ask you not to regard them as the fruit of idle speculation. They are based on a laborious individual examination of patients which has in most cases taken up a hundred or more hours of work. What is even more important to me than the value you put on my results is the attention you give to the procedure I have employed. The procedure is new and difficult to handle, but it is nevertheless irreplaceable for scientific and therapeutic purposes. You will realize, I am sure, that one cannot properly deny the findings which follow from the modification of Breuer's procedure so long as one puts it aside and uses only the customary method of questioning patients. To do so would be like trying to refute the findings of histological techniques by relying on macroscopic examination. The new method of research gives wide access to a new element in the psychical field of events, namely, to processes of thought which have remained unconscious - which, to use Breuer's expression are "inadmissible to consciousness". Thus it inspires us with the hope of a new and better understanding of all functional psychical disturbances. I cannot believe that psychiatry will long hold back from making use of this new pathway to knowledge.

The Meagre Menu of Mawson and Mertz:

a re-evaluation of their sufferings during their ill-fated expedition to the Antarctic in 1911-1914

(This material comprised Appendix III of my Master's thesis, 2004. An abbreviated version was published by the Medical Journal of Australia (2005):

Mawson and Mertz: a re-evaluation of their ill-fated mapping journey during the 1911-1914 Australian Antarctic Expedition.)

Introduction

In 1969 Cleland and Southcott (1969a) proposed that Mawson had suffered and Mertz had died from the effects of vitamin A toxicity as a result of eating huskie liver following the loss of most of their food supplies down a crevasse. This hypothesis was supported by Shearman (1978) but is here questioned. It is proposed that Mawson suffered from the effects of severe food deprivation and Mertz died, being unable to tolerate the change of diet, as he was by habit vegetarian. It is further suggested that Mertz' condition was aggravated by psychological stress brought on by being forced to eat the dogs he had cared for for 18 months.

Background

It is now ninety years since Douglas Mawson led the Australasian Expedition to the Antarctic with the intention of mapping as much as possible of the Australian quadrant of the Antarctic coastline. After setting up camp in January 1911, the party wintered at Cape Denison, being forced to wait until November of that year before the weather moderated sufficiently to allow sledging. The expedition members broke up into small groups, most of which were to map different areas, with one group remaining at Base Camp. Mawson and his companions, Ninnis and Mertz, took the dogs to enable them to travel rapidly over the first part of their course, which would be mapped by another party, so that they could concentrate on an area which would not otherwise be reached. They therefore travelled further from Base Camp than any of the other parties.

On 14th December, 1911, Ninnis was killed when he fell into a crevasse. With him was lost the sledge carrying most of the food supplies which, being the heaviest, was being pulled by the strongest dogs. Mawson and Mertz were faced with the daunting prospect of making their way back to Base Camp on very reduced rations. If they were to survive, they would have no option but to kill and eat the six remaining dogs.

The tragedy occurred as Mawson's party approached the end of their proposed outward journey. Mawson had allocated 700 lb. (just over 300 kg) of food for the dogs (dried seal meta and blubber) but it would seem that this was not adequate for the dogs while they were working hard, pulling the sledges against strong winds and over very rough ground. Before completing the proposed outward journey, Mawson had been obliged to divide the dogs into two teams – the strongest and fittest pulling the heaviest load, which included all the dog food and most of the human food, while the weaker dogs pulled the lighter sledge. "All the dogs which had perished were big and powerful ... We had fully anticipated that those at least would come back alive, at the expense of the six dogs in my sledge (Mawson 1915: 243-244). It is not clear whether Mawson had always anticipated that some of the dogs would not return or whether this was the conclusion he was reaching towards the end of the outward journey. As it transpired, the first of the dogs to die did so on the first day of the premature return journey, suggesting that some of the dogs at least would not have survived the outward journey, let alone the return journey, even if it had been completed as planned.

Whatever the original intention, it is clear that the dogs were suffering severe malnutrition and those that did not perish with Ninnis down the crevasse eventually succumbed to starvation. The dogs continued working until they dropped, because that was their nature. They were then carried on the sledge in a comatose condition until they were shot and used for food for both man and dog. Mawson described the dog meat as "tough, stringy and without a vestige of fat" (Mawson 1915: 247).

Early on 8th January, 1912, Mertz died. One month later, on 8th February, 1912, Mawson reached Base Camp having survived one of the most incredible journeys under conditions of extreme hardship ever recorded.

In 1969 an article by Cleland and Southcott (1969a) was published suggesting that both Mawson and Mertz had suffered hyervitaminosis A as a result of eating the livers of the dogs. Cleland and Southcott's hypothesis is here re-investigated and the conclusion reached that both explorers suffered from starvation, not vitamin A toxicosis.

Discovery of vitamins

As the world moved into the age of science and human nutrition became subject to investigation, it was considered that proteins, carbohydrates, fats, minerals and water were all that were necessary to maintain health. Persistent sickness by sailors who were provided with these five 'essentials' made it clear that some other substances were necessary. Since these vital substances were believed to be amines, in 1912 the term 'vitamines' (i.e. vital amines) was used to describe this new class of nutrients (Rosenberg 1945: 10). Subsequent research showed that these nutrients were not, in fact, amines and the name was shortened to that by which we now know them, 'vitamins'. The following three years saw the discovery of the fat soluble vitamin 'A' in cod-liver oil and in butter, its structure being determined in 1931, the year I which it was first synthesized (Rosenberg 1945: 37).

Other vitamins were discovered and named, but it was not until after the Second World War that vitamin preparations became widely available and were consumed in large quantities by the general public, both on medical advice and of their own volition. Until this time, scientific work had concentrated on the effects of vitamin deficiency, but with the consumption of large quantities of vitamin A, taken for its curative effect in skin disorders, reports of toxic effects began to surface, especially during the 1950s, 1960s and 1970s. Attention began to be focused on the possible effects of over doses of these readily available vitamins. Vitamins B and C, being water soluble and readily excreted in the urine, proved to have very low toxicity but vitamin A, being fat soluble and stored in the body to some extent, did exhibit toxicity at very high dosages taken over long periods of time.

Acute Vitamin A toxicity

Anecdotal evidence of people in Arctic regions becoming sick after eating the liver of Arctic fish/mammals has been reported for several centuries (Kane 1856; Rodahl and Moore 1943). Kane reported mixed results from the eating of bear liver. He 'ate it freely and succeeded in making it a favourite dish with the mess'. However, he once became sick after eating the liver of a bear cub and on two other occasions all of his party became sick after eating the liver of bears which were old and fat (Kane 1845: 392-393). Because the liver of some bears and seals had been found to be a rich source of vitamin A, it was assumed that the cause was acute vitamin A poisoning. Symptoms reported included abdominal pains, nausea, vomiting, diarrhea, headache (at times severe), dizziness, irritability, sluggishness and a desire to sleep, starting within a few hours of the meal, followed by a rapid recovery.

Cleland and Southcott (1969b) reported a case of two fishermen who in 1935 became ill after eating the dried liver of a seal taken off Kangaroo Island, South Australia, which they hypothesized was caused by vitamin a poisoning. However, this liver had been stored for about 18 hours in a kerosene tin used for swabbing decks so a question mark must be placed over which symptoms may have been those of vitamin A poisoning and which may have been related to kerosene, since the symptoms (headache, vomiting, abdominal pain diarrhea) are generic food poisoning symptoms and only subsequent peeling of the skin and some numbness may be considered unusual.

It is unfortunate that most of the cases of assumed vitamin A poisoning have occurred in places far removed from medical attention so that no analysis of the patient or remains of the food consumed was possible. An exception is a case cited by Lonie (1950) in which seven members of a family of Chinese in New Zealand experienced food poisoning after eating fresh shark liver. No pathogenic organisms or their toxins were found in the remains of the cooked liver, which had broken up during cooking, having been mixed with soy beans to form a pasty mix. A second shark, caught at the same time, was still on the beach and its liver was found to contain 100,000 IU of vitamin A per gram. Consequently, vitamin A toxicity was deemed to be the cause of the sickness. No tests to determine the possible presence of other toxic substances were performed.

An unusual case of acute intoxication was that of a 28 year old female who was sensitive to exposure to the sun. As part of an experiment, she took two tablets totaling 50,000 IU vitamin A daily for three days prior to test sun-exposure and a similar dose every 30 minutes when symptoms of sunburn began, continuing for 27 hours despite worsening of symptoms, by which time she had ingested 1,500,000 IU vitamin A. Her symptoms were intense and included headache blurred vision and inability to stand or sit because of diarrhea and vertigo On the fourth day she started to exfoliate, with skin peeling off in sheets from limbs and trunks. There was slight hair loss from the head over the next two weeks (Furman 1973).

Most of the evidence for acute vitamin a poisoning is anecdotal, the assumed acute symptoms of one case being used to support the diagnosis of the next. Whatever future scientific work may show to be the exact symptoms of acute hypervitaminosis A from ingestion of food, all presumed accounts to date showed rapidity of onset and rapidity of recovery.

Southcott and Chesterfield (1971) tested the vitamin A levels of ten Antarctic huskies and found considerable variation, the range being from 3,700 IU/gm to 24,00 IU/gm. However, those levels were obtained from healthy dogs and not from those dying of starvation. It is extremely unlikely that the dog and the livers eaten by Mawson and Mertz contained anything like these amounts of vitamin A, since the body utilizes its reserves of all nutrients, fats, proteins, carbohydrates, minerals, vitamins and water, before it expires.

Chronic vitamin A toxicity

The symptoms of chronic hypervitaminosis A are well recorded in the literature. Most have resulted from the ingestion of large amounts of the vitamin in tablet form over extended periods of time (usually several years). Symptoms include coarseness and sparseness of hair of the scalp, eyebrows and other parts of the body, hepatosplenomegally, hyperostosis, cessation of menstruation, blurred vision, bone tenderness/joint pain, especially of the distal extremities, which may be accompanied by weakness, making walking difficult (in humans) and sitting/walking difficult (in animals), dizziness, myalgia, hæmorrhagic tendency, œdema both of the brain and the legs, dryness of the skin, ulceration, desquamation, increased intercranial pressure causing bulging of the fontanelles in infants and severe headache in adults, irritability and depression, loss of appetite/anorexia, diarrhea (Bergen and Roels 1965; Di Benedetto 1967; Feldman and Schlezinger 1970; Frame et al. 1974; Hillman 1956; Jowsey and Riggs 1968; Katz and Tzagournis 1972; Morrice et al. 1960; Muenter et al.1971; Rodahl and Moore 1943; Russell et al. 1974; Shaw and Niccoli1953; Smith and Goodman 1976; Stimson 1961; Sulzberger and Lazar 1951).

Osteological changes have been noted by a number of authors in both humans and animals at the cellular level, extending to include retardation of growth and early closure of the epiphyses (Belanger and Clark 1967; Frame et al. 1974; Gerber et al. 1954; Jowsey and Riggs 1968; Pease 1962; Ruby and Mital 1974).

Only one instance of chronic hypervitaminosis A appears to have been caused by the natural ingestion of food. In this case a mother had fed her twin female infants chicken liver twice daily for four months, starting at the age of 5 months, because she believed it to be nutritious and easy to cook, homogenize and freeze. Symptoms were dryness of the skin, bulging of the anterior fontanelles, irritability and vomiting (Mahoney et al. 1980).

Mawson's account

In 1915, Mawson published a two volume account of the Expedition, followed in 1930 by an abridged one volume version. These books are interesting, not simply because of the account given of the Expedition itself, but because of the large amount of detail given in relation to weather, physical conditions, work undertaken and any ill health or accidents suffered by any member of the3 expedition. An extraordinarily small amount of personal information was given regarding the character of the individual members of the Expedition, the friendships they formed or any conflicts which may have arisen among them. One thing we do know is that Mertz and Mawson were friends. It was they who travelled from England on the Aurora in charge of the dogs, which remained in their care for the rest of their all to brief lives. It is interesting to note that the first illness and death to be recorded was that of some of the dogs during their outward voyage. They were being fed the finest 'dog cakes' but "... undoubtedly felt the need of fresh meat and fish ... the rough weather ... so lowering their vitality that a number died from seizures" (Mawson 1915: 20-21). As pointed out above, the role of fresh food and vitamins in the diet of both humans and animals was yet to be fully understood.

Shearman (1978: 285) suggested that "Mertz may have found the liver less repulsive and they may have struck a bargain on the issue". However, Mawson makes no mention of such an arrangement in his autobiography. In the absence of any evidence to the contrary, therefore, it may be assumed that Mawson and Mertz shared all rations equally until that occasion when Mertz, finding the dog meat difficult to stomach, requested an extra portion of Glaxo (dried milk) while Mawson took an extra ration of dogs' meat in exchange (Mawson 1915: 257).

If this was the case, then in the three weeks before he died, Mertz could have eaten no more than three huskie livers – one a week or one-seventh per day. It is difficult to form an estimate of the likely vitamin A level in the livers of these emaciated dogs, but it will have been very low, simply because they starved to death and their reserves would have been utilized before they collapsed. It is also questionable how much vitamin A, assuming it was available, would actually have been absorbed, since the fat necessary for its metabolism was absent from their diet.

It an overdose of vitamin A was present in the livers such as to cause toxicity, it would have been expected that some acute symptoms would have been noticed – headache, nausea, dizziness – but none were mentioned. It must be assumed that they did not occur.

The symptoms of chronic toxicity are well recorded and many stem from elevated intracranial pressure – the bulging fontanelles in infants and severe headache in adults, the dizziness and vertigo, the pain and tenderness of the distal extremities as well as the weakness of the (back) legs. Irritability and extreme fatigue are frequently present. Again, these prominent symptoms are not recorded by Mawson. What is recorded is desquamation, which is definitely a symptom of vitamin A overdose, but not necessarily a symptom exclusive to it.

The only change of clothing undertaken by Mawson, Mertz and Ninnis was that of their socks, since the wearing of wet socks overnight would have led to frostbite. Their other clothes they ever removed. There was a buildup of dead skin within clothing "... As we never took off our clothes, the peeling of hair and skin from our bodies worked down into our under-trousers and socks and regular clearances were made" (Mawson 1915: 258). On 6th January Mawson recorded that the skin was peeling off their bodies and that Mertz lifted from his ear a perfect skin cast. On 11th January, he suffered from separation of a thickened layer of skin from the soles of his feet. Casts of this nature have been reported in vitamin a toxicity (Lonie 1950). However, their loss of weight would have made their clothes ill-fitting and continually damp clothes and sleeping bags may well have caused the skin to become sodden and to separate, the cast helped to keep its perfect shape by being frozen.

In their differential diagnosis between hypervitaminosis A and scurvy, Cleland and Southcott (1969a) point out that the symptoms of scurvy include swelling of the gums, ecchymosis and petechiæ due to fragility of the capillaries. None of these were reported by Mawson as would have been expected in view of the many falls experienced. Mawson, in particular, had several falls down crevasses, and records being saved by the rope attached to his sledge, the latter having jammed cross the top of the crevasse. After struggling to haul himself out, Mawson made himself a rope ladder. It is doubtful that he would have been able to do all the things he did without hæmorrhaging had he been suffering from scurvy. In respect to bruising resulting from fragility of the capillaries, this is a symptom of both excess intake of vitamin A and a deficiency of vitamin C (scurvy).

Not only did Mawson haul himself out of crevasses on more than one occasion, he also cut one of the sledges in half using a pocket knife. He would not have been able to do these things if he had been suffering from the sore and painful extremities associated with chronic vitamin A toxicity or, for that matter, from the dizziness and vertigo.

Why did Mertz die?

Mertz ate very little meat, being a 'near vegetarian' (Bickel 1977; Shearman 1978). His reputation in omlette making was undoubtedly a result of his near vegetarian habit. He accepted the need to eat the pemmican biscuits, make from dried, powdered beef, as part of the sledging rations, but this is a far cry from being forced to eat the flesh of his beloved dogs. "One must reckon with the psyche factor when eating" (Mawson 1915: 187).

Was the 'psyche' (psychological) factor the reason why Mertz died while Mawson survived? Mawson suffered the loss of a companion and a member of the party for whom he was ultimately responsible, but Mertz lost a friend when Ninnis died. Indeed, he lost seven friends, one human and six animal, since, as pointed out above, the dogs had been in the care of Ninnis and Mertz since the time they left England. Not only did Mertz loose these fiends, but the remaining dogs were dying, one by one. In addition to witnessing their suffering, he then had to assist in their final killing, and to eat their flesh, with himself a near vegetarian! Is it any wonder that both his mind and his body rebelled against the situation? A sudden change of diet to one consisting mainly of meat, at a time when he would have been stricken with grief, must have taken a huge toll upon him at a time when he needed all his psychological strength to overcome the difficulties with which he and Mawson were faced. In addition to this psychological aspect of the forced change of diet, a sudden change from a mixed diet to a primarily meat based diet leads to asymptomatic ketosis and ketonuria (Draper 1977).

A week before he died, Mertz felt that he could no longer tolerate eating the dog meat and swapped his ration of meat for Mawson's ration of Glaxo dried milk powder. He thus had no fresh food for the last days of his life and may have suffered from a lack of the vitamin B group, which is water soluble and is only stored in limited amounts in the body. The amount present is enough to maintain proper life for only a few days, the body absorbing only as much of the vitamin B group as is needed for the time being, the excess being excreted in the urine (Rosenberg 1945: 143-135). Vitamin B has become known as the 'stress' vitamin because during the times of psychological stress the body calls upon the reserves of this vitamin group, especially vitamin B1, thiamine, to support the nervous system. Thiamine is readily available from a large variety of seeds, beans and nuts, as well as eggs, which Mertz would have taken as part of his normal vegetarian diet, but this source was not now available to him. Vitamin B is also available from meat, since it occurs in all living cells, but for the last few days of his life, Mertz avoided this source of the vitamin as well. Lack of vitamin B results in loss of appetite, unusual susceptibility to fatigue, muscular weakness and œdema of the ankles, all suffered by Mawson later. Severe avitaminosis B causes convulsions and paralysis of the lower extremities (Rosenberg 1945). Mertz suffered 'fits' during the last twenty-four hours of his life.

Hypovitaminosis B6 (pyridoxine) in rats causes dermatitis, primarily of the peripheral parts of the body, such as paws, mouth tail, ears and nose (Rosenberg 1945). If these symptoms are extrapolated to humans, they might have contributed to the exfoliation of Mertz ears and soles. In 1951 an epidemic of convulsions among infants in the United States was traced to excessive destruction of vitamin B6 during preparation of a dried milk infant formula (Best and Taylor 1961). In this connection, it will be recalled that the dogs which died during the trip from England to Australia, who were fed dried food, also fitted before they died. Cramping pains in the stomach have been observed in humans (Rosenberg 1945). Both Mawson and Mertz suffered stomach pains, which Mawson attributed to stomach acid in the absence of food, and this simple explanation is still the most likely, but shortage of vitamin B6 may have contributed.

Lack of vitamin B6 leads to dermatitis, especially of body parts exposed to sunlight or to friction and hypovitaminosis B3 (pantothenic acid) has caused dermatitis, loss of hair and sloughing of the skin in chicks (Best and Taylor 1961; Roseberg 1945).

Discussion

Notwithstanding the occurrence of loss of hair and skin by Mawson and Mertz it is not felt that these symptoms alone are sufficient to support the theory that Mawson and Mertz suffered from hypervitaminosis A, as proposed by Cleland and Southcott (1969a) and supported by Shearman (1978). There are insufficient supporting symptoms. No mention is made by Mawson of possible acute poisoning after eating any of the livers (headache, nausea, vomiting) nor did he mention most of the chronic symptoms of vitamin a poisoning (severe headache, pain and tenderness of the limbs, hæmorrhaging, dizziness and vertigo and weakness of the limbs). Indeed, it is difficult to comprehend how Mawson could have completed his journey, involving as it did extreme physical exertion, had he actually been suffering vitamin a toxicity. The lack of hæmorrhaging is particularly significant, considering the falls into crevasses Mawson experienced. Further, his ability to pull himself out contra indicates weakness and tenderness of the extremities, which again would have been expected to have been present.

The symptoms suffered by Mawson and Mertz may readily be explained by generalized malnutrition a the result of semi-starvation. Mertz' death probably saved Mawson's life, since it made available to him a double ration of the remaining food. Although the nutritional value of the dog meat would have been low, since the animals themselves starved to death, such little as there was undoubtedly contribured to Mawson's survival.

That Mawson survived while Mertz died may also be attributed to Mertz' dietary preference. It is likely that Mertz suffered more psychologically than did Mawson because he lost a close friend, Ninnis, and also all of his beloved dogs. Not only was he forced to betray his principles by eating meat, the meat he was forced to eat was that of the dogs for which he had cared for 18 months. Psychological stress depletes the body's reserves of the B group of vitamins which may have been supplied by the very dog meat which Mertz felt himself unable to stomach! It is ironic that his obvious compassion for animals should have prevented him from receiving from his canine friends one last gift which may have saved his life.

Conclusion

It is the conclusion of this reassessment of the death of Mertz and sickness of Mawson that they did not suffer from vitamin A toxicosis. While it is proposed that these two people did suffer from the effects of a lack of vitamin B, it is not the conclusion that Mawson and Mertz suffered exclusively from hypovitamiosis B. The simple explanation of starvation is preferred. However, it is thought that the fits which Mertz threw shortly before his death may indicate that he was particularly deprived of the vitamin B group, possibly as a result of his near-vegetarian diet over many years. It is also suggested that the psychological and physiological effects of his being forced to change to a diet which consisted mostly of meat, during a time of great physical and emotional stress, would have made Mertz more vulnerable than Mawson and precipitated his demise.

References

Adams, M. (1980) Sergei Chetverikov, the Kol'stov Institute, and the Evolutionary Synthesis. In E. Mayr and W. Provine (eds.). *The Evolutionary Synthesis: Perspectives on the Unification of Biology.* Cambridge, Mass: Harvard University Press

Adcock, G.J., Denniss, E.S., Easteal, S., Huttley, G.A., Jermin, L.S., Peacock, W.J. and Thorne, A. (2001) Mitochondrial DNA sequences in ancient Australians: Implications for modern human origins. *Proceedings of the National Academy of Science* 98(2): 537-542

Aderaye, G. (1996) Hypertrophic pulmonary osteoarthropathy n pulmonary metastasis - a case report. *Ethiopian Medical Journal* 4: 251-254

Aiello, L. (1993) The fossil evidence for modern human origins in Africa: A revised view. *American Anthropologist:* 95(1): 73-96

Andersson, J.G. (1934) *Children of the Yellow Earth.* Cambridge, Mass: The MIT Press

Arsuaga, J-L., Martinez, I., Gracia, A., Carretero, J-M. and Carboneli, E. (1993) Three new human skulls from the Sima de los Huesos Middle Pleistocene site in Sierra de Atapuerca, Spain. *Nature* 362: 534-537

Atchley, W. (1981) M.J.D. White: The Scientist and the Man. In W. Atchley and D. Woodruff (eds.), *Evolution and Speciation:* Essays in honor of M.J.D White ,pp.3-22. Cambridge: Cambridge University Press

Atkins, R.C. (1999) *Dr. Atkins' New Diet Revolution.* New York. Aon

Bahn, P. (1996) Archæology: *Theories, Methods and Practice.* London: Thames and Hudson

Barber, L. (1980) *The Heyday of Natural History.* London: Jonathon Cape

Barham, J. (2004). The emergence of biological value. In W. Dembski and M. Ruse (eds.) *From Darwin to DNA: Debating Design.* Cambridge: Cambridge University Press

Barlow, N. (1967) *Darwin and Henslow: The Growth of an Idea.* London: John Murray

Barrett, P., Gautrey, P., Herbert, S., Kohn, D. and Smith, S. (eds.) (1987)

Charles Darwin's Notebooks, 1836-1844. New York: Cornell University Press

Barrow, J. and Tipler, F. (1986) *The Anthropic Cosmological Principle.* Oxford: Oxford University Press

Bateson, W. (1909a) *Mendel's Principles of Heredity.* Cambridge: Cambridge University Press

Bateson, W. (1909b) Heredity and Variation in Modern Lights. In A.C. Seward (ed.), *Darwin and Modern Science*, pp.85-101. Cambridge: Cambridge University Press

Bednarik, R. (1992) Palæoart and archæological myths. *Cambridge Archæological* Journal 2(1): 27-43

Bednarik, R. (2001) An Acheulian figurine from Morocco. *Rock Art Research*: 115-116

Beebe, W. (1944) *The Book of Naturalists: An Anthology of the Best Natural History.* London: Robert Hale Ltd.

Beer, de, G. (1958) *Evolution by Natural Selection: Charles Darwin and Russel Wallace.* Cambridge: Cambridge University Press

Beer, de, G. (1963/1976) *Charles Darwin: A Scientific Biography.* New York: Anchor Books

Behe, M. (1996) *Darwin's Black Box.* New York: Simon and Schuster

Behe, M. (2000) Self-organization and Irreducibly Complex Systems: A reply to Shanks and Joplin. *Philosophy of Science* 67(1): 155-162

Behe, M. (2001) Reply to my critics: A response to reviews of Darwin's Black Box. *Biology and Philosophy* 16: 685-709

Behe, M. (2004) Irreducible Complexity: Obstacle to Darwinian Evolution. In W. Dembski and M. Ruse (eds.) *Debating Design: From Darwin to DNA.* Cambridge: Cambridge University Press

Belanger, L. F. and Clark, I. (1967) Alpharadiographic and histological observations on the skeletal Effects of hypervitaminosis A and D in the rat. *The Anatomical Record* 158: 443-452

Benazzi, S., Douka, K., Fornai, C., Bauer, C., Mullmer, O., Svoboda, J., Pap, I., Mallegni, F., Bayle, P., Coquerelle, M., Condemi, S., Ronchitelli, A., Harvati, K. and Weber G. (2011) Early dispersal of modern humans in Europe and implications for Neanderthal behaviour. *Nature* 479: 525-528

Bergen, S. and Roels, A. (1965) Hypervitaminosis A: A Report of a case. *American Journal of Clinical Nutrition* 16: 265-269

Best, C. H. and Taylor, N. B. (1961) *The Physiological Basis of Medical Practice*. Baltimore: Williams and Wilkins

Bickel, I. (1977) *The Accursed Land*. Melbourne: MacMillan

Binford, L. (1978) *Nanamiut Ethnoarchæology*. New York: Academic Press

Binford, L. (1985) Human ancestors: changing views of their behaviour. *Journal of Anthropological Archæology* 4: 292-327

Binford, L. and Ho. C. (1985) Taphonomy at a Distance: Zhoukoudian – "The Cave Home of Beijing Man"? *Current Anthropology* 26(4): 413-442

Binford, L. and Stone, N. (1986) Zhoukoudian: A Closer Look. *Current Anthropology* 27(5): 453-475

Birdsell, J. (1975) *Human Evolution: An introduction to the New Physical Anthropology*. Chicago: Rand McNally College Publ.

Blyth, E. (1835) An attempt to Classify the 'Varieties' of Animals. (The Magazine of Natural History (London) 8: 40-53) In L. Eiseley (1979) *Darwin and the Mysterious Mr. X*. London: J.M. Dent and Ass.

Blyth, E. (1836) Observations on the Various Seasonal and Other Changes which Regularly Take Place in Birds. (The Magazine of Natural History (London), 9: 399) In L. Eiseley (1979) *Darwin and the Mysterious Mr. X*. London: J.M. and Ass.

Blyth, E. (1837) On the Psychological Distinctions Between Man and All Other Animals. (The Magazine of Natural History (London), 1: 1-9, 77-85, 131-141) In L. Eiseley (1979) *Darwin and the Mysterious Mr. X*. London: J.M. and Ass.

Boakes, R. (1984) *From Darwin to Behaviourism*. Cambridge: Cambridge University Press

Bogin, B. (1999) *Patterns of Human Growth*. Cambridge: Cambridge University Press

Boule, M. (1923) *Fossil Men: Elements of Human Palæontology*. Edinburgh: Oliver and Boyd (Translated by J. and J. Ritchie)

Boule, M. and Vallois, H.V. (1957) *Fossil Men: A Textbook of Human Palæontology* (4th edn.) London: Thomas and Hudson

Bowler, P.J. (1984) *Evolution: The History of an Idea*. Berkeley: University of

California Press

Brace, C.L. (1964) The fate of the "classic" Neanderthals: A consideration of hominid catastrophism. *Current Anthropology* 5(1): 3-37

Brackman, A. (1980) *A Delicate Arrangement.* New York: Times Books

Brooks, J. (1984) *Just before The Origin: Alfred Russel Wallace's Theory of Evolution.* New York: Columbia University Press

Brown, A. (1999) *The Darwin Wars.* London: Simon and Schuster

Brown, P., Sutikna, T., Morwood, M.J., Soejono, R.P., Jatmiko, E., Wayhy Saptome, E. and Rokis Awe Due. (2004) A new small-bodied hominin from the Late Pleistocene of Flores, Indonesia. *Nature* 431: 1055-1061

Bruemmer, F. (1993) *Arctic Memories.* Toronto: Key Porter Books

Brunet, M., Guy, F., Pilbeam D., Mackaye, H., Likius, A., Ahounta, D., Beauvilain, A., Blondel, C., Bocherensk, H., Boisserie, J-R., de Bonis, L., Coppens, Y., Deja, J., Denys, C., Duringer, P., Eisenmann, V., Gongdibe, F., Fronty, P., Geraads, D., Lehmann, T., Lihoreau, F., Louchart, A., Mahamat, Merceron, G., Mouchelin, G., Otero, O., Campomanes, P., Ponce de Leon, M., Rage, J-C., Sapanetkk, M., Schuster, M., Sudre, J., Tassy, P., Vaientin, X., Vignaud, P., Viriot, L., Zazzo. A., and Zollikofer,C (2002) A new hominid from the Upper Miocene of Chad, Central Africa. *Nature* 418: 145-151

Bryant, V and Williams-Deane, G. (1975) The coprolites of Man. *Scientific American* 232(1): 100-108

Buffon, G.L. (1781/1834) *Natural History, General and Particular.* London: Thomas Kelly

Bullock, A. (1985) *The Humanist Tradition in the West.* London: Thames and Hudson

Burr, V. (1995) *An Introduction to Social Constructionism.* London: Routledge

Burton B. (1976) *Human nutrition.* New York: McGraw Hill

Cann, R.L. (1987) In search of Eve. *The Sciences* 27: 30-37

Cann R.L. (1988) DNA and Human Origins. *Annual Review of Anthropology* 17: 127-143

Cann, R.L., Stoneking, M. and Wilson, A.C. (1987) Mitochondrial DNA and human evolution. *Nature* 325: 31-36

Cann, R., Rickards, O. and Koji Lum, J. (1994) Mitochondrial DNA and Human

Evolution: Our One Lucky Mother. In M. Nitecki and D. Nitecki, (eds.), *Origins of Anatomically Modern Humans*, pp.136-148 New York: Plenum Press

Carrington-Smith, D. (2004a) Behaviourally induced pathologies in Pleistocene Human Populations. *Unpublished Master's thesis.* School of Anthropology, Archæology and Sociology. Cairns: James Cook University

Carrington-Smith, D. (2004b) Mawson and Mertz: A re-evaluation of their ill-fated mapping journey during the 1911-1914 Australian Antarctic Expedition. *Medical Journal of Australia* 11/12: 638-641

Carrington-Smith, D. (2015) *Outshining Darwin: Lamarck's Brilliant Idea.* Mossman: Mossman Print

Cassim, B., Mody, G. and Rubin, D. (1990) The prevalence of diffuse idiopathic skeletal hyperostosis in African blacks. *British Journal of Rheumatology* 29: 131-132

Cavalli-Sforza, L., Menozzi, P. and Piazza, A. (1994) *The History and Geography of Human Genes.* Princeton: Princeton University Press

Cavallo, J. (1990) Cat in the human cradle. *Natural History* 2: 52-60

Csányi, V. (1982) *General Theory of Evolution.* Budapest: Akadémiai Kiadó

Chakravarty, K. and Bednarik, R. (1997) *Indian Rock Art.* Delhi: Motdal Barnarsidass Publishers

Chambers, R. (1844/1994) *Vestiges of the Natural History of Creation.* Chicago: University of Chicago Press

Chambers, R. (1845/1994) *Explanations: A Sequel to Vestiges of the Natural History of Creation.* Chicago: University of Chicago Press

Chyba, C.F. and Hand, K.P. (2005) Astrobiology: the study of the living Universe. *Annual Review of Astronomy and Astrophysics* 43: 31-74

Cleland, J. and Southcott, R. (1969a) Hypervitaminosis A in the Antarctic in the Australian Antarctic Expedition of 1911-1914: A possible explanation for the illnesses of Mawson and Mertz. *The Medical Journal of Australia* 26: 1337-1342

Cleland, J. and Southcott, R. (1969b) Illnesses following the eating of seal liver in Australian waters. *The Medical Journal of Australia* 1: 760-763

Coleman, W. (1964) *Georges Cuvier, Zoologist.* Cambridge, Mass: Harvard University Press

Coon, C.S. (1962) *The Origins of Races.* London: Jonathan Cape

Creswell, J. (2007) *Qualitative Enquiry and Research Design: Choosing among five Approaches* (2nd ed.) London: Sage Publications

Crookes, W. (1864) The reputed fossil man of the Neanderthal. *The Quarterly Journal of Science* 1: 88-97

Crubézy, E. (1990) Diffuse Idiopathic Skeletal Hyperostosis: Diagnosis and importance in Paleopathology. *Journal of Paleopathology* 3(2): 107-118

Crubézy, E. and Trinkaus, E. (1992) Shanidar 1: A case of hyperostotic disease (DISH) in the Middle Palæolithic. *American Journal of Physical Anthropology*: 89: 411-420

Cuvier, G. (1798/1817/1978) *Essay on the Theory of the Earth.* (3rd edn.) New York: Arno Press

Cuvier, G. (1863/1969) *The Animal Kingdom.* London: Henry G. Bonn

Dalton, R. (2005) More evidence for hobbit unearthed as diggers are refused access to cave. *Nature* 437: 934-935

Dalton, R. (2006) Decoding our Cousins. *Nature* 422: 238-240

Darlington, C. (1959) *Darwin's Place in History.* Oxford: Blackwell

Darwin, C. (1839/1979) *The Journal of a Voyage in the H.M.S. Beagle.* Facsimile edition of volume 3. Guildford: Genesis Publications in association with Australia and New Zealand Book Co.

Darwin, C. (1845) *Journal of Researches into the Natural History and Geology of the Various Countries Visited During the Voyage of H.M.S. Beagle Round the World.* London: John Murray

Darwin, C. (1859/1998) *The Origin of Species by Means of Natural Selection, or the Preservation of Favoured Races in the Struggle for Life.* (1st edn. with additions from 3rd edn.) Ware, Herts: Wordsworth Editions

Darwin, C. (1868/1893) *The Variation of Animals and Plants under Domestication.* London: John Murray.

Darwin, C. (1871/1906) *The Descent of Man.* London: John Murray

Darwin, C. (1887) *The Life and Letters of Charles Darwin.* Edited by Francis Darwin. London: John Murray

Darwin, C. (1903) *More Letters of Charles Darwin.* Edited by Francis Darwin. London: John Murray

Darwin, C. (1909/1969) *The Foundations of the Origin of Species: Two Essays written in 1842 and 1844 by Charles Darwin*. (F. Darwin ed.) Reprinted in 1969 by Kraus Reprint Company, New York (First published in 1909 by Cambridge University Press)

Darwin, E. (1791) *The Botanic Garden. Part I: The Economy of Vegetation; Part II: The Loves of the Plants*. London: J. Johnson

Darwin, E. (1794) *Zoonomia: The Laws of Organic Life*. London: J. Johnson

Darwin, F. (ed.) (1887) *The Life and Letters of Charles Darwin* (3 vols.) London: John Murray

Darwin, F. (ed.) (1902/1995) *The Life of Charles Darwin*. Twickenham, Middlesex: Senate

Darwin, F. (1929) *Autobiography of Charles Darwin*. London: Watts & Co.

Darwin, F. and Seward, A. (eds.) (1903) *More Letters of Charles Darwin* (2 vols.) London: John Murray

Darwin, G.P. (1839/1979) Introduction. In C. Darwin. *The Journal of a Voyage in H.M.S. Beagle*. Facsimile edition of volume 3 (1979) by Genesis Publications in association with Australia & New Zealand Book Co.

Davies, P. (2004) Emergent Complexity, *Teleology and the Arrow of Time*. In W. Dembski and M. Ruse Ruse (eds.) *From Darwin to DNA: Debating Design*. Cambridge: Cambridge University Press

Davies, R. (2008) *The Darwin Conspiracy: Origins of a Scientific Crime*. London: Golden Square Books

Davies, R. (2012) How Charles Darwin received Wallace's Ternate paper 15 days earlier than he claimed. *Biological Journal of the Linnean Society* 105: 472-477

Dawkins, R. (1976) *The Selfish Gene*. Oxford: Oxford University Press

Dawkins, R. (1982) *The Extended Phenotype*. Oxford: Oxford University Press

Dawkins, R. (1986) *The Blind Watchmaker*. London: Penguin Books

Dawkins, R. (1995) *River Out of Eden*. London: Weidenfeld and Nicolson

Dawkins, R. (1996) *Climbing Mount Improbable*. London: Viking

Dawkins, R. (1998) *Unweaving the Rainbow*. London: Penguin

Dawkins, R. (2003) *A Devil's Chaplain*. London: Weidenfeld and Nicolson

Dawkins, R. (2004) *The Ancestors' Tale: A Pilgrimage to the Dawn of Life*. London: Weidenfeld and Nicolson

Dawkins, R. (2006) *The God Delusion*. Boston: Houghton Mifflin Co.

Dawkins, R. (2009) *The Greatest Show on Earth*. New York: Free Press

Day, M. (1965) *Guide to Fossil Man*. London: Cassell

de Lumley, H. (1969a) A Palæolithic camp at Nice. *Scientific American* 200(5): 42-50

de Lumley, H. (ed.) (1969b) Une cabane achuléenne dans La Grotte du Lazaret (Nice). Paris: *Société Préhistorique Française*

de Lumley, H. and Boone, Y. (1976) Les structure d'habitat au Paléolithique inferior. In La Préhistoire Française (1): 625-643. H. de Lumley (ed.) Paris: *Centre National de la Recherche Scientifique*

Delporte, H. (1979) L'Image de La Femme dans l'art préhistorique. Paris: Pic*ard Press*

Dembski, W. and Ruse, M. (eds.) (2004) Debating Design: From Darwin to DNA. Cambridge: Cambridge University Press

Denko, C. Boja, B. and Moscowitz, R. (1994) Growth promoting peptides in osteoarthritis and diffuse idiopathic skeletal hyperostosis - insulin, insulin-like growth factor-1, growth hormone. *Journal of Rheumatology* 21(9): 1725-1730

Dennett, C. (1995) *Darwin's Dangerous Idéa*. New York: Simon & Schuster

Desmond, A. and Moore, J. (1991) *Darwin*. London: Penguin Books

Di Benedetto, J. (1967) Chronic hypervitaminosis A in an adult. *Journal of the American Medical Association* 201: 700-702

Dobzhansky, T. (1937/1951) *Genetics and the Origin of Species*. (3rd edn.) New York: Columbia University Press

Dobzhansky, T. (1959) *Genetics and the Origin of Species*. (3rd edn. revised) New York: Columbia University Press

Dobzhansky, T. (1962) *Mankind Evolving*. New Haven: Yale University Press

Dobzhansky, T. (1964) *Heredity and the Nature of Man*. New York: Signet

Dobzhansky, T. (1967) *The Biology of Ultimate Concern*. New York: The New American Library

Dobzhansky, T. (1970) *Genetics and the Evolutionary Process*. New York: Columbia University Press

Draper, H. H. (1977) The aboriginal Eskimo diet in modern perspective.

American Anthropologist 79: 309-316

Duhard, J. (1991) The shape of Pleistocene women. *Antiquity* 83: 552-561

Eiseley, L. (1957) *The Immense Journey.* New York: Vintage Books

Eiseley, L. (1961) *Darwin's Century: Evolution and the Man Who Discovered It.* New York: Anchor Books

Eiseley, L. (1979) *Darwin and the Mysterious Mr. X.* London: J.M. Dent and Ass.

Eldredge, N. (1979) Alternative approaches to evolutionary theory. *Bulletin of Carnegie Museum of Natural History* 13: 7-19

Eldredge, N. (1985a) *Time Frames.* New York: Simon and Schuster

Eldredge, N. (1985b) *Unfinished Synthesis.* Oxford: Oxford University Press

Eldredge, N. (1991a) The Miner's Canary: Unravelling the Mysteries of Extinction. Princeton: Princeton University Press

Eldredge, N. (1991b) *Fossils: The Evolution and Extinction of species.* Princeton: Princeton University Press

Eldredge, N. (1995) *Reinventing Darwin.* London: Weidenfeld and Nicolson

Eldredge, N. (1999) *The Pattern of Evolution.* New York: W.H.Freeman

Eldredge, N. and Gould, S.J. (1972) Punctuated Equilibrium: An Alternative to Phyletic Gradualism. In T.J.M. Schopf (ed.), *Models in Paleobiology,* pp.82-225 San Francisco: Freeman, Cooper

Eldredge, N. and Gould, S.J. (1993) Punctuated Equilibrium Comes of Age. *Nature* 366: 223-227

Elton, G. (1990) Humanism in England. In A. Goodman and A. Mackay (eds.) *The Impact of Humanism in Western Europe.* pp.259-278 London: Longman

Elshakry, M. (2009) Global Darwin: Eastern Enchantment. *Nature* 46(1): 1200-1201

Errico, d'., F., Zilhao, J., Jullen, M., Baffier, D. and Pelegrin, J. (1998) Neanderthal acculturation in Western Europe. *Current Anthropology* 39: S1-S44

Evans-Pritchard, E. (1940) *The Nuer: A description of the modes of livelihood and political institutions of a Nilotic people.* Oxford: Clarendon Press

Falk, D. (2011) *The Fossil Chronicles: How two consecutive discoveries changed our view of human evolution.* California: University of California Press

Farizy, D. (1994) Behavioural and Cultural Changes at the Middle to Upper Paleolithic Transition in Western Europe. In M. Nitecki and D. Nitecki (eds.) *Origins of Anatomically Modern Humans,* pp.93-100 New York: Plenum Press

Feldman, M. and Schlezinger, N. (1970) Benign intercranial hypertension associated with hypervitaminosis A. *Archives of Neurology* 22: 1-7

Femoral Fennell, K. and Trinkaus, E. (1907) Bilateral femoral and tibial periostosis in the La Errassie 1 Neanderthal. *Journal of Archæological Science* 24(1): 985-995

Fernández-Armesto, F. (2004) *So You Think You're Human?* Oxford: Oxford University Press

Finlayson, C. (2009) The Humans who went extinct: Why Neanderthals died out and we survived. Oxford: Oxford University Press

Finlayson, C. (2010) *Humans who went extinct: Why Neanderthals died out and we survived.* (2nd ed.) Oxford: Oxford University Press

Fisher, R. (1929/1958) *The Genetical Theory of Natural Selection.* New York: Dover Publications

Fisher, R. (1936) Has Mendel's work been rediscovered? *Annals of Science* 1(2): 115-137

Foder, J. and Piatelli-Palmarini, M. (2010) *What Darwin Got Wrong.* New York: Farrar, Straus and Giroux

Ford, E. (1931) *Mendelism and Evolution.* London: Methuen

Ford, E. (1949/1963) Early Stages in Allopatric Speciation. In G. Jepson, E. Mayr and G.Simpson (eds.). *Genetics, Paleontology and Evolution,* pp.309-314. New York: Atheneum

Forestier, J. and Lagier, R. (1971) Ankylosing hyperostosis of the spine. *Clinical Orthopædics and Related Research* 74: 65-83

Forrest, B. (2004) *Understanding the Intelligent Design Creationist Movement - Its True Nature and Goals.* A position Paper from The Cere from the Centre for Enquiry Office of Public Policy. Washington D.C.

Forrest, B. and Gross, R. (2004) *Creationism's Trojan Horse.* Oxford: Oxford University Press

Foucault, M. (1972) *The Archæology of Knowledge.* London: Tavistock Publication

Frame, R., Jackson, C., Reynolds, W. and Umphrey J. (1974) Hypercalæmia and skeletal effects in Chronic hypervitaminosis A. *Annals of Internal Medicine* 80: 44-48

Frayer, D., Wolpoff, M., Thorne, A., Smith, F. and Pope, G. (1993) Theories of Modern Human Origins: The Paleontological Test. *American Anthropologist* 95(1): 14-50

Furman, K. (1973) Acute hypervitaminosis A in an adult. *American Journal of Clinical Nutrition* 26: 575-577

Gabunia, L. and Vekus, A. (1995) A Plio-Pleistocene hominid from Dmanisi, East Georgia, Caucasus. *Nature* 373: 509-512

Galton, F. (1892/1962) *Hereditary Genius.* London: Meridian Books

Ganong, W. (1999) *Medical Physiology.* Stamford: Appleton & Lange

Garrow, J., Jams, W. and Ralph, A. (1998) *Huan Nutrition and Dietetics.* Edinburgh: Churchill Livingtone

Gassman, D. (1971) *The scientific origins of National Socialism: social Darwinism in Ernst Haeckel and the German Monist League.* London: Macdonald & Co.

Gerber, A., Raab, A. and Sobel A. (1954) Vitamin A poisoning in adults. *American Journal of Medicine* 16: 729-745

Gimbutas M. (1979) *The Gods and Goddesses of Old Europe.* London: Thames and Hudson

Ginger, R. (1958) *Six Days or Forever: Tennessee v. J. T. Scopes.* Boston: Beacon Press

Glass, B. (1955) Maupertuis, a forgotten genius. *Scientific American* 193: 100-110

Glass, B., Temkin, O., Straus, W. (eds.) (1959) *Forerunners of Darwin: 1745-1859.* Baltimore: John Hopkins Press

Goldschmidt, R. (1938) *Physiological Genetics.* New York: McGraw-Hill Book Co.

Goldschmidt, R. (1940) *The Material Basis of Evolution.* New Haven: Yale University Press

Goode, E. (1988) *Diet advice from our ancestors.* Mailer-Daemon@email-delivery.galegroup.com

Gould, S.J. (1977) *Ontogeny and Phylogeny.* Cambridge, Mass: Harvard University Press

Gould, S.J. (1991) *Bully for Brontosaurus.* London: Penguin Books

Gould, S.J. (2002) *The Structure of Evolutionary Theory*. Cambridge, Mass: Harvard University Press

Grabiner, J. and Miller, P. (1974) Effects of the Scopes Trial. *Science* 185: 832-837

Gray, A. (1876/1963) *Darwiniana: Essays and Reviews Pertaining to Darwinism*. Cambridge, Mass: Belknap Press

Gray, J.L. (1894) *Letters of Asa Gray*. Boston: Houghton, Miffly and Co.

Green, R., Krause, J., Ptak, S., Briggs, A., Ronan, M., Simons, J., Du, L., Egholm, M., Rothberg, J., Paunovic, M. and Pääbo, S. Analysis of one million base pairs of Neanderthal DNA. (2006) *Nature* 444: 330-336

Green, R., Krause, J., Briggs, A., Maricic, T., Stenzel, U., Kircher, M., Patterson N., Lii, H., Zhai, W., Fritz. M., Hansen, F., Durand, E., Malaspinas, A.-S., Jensen, J., Marques-Bonet, T., Alkan, C., Prüfer, K.Meyer, M., Barbano, J., Good, J., Schultz, R., Aximu-Petri, A., Butthof, A., Höffner, B., Siegemund, M Erihmann, A., Nushaum, C., Lander, E., Russ, C., Novod, N., Affourtit, J., Egholm M., Verna, C., Rudan, P., Brajkovic, D., Kukan, A., Gusic, I., Doronichev, V., Godovanova, L., Lalucza-Fox, C., de la Rasilla, M., Forica, J., Rosas, A., Schmmitz, R., Johnson, P., Eichler, E., Falush, D., Birney, E. Mullikin, J., Slatkin, M., Nielsen, R., Kelso, J., Lachmann, M., Reich D. and Pääbo, S. (2010) A draft sequence of the Neanderthal Genome. *Science* 328: 710-722

Gruber, H.. (1974) A Psychological Study of Scientific Creativity. In H. Gruber and P. Barrett. *Darwin on Man*. New York: E.P. Dutton

Grün, R., Spooner, N., Thorne, A., Mortimer, G., Simpson, J., McCulloch, M., Taylor, L. and Curnoe, D. (2000) Age of the Lake Mungo 3 skeleton. *Journal of Human Evolution* 38: 733-741

Gutmann, J. (1936) *Schelling: Of Human Freedom. A translation of F.W.J.Schelling's Philosophische Untersuchungen first published in 1809*. Chicago: The Open Court Publishing Company

Haldane, J. (1932) *The Causes of Evolution*. New York: Cornell University Press

Haldane, J. (1938) *Heredity and Politics*. London: George Allen & Unwin

Haldane, J. (1951) *Everything Has a History*. London: George Allen & Unwin

Haldane, J. (1954) *The Biochemistry of Genetics*. London: George Allen & Unwin

Hamilton, W. (1964) The genetical evolution of social behaviour. *Journal of Theoretical Biology* 7: 1-51

Hammer, H. (1995) A recent common ancestry for human Y chromosome. *Nature* 378: 376-378

Harrold, F. (1983) The Châtelperronian and the Middle-Upper Paleolithic Transition. In E. Trinkaus (ed.), *The Mousterian Legacy*, pp.123-140 Oxford: BAR International Series 164

Hawks, J. and Wolpoff, M. (2001) Paleoanthropology and the population genetics of ancient genes. *American Journal of Physical Anthropology* 114: 269-272

Hawks, J. and Wolpoff, M. (2001) The accretion model of Neanderthal evolution. *Evolution* 55: 1474-1485

Hedges, S., Kumar, S., Tamura, K. and Stoneking, M. (1992) Human origins and analysis of mitochondrial DNA sequences. *Science* 255: 737-739

Higham, T., Compton, T., Stringer, C., Roger, J., Shapiro, B., Trainkaus, E., Chandler, B., Grööning, F., Collins, C., Hillson, S., O'Higgins, P., Fitzgerald, C. and Fagan, M. (2011) The earliest evidence for anatomically modern humans in northwestern Europe. *Nature* 479: 521-524

Hillman, R. (1956) Hypervitaminosis A: Experimental induction in the human subject. *American Society for Clinical Nutrition* 4: 603-608

Hollick, M. (2006) *The Science of Oneness*. Winchester, U.K: O Books

Honeywell, R. (2008) *Lamarck's Evolution: Two Centuries of Genius and Jealousy*. Miller's Point, N.S.W: Murdoch Books

Howell, F.C. (1952) Pleistocene glacial ecology and the evolution of 'Classic Neanderthal' man. *Southwestern Journal of Anthropology*. 6(4): 377-400

Howells, W. (1959) *Mankind in the Making: The Story of Human Evolution*. New York: Doubleday

Hublin, J. (1985) Human fossils from the North African Middle Pleistocene and the origins of Homo sapiens. In E. Delson (ed.), *Ancestors: The Hard Evidence*, pp.283-288 New York: Alan R. Liss

Hudson, W.H. (1892-1893) Patagonian Memories. In W. Beebe. (1944: 178-180) *The Book of Naturalists: An Anthology of the Best Natural History*. London: Robert Hale Limited.

Hull, D. (1973) *Darwin and His Critics*. Cambridge, Mass: Harvard University

Press

Hunter, J., Roberts, C. and Martin, A. (1996) *Studies in Crime: An Introduction to Forensic Archæology.* London: Routledge

Hutton, J. (1788/1794/1970) Systems of the Earth (1788), Observations on Granite (1794). In G.W. White (ed.) *Contributions to the History of Geology.* Vol.5. Facsimile reproduction. Darien, Conn: Hasner

Huxley, J. (1942) *Evolution: The Modern Synthesis.* London: Allen and Unwin

Huxley, L. (1900) *Life and Letters of Thomas Huxley* (2 vols.) London: Farnborough Gregg International

Huxley, L. (1918) *Life and Letters of Sir Joseph Dalton Hooker* (2 vols.) London: John Murray

Huxley, T.H. (1863/1959) *Man's Place in Nature.* Michigan: University of Michigan Press

Huxley, T.H. (1893-1917/1968) *Collected Essays* (10 vols.) New York: Greenwood Press

Irvine, W. (1955) *Apes, Angels and Victorians.* London: Weidenfeld and Nicolson

Jablonski, N. and Chaplin G. (2000) The evolution of human skin colouration. *Journal of Human Evolution* 39: 57-106

Jepson, G., Mayr, E. and Simpson, G.G. (eds.) (1949/1963) *Genetics, Paleontology and Evolution.* York: Atheneum

Jia Lanpo and Huang Weiwen. (1990) *The Story of Peking Man: From Archæology to Mystery.* Bejing: Foreign Language Press

Johanson, D. and Edgar, B. (1996) *From Lucy to Language.* London: Weidenfeld and Nicolson

Jones, S. (1993) *The Language of Genes.* London: Harper Collins

Jones, S. (1999) *Almost Like a Whale.* London: Doubleday

Jones, S. (2002) *Y: The Descent of Men.* London: Abacus

Joravsky, D. (1970) *The Lysenko Affair.* Cambridge, Mass: Harvard University Press

Jordan, P. (1999) *Neanderthals: Neanderthal Man and the Story of Human Origins.* Stroud: Sutton Publishing

Jordanova, L.J. (1984) *Lamarck.* Oxford: Oxford University Press

Jørgensen, M. and Phillips, L. (2002) *Discourse Analysis as Theory and Method.* London: Sage Publications

Jowsey, J. and Riggs, B. (1968) Bone changes in a patient with hypervitaminosis A. *Journal of Clinical Endocrinology and Metabolism* 28: 1833-1835

Jurmain, R., Nelson, H., Kilgore, L and Trevathqn, W. (1997) *Introduction to Physical Anthrolpology.* Belmot, CA: Wadsworth Publishing

Kammerer, P. (1924) *The Inheritance of Acquired Characteristics.* New York: Boni and Liveright

Kane, e.K. (1856) *Arctic Explorations: The Second Grimmell Expedition in Search of Sir John Franklin, 1853, '54, '55.* Vol.1). Philadelphia: Childs and Peterson

Katz C and Tzagournis M. (1972) Chronic adult hypervitaminosis A with hypercalcæmia. *Metabolism* 21: 1171-1176

Keene, A. (1985) Nutrition and economy: models for the study of prehistoric diet. In R. I. Gilbert and H. Mielke (eds.), *The Analysis of Prehistoric Diets* pp.55-190 New York: Academic Press

Keith, A. (1915) *The Antiquity of Man.* London: Williams and Northgate

Keith, A. (1948) *A New Theory of Human Evolution.* London: Watts and Co.

Keith, A. (1955) *Darwin Revisited.* London: Watts and Co.

Kemp, M. (2005) A hobbit-forming image. *Nature* 437: 555

Kennet, J.P. (1995) A review of Polar Climate: Evolution during the Neogene, based on The Marine Sediment Record. In E. Vbra, G. Denton, T. Partridge and L. Burckle (eds.), *Paleoclimate and Evolution, with Emphasis on Human Evolution,* pp.46-64 New Haven: Yale University Press

King, W. (1864) The reputed fossil man of the Neanderthal. *The Quarterly Journal of Science* 1: 88-97

King-Hele, D. (1963) *Erasmus Darwin.* New York: Charles Scribner's Sons

King-Hele, D. (1968) *The Essential Writings of Erasmus Darwin.* London: MacGibbon and Kee.

King-Hele, D. (1999) *Erasmus Darwin: A Life of Unequalled Achievement.* London: Giles de la Mare Publishers

Kirschymann, G. (1998) *Nutrition Almanac* (4th ed.) Singapore: McGraw-Hil

Klein, J. and Takahata, N. (2002) *Where Do We Come From? The Molecular Evidence for Human Descent.* New York: Springer

Klein, R. (1994) The Problems of Modern Human Origin. In M. Nitecki and D. Nitecki (eds.), *Origins of Anatomically Modern Humans,* pp.3-17 New York: Plenum Press

Klein, R. (1998) Why anatomically modern people did not disperse from Africa 100,000 years ago. In T. Akazawa, K. Aoki. and O. Bar-Yosef (eds.), *Neanderthals and Modern Humans in Western Asia,* p.33 New York: Plenum Press

Kœnigswald, G.H. von (1956) *Meeting Pre-historic Man.* London: Thames and Hudson

Kœstler, A. (1971) *The Case of the Midwife Toad.* London: Pan Books

Kohn, D. (1981) On the origin of the Principle of Diversity. *Science* 213: 1105-1108

Krauss, L. and Dawkins, R. (2012) *A Universe from Nothing.* London: Free Press

Krings, M., Capelli, C., Tschentscher, F., Gelsert, H., Meyer, s., von Hæselelr, A., Grosschmidt. K., Possnert, G., Paunovic, M and Pääbo, S. (2000) A view of Neanderthal genetic diversity. *Nature Genetics* 26: 144-146

Krings, M., Stone, A., Schmitz, R.W., Kraintzki, H., Stoneking, M. and Pääbo, S. (1997) Neanderthal DNA sequences and the origin of modern humans. *Cell* 90: 19-30

Kroher, M. (2000) Morphological chimeras of larvæ and adults in hydrozoans. *The International Journal of Developmental Biology.* 44(8): 861

Kuhn, T. (1962/1970) *The Structure of Scientific Revolutions.* Chicago: Chicago University Press

Lack, D. (1949/1963) The significance of Ecological Isolation. In G. Jepson, E. Mayr and G.G. Simpson (eds.), *Genetics, Paleontology and Evolution,* pp.229-308. New York: Atheneum

LaHaye, T. (1980) *The Battle for the Mind.* Old Tappan, N.J.: Fleming H. Revell Company

Lahr, M. and Foley, R. (1994) Multiple dispersals and modern human origins. *Evolutionary Anthropology* 3: 48-60

Lakatos, I. (1970) Falsification and the Methodology of Scientific Research

Programmes. In I. Lakatos and A. Musgrave (eds.), *Criticism and the Growth of Knowledge*, pp.91-196 Cambridge: Cambridge University Press

Lam, V. (2010) The Truth about Chimeras. *The Science Creative Quarterly.* Issue 4; http.//www.scq. ca/the-truth-about-chimeras. Accessed 14/09/2010

Lamarck, J.B. (1809/1963) *Zoological Philosophy.* Translated by H. Elliot. New York: Hafner Publishing Co.

Lamont, C. (1965) *The Philosophy of Humanism.* New York: Ungar

Larsen, C. S. (1996) Biological changes in human populations with agriculture. *Annual Review of Anthropology* 24: 185-213

Leakey, R.E.R., Butzer, K.W. and Day, M.H. (1969) Early Homo sapiens remains from the Omo River region of southwest Ethiopia. *Nature* 222: 1132-1138

Leakey M. and Walker, A. (2003) Early hominid fossils from Africa. *Scientific American* 13(2): 14-19

Lepre, C., Roche, H., Kent, D., Harmand, S., Quinn, R., Brugai, J-P., Texier, J., Lenoble, A.and Craig, F. (2011) An earlier origin for the Acheulian. *Nature* 477: 82-85

Levins, R. and Lewontin, R. (1985) *The Dialectical Biologist.* Cambridge, Mass: Harvard University Press

Lewin, R. (1987) *Bones of Contention: Controversies in the Search for Human Origins.* New York: Simon and Schuster

Liddell, R. (1991) Cape York: *The Savage Frontier.* Dubbo: CS Printers Pty. Ltd.

Lieberman, D. (2005) Further fossil finds from Flores. *Nature* 437: 957-958

Linden, E. (1974) *Apes, Men and Language.* New York: E. P. Dutton & Co.

Lips, J. (1949) *The Origin of Things.* London, George G. Harrap and co. Ltd

Lonie, T. (1950) Excess Vitamin A as a cause of food poisoning. *New Zealand Medical Journal* 49: 680-685

Lookingbill,D., Lookingbill G and Leppard B. (1995) Active damage and skin cancer in albinos in northern Tanzania. *Journal of the American Academy of Dermatology* 32(4): 653-658

Lovejoy, A. (1936/1964) *The Great Chain of Being.* Cambridge, Mass: Harvard

University Press

Lovejoy, A. (1959) Buffon and the Problem of Species. In B. Glass, O. Temkin and W. Straus (eds.), *Forerunners of Darwin: 1745-1859*, pp.84-113 Baltimore: The John Hopkins Press

Lyell, C. (1830-1833/1997) *Principles of Geology*. (Vols.1-3) Abridged edition. (James A.) (2nd ed.) London: Penguin

Lyell, C. (1863) *The Geological Evidences of the Antiquity of Man*. London: John Murray

Macgilivary, J. (1852) *Narrativve of the Voyage of HMS Rattlesnake*. London: T. & W. Boone

McBrearty, S. and Brooks, A. (2000) The Revolution that Wasn't. *Journal of Human Evolution* 39: 453-563

McBride S. and Leppard, B. (2002) Attitudes and beliefs of an albino population towards sun avoidance. *Archives of Dermatology* 138: 620-632

McCalman, I. (2009) *Darwin's Armada*. London: Viking

McGrath, A. (2005) *Darwin's God: Genes, Memes and the Meaning of Life*. Malden, Mass: Blackwell Publishing

McKinney, H. L. (1996) Alfred Russel Wallace and the Discovery of Natural Selection. *Journal of the History of Medicine and Allied Sciences*. October: pp.353-357

McLellan, D. (1973) *Karl Marx: His Life and Thought*. St. Albans (Herts.): Paladin

Mahoney, C., Margolis, M., Knauss T. and Labem R. (1980) Chronic vitamin A intoxication in infants fed chicken liver. *Pediatrics* 65: 893-896

Malthus, R. (1816/1890) *An Essay on the Principle of Population*. London: Ward, Lock & Co.

Malthus, R. (1820/1989) *Principles of Political Economy* (2 vols.) Cambridge: Cambridge University Press

Marshack, A. (1990) Early hominid symbol and evolution of the human capacity. In P. Mellars (ed.) *The Emergence of Modern Humans*, pp.497-493 Edinburgh: Edinburgh University Press

Marshack, A. (1996) A Middle Palæolithic composition from the Golan heights: the earliest known depicted Image. *Current Anthropology* 37: 357-365

Marshack A. (1997) The Berekhat Ram figurine: a late Acheulian carving from the Middle East. *Antiquity* 71: 327-337

Matthew, P. (1831) *On Naval Timber and Aboriculture*. London: Longman

Maupertuis, P.L-M. de (1753/1966) *The Earthly Venus*. Translated by S. Boas, New York: Johnson Reprint Corporation

Mawson, D. (1915) *The Home of the Blizzard*. London: William Heinemann

Maynard Smith J. (1956) *The Theory of Evolution*. London: Penguin Books

Maynard Smith, J. (1982) *Evolution Now: A Century after Darwin*. San Francisco: Freeman Press

Maynard Smith, J. (1984) Palæontology at the High Table. *Nature*: 309: 401-402

Maynard Smith, J. (1987) Darwinism stays unpunctured. *Nature* 330: 516

Maynard Smith, J. (1988) *Games, Sex and Evolution*. New York: Harvester Wheatsheaf

Mayr, E. (1942) *Systematics and the Origin of Species*. New York: Columbia University Press

Mayr, E. (1949/1963) Speciation and Systematics. In G. Jepsen, E. Mayr and G.G. Simpson (eds.), *Paleontology and Evolution*, pp.281-295 New York: Atheneum

Mayr, E. (1963) *Animal Species and Evolution*. Cambridge, Mass: Harvard University Press

Mayr, E. (1972) The Nature of the Darwinian Revolution. *Science* 176: 981-989

Mayr, E. (1976) *Evolution and the Diversity of Life*. Cambridge, Mass: Harvard University Press

Mayr, E. (1977) Darwin and Natural Selection. *American Scientist* 65: 321-377

Mayr, E. (1982) *The Growth of Biological Thought*. Cambridge, Mass: Harvard University Press

Mayr, E. (1988) *Towards a new philosophy of biology*. Cambridge, Mass: Harvard University Press

Mayr, E. (1991) *One Long Argument*. London: Penguin Press

Mayr, E. (2001) *What Evolution Is*. New York: Basic Books

Mayr, E. and Provine, W.B. (1980) *The Evolutionary Synthesis: Perspectives on the Unification of Biology*. Cambridge, Mass: Harvard University Press

Medvedev, Z. (1969) *The Rise and Fall of T.D. Lysenko.* New York: Columbia University Press

Mendel, G. (1865/1966) Experiments on Plant Hybrids. In C. Stern and E. Sherwood (eds.), *The Origin of Genetics: A Mendel Source Book*, pp.1-48 San Francisco: W.H. Freeman

Menuge, A. (2004) Who's afraid of ID? A Survey of the ID Movement. In W. Dembski and M. Ruse (eds.), *Debating Design: From Darwin to DNA.* Cambridge: Cambridge University Press

Meyer, S. (2004) The Cambrian information explosion: Evidence for Intelligent Design. In W. Dembski and M. Ruse (ed.), *Debating Design: From Darwin to DNA.* Cambridge: Cambridge University Press

Miller, K. (1999) *Finding Darwin's God: A scientific Search for Common Ground between God and Evolution.* New York: Harper Collins

Miller, K. (2004) The Flagellum Unspun. In W. Dembski and M. Ruse (ed.), *Debating Design: From Darwin to DNA.* Cambridge: Cambridge University Press

Mithen, S. (2005) *The Singing Neanderthal: The Origins of Music, Language, Mind and Body.* London: Weidenfeld and Nicolson

Molleson, T. and Cox, M. (1993) *The Spitalfields Project (Vol.2) The Anthropology: The Middling Sort.* London: Council for British Archæology

Montgomery, P., Williams H., Reading, N. and Stringer, C. (1994) An assessment of the temporal bone Lesions of the Broken Hill cranium. *Journal of Archæological Science* 21: 331-337

Montgomery, P., Williams, H., Reading, N. and Stringer, C. (1994) An assessment of the temporal bone lesions of the Broken Hill cranium. *Journal of Archæologial Science* 21: 331-337

Morean, C.W. and Kim, S.Y. (1998) Mousterian large mammal remains from Kobeh Cave. *Current Anthropology* 39: S79-S113.

Morrice G., Havener, W. H. and Kapetansky, F. (1960) Vitamin A intoxication as a cause of pseudo-tumor cerebri. *Journal of the American Medical Association* 175: 1802-1805

Morwood, M.J., O'Sullivan, P., Azizi F. and Raza, A. (1998) Fission-track ages of stone tools and fossils on the east Indonesian island of Flores. *Nature* 392: 73-176

Morwood, M.J., Azizi, F., O'Sullivan, P., Nasruddin, Hobbs, D.R. and Raza, A. (1999) Archæological and palæontological research in central Flores, east Indonesia: Results of fieldwork 1997-98. *Antiquity* 73: 273-286

Morwood, M.J., Soejono, R.P., Roberts, R.G., Sutikana, T. Curney, C.S., Westaway, K.E., Rink, W.J., Zhao, J., van der Bergh, G.D., Rokus Awe Due, Hobbs, D.R., Moore, M.W., Bird, M.I. and Fifield, L.K. (2004) Archæology and age of a new hominin from Flores in eastern Indonesia. *Nature*: 431: 1067-1091

Morwood, M.J., Brown, P. Jatmiko, T., Sutikana, T., Wahyu Saptomo, E., Westaway, K.E.,Rokis Awe Due, Roberts R.G., Maeda, T., Wasisto, S. and Djubiantono, T. (2005) Further evidence for small- bodied hominins from the Late Pleistocene of Flores, Indonesia. *Nature* 437: 1012-1017

Morgan, E. (1972) *The Ascent of Woman*. London: Souvenir Press

Muenter, M., Perry H. and Ludwig J. (1971) Chronic vitamin A intoxication in adults. *America Journal of Medicine* 50: 129-136

Muller, H. (1949/1963) Re-integration of the Symposium on Genetics, Paleontology and Evolution. In G. Jepsen, E. Mayr and G.G. Simpson (eds.), *Paleontology and Evolution*, pp.421-447 New York: Atheneum

Neuman, W. (1994) *Social Research Methods: Qualitative and Quantitative Approaches*. (2nd ed.) Boston: Allyn and Bacon

Numbers, R. (1982) Creationism in 20th century America. *Science* 218: 538-544

Oakley, K., Campbell, B and Molleson, T. (1971) *Catalogue of Fossil Hominids* (Vol.2) London: British Museum (Natural History)

Oldroyd, D.R. (1980) *Darwinian Impacts*. Kensington: New South Wales University Press

Ortner, D. and Putschar G. (1981) *Identification of pathological conditions in Human Skeletal remains*. Washington: Smithsonian Institution Press

Osler, W. (1892) *The Principle and Practice of Medicine*. New York. Appleton & Co.

Ovchinnikov, I.V., Gotherstrom, A., Romanova, G.P. (2000) Molecular analysis of Neanderthal DNA from the northern Caucasus. *Nature* 404: 490-493

Owen, R. (1840/1890) *Paleontology: or A Systematic Summary of Extinct Animals and their Geological Relations*. Edinburgh: Adam and Charles Black

Paley, W. (1785/1791/1794/1802/1833) *The Works of William Paley*, D.D. Edinburgh: Peter Brown and Thomas Nelson

Partridge, D. (2015a) Further details concerning the Darwin-Wallace presentation to the Linnean Society in 1858, including its submission on 1 July, not 30th June. *Journal of Natural History* http://www.tandfonline.com./loi/tnah20

Partridge, D. (2015b) 1 July 1858 and the 1844 essay: what Lyell and Hooker decided and what Darwin did not want and did not know. *Biological Journal of the Linnean Society* 116: 247-251

Pavlov, P., Svendsen, J. and Indrelid, S. (2001) Human presence in the European Arctic nearly 40,000 years ago. *Nature* 413: 64-67

Pease C. (1962) Focal retardation and arrestment of growth of bone due to vitamin A intoxication. *Journal of the American Medical Association* 182: 980-985

Phillips, N. and Hardy C. (2002) *Discourse Analysis: Investigating Processes of Social Construction*. London: Sage Publications

Peake, H. and Fleure, H.J. (1927) *Apes and Men*. Oxford: Clarendon Press

Peckham, M. (1959) "Darwinism and Darwinisticism". In M. Peckham (1970), *The Triumph of Romanticism: Collected Essays*. Columbia: University of South Carolina Press

Pickford, M., Senut, B., Gommercy, D. and Treil, J. (2002) Bipedalism in Orrorin tugenensis revealed by its femora. *Palevol* 1(4): 191-203

Playfair, J. (1802/1956) *Illustrations of the Huttonian Theory of the Earth*. Edinburgh: William Creech. (Facsimile reproduction with Introduction by G.W. White.)

Popper, K. (1972) *Conjectures and Refutations*. London: Routledge and Kegan Paul

Price J. and Molleson, T. (1974) A radiographic examination of the left temporal bone of Kabwe Man Broken Hill Mine, Zambia. *Journal of Archæological Science* 1: 285-289

Pritchard, Radovcic, J. (1985) Neanderthals and their Contemporaries. In E. Delson (ed.), *Ancestors: The Hard Evidence*, pp.310-318 New York: Alan R. Liss

Raphael, D. (1985) *Adam Smith*. Oxford: Oxford University Press

Reich, D., Green R., Kircher, N., Krause, J., Patterson, N., Durand, E., Viola, B.,

Briggs, A., Stenzel, U,. Johnson, P., Maricic, T., Good, J., Marques-Bonet, T., Alkan, C., Fu, Q.,Mallick, S., Li H., Meyer, M., Eichler, E.,Stoneking, M., Richards, M., Talomo, S., Shunkov, K., Derevianko, A., Hublin, J., Kelso, J., Slatkin, M. and Pääbo, S. (2010) Genetic history of an archaic hominin group from Denisova Cave, Siberia. *Nature* 476: 136-137

Reynolds, T. (1990) The Middle-Upper Paleolithic Transition in Southwestern France: Interpreting the Lithic Evidence. In P. Mellars (ed.), *The Emergence of Modern Humans*, pp.262-273 Edinburgh: Edinburgh University Press

Rightmire, G.P. (1990) *The evolution of Homo erectus*. Cambridge: Cambridge University Press

Ritvo, L. (1990) *Darwin's Influence on Freud: A Tale of Two Sciences*. New Haven: Yale University Press

Roberts, C. and Manchester, K. (1997) *The Archæology of Disease*. New York: Cornell University Press

Roberts, R., Jones, R. and Smith, M. (1990) Thermoluminescence dating of a 50,000-year-old human occupation site in northern Australia. *Nature* 345: 153-156

Roberts, M., Stringer, C. and Parfitt, S. (1994) A hominid tibia from Middle Pleistocene Sediments at Boxgrove, U.K. *Nature* 369: 311-313

Rodahl, K. and Moore, T. (1943) The Vitamin A content and toxicity of bear and seal liver. *Biochemical Journal* 37: 166-168

Rogers, J., Watt, I., and Dieppe P, (1985) Palæopathology of spinal osteophytosis, vertebral ankylosis, ankylosing spondylitis and vertebral hyperostosis. *Annals of the Rheumatic Diseases* 44: 113-120

Romanes, E. (ed.) (1896) *The Life and Letters of George John Romanes*. London: Longmans, Green and Co.

Romanes, G.J. (1882) *Animal Intelligence*. London: Kegan, Paul, Trench and Co.

Romanes, G.J. (1884) *Mental Evolution in Animals*. New York: Appleton and Co.

Romanes: G.J. (1888) *Mental Evolution in Man*. London: Kegan, Paul, Trench and Co.

Romanes, G.J. (1892-1897) *Darwin, and after Darwin* (3 vols.). London: Longmans Green and Co.

Romanes, G.J. (1893) The Spencer-Weismann Controversy. *The Contemporary*

Review 64: 50-53

Rosenberg, H.R. (1945) *Chemistry and Physiology of the Vitamins*. New York: Interscience Publishers Inc.

Roth, H. L. (1899) *The Aborigines of Tasmania*. Halifax: F. King and Sons

Ruby, I. and Mital, M. ((1974) Skeletal deformities following chronic hypervitaminosis A. *Journal of Bone and Joint Surgery* 56: 1283-1287

Ruse, M. (1999) *The Darwinian Revolution: Science Red in Tooth and Claw*. Chicago: Chicago University Press

Ruse, M. (2006) *Darwinism and its Discontents*. Cambridge: Cambridge University Press

Russell, R., Boyer J., Bagheri S. and Hruban Z. (1974) Hepatic injury from chronic hypervitaminosis A Resulting in portal hypertension and ascites. *New England Journal of Medicine* 291: 435-440

Sahlins, M. (1972) *Stone Age Economics*. London: Tavistock Publications

Sarantakus, S. (1993) *Social Research*. Melbourne: Macmillan

Schopf, T. (ed.) (1972) *Models in Paleobiology*. San Francisco: Freeman Cooper

Scopes, J. and Presley, J. (1967) *Centre of the Storm: Memoirs of John T. Scopes*. New York: Holt, Rinehart and Winston

Scott, E. (2004) *Evolution v. Creationism*. Westport, Conn: Greenwood Press

Seward, C. (ed.) (1909) *Darwin and Modern Science*. Cambridge: Cambridge University Press

Shackley, M. (1980) *Neanderthal Man*. London: Duckworth

Shapiro, R. (1986) *Origins: A Skeptics Guide to the Creation of Life on Earth*. Toronto: Bantam Books

Shaw E. and Niccoli J. (153) Hypervitamonisos A: Report of a case in an adult male. *Annals of Internal Medicine* 39: 131-134

Shea,J.J. (1998) Neandertal and early modern behavioural variability. *Current Anthropology* 39: S45-S78

Shearmn, D.J. (1978) Vitamin A and Sir Douglas Mawso. *British Medical Journal* 1: 283-285

Sheldrake, R. (1988) *The Presence of the Past*. London: Fontana

Shermer, M. (2006) *Why Darwin Matters*. New York: Times Books

Simpson, G.G. (1944) *Tempo and Mode in Evolution.* New York: Columbia University Press

Simpson, G.G. (1950) *The Meaning of Evolution.* London: Oxford University Press

Simpson, G.G. (1953) *Life in the Past: An Introduction to Paleontology.* New Haven: Harvard University Press

Smith, A. (1776/1937) *An Inquiry into the Nature and Causes of the Wealth of Nations.* New York: The Modern Library

Smith, C. H. (ed.) (1991) *Alfred Russel Wallace: An Anthology of his Shorter Writings.* Oxford: Oxford University Press

Smith F. and Goodman D-W.S. (1976) Vitamin A transport in human vitamin A toxicity. *New England Journal of Medicine* 294: 805-808

Smuts, J. (1926) *Holism and Evolution.* London: Macmillan and Co.

Solecki, R. (1971) *Shanidar: The First Flower People.* New York: Alfred A. Knopf

Sollas, W. (1924) *Ancient Hunters and their Modern Representatives.* London: Macmillan & Co.

Southcott, R. V. and Chesterfield, N. J. (1971) Vitamin A content of the livers of huskies and some seals from Antarctic and subantarctic regions. *The Medical Journal of Australia* 1: 311-313

Spencer, H. (1852a) *A Theory of Population deduced from the General Law of Animal Fertility.* Westminster Review.

Spencer, H. (1852b) *The Developmental Hypothesis. The Haythorne Papers, No.2.* The Leader (London)

Spencer, H. (1861/1966) *Essays in Education.* London: Dent

Spencer, H. (1863) *First Principles of a New System of Philosophy.* London: Dent

Spencer, H. (1879) *The Evolution of Society.* Chicago: University of Chicago Press

Spencer, H. (1893) The Inadequacy of Natural Selection. *Contemporary Review* 63: 153-166

Speth, J.D. (2004) News flash: negative evidence convicts Neanderthals of gross mental incompetence. *World Archæology* 36: 519-526

Speth, J.D. and Tchernov, E. (1998) The Role of Hunting and Scavenging in Neandertal Procurement Strategies. In T. Akazawa and O. Bar-Yosef (eds.), *Neandertal and Modern Humans in Western Asia*, pp.223-240 New York: Plenum Press

Steadman, D. and Zoussmer, S. (1988) *Galapagos: Discovery on Darwin's Islands.* Washington: Smithsonian Institution Press

Stebbins, G. and Ayala, F. (1981) Is a New Evolutionary Synthesis Necessary? *Science* 213: 967-971

Steele, E.J. (1979) *Somatic Selection and Adaptive Evolution: On the Inheritance of Acquired Characters.* Toronto: Williams-Wallace International

Steele, E. J., Lindley, R.A. and Blanden, R.V. (1998) *Lamarck's Signature: How Retrogenes are Changing Darwin's Natural Selection Pa*radigm. Sydney: Alan and Unwin

Stern, C. (1949/1963) Gene and Character. In G. Jepsen, E. Mayr and G.G. Simpson (eds.), *Genetics, Paleontology and Evolution*, pp.13-22 New York: Atheneum

Stimson, W. (1961) Vitamin A intoxication in adults. *The New England Journal of Medicine* 265: 369-373

Strange L., Caristrom K. and Eriksson M. (1978) Hypervitaminosis A in early human pregnancy and malformation of the central nervous system. *Acta Obstrreticia et Gynecologica Scandinavia* 57: 289-291

Stringer, C. (1985) Middle Pleistocene Hominid Variability and the Origin of Late Pleistocene Humans. In E. Delson (ed.), *Ancestors: The Hard Evidence*, pp.289-295 New York: Alan R. Liss

Stringer, C. (1988) The Dates of Eden. *Nature* 331: 565-566

Stringer, C. (1994) Out of Africa – A Personal History. In M. Nitecki and D. Nitecki (eds.) *Origins of Anatomically Modern Humans*, pp. 150-172 New York: Plenum Press

Stringer, C. (1996) *African Exodus.* London: Jonathan Cape

Stringer, C. (1998) Chronological and biographic prospective on later human evolution. In T. Akazawa, K. Aoki and O. Bar-Joseph (eds.), *Neanderthals and Modern Humans in Western Asia*, pp.30-38 New York: Plenum Press

Stringer, C. (2011) *The Origin of our Species.* London: Allen Lane

Stringer, C. and Andrews, P. (1988) Genetic and Fossil Evidence for the Origin of Modern Humans. *Science* 239: 1263-1265

Stringer, C. and Davies, W. (2001) Those Elusive Neanderthals. *Nature* 309: 701-703

Stringer, C. and Gamble, C. (1993) *In Search of the Neanderthals.* London: Thames and Hudson

Stringer, C. and Grun, R. (1991) Time for the last Neanderthal. *Nature* 351: 701-702

Stringer, C. and McKie, R. (1996) *African Exodus.* London: Jonathan Cape

Sulzberger, M. and Lazar P. (1951) Hypervitaminosis A. Report of a case in an adult. *Journal of the American Medical Association* 146: 788793

Swift, J. (1735/1961) *Gulliver's Travels.* New York: W.W. Norton and Co.

Swisher, C., Curtis, G.H., Jacob, T., Getty, A.G., Suprijo, A. and Widiasmoro. (1994) Age of the earliest known hominids in Java, Indonesia. *Science* 263: 1118-1121

Swisher, C., Rink, W., Antón, S., Schwarcz, H., Curtis, G., Suprijo, A. and Widiasmoro. (1996) Latest Homo erectus from Java: Potential Contemporaneity with Homo sapiens in Southeast Asia. *Science* 274: 1870-1874

Sykes, B. (2001) *The Seven Daughters of Eve.* London: Bantam Press

Tanaka, J. (1976) Subsistence Ecology of Central Kalahari San. In R. Lee and I. De Vere (eds.) *Kalahari Hnter-Gatherers.* Cambridge, Mass: Harvard University Press

Tattersall, I. (1998) *Becoming Human: Evolution and Human Uniqueness.* San Diego: Harcourt Brace

Tattersall, I. (1999) *The Last Neanderthal.* New York: Nevraumont Publishing

Tattersall, I. (2002) *The Monkey in the Mirror.* Oxford: Oxford University Press

Tattersall, I. and Schwartz, J.H. (2001) *Extinct Humans.* New York: Nevraumont Publishing

Teilhard de Chardin, P. (1951/1955) *The Phenomenon of Man.* New York: HarperTorch Books

Teillhard de Chardin, P. (1956/1965) *The Appearance of Man.* London: Collins

Teilhard de Chardin, P. (1956/1966) *Man's Place in Nature.* London: Collins

Templeton, A.R. (1992) Human Origins and Analysis of Mitochondrial DNA sequences. *Science* 255: 737-739

Templeton, A.R. (1993) The 'Eve' hypothesis: A genetic critique and reanalysis. *American Anthropologist* 95: 51-72

Templeton, A.R. (2002) Out of Africa again and again. *Nature* 416: 45-51

Tennant, F. (1930/1968) *Philosophical Theology* (Vol.II). Cambridge: Cambridge University Press

Thaxton, C., Bradbury, W. and Olsen R. (1984) *The Mystery of Life's Origins: Re-assessing Current Theories. New York: Philosophical Library*

Thoma, A. (1973) New evidence for the polycentric evolution of Homo sapiens. Journal of Human Evolution 2: 529-539

Thompson, D.D. and Trinkaus, E. (1981) Age determination for the Shanidar 3 Neanderthal. *Science* (2)2: 375-577

Thorne, A. and Macumber, O. (1972) Discoveries of Late Pleistocene Man at Kow Swamp, Australia. *Nature* 238: 316-319

Thorne, A. and Wolpoff, M. (1992a) Regional continuity in Australian Pleistocene Hominid Evolution. *American Journal of Physical Anthropology* 55: 337-349

Thorne, A. and Wolpoff, M. (1992b) The Multiregional evolution of humans. *Scientific American* 266: 28-33

Thorne, A., Grun, R. and Mortimer, G. (1999) Australia's oldest human remains: age of the Lake Mungo 3 skeleton. *Journal of Human Evolution* 36: 591-612

Tocheri, M., Orr, C., Larson, S., Sutikna, T., Jatmiko, E. Saaptomo W., Rokus, A., Djubiantono, T., Morwood, M. and Jungers, W. (2007) The primitive wrist of Homo floresiensis and its implications for Hominin evolution. *Science* 317: 1743-1745

Todd, M. (1987) *Christian Humanism and the Puritan Social Order.* Cambridge: Cambridge University Press

Trinkaus, E. (1983a) *The Shanidar Neanderthals.* New York: Academic Press

Trinkaus, E. (1983b) *The Mousterian Legacy* (ed.) Oxford: BAR

Trinkaus, E. (1989) *The Emergence of Modern Humans.* Cambridge: Cambridge University Press

Trinkaus, E. (1995) Neanderthal Mortality Patterns. Journal of Archæological Science 22: 121-142

Trinkaus, E. and Shipman, P. (1992) *The Neanderthals: Changing the Image of*

Mankind. New York: Alfred A. Kopf

Trinkaus, E. and Smith, F. (1985) The fate of the Neanderthals. In E. Delson (ed.), *Ancestors: The Hard Evidence*, pp.325-333 New York: Alan R. Liss

Trinkaus, E., Zilhäo, J. and Duate, C. (2001) *The Lapedo Child: Lagar Vello 1 and our perceptions of the Neanderthals*. http:/w.w.w.med.abaco-mac-it/issue-1/articles/doc/013.htm Accessed 28/10/01

Turnbull, C. (1994) (1994) *The Mountain People*. London: Pimlico

Tynan, E, (1994) *Survival of the fittest immune system: Darwin and Lamarck were both right*. ANU Reporter. Canberra: The Australian National University

Ustinger, P.D. (1985) Diffuse idiopathic skeletal hyperostosis. *Clinics in Rheumatic Diseases* 11(2): 325-351

Vallois, H. (1937) La durée de la vie chez l'homme fossile. *L'Anthropologie* 47: 499-532

Van Dijk, T. (2003) Critical Discourse Analysis. In D. Schiffrin, D. Tanner and H. Hamilton (eds), *The Handbook of Discursive Analysis*, pp. 351-367 Malden, M.A: Blackwell Publishing

Van Wyhe, J. and Rookmacker, K. (2012) A new theory to explain the receipt of Wallace's Ternate Essay by Darwin in 1858. *Biological Journal of the Linnean Society* 105: 249-252

Vigilant, L., Stoneking, M., Harpending, H., Hawkes, K. and Wilson, A.C. (1991) African populations and the evolution of human mitochondrial DNA. *Science* 253: 1503-1507

Vrba, E. (1995) On the connection between paleoclimate and evolution. In E. Vrba, G. Denton, T. Partridge and L. Burckle (eds.), *Paleoclimate and Evolution, with Emphasis on Human Evolution*, pp.24-48 New Haven: Yale University Press

Vorzimmer, P. (1970) *Charles Darwin: The Years of Controversy*. Philadelphia: Temple University Press

Walker, A. (1984) Extinction in Hominid Evolution. In M. Nitecki (ed.), *Extinctions*, pp. 119-152 Chicago: University of Chicago Press

Walker, A. and Shipman, P. (1996) *The Wisdom of Bones: In Search of Human Origins*. London: Phoenix

Walker, A., Zimmerman, M. and Leakey, R. (1982) A possible case of hypervitaminosis A in Homo erectus. *Nature* 296: 248-250

Wallace, A.R. (1855) On the law which has regulated the introduction of new species. *Annals and Magazine of Natural History* 16(93): 184-196

Wallace, A.R. (1856a) On the habits of the orang-utan in Borneo. *Annals and Magazine of Natural History* 17(103): 26-32

Wallace, A.R. (1856b) Attempts at a natural arrangement of birds. *Annals and Magazine of Natural History* 2nd series 18: 93-216

Wallace, A.R. (1856c) Splendours of the eastern archipelago: observations on the geology of Borneo. *Zoologist*: 5116-5117

Wallace, A.R. (1857a) On the great bird of paradise of the Malays. *Annals and Magazine of Natural History* 2nd series 20: 411-416

Wallace, A.R. (1857b) On the natural history of the Aru islands. *Annals and Magazine of Natural History* 20 supplement: 473-485

Wallace, A.R. (1858a) Note on the theory of permanent and geographical varieties. *Zoologist* 16: 5887-5888

Wallace, A.R. (1858b) On the habits and transformation of a species of Ornithoptera, allied to O. Priamus, inhabiting the Aru islands near New Guinea. *Transactions of the Entomological society of London* 4: 272-273

Wallace, A.R. (1858c) On the tendency of varieties to depart indefinitely from the original type: instability of varieties supposed to prove the permanent distinction of species. *Journal of the Proceedings of the Linnean Society of London* 3: 53-62

Wallace, A.R. (1870/1973) *Contributions to the Theory of Natural Selection.* New York: AMS Press

Wallace, A.R. (1889a) *Travels on the Amazon and Rio Negro.* London: Ward, Lock and Co. Ltd.

Wallace, A.R. (1889b/1975) *Darwinism: An Exposition of the Theory of Natural Selection.* New York: AMS Press

Wallace, A.R. (1896) *Miracles and Modern Spiritualism.* London: Macmillan

Wallace, A.R. (1905/1969) *My Life.* London: Chapman and Hall

Wallace, A.R. (1910) *The World of Life.* London: Chapman Hall

Watson, J. and Crick, F. (1953) Molecular structure of nucleic acids. *Nature* 171: 737-738

Watson, L. (1974) *Supernature: A Natural History of the Supernatural.* London:

Coronet Books

Weidenreich, F. (1939) On the earliest representatives of modern mankind recovered on the soil of East Asia. Peking Natural History Bulletin Vol.1313. In F. Weidenreich (1949), *Anthropological Papers of Franz Weidenreich 1939-1948*; pp.205-224 New York: The Viking Fund Inc.

Weidenreich, F. (1940) The external tubercle of the human tubercalcanel. American Journal of Physical Anthropology, Vol.26. In F. Weidenreich (1949) *Anthropological Papers of Franz Weidenreich 1939-1948*; pp.99-123. New York: The Viking Fund Inc.

Weidenreich, F. (1947) Facts and speculations concerning the origins of Homo sapiens. *American Anthropologist* 49: 187-203

Weidenreich, F. (1949) *Anthropological Papers of Franz Weidenreich: 1939-1948.* New York: The Viking Fund Inc.

Weismann, A. (1893) The All-Sufficiency of Natural Selection. Contemporary Review 64: 309-338

Weismann, A. (1904) *The Evolutionary Theory* (2 vols.) (Translated by J.A. and M.R. Thomson). London: Edward Arnold

Wells C. (1964) *Bones, Bodies and Disease.* London: Thames and Hudson

White: M. (1937) *The Chromosomes* (1st edn.). London: Methuen

White, M. (1973a) *The Chromosomes* (6th edn.). London: Chapman and Hall

White, M. (1973b) *Animal Cytology and Evolution.* Cambridge: Cambridge University Press

White, M. (1978) *Modes of Speciation.* San Francisco: W.H. Freeman and Co.

White, T.D., Asfaw, B., DeGusta, D., Gilbert, H., Richards, G.D., Suwa, G. and Clark Howell, F. (2003) Pleistocene Homo sapiens from the Middle Awash, Ethiopia. *Nature* 423: 742-747

Williams, G. (1966/1992) *Plan and Purpose in Nature.* London: Weidenfeld and Nicolson

Wilson, A.C. and Cann, R.L. (1992) *Adaptation and Natural Selection.* Princeton: Princeton University Press

Wilson, E.D. (1975) *Sociobiology:* The New Synthesis. Cambridge, Mass: Harvard: Harvard University Press

Wolf, F. (1999) *The Spiritual Universe.* Needham, Mass: Moment Point Press

Wolinsky, H. (2007) A mythical beast. Increased attention highlights the hidden wonders of Chimeras. *EMBO Reports* 8: 212-214

Wolpoff, M. (1989) The place of the Neanderthals. In E. Trinkaus (ed.), *The Emergence of Modern Humans.* Cambridge: Cambridge University Press

Wolpoff, M. and Caspari, R. (1997) *Race and Human Evolution.* Boulder: Westview Press

Wolpoff, M. and Thorne, A. (1991) The Case against Eve. *New Scientist* 6: 33-37

Wolpoff, M., Thorne, E., Smith, F., Frayer, D. and Pope, G. (1994) Multi-regional Evolution - A World-wide Source for Modern Human Populations. In M. Nitecki and D. Nitecki (eds.), *Origins of Anatomically Modern Humans,* pp.175-199 New York: Plenum Press

Wong, K. (2003) An ancestor to call our own. *Scientific American* 13(2): 4-13

Wright, S. (1949-1963) Adaptation and Selection. In G. Jepsen, E. Mayr and G.G. Simpson (eds.), *Genetics, Paleontology and Evolution,* pp.365-389 New York: Atheneum

Wu, R. and Lin, S. (1983) Peking Man. *Scientific American* 249(6): 78-86

Wynn, T. and Coolidge, F. (2011) *How to think like a Neanderthal.* Oxford: Oxford University Press

Yearsley, M. (1928) The pathology of the left temporal bone of the Rhodesian Skull. In *Rhodesian Man and Associated Remains.* London: British Museum (Natural History) pp. 59-63

Yukteswar, J. (1949/1990) *The Holy Science.* Los Angeles: Self-Realization Fellowship

Ziegler, R. (1992) Diabetes mellitus and bone metabolism. *Hormone Metabolism Research Supplement* 26: 90-94

Zimmer, C. (2001) *Evolution: The Triumph of an Idea.* New York: Harper Collins

Glossary

acrocentric : chromosome with unequal length arms; 'J' shaped

adiabatic expansion : cooling of a gaseous substance (air) during expansion without transfer of heat to another substance

allopatric : separate habitats, reproductive isolation

amixis : fertilization

ammonites : extinct shelled sea creatures

amphimixis : mixing of generative materials

anagenesis : one species gradually changing into another

anastomoses : interconnecting pathway allowing diffusion

atavism : expression of an ancestral line not observed in more recent progenitors

bases : subcomponents of DNA molecules

bimanus : having two hands; humans

biofor : submicroscopic particle responsible for reproducing life

B.P.: Before Present. Scientists needed to adjust dating calculations to allow for artificially produced radiation after 1945 A.D. For convencience, 1950 A.D. was chosen to mark the new Era. 10,000 B.C. is ~12,000 B.P.

brachydactylism : short fingers, short stature

calvaria : top portion of skull

centromere : region of chromosome to which spindle fibre attaches during cell division

cervid species : deer

cladogenesis : the divergence of one species into two

CSIRO : Commonwealth Scientific and Industrial Research Organisation

determinant : microscopic particles responsible for development of cells and body parts

dimorphism : different physical appearance within same species, especially between sexes

DNA : deoxyribonucleic acid; a large molecule which carries hereditary instructions that determine the formation of all living organisms

eukaryotic : species in which both the male and female contribute genetic material during the process of reproduction

evolutionary synthesis : synthesis of Darwinian theory of evolution by natural selection with Mendelian genetics

foramen : space in bone for passage of blood vessels and nerve fibres gemmiparous reproducing asexually

gemmules : particles which Darwin postulated circulated throughout the body carrying information about acquired change to the reproductive material, thus bringing about change to the next generation

germ-plasm : reproductive material

gynæcomastia : enlarged breast tissue in male

heterozygote : having both X and Y chromozomes (female)

hybridization : crossing of species/varieties other than that which normally occurs in nature

imago : adult form of caterpillar, either butterfly or moth

infusoria : amorphous animals reproducing by fission or budding with no special organs, even of digestion

karyotypes : number and arrangement of chromosomes

metacentric : chromosome with two arms of (approximately) equal length, 'Y' shaped

metaphase : stage of cell division at which spindle fibres attach to chromosomes

microchromosomes : very small chromosomes with no known function

monotreme : egg-laying mammal

mtDNA : DNA found in the mitochondria of cells, inherited through the maternal line

multi-regional evolution : theory that humans had evolved in many areas of the world after an initial diaspora from Africa about one million years ago

Neanderthals : occupants of Europe and western Asia 120,000-30,000 BP

noosphere : Teilhard's term for the sphere of the reasoning mind

orthoganous : jaw in 'normal' alignment, neither protruding nor receding

os lunatum : a bone from the wrist

Out of Africa : theory that the species Homo sapiens evolved in Africa approximately 150,000 ya before spreading to the rest of the world

oviparous : egg-laying

pleitropism : single gene responsible for more than one characteristic; single characteristic influenced by more than one gene

polydactyly : more than five digits on hand or foot

polyploidy : an extra one, or more, complete chromosomes produced during cell division

polyps : gelatinous animals which reproduce by budding; no internal organs other than an alimentary canal

prognathous : having jaw that projects forward to a marked degree

prokaryote : single-celled organism with asexual reproduction

protoplasm : most simple form of organic material capable of sustaining life

quadrumana : having four hands; primates

radiarians : suboviparous animals with regenerating bodies; no head, eyes; or jointed legs; mouth on inferior surface; star fish, sea urchin

radiometric dating : absolute dating method that measures the decay of radio-active material

saltation : evolutionary change by 'jumps' or 'leaps'; sudden change

seminiferous tubules : tubules for the passage of semen

skeletal hyperostosis : overgrowth of bone

speciation : establishment of a new species

spontaneous generation : life form which appears without parent 'sport' distinct variation between parents and offspring, usually in plants; for example, double the number of expected petals

sympatric : living in the same area; two or more species whose habitat partly or largely overlap

taphonomy : history of fossil bones in the ground before discovery

taurodontism : enlarged pulp cavities found in some fossil human teeth

teleocentric : chromosomes with one arm; 'rod' shaped

thelytokous :   process of reproduction in which males play no part

Tory : British Conservative Party supported principally by the aristocracy and landed gentry

trilobite : extinct sea-floor dwelling species, with multiple eyes and a body segmented into three sections

tylosis : thick soles and palms

vernalisation : increasing growth and reproduction by increasing warmth and light

viviparous : giving birth to life young

Whig : British Liberal Party, principally supported by financial and mercantile interests

Aristotle, 4-7, 136, 268, 274

art, 3, 62, 144, 203, 242, 245, 272, 350, 376, 390, 391, 418-422, 446, 458, 459, 543, 545

Atapuerca, 365, 381

aura vitalis, 46, 47, 58

Australopithecus, 300, 344, 354, 360, 375, 386, 387, 389, 412, 422, 429, 437, 472, 537

Aurignacian, 306, 308, 416, 418, 446

aversions, 30, 48

Azilian, 301, 446

Bacon, F. 6, 269

bacteria, 105, 338, 485, 488, 489, 530, 540

bacterial flagellum, 485

basicranial fllexion, 440

Bates, W.143, 156-159, 172, 179

Bateson, W. 244, 245, 286-288, 290, 320

Baudin, 38

Beagle, 9, 27, 79, 80, 89, 98, 100, 113, 195, 212

Behe, 483-494, 515, 518, 543, 545

bimanus, 58

biophors, 236-240, 335

biosphere, 342, 344

biped, 6, 59, 138, 429, 430, 440

blending, 82, 124, 216, 225, 226, 242, 244, 248, 310, 318, 505

Blyth, E., 77-83, 88, 105, 107, 113, 152, 158, 162

Border Cave, 368, 381

Boule, M., 301-3-3, 307, 308, 353, 363, 421

Bowler, P.J., 58

Brace, C. L., 366, 367, 376, 427

Brackman, A., 141, 145, 151, 152, 154, 159-161, 163, 171, 173, 178

brain, 29, 32, 55, 56-58, 135, 197, 220, 278, 289, 290, 302, 343, 352, 361, 392, 421, 435, 436, 438, 441, 456, 478, 482, 507, 509, 536

Broken Hill, 307, 367, 376, 378, 389, 425, 469, 470

Brooks, J. 141, 143, 151, 152, 154, 159, 162, 169, 171-173, 187, 193 416

Bryan, W. J., 292, 293, 295-297, 403, 457, 481, 496

Buffon, G., 19-26, 30, 33, 37, 43-45, 47, 49, 61, 62, 64, 75, 77, 84, 86, 88-90, 103, 105, 133, 136, 271

Eldredge., N., 381, 392-402

Elliot, H., 35, 36, 38, 40-43, 52, 58

embryo, 16-18, 31, 87, 122, 245, 247, 253, 286, 290, 405, 412, 511, 534

Engels, F., 251, 252

entropy, 532

eosinophilic granuloma, 470, 471

Eskimo, 309, 353, 423, 474

eugenics, 148, 254, 280, 292, 316, 320, 321, 345

Europe, 4-10, 12, 13, 22, 39, 48, 66, 69-72, 89, 90, 101, 103, 110, 124, 137, 181, 204, 205, 246, 250, 251, 253, 254, 256, 257, 266, 268, 277, 278, 280, 286, 291, 293, 300, 301, 304, 308, 309, 312, 346, 348-350, 353, 354, 363, 364, 366-369, 372-383, 387-391, 396, 397, 414-425, 428, 429, 431, 433, 441, 446, 447, 450-452, 457-462, 466-469, 473, 474, 491, 492, 501, 502, 525, 527, 535, 539

Eve, 35, 269, 369, 370, 377

Evolution, 3-5, 8, 12, 13, 15, 16, 19, 20, 22, 23, 27, 28, 30, 31-34, 38-40, 43-45, 49, 52, 54, 55, 56-62, 64, 66, 67, 70, 74, 77-79, 83-89, 91, 92, 97, 101, 102, 105, 108, 109, 112, 113, 115, 124-128, 133, 134, 126, 138, 141, 143, 144, 146-149, 151-153, 156, 170, 171, 174, 194, 197, 198, 201, 203, 205-206, 211, 213-217, 219-229, 241-253, 256-258, 261, 265, 267, 270, 271-276, 279, 289, 282, 285, 290-301, 307, 308, 310, 311, 313-324, 327, 328, 333-348, 360, 363-369, 371, 373, 375-381, 385-388, 390. 392-413, 419, 422, 425-427, 431, 433, 434, 436, 438-440, 469, 472, 473, 476, 483-493, 496-502, 504, 506, 507, 509- 517, 518, 521-524, 526, 528-532 535-544, 547

Evolutionary, 12, 33, 241, 242, 244, 248, 249, 259, 253, 271, 273, 274, 279, 280, 282, 285, 291, 300, 301, 307, 311, 312, 313, 315, 318, 319, 321,323, 324, 327, 328, 333, 335, 336-339, 364, 365, 371, 373, 377, 379, 380, 385, 392, 395, 396, 398, 400-408, 410, 419, 436, 439, 472, 473, 488, 490, 491, 498, 505-507, 509, 511, 512, 516, 523, 526, 529, 532, 535, 539

Feldhofer, 373, 374, 524, 525

First Amendment, 498, 501, 544

First Cause, 26, 85, 91, 92, 252, 271

Fisher, R., 244, 317, 318, 321, 323, 406, 505, 529

Fitzroy, Captain R., 27, 99, 100, 113, 195, 212

Fleiss, W., 259, 262-264

F(f)lood, 4, 20, 25, 39, 40, 42, 44, 64, 111, 250, 529

Flores, 535-537

floresiensis, 535, 536

Ford, E., 244, 247, 314, 318, 320, 321

forethought, 105, 110, 116, 120, 136, 138, 174, 541

fossil, 3, 9, 39, 43, 48, 59, 60, 62-65, 86, 88-92, 109, 111, 125, 143, 151, 153, 204, 205, 207, 221, 223, 224, 228, 239, 240, 298, 300, 309, 341, 346, 347, 349-351, 353-358, 360, 366-368, 370, 372-

379, 381-383-389, 392-394, 396-398, 401, 404, 414, 415, 427, 419-429, 432, 456, 469, 470, 472, 474, 475, 527, 536, 537, 539, 540

Foundations, 97, 104, 112, 113, 115, 116, 118-120, 123, 132, 134, 136, 140, 151, 154, 157, 182, 198, 251, 528

Freud, Sigmund, 257-264, 273, 277

Galápagos, 100-103, 109, 110, 113, 114, 125, 196, 262, 313

Galileo, 5, 20, 338

Galton, Sir Francis, 27, 254, 320

ganglia, 55, 219

gelatinous, 41, 46, 48, 51, 87, 285

gemmiparous, 54

gemmules, 1134, 135, 136, 234

Genesis, 9, 20, 21, 33, 39, 40, 47, 65, 212, 225, 229, 271, 292, 298, 499

Gibraltar, 300, 303

Giganthropus, 358

God, 4-11, 15, 16, 21, 33, 66, 70, 73, 74, 85, 92, 110, 113, 119, 120, 127, 136, 147, 150, 153, 164, 203, 206, 210, 224, 225, 229, 240, 250, 252, 253, 266, 269-272, 276, 291, 292, 296, 298, 483, 484, 485, 491-494, 498, 499-501, 503, 513-522, 531, 542, 543

Goldilocks planet, 542

Goldschmidt, R., 245, 247, 312, 314, 315, 321-324, 328, 335, 399, 510

Goodness, 5, 9, 15, 119

gorilla, 10, 204, 351, 408, 511, 534, 544

Gould, S., 50, 83, 133, 210-212, 333, 381, 392-394, 399-403

gradualism, 140, 204, 328, 392, 393, 397, 399, 400, 438, 489, 510

Gray, A., 151, 152, 155, 161, 163, 164, 166, 167-169, 171, 174, 175, 177-187, 192

Gray, J., 177, 191

Greeks, 3, 4, 5, 265, 268

Hækel, E., 252-255, 257, 541

hæmophilia, 358, 368, 369

Haldane, J., 239, 245, 247, 316-321, 323, 333, 335, 406, 505, 523

Hardy/Weinburg Law, 316, 319, 362

Henslow, J., 99-101, 194-196, 198, 222

Hobbit, 535, 537

Hollick, M., 527, 528, 543

hominid, 344, 347, 351, 355-357, 361-363, 366-369, 378, 393, 396, 397, 424, 427, 430, 469, 470, 476

hominoid, 344, 387, 472

Homo erectus, 3, 300, 303, 307, 309, 346, 347, 349, 350, 354, 355, 357, 360, 363, 364, 366, 368, 369, 371, 373, 374, 378, 379, 381, 382, 385-390, 396, 404, 424, 426-429, 438, 440, 458, 469, 472, 475, 476, 479-482, 535, 536, 541

Homo heidelbergensis, 301, 375, 381, 382, 386, 388, 390, 440, 469, 537

Homo neanderthalensis, 301, 381, 382, 387, 428

HOMO RHODESIENSIS, 301, 307, 469

Homo sapiens, 3, 301, 308, 347, 349, 364, 366, 367-370, 372-6, 380, 382-384, 386, 387-390, 396, 407, 416, 421, 428, 433, 439, 469, 470, 474, 526, 535, 536

Homo sapiens idaltu, 367

Honeywill, R., 38, 40

Hooker, Sir J., 141, 151-153, 157, 158, 160-169, 171, 173-175, 178, 179, 181-186, 188, 189, 191, 192, 194, 196-199, 211, 212, 214, 222, 224

Hooker, Sir William., 152, 196

Howell, F., 389, 474

Howells, W., 365, 425

Hox genes, 511, 534

HPO (hypertrophic pulmonary osteoarthropathy), 464-466

humanism, 258, 264, 265-276, 278, 318, 483, 498, 507, 543

humanist, 15, 120, 250, 259, 264, 267, 269, 270, 272, 273-276, 298, 339, 498, 499, 504

Hutton, J., 24-26, 28, 32, 33, 47, 62, 66, 67, 77, 84, 89, 90, 109, 199

Huxley, Sir J. 246, 271, 273, 311, 315, 318, 319, 344, 384

Huxley, L., 153, 162, 211,

Huxley, Prof. T., 8, 120, 128, 137, 148, 153, 163, 171, 174, 194, 199-212, 214, 221-223, 227, 269, 273, 274, 277, 278, 298, 300, 333,

hybrid, 23, 106, 107, 117, 123, 124, 130, 133, 134, 204, 227, 242, 243, 280, 285, 286, 302, 313, 326, 379, 380, 392

hypocephalon, 56, 57

ids, 235, 237, 239, 240

IDS (infant directed speech), 434

incipient species, 116, 314, 323

inferior, 37, 38, 55, 87, 138, 139, 209, 320, 421, 422, 426, 501

Middle Ages, 6, 7, 9, 303, 308, 420

Middle Eastern, 377, 428

Mithen, S., 426, 434, 435, 440

mitochondrial, 239, 369-372, 532, 533

mitosis, 234, 240, 311, 324-328, 398, 410, 413, 483

Mivart, G., 221, 222, 225, 227, 228

Mojokerto, 354

Monist League, 253, 254, 292, 541

Monkey Trial, 291

monkeys, 10, 62, 396, 397, 429, 436, 528

Moon, 5, 7, 12, 14, 20, 21, 27, 45, 271

morphic field, 527-529

Morgan, 361-363

Morris, D., 363

Mount Carmel, 308

Mousterian, 302, 389, 416, 417, 525

mtDNA, 239, 369-374, 379, 424, 524-527, 533

mucilaginous, 41, 46, 48, 51

mule, 22, 81, 106, 123, 134-136, 204, 234, 542

Multiregional, 353, 369, 370, 375,376, 378, 381, 385, 425, 469, 526, 527, 537, 540

mutation, 16, 79, 87, 92, 240, 247, 311, 312, 314-319, 323-325, 327, 328, 334, 335, 369, 374, 400, 412, 434, 487, 489, 507, 509-511, 517, 524, 527, 529-531, 533, 538, 540

Nariokotome, 389, 426, 429

natural selection, 3, 15, 30, 70, 78, 79, 83, 92, 97, 101, 109, 115, 116, 118-121, 126-130, 133, 136, 139, 141, 144, 146, 148, 150- 154, 158, 162, 167, 169, 171, 174, 178, 179, 181, 182, 186, 190, 198, 203, 205, 207,208, 214, 216, 218, 219, 221, 227, 229, 237-239, 241, 243, 245, 247, 250, 271, 276, 293, 296, 310, 312, 316, 319-321, 328, 336, 338, 345, 398, 399, 402, 406, 434, 439, 438, 470, 484-486, 488, 490. 491, 498, 504, 506, 508, 514, 516, 517, 519, 521, 522, 529, 530, 531, 538, 541-544, 547

natura non facit saltum, 10, 128, 541

Neanderthals (see also Homo neanderthalensis, 3, 91, 204, 300-303, 305-309, 333, 344, 346, 347, 349, 363, 366, 367, 369, 370, 372, 374, 375-380, 382, 386-390, 397, 404, 414-429, 431, 433, 434, 435, 440, 441, 453, 454, 456-458, 461, 462, 468, 469, 473, 524-526, 535-537

Negroid, 308, 309, 353, 425, 472, 473, 474

Neo-Darwinism, 120, 137, 214, 216, 241, 246, 290, 328, 399, 400, 401, 485, 489, 491, 492, 542, 545

Neo-Lamarckist, 246

Sedgwick, A., 99, 100, 198, 221, 245

Shanidar, 306, 375, 415, 416, 422, 426, 452-454, 456, 457, 459, 462, 464, 468

Sheldrake, R., 528

Shipman, P., 367, 377-380, 426-429

Simpson, G., 244-247, 273, 314, 317, 319, 333, 364, 399, 524

Sinanthropus, 303, 307, 346, 347, 351-353, 356-358, 363

Skhul, 376, 378, 419

Smith, A., 78, 79, 80, 81, 82

Smith, J., 164

Smith, M., 156, 381

Social Darwinism, 207, 209, 241, 279, 540

Sociobiology, 505, 506

Socrates, 5

Solecki, R., 415, 416, 452, 453, 457

Sollas, W., 301-303, 305-309

Solo, 346, 354, 363, 446

somatic, 137, 208, 215, 234, 247, 263, 311, 327, 328, 335, 410, 485, 511

soul, 6, 7, 10, 58, 147, 150, 168, 253, 274, 292, 416, 520

South-East Asia, 8, 64, 349, 354, 374 378, 387

Spain, 4, 69, 254, 306, 365, 474

Special Creation, 41, 42, 47, 113, 153, 206, 500

species, 17, 19, 22, 23, 30, 32 33, 37, 39, 40,, 42, 43, 48, 49, 50, 51, 58, 59, 62-65, 73, 74, 78-82, 87, 91, 92, 100-125, 129-131, 133, 137, 139, 141, 143-147, 153-159, 164-166, 170, 174, 180, 184, 186, 195, 197, 204-206, 216-218, 221, 222, 224-227, 235-237, 239, 240, 246, 250, 254, 257, 273, 290, 301-319, 322-328, 334, 337, 341, 342, 344, 346, 347, 349, 350, 353-356, 360, 364, 367, 368, 370, 372, 374-376, 379-390, 393-396, 398, 399-402, 405-408, 414, 415, 417, 422, 428, 429, 445, 463, 469, 472, 475, 476, 479, 481, 491, 499, 502, 505-509-512, 526, 529, 535, 537, 540, 541

speech, 209, 211, 212, 216, 299, 307, 359, 418, 426, 433-435, 437, 439, 441, 502, 545

Spencer, H., 16, 73, 136, 161, 194, 205,, 206, 209, 214, 269, 340

sphere, 47, 88, 275, 342,

spiritualism, 147, 214, 521

Spitalfields, 303

spontaneous generation, 41, 44, 47, 51, 52, 54, 58, 87, 88, 91, 221, 224, 225, 233, 237, 238

stars, 5, 13, 32, 45, 86, 340, 493, 542

Steele, E., 137, 290

About the Author

Born in London in 1937, Denise Carrington-Smith came to Australia in 1967, raising her family in Melbourne, where she lived for nearly thirty years.

Denise became interested in healing, training first as a natural therapist and homœopath and then as a psychologist and clinical hypnotherapist.

For some years, Denise lectured in herbalism, the Bach Flower Remedies and homœopathy before becoming Principal of the Victorian College of Classical Homœopathy. She also served as President of the Australian Federation of Homœopaths.

In 1995, Denise retired to Far North Queensland, returning to University where she took up the study of archæology, specializing in the study of evolutionary theory, which was the subject of her doctoral thesis.

Denise has seven children (each born on a different day of the week!), eighteen grandchildren and a smattering of great-grand children..

www.ingramcontent.com/pod-product-compliance
Lightning Source LLC
Chambersburg PA
CBHW081210220326
41598CB00037B/6734